Time Series Modeling of Neuroscience Data

CHAPMAN & HALL/CRC
Interdisciplinary Statistics Series

Series editors: N. Keiding, B.J.T. Morgan, C.K. Wikle, P. van der Heijden

Published titles

Published titles

Chapman & Hall/CRC
Interdisciplinary Statistics Series

Time Series Modeling
of Neuroscience Data

Tohru Ozaki

CRC Press
Taylor & Francis Group
Boca Raton London New York

CRC Press is an imprint of the
Taylor & Francis Group, an **informa** business

A CHAPMAN & HALL BOOK

CRC Press
Taylor & Francis Group
6000 Broken Sound Parkway NW, Suite 300
Boca Raton, FL 33487-2742

Printed in the United States of America on acid-free paper
Version Date: 20120104

International Standard Book Number: 978-1-4200-9460-2 (Hardback)

Library of Congress Cataloging-in-Publication Data

Ozaki, Tohru, 1944-
    Time series modeling of neuroscience data / Tohru Ozaki.
       p. ; cm. -- (Chapman & Hall/CRC interdisciplinary statistics)
    Includes bibliographical references and index.
    ISBN 978-1-4200-9460-2 (hardcover : alk. paper)
    I. Title. II. Series: Interdisciplinary statistics.
    [DNLM: 1. Diagnostic Techniques, Neurological--statistics & numerical data.
    2. Brain Mapping--statistics & numerical data. 3. Data Interpretation, Statistical. 4.
Models, Neurological. 5. Time Factors. WL 141]

    616.8'0475--dc23   2011046671

Visit the Taylor & Francis Web site at
http://www.taylorandfrancis.com

and the CRC Press Web site at
http://www.crcpress.com

To the memory of my teacher and friend, Hiro Akaike.

Contents

Part III State Space Modeling

Preface

The aim of this book is to provide a comprehensive picture of fast-growing state space modeling approaches in neuroscience data analysis (such as EEG, MEG, fMRI, and NIRS) together with cutting-edge time series modeling techniques (developed first in Hirotugu (Hiro) Akaike's laboratory and later in Tohru Ozaki's laboratory at the Institute of Statistical Mathematics) to scientists and graduate students in neuroscience and statistics as well as in experimental science and engineering.

Recent developments of measurement technology in brain sciences have produced massive time series data. Typical examples are MEG/EEG data, which measure the dynamically changing electrical and magnetic activity inside the working brain. Here, dynamic change of electromagnetic field is measured at 20–100 locations on the surface of the brain every 10 ms. Another example is functional magnetoresonance imaging (fMRI) data, which measure the hemodynamics inside the working brain, where about 140,000 dimensional time series are generated every 2–3 s. To analyze such massive data, efficient computational and statistical methods are required.

Regrettably, most of the commonly used techniques in this field are either nondynamic or are not time series oriented. For example, in EEG/MEG inverse problems, solutions are derived at each slice of time without taking into consideration the temporal correlations contained in the data. Another example is the independent component analysis (ICA) technique for signal decomposition of EEG, MEG/fMRI time series. Obviously, the dynamic characteristics of the signal in the data are ignored in the ICA treatment, yielding less efficient results than the time series approach.

The main point of interest in this book is to show that the same data can be treated using both a dynamic and time series approach so that the neural and physiological information can be extracted more efficiently. Of course, time series modeling is valid not only in neuroscience data analysis but also in many other sciences and engineering where the statistical inference from the observed time series data plays an important role. The modeling methods presented in this book are also valid in many other fields of science and engineering.

One of the characteristic points in the time series modeling approach is "causal modeling" of the dynamic phenomena through dynamical system models, where the past state can influence and cause the future state, but the future state can never change the past state. Here, the direction of the "arrow of time" from the past to the future plays an essential role.

In the causal modeling of neuroscience data, we use several state variables in order to characterize the neural dynamic phenomena, either in continuous-time dynamical system models or discrete-time dynamical system models.

Thus, in the causal dynamic modeling of neuroscience data, we need to deal with huge dimensional multivariate time series data.

When we take into account the merit of causal dynamic models in real-world science and engineering, it becomes important to pay attention to the observability of the variables—whether they are "observable" or "non-observable" as time series data. In neuroscience, for example, we need to guess what is going on using the measured time series data because we cannot measure all the important variables from the live brain. However, it is very likely that dynamics of observed time series variables are influenced by some other unobserved variables of the live functioning brain. Therefore, state space modeling becomes essential for describing the dynamics and predicting the observed time series variables using some unobserved variables in the state dynamic equation.

In the state space model, the dimension of the observed time series x_t could be smaller than the dimension of the state z_t of the dynamic model. Also the observed data x_t is, in general, contaminated by the observation error e_t. This makes the inference problem of the state space models more complex, and therefore two types of inference problems need to be considered: (1) "inference on state variables," where we assume the parameters of the state space model are specified, and (2) "inference on the model structure" of the state space model. In real applications, what we know are the observed time series, $x_1, x_2, \ldots, x_N$, and we are required to estimate, from time series data, both the unobserved state variables and the state space model itself at the same time. We will see that this innovative approach will play a central role in solving both these inference problems and in developing numerical techniques for estimating the states and the model parameters from time series data.

Another important idea of time series modeling in neuroscience data is heteroscedasticity. The time-varying scale of noise variance is typically seen in time series data of neural activity. We show that some of the time series modeling techniques of heteroscedasticity developed in financial data analysis can be efficiently used to solve similar heteroscedastic problems in neuroscience data analysis by generalizing the techniques into a state space framework.

Using the causal dynamic modeling approach, significant improvement over conventional (mostly static or dynamic but deterministic) methods are expected in the following examples: (1) the inverse solution of EEG/MEG: conventional nondynamic inverse solutions is shown to be improved by extending the model from static model to dynamic model by using the state space modeling approach; (2) Detection of consciousness from EEG of anaesthesized patient: rise and decrease of delta rhythm in EEG are shown to be detectable online using dynamic signal decomposition of EEG using the heteroscedastic state space modeling approach; (3) fMRI activation and connectivity analysis: activation and connectivity between different cortical regions are shown to be specified by the stochastic nearest-neighbor autoregressive

(NN-AR) modeling of spatial time series data of fMRI; (4) Identification of nonlinear dynamic models for EEG: nonlinear differential equation models, such as the Hodgkin–Huxley model, the Zetterberg model, the Balloon model, etc., are shown to be identified from time series data using the state space modeling approach; and (5) Causality analysis of fMRI data: the causality between some chosen visual cortical regions (V1, V5, PP) are shown to be identified using the Akaike causality method.

Among these five topics, I have focused on causality in this book because the method for causality analysis may well turn out to be one of the most useful and influential techniques in science and engineering. (In fact, I was hoping to include all five topics of applications, each with numerical examples and solutions. However, due to space limitations, I had to choose only one. For those who are interested in the other four topics, the papers referenced at the end of the book can be of use.)

"Causality" is a concept that is associated with spatial and temporal "direction." In this respect, causality is very different from the concept of "correlations" in statistics. In the study of dynamic brain functioning, for example, what we are interested in is (1) to find a causal "direction" (from which cortical region to which cortical region?), and (2) to assess the "strength" of the causal influence in relation to some specific brain functions (such as a visual cognitive task) using the measured time series (e.g., fMRI data) from the live working brain. In Chapter 14, we show how Wiener and Granger's original idea of pair-wise causality methods is extended into the more powerful Akaike's total causality method using multivariate AR models as well as state space models.

In this book, I have tried to discuss both continuous-time dynamical system models and time series models (i.e., discrete-time dynamical system models) as useful examples of prediction models. This is rather unusual for standard time series textbooks because both these model families are studied in entirely different branches of mathematics even though they share many similar characteristics. I have tried to do this only because I find both model families useful for real-world time series data analysis.

My goal is not to present the development of "mathematical" time series theory but rather to introduce and use just as little mathematics as possible to explore some of its basic concepts and their derivatives as useful tools for time series analysis. I believe the goal of mathematics is discovery, not proof. Certainly, I belong to the school that holds ideas and exposition to be more important than mere results.

My references to the literature are not systematic. I have tried to give credit for all the work for which I knew an attribution. However, there are many sections in which I have worked out my own treatment, very possibly in ignorance of existing work. Let me apologize in advance to authors thus unwittingly overlooked and affirm my readiness to correct the record at the first opportunity.

While the materials in this book may include many of the standard topics in ordinary time series textbooks, the chapters in Parts I and II are shaped considerably by the ideas covered in Part III.

I hope that the readers not only learn already achieved innovative methods but also learn to develop more innovative methods based on their own problems and necessities. This is exactly the way I was trained by Professor Akaike in the early stages of my career, and I hope that current scientists and engineers pass this on to future generations.

Acknowledgments

I would first like to express my thanks to two of my best friends, the late Hiro Akaike and Pedro Valdes. Their direct and indirect influence on my work has been very strong and can never be expressed in words.

I started my career as a time series analyst at the Institute of Statistical Mathematics (ISM), Tokyo, in 1973, under the supervision and guidance of Hiro Akaike. His influence on my work and career has been immense.

I first met Pedro Valdes in Tokyo in 1994, when I was beginning to be interested in neuroscience time series modeling. Since then he has been a constant source of inspiration and has assisted me in my work on neuroscience time series analysis.

I would like to thank all my friends and colleagues, Rolando Biscay, Jorge Bosch, Mitsunori Iino, Juan Carlos Jimenez, Ryuta Kawashima, Fumi Miwakeichi, the late Roy John, Jorge Riera, Norihiro Sadato, Isao Shoji, Yukihiro Toyoda, Mitchell Valdes, Kevin K.F. Wong, the late Keith Worsley, Okito Yamashita, Atsushi Yoshimoto, and my wife Valerie Haggan-Ozaki, who have helped and inspired me in my work on time series analysis and its applications in science and engineering. Their encouragement and constant support has been of great help. Special thanks go to Andreas Galka, who corrected errors in the first draft.

Hiro Akaike passed away in the summer of 2009 while I was writing this book. I find it difficult to express my grief and sense of loss. I can no longer see him but he will always be there in my heart.

Author

Tohru Ozaki is a mathematician and statistician. He received his BSc in mathematics from the University of Tokyo in 1969. He then joined the Institute of Statistical Mathematics (ISM), Tokyo, in 1970 and studied and worked with Hiro Akaike. He received his DSc from Tokyo Institute of Technology in 1981 under the supervision of Akaike. From 1987 to 2008, he was a professor at ISM and, after Akaike's retirement, served as the director of the prediction and control group. His major research areas include time series analysis, nonlinear stochastic dynamic modeling, predictive control, signal processing and their applications in neurosciences, control engineering, and financial engineering.

While he was at ISM, Ozaki was engaged in various projects in applied time series analysis in science and engineering: EEG dynamic inverse problems, spatial time series modeling of fMRI data, causality analysis in behavioral science, modeling nonlinear dynamics in ship engineering, predictive control design in fossil power plant control, seasonal adjustment in official statistics, heteroscedastic modeling and risk-sensitive control in financial engineering, nonlinear dynamic modeling in macroeconomics, spectral analysis of seismology data, point process modeling of earthquake occurrence data, river-flow prediction in stochastic hydrology, etc.

Ozaki retired from ISM in 2008. Since then he has been a visiting professor at Tohoku University, Sendai, Japan, and at Queensland University of Technology, Brisbane, Australia. He has been involved in supporting several research projects (in dynamic modeling of neuroscience data, fossil power plant control design, and risk-sensitive control in financial engineering) in universities and industry. He has also led, through his international research network, a time series research group called Akaike Innovation School from his office in Mount Fuji and organizes seminars every summer.

1

Introduction

1.1 Time Series Modeling

Time series modeling aims to provide methods for "predicting" and "understanding" dynamic phenomena from observed time series data through modeling. In this book, we are especially interested in time series modeling methods that are useful for neuroscience data, observed from the functioning of the "dynamic brain." Typical examples of neuroscience time series data are EEG (electroencephalography), MEG (magnetoencephalography), fMRI (functional magnetic resonance imaging), and NIRS (near-infrared spectroscopy).

However, the methods are essentially valid for any time series data from any kind of dynamic phenomena in sciences and engineering. Examples of time series produced by dynamic processes are known not only in the neurosciences but also in many research areas in sciences and engineering. Examples are mechanical engineering processes, chemical and industrial engineering processes, geophysical processes, meteorological processes, macroeconomic processes, and financial market processes.

One of the main purposes of time series modeling is to predict "dynamic phenomena." Closely related to prediction is the task of "understanding and explaining" the mechanism of dynamic phenomena, such as the causality between several variables in the phenomena.

From time immemorial, the problem of predicting the future has captured the imagination of statisticians, mathematicians, engineers, and scientists from many disciplines, not to mention humankind in general. The scope and depth of applications of prediction theory are boundless. Leaving aside the crystal ball approach, any prediction method based on scientific principles first needs some kind of "model" that closely describes and to some extent explains the observed dynamic phenomena.

When we predict the dynamic phenomena by a model, we use some observed time series data, which are employed as a kind of objective scientific evidence of the dynamic phenomena. The model we use for the prediction is often revised by using objective evidence and reinforced to make the prediction as precise as possible. Naturally, we try to predict future phenomena, as accurately as possible, by using this reinforced model.

Time series modeling is an inferential task of finding a model for prediction from the observed time series data. In contrast to this, the main task of simulation studies, commonly used for predictions in modern sciences, is to find what will happen (i.e., what kind of trajectory will occur) with a fixed hypothetical model. In sciences and engineering, the decision as to which model to choose for the simulation study is sometimes not obvious, but it is a rather delicate and critical issue. The inferential problem of finding a proper model from observed data is called an "inverse problem," in contrast to the problem of finding possible future events from a fixed model, which is called a "forward problem." Obviously, time series modeling represents an inverse problem.

A prediction model is classified as nondeterministic (stochastic) or deterministic according to whether it does or does not contain random variables. If our prediction is exact and the future value of the process is exactly specified by the initial value of the process, the model is deterministic. However, if the prediction model includes some kind of inexplicable term affecting the future behavior of the process, the inexplicable term is usually treated as a random variable and the model is classified as a stochastic model. Scientists know that deterministic models are possible only in the ideal hypothetical world. In the real world, where our observed time series are exposed to all sorts of external disturbances and interferences, stochastic models are indispensable, and we need to deal with them whether we like it or not.

For the prediction of nondeterministic dynamic phenomena, many types of prediction models have been used in various fields of science and engineering. Before examining these prediction models, we first point out one of the most important relations found by Kolmogorov (1939) (and later by Whittle 1953 for multivariate cases), i.e., the relation between the prediction error variance σ_ε^2 and the power spectrum $p(f)$ of a stationary process:

$$\sigma_\varepsilon^2 = \exp\left\{ \int_{-1/2}^{1/2} \log p(f) df \right\}.$$

The implications of this relation should not be underestimated. It implies that if $p(f) = 0$ for the interval $f_0 \leq f \leq f_0 + \varepsilon$, then $\sigma_\varepsilon^2 = 0$. This means that if the power spectral density of the time series process is zero for any small frequency band $[f_0, f_0 + \varepsilon]$, the process is deterministic and the prediction error variance is zero. It also implies the presence of a continuum of all frequencies in the power spectrum of any nondeterministic process. One of the typical stochastic processes with all the frequencies in its power spectrum is the Gaussian white noise process, and the theorem implies the importance and the fundamental role played by Gaussian white noise in the modeling of the stochastic prediction model. The theorem can be proved under a very mild (weak) condition (namely, stationarity), so it is relevant to most dynamic phenomena studied in science and engineering. Because of this, we will first look into the classic method of characterizing a stationary process by the power spectrum and by its equivalent counterpart, autocovariance functions, in Chapter 2.

Although Kolmogorov's theorem indicates a useful theoretical relation between the prediction error variance and the power spectral density, it does not indicate how to obtain the predictor, prediction errors, and the power spectrum from the observed time series. Since our goal is an actual prediction of the time series, we need to find a model that gives rise to the smallest prediction error variance from the time series data.

1.2 Continuous-Time Models and Discrete-Time Models

Originally, in the era of Kolmogorov and Wiener, time series analysis aimed at predicting and controlling stochastic processes both in discrete time and continuous time. Unfortunately, people working in this area seem to have gradually lost interest in continuous-time modeling problems, and nowadays time series models and model estimation methods are restricted mainly to discrete-time cases.

On the other hand, many well-known useful dynamic models in science and engineering are formulated in continuous time. A question is whether we could use the "continuous-time" model or the "discrete-time" model for the actual prediction of the phenomena from which our time series data are sampled. In this book, we try as much as possible to take advantage of both kinds of models for our prediction purpose.

Although, on the face of it, continuous-time dynamic models and discrete-time dynamic models look quite different, they do in fact have many similarities and may in some ways be thought of as equivalent if we look at the process from the point of view of prediction. The Doob decomposition theorem is a useful theoretical tool to look into the similarities of various types of (mostly nonlinear) dynamic models for stochastic processes or time series. It suggests splitting the process or time series, using a prediction model, into two parts, a predictable part and an unpredictable part. For example, in the discrete-time case, the time series x_t is decomposed by a prediction model M as

$$x_t = y_t^{(M)} + W_t^{(M)}$$

$$= \sum_{k=1}^{t} \Delta y_k^{(M)} + \sum_{k=1}^{t} \Delta W_k^{(M)} \quad t = 1, 2, \dots, \tag{1.1}$$

where
$$\Delta y_k^{(M)} = E^{(M)}[\Delta x_k | x_{k-1}, \dots, x_0] = E^{(M)}[x_k | x_{k-1}, \dots, x_0] - x_{k-1}$$
$$\Delta W_k^{(M)} = x_k - E^{(M)}[x_k | x_{k-1}, \dots, x_0]$$

Here $E^{(M)}[x_k|x_{k-1}, ..., x_0]$ is the predictor of x_k by the model M. We sometimes use $w_k^{(M)}$ for $\Delta W_k^{(M)}$, i.e., $w_k^{(M)} = \Delta W_k^{(M)}$. This decomposition theorem (1.1) is discussed in more detail in Chapter 7.

A disadvantage of continuous-time models is that when we try to estimate the model from time series data, the estimation is not as easy as in the case of discrete-time models. This is because the time series data are usually given at discrete time points. However, continuous-time models offer in some cases advantages over the discrete-time models. One example is the characterization of the non-Gaussian distributional properties of stochastic nonlinear dynamic models.

Sometimes the distribution of time series is non-Gaussian, and this is when nonlinear dynamic models can give rise to better prediction performance than that of linear dynamic models. Although many discrete-time nonlinear dynamic models are known in science and engineering (see, e.g., Pearson 1999), a general characterization of the non-Gaussian distribution of the discrete-time process is almost impossible. We can use these discrete-time nonlinear models to predict better than linear models, but we cannot specify their stochastic distributional properties by analytical methods explicitly. These can be investigated only by numerical methods, including Monte Carlo simulations.

Moreover, a nice analytical tool has been established (Kolmogorov 1931, Wong 1963) for the characterization of the close relationship between non-Gaussian distributions and nonlinear dynamics for the continuous-time case. This means that the distributional properties of some of the discrete-time nonlinear dynamic models can be characterized approximately by the continuous-time nonlinear dynamic model, which is the closest to the discrete-time nonlinear dynamic model.

To understand this, in Chapter 8 we first look into the beautiful work of Kolmogorov and Wong's method for characterizing the non-Gaussian distributional properties of continuous-time Markov diffusion process models. These relations show that many non-Gaussian distributions are characterized as limiting marginal distributions of Markov diffusion processes. Considering that Markov diffusion processes are characterized by nonlinear stochastic differential equation models, and that the processes defined by stochastic differential equations are generated from Gaussian white noise, we can see that many non-Gaussian distributions are in fact generated from Gaussian white noise. Incidentally, if we ignore the temporal dependence of the generated non-Gaussian distributed trajectories of nonlinear stochastic differential equation models, this can be an efficient algorithm of a Metropolis–Hastings type for generating random numbers with a non-Gaussian distribution, to be used in MCMC (Monte Carlo Markov chain) studies. (In fact, non-Gaussian random number generation was one of the motivations of Ozaki's (1985) introduction of the local linearization scheme.)

The above-mentioned results by Kolmogorov and Wong apply only to continuous-time dynamic models. Naturally, we wonder whether we can have similar access to non-Gaussian distribution families from discrete-time dynamic models by making a connecting bridge between discrete-time dynamic model families and continuous-time dynamic model families. The important point to remember here is that these relations are interesting only for non-Gaussian distributions and "stationary" (i.e., finite-variance) stochastic processes. If the process is of infinite variance and is nonstationary, it is meaningless to consider its marginal distribution. Naturally, we are interested in a connecting bridge between continuous-time dynamic models yielding stationary stochastic processes and discrete-time dynamic models yielding stationary time series.

In fact, in the last decades, discretized versions of continuous-time models have been used implicitly in the computational sciences. In physics, biology, and related sciences, continuous-time differential equation models (whether partial or ordinary), in most cases highly nonlinear, have been used in large-scale computer simulations, where the time (and space as well) is discretized by a small interval. Solutions of the differential equations are given on these discrete-time points. Then these solutions can be regarded as trajectories of an approximate nonlinear discrete-time model derived from nonlinear continuous-time models by a discretization scheme. This observation suggests an idea of connecting the two types of model families, one in continuous time and the other in discrete time, by a numerical discretization scheme. The idea is valid for deterministic cases, whereas for stochastic cases it encounters an intrinsic difficulty caused by the Gaussian white noise. As a simple example, consider the following nonlinear stochastic differential equation model:

$$\dot{x} = -x^3 + w(t),$$

which is driven by Gaussian white noise $w(t)$. If we discretize the model by the well-known Euler scheme with a time interval Δt, we have a discrete-time nonlinear stochastic difference equation model (i.e., nonlinear autoregressive model):

$$x_{t+\Delta t} = x_t - \Delta t x_t^3 + \sqrt{\Delta t}\, w_{t+\Delta t}.$$

The problem with this discretized model is that it does not represent a "finite-variance" process, no matter how small a time interval we choose, as long as the variance of Gaussian white noise w_t is nonzero. Apart from the Euler scheme, more sophisticated discretization schemes, such as the Milshtein method and the Runge–Kutta method, are available in numerical mathematics (Kloeden and Platen 1995). These schemes are better than the

Euler scheme in the sense that they converge for $\Delta t \to 0$ faster to the original continuous-time model than the Euler scheme. Unfortunately, however, they also represent an infinite-variance (explosive) process when Δt is fixed to some small constant, no matter how small it is.

For our purpose of connecting the land of stationary continuous-time dynamic model families and the land of stationary discrete-time dynamic model families, we need a proper discretization scheme that transforms a stationary continuous-time process into a stationary discrete-time process for a "fixed" nonzero Δt. Such a scheme is called an "A (absolute)-stable" scheme in numerical mathematics. In Chapter 9, we introduce an A-stable discretization scheme that establishes a kind of "proper" bridge connecting the land of stationary continuous-time dynamic model families and the land of stationary discrete-time dynamic model families. The bridge provides us with a an approximate discrete-time counterpart of a continuous-time non-linear dynamic model for predicting time series and generating prediction errors, which are required for writing down the likelihood of continuous-time models. This is discussed in Chapter 10.

Incidentally, A-stable schemes are useful not only for obtaining the prediction errors for writing down the likelihood of stationary stochastic differential equation models but also for accelerating the speed of computer simulations of highly nonlinear differential equation systems. This is because we do not need to make the interval of time discretization very small to prevent the trajectories from exploding in the simulations. If we can increase the discretization time interval from $\Delta t = 10^{-6}$ to $\Delta t = 10^{-3}$, we can accelerate the computation by a factor of 1000. With the A-stable schemes, the numerical solutions are guaranteed to be nonexplosive irrespective of whether the noise variance is zero, small, or very large.

When we try to use dynamic models for predicting the dynamic systems in science, naturally difficulties may arise, because the systems studied may be known only approximately or partially; i.e., some of the parameters of the model for the system may not be known precisely, or some variables may be missing from the model. The interpretation and prediction of the phenomena based on these imprecise dynamic models bring into question the validity of the scientific conclusion. What we can do to improve the accuracy of the prediction by the model for the phenomena is to take advantage of the time series data of the past together with statistical model identification methods.

An important point of time series modeling in real-world applications is that we do not really know the true model, and we have, for the same time series data, several candidate models, M_i, $(i = 1, 2, \ldots)$. In some cases, these models belong to the same model family but with slightly different parameter values. In other cases, they belong to different model families (such as stochastic differential equation models, autoregressive models, and neural network models). Each dynamic model M_i specifies its own predictor,

$E^{(M_i)}[x_k | x_{k-1}, ..., x_0]$, yielding its own way of decomposing x_t into a predictable part, $y_t^{(M_i)}$, and an unpredictable part, $W_t^{(M_i)}$, i.e.,

$$x_t = y_t^{(M_i)} + W_t^{(M_i)} = \sum_{k=1}^{t} \Delta y_k^{(M_i)} + \sum_{k=1}^{t} w_k^{(M_i)} \quad t = 1, 2, ...$$

Note that the Doob decomposition gives a way of whitening the dependent time series $x_1, x_2, ..., x_N$ into independent series $w_1^{(M_i)}, w_2^{(M_i)}, ..., w_N^{(M_i)}$ for each M_i. Finding a dynamic model M_i attaining the best "predictor," $E^{(M_i)}[x_k | x_{k-1}, ...]$, and the best "whitening method" of the time series, $x_1, x_2, ..., x_N$, is one of the most important and challenging themes in time series analysis. Since we do not know the true model, we can introduce a few candidate models, M_i ($i = 1, 2, ...$), or try to use some commonly known general prediction models, estimate these models from the time series data, and choose the best model from the candidate models.

In classic statistical modeling, the number of models to be considered is not large (usually one or two), and often statistical testing methods have been used. In the case of time series modeling, however, there are usually many candidate models. In real applications, we do not know how many variables are involved in generating the time series. In addition, the dynamics of the generating mechanism may be nonlinear, and the true nonlinear dynamic model may not be included in our set of candidate models for the data analysis. In such situations, the prediction performance of the model improves when the model includes many terms, including high-order lags of the data, $x_{t-1}, x_{t-2}, ..., x_{t-k}$, and their nonlinear functions, such as $x_{t-1}^2, x_{t-1}^3, ..., x_{t-2}^2, x_{t-2}^3, ...$ The problem has often been treated in a framework of statistical model identification problems rather than a framework of statistical testing. The statistical identification problem is solved by a strategy of choosing a model that is closest, in terms of a natural measure, to the true model. That is a strategy of finding the best approximate model rather than one of finding the unique true model. Here, the statistical assessment of the deviation of each model from the true model, in terms of the natural measure, is a critical problem for the validity of the identification method. This topic is discussed in Section 10.5.

1.3 Unobserved Variables and State Space Modeling

Many well-known and useful dynamic models in science and engineering are formulated with several variables. When these dynamic models are employed for time series modeling problems, it is often the case that some of the important variables in the dynamic model are not available as observed data.

If a value x_t within a time series is determined not only by its own past values but also by past values of some other variables that are not directly observed as time series data, difficulties may arise. Moreover, if the variable z_t of a dynamic model is observed with an additive observation error ε_t, the prediction of the time series, x_t, is a little more complicated than that in cases where all the important variables of the dynamic system are directly observed without observation errors. Here, the calculation of the predictor of x_t, i.e., $E[x_t|x_{t-1}, x_{t-2}, \ldots]$, for the derivation of the prediction error,

$$w_t = x_t - E[x_t|x_{t-1}, x_{t-2}, \ldots],$$

is not as simple as that in the case with the usual dynamic models such as an autoregressive model.

Such a situation is typically described by the following linear state space model:

$$\begin{aligned} z_t &= Az_{t-1} + Bw_t \\ x_t &= Cz_t + \varepsilon_t, \end{aligned} \tag{1.2}$$

where $r = \dim(x_t) < \dim(z_t) = k$ and C is an $r \times k$ rectangular matrix, and both w_t and ε_t are Gaussian white noise. Here the question arises about which is the best predictor of the time series x_t, and whether it is possible to identify and estimate the dynamic model, $z_t = Az_{t-1} + Bw_t$, and the unobserved state variable, z_t, from the observed time series, $x_1, x_2, \ldots, x_N$. If this is possible and if appropriate statistical methods are available for solving the problems, a further question is whether we can apply the methods in nonlinear situations where the state dynamic models are either discrete-time nonlinear as

$$\begin{aligned} z_t &= f(z_{t-1}) + Bw_t \\ x_t &= Cz_t + \varepsilon_t, \end{aligned} \tag{1.3}$$

or continuous-time nonlinear as

$$\begin{aligned} \dot{z}(t) &= f(z(t)) + Bw(t) \\ x_t &= Cz_t + \varepsilon_t, \end{aligned} \tag{1.4}$$

where $f(\cdot)$ is a k-dimensional nonlinear function vector.

In such situations, prediction and estimation problems are more complicated, as the available information (i.e., data) is limited compared with other situations where all the variables are observed as time series without observation errors. In the present situation, the following two problems need to be solved for the given time series data $x_1, x_2, \ldots, x_N$:

1. To obtain the best predictor of the state, i.e., z_{t+1}, and the observation, x_{t+1}, from $x_t, x_{t-1}, x_{t-2}, \ldots$, in order to calculate prediction errors
2. To obtain the best estimate of the state space model, e.g.,

$$\dot{z}(t) = f(z(t)) + Bw(t)$$

$$x_t = Cz_t + \varepsilon_t,$$

from $x_1, x_2, \ldots, x_N$.

To solve (1), we need to solve (2), but to solve (2), we need to solve (1), because small prediction errors represent one of the most reliable sources of objective evidence for the goodness of the estimated dynamic model. Therefore, these two problems are inseparable. However, to give a complete solution to the inference problem of the state space models, and to clarify the problems involved, we solve the following two subproblems:

1'. To obtain the best predictor of the state, i.e., z_{t+1}, and the predictor of the time series, x_{t+1}, from $x_t, x_{t-1}, x_{t-2}, \ldots$ to calculate prediction errors of time series $x_1, x_2, \ldots, x_N$, under the assumption that the state space model is specified and given
2'. To obtain the best estimate of the state space model, e.g.,

$$\dot{z}(t) = f(z(t)) + Bw(t)$$

$$x_t = Cz_t + \varepsilon_t,$$

from $x_1, x_2, \ldots, x_N$, using the method of (1').

Here, the inference problem of the state space model identification is divided into two subproblems. We call the first subproblem (1') "inference problem (*a*)" and the second subproblem (2') "inference problem (*b*)." We discuss inference problem (*a*) in Chapter 11 and inference problem (*b*) in Chapter 12.

Topics relating to the problem of maximizing the likelihood (or, equivalently, minimization of (–2)log-likelihood) of state space models are discussed in Chapter 13. The basic strategy of the maximum likelihood method for state space models is the same as that in the case of ordinary nonstate space models. We simply write down the likelihood of the model and maximize the log-likelihood (or minimize the (–2)log-likelihood) with respect to the model parameters. For the maximization of the likelihood, we need to use a numerical nonlinear optimization technique.

The difficulty we face with state space modeling originates mostly from the numerical problems associated with this minimization of the (–2)log-likelihood. As examples for typical numerical difficulties, we mention the divergence (and explosive trajectory) problem associated with the "instability" of the model and the slow "convergence problems" associated with the iterative procedure for the minimization of the (–2)log-likelihood. The situation is similar to the one we face in the minimization of the (–2)log-likelihood of autoregressive moving average (ARMA) models. However, the situation is more complicated with state space models because the predictors and the prediction errors of the state space models are calculated indirectly through the intermediate state variables, and the minimization of (–2)log-likelihood, which is essentially equivalent to the minimization of the prediction error variance, is not as easy as the case for ARMA models.

People may think that the slow convergence problem could be solved by using more sophisticated iterative optimization techniques. However, the problem associated with the maximization of the likelihood of state space models is not so simple. In fact, the cause of the numerical difficulties of slow convergence may come from the model itself, i.e., from improper parameterization of the state space model. If the parameterization of the model is improper (such as redundantly parameterized or wrongly structured), we cannot solve the slow convergence problem just by replacing the numerical optimization method with a different method that is more sophisticated and more efficient. To make things worse, even if the state space model is properly structured and parameterized, it might still not be suitable for the given data, and in this case the maximization of the likelihood may fail.

An Innovation approach is necessary for checking and diagnosing such delicate problems. The innovation approach directs us to find a model M_i that yields the smallest possible prediction error sequence, $w_1^{(M_i)}, w_2^{(M_i)}, \ldots, w_N^{(M_i)}$, satisfying the following three properties: (1) "independence," (2) "homogeneity," and (3) "Gaussianity." Here we have

$$w_t^{(M_i)} = x_t - E^{(M_i)}[x_t | x_{t-1}, x_{t-2}, \ldots],$$

and $E^{(M_i)}[x_t | x_{t-1}, x_{t-2}, \ldots]$ is a predictor of x_t provided by the use of model M_i.

Checking the innovations often helps in updating and improving the model structure. Even though the parameterization of the dynamic model is proper and the residuals are independent after the optimization of the likelihood by adjusting the parameter values of the dynamic model, the residuals may still appear to be inhomogeneous and non-Gaussian. However, in such circumstances, there are still certain ways to improve the model, leading to an improved likelihood, thus bringing us some more useful and more reliable information about the dynamic characteristics of the phenomena behind the time series data.

The "inhomogeneity" of the innovations of the state space model is caused by either inappropriate initial state values or the intrinsic time-varying nature of the driving noise of the state dynamics. If the cause is inappropriate initial state values, we can reduce the variance of the innovations of some period immediately after the starting point, leading to an "improvement" of the likelihood. If the inhomogeneity of the innovation variance is not restricted to the beginning part of the time series, the inhomogeneity is likely to be caused by the latter effect, i.e., by the intrinsically time-varying nature of the driving noise. A method of generalizing a stationary homoscedastic (i.e., with driving noise of a constant variance) state space model into a "heteroscedastic state space model" is introduced in Chapter 13.

The non-Gaussianity of the innovations we may encounter in time series modeling is mostly given by fat-tailed distributions caused by occasional outlying large prediction errors. This kind of non-Gaussianity can be regarded as a sign of the Levy-type driving noise, where Poisson-type shot noise (which is also called jump noise) is combined with the Gaussian white noise. Generalizing the homoscedastic state space model into a state space model with Levy-type driving noise (which we call the local Levy state space model) is one possible way of dealing with and "improving" the likelihood of the state space model with non-Gaussian innovations. A method of identifying jumps and a state space model with Levy-type driving noise is introduced and discussed in Chapter 13.

In Chapter 14, we see how these time series modeling techniques, especially the innovation approach to multivariate autoregressive modeling and to state space modeling, together with the innovation representation of their power spectrum as introduced by Kolmogorov (1939) and Whittle (1953), become useful for solving problems of finding and understanding the causality pattern hidden in the multivariate time series of fMRI data.

Finally, a brief concluding discussion is given in Chapter 15.

Part I

Dynamic Models for Time Series Prediction

Dynamic models play an essential role in explaining, understanding, and predicting the dynamic phenomena behind time series data. Many kinds of dynamic models have been introduced in science and engineering. Some of them are continuous-time dynamic models defined by differential equations. Some of them are defined in discrete time. Some are deterministic models and some are stochastic models, and they are often treated in different branches of mathematics. However, as far as time series data analysis is concerned, they are all useful in one way or another. There is no reason why we should restrict ourselves to one type of dynamic model in our study of time series data analysis. In Part I, we study all these dynamic models from the viewpoint of statistical modeling of time series data in neuroscience.

2

Time Series Prediction
and the Power Spectrum

2.1 Fantasy and Reality of Prediction Errors

2.1.1 Laplace's Demon

From the beginning of human history, the prediction of important dynamic phenomena in everyday life, such as the trajectory of the sun and the moon, attracted people's (not only scientists') strong attention. Naturally, people are interested in finding an explanation and the rule of dynamic change so that they can predict the future of the phenomena exactly. The philosophical proposition that every event, decision, and action, including human cognition and behavior, is causally determined by an initial state of the universe and common universal rules is called determinism. The opposite view is called indeterminism. Laplace strongly believed in causal determinism, which is translated in the following quotation from the introduction to his original essay (Laplace 1814, p. 4):

> We may regard the present state of the universe as the effect of its past and the cause of its future. An intellect which at a certain moment would know all forces that set nature in motion, and all positions of all items of which nature is composed, if this intellect were also vast enough to submit these data to analysis, it would embrace in a single formula of the movements of the greatest bodies of the universe and those of the tiniest atom; for such an intellect nothing would be uncertain and the future just like the past would be present before its eyes.

This intellect is often referred to (later by biographers) as Laplace's demon. Laplace must have recognized that an enormous calculating power and a significant progress in science would still be needed to compute it all in a single instant.

2.1.2 Reality of Prediction Errors

In modern statistical time series analysis, the debate between determinists and nondeterminists could go back at least as far as Wold (1938), who showed that any stationary (even non-Gaussian or deterministic) process has a decomposition,

$$z_t = \xi_t + \eta_t$$

where ξ_t is a deterministic process and η_t is a purely random process represented in the form

$$\eta_t = b_0 v_t + b_1 v_{t-1} + b_2 v_{t-2} + \cdots$$

where v_t is an uncorrelated (i.e., white noise) sequence. The dichotomy of determinism and nondeterminism was further studied in Wiener and Masani (1957) and Cramer (1960). What determinists see from Wold's theorem is the idea that η_t could be zero. What nondeterminists see from the same theorem is naturally ξ_t could be zero. In fact, no one can mathematically prove either of these. The Wold theorem is, in any case, an existence theorem and does not show how to obtain the components ξ_t and η_t. In practice, we anticipate that, from whichever point of view we start, we would finally find ourselves in the same place, that is, with the deterministic process ξ_t and the stochastic process η_t.

In the history of science, the idea of determinism has repeatedly returned through a new fashion. The study of chaos in the 1960s–1980s promoted by scientists in meteorology, zoology, physics, and applied mathematics (Lorenz 1963, Ruelle and Takens 1971, May 1976, Stewart 1989) is a prominent example. According to Stewart (1989), the definition of chaos, proposed at a prestigious international conference on chaos held by the Royal Statistical Society in London in 1986, is stochastic behavior occurring in a deterministic system. Some statisticians (see Tong 1994) thought that the subject would bring a revolution in modeling and data analysis in statistics. Indeed, it would truly be revolutionary if all the noise in stochastic models were replaceable by deterministic chaos. Everyone would naturally prefer to have a prediction model with less uncertainty.

All statisticians would agree that deterministic prediction, if possible, is preferable to stochastic prediction. Time series analysts would be extremely happy to have time series models with zero system noise variance if possible, so that the innovations v_t,

$$v_t = z_t - E[z_t \mid z_{t-1}, z_{t-2}, \dots]$$

defined by the model become zero. The point is whether such deterministic prediction can really work and whether the prediction error could really be zero in practice. Although some time series analysts may still be reluctant to accept leaving behind the idea of determinism (see how many time series analysts had conceived an overly rosy view of chaos in the discussion paper by Bartlett 1990), there are several reasons why we could say that determinism is lost (see Casdagli et al. 1991 or Ozaki et al. 1997).

Since the prediction is not given in a deterministic way and prediction errors are unavoidable, we are forced to consider the stochastic distribution of errors or the conditional predictive distributions, whether we like it or not. This is the nondeterminists' view. By the 1990s, a similar conclusion had been reached from the determinists' side (Takens 1994), where it was shown that small system noise and observation noise are not distinguishable in practice.

If the prediction cannot be deterministic, our second concern should be unbiased prediction and smallest prediction error variance, which has all along been one of the main objectives of conventional time series analysis. In effect, this means there is nothing left to distinguish nonstochastic chaos studies from conventional stochastic nonlinear time series analysis.

Some statisticians still do not seem to notice the gap between the "fantasy" and reality of the prediction error. Tong (1994) has introduced noise into the original chaos model as $z_t = f(z_{t-1}) + n_t$, where $f(.)$ is a nonlinear function and n_t is a Gaussian white noise, and then called the result a noisy chaos model, which surely is a contradiction in terms. Then natural questions arise such as "exactly what is the difference between a conventional nonlinear stochastic dynamical system model and a chaos model?" and "exactly what is revolutionary about a chaos model if the noise in a stochastic model cannot really be replaced by a purely deterministic model?" But answers to these questions are none (with apologies to Carroll 1872), at least no satisfactory ones.

2.1.3 Prediction Error Variance and the Power Spectrum

Having realized that in real data analysis prediction using any kind of model cannot be completely free from prediction errors, we would naturally look for a predictor with zero-mean prediction error, that is, $E[v_t] = 0$, where $v_t = z_t - f(z_{t-1}, ..., z_{t-d})$. Also, we would naturally prefer that the prediction error variance, $\sigma^2 = E[v_t^2]$, should be as small as possible.

The prediction theory of dynamic phenomena has been systematically studied since the 1930s (Wold 1938, Kolmogorov 1939, Wiener 1949). The theory has been developed for both continuous time dynamical systems and for discrete-time dynamical systems. Since the expression of prediction errors is straightforward for the case of discrete-time dynamic models, we look at typical discrete-time dynamical system models first. Among many theorems derived for linear discrete-time prediction theory, the following theorem by Kolmogorov (1939) is one of the most remarkable and intriguing (Hannan 1960).

The theorem states that if the best linear predictor of a stationary process has non-zero prediction error ε_t with the prediction error variance, σ_ε^2, then we have

$$\int_{-1/2}^{1/2} \log p(f)df > -\infty \tag{2.1}$$

where $p(f)$ denotes the power spectral density of the process, and the following relation holds:

$$\sigma_\varepsilon^2 = \exp\left\{\int_{-1/2}^{1/2} \log p(f)df\right\}. \tag{2.2}$$

The multivariate version of the theorem was given by Whittle (1953). Here we assume that the sampling interval Δt is fixed, and so the frequency f is related to the original frequency ω by $2\pi f = \omega$, where $-1/(2\Delta t) < \omega < 1/(2\Delta t)$.

Since this book is aimed mainly at applied time series analysts, we skip this of the theorems. However, we note that the implication of the present theorem is important and is rather surprising. It implies in particular that if the spectrum of a stationary process has zero power for a certain frequency band, $[f, f + \Delta f]$, the stationary process is deterministic. This is seen from the formulas, where the integral is minus infinity as long as $\log p(f)$ is minus infinity for the interval $[f, f + \Delta f]$, no matter how small Δf is.

2.2 Power Spectrum of Time Series

2.2.1 Power Spectrum of Nondeterministic Processes in Continuous Time

The power spectrum of deterministic process $x(t)$ is defined as the square of a Fourier transformed process, that is,

$$p(f) = \left|\int_{-\infty}^{\infty} \exp(-i2\pi ft)x(t)dt\right|^2 .$$

For the integral

$$\int_{-\infty}^{\infty} x(t)\exp(-i2\pi ft)dt$$

to exist and to take a finite value, $x(t)$ needs to satisfy

$$\int_{-\infty}^{\infty} |x(t)|\, dt < \infty.$$

This means $x(t)$ needs to converge to zero, that is, $x(t) \to 0$, for $t \to \pm\infty$. If the process $x(t)$ is a nondeterministic stationary stochastic process, $x(t)$ is fluctuating all the time, and we cannot expect $x(t)$ to have this property. Mathematically speaking,

$$\lim_{T \to \infty} \int_{-T}^{T} \exp(-i2\pi ft)x(t)\, dt$$

does not exist for a nondeterministic stochastic process. In order to obviate this technical difficulty, the power spectrum of a nondeterministic stochastic process is usually defined by the following Fourier Analysis approach, which was introduced by Wiener.

Let us define $x_T(t)$ by

$$x_T(t) = \begin{cases} x(t) & -T \le t \le T \\ 0 & \text{otherwise} \end{cases}$$

then we have $x_T(t) \to 0$ for $t \to \pm\infty$. Then it holds that

$$\int_{-\infty}^{\infty} |x_T(t)|\, dt < \infty,$$

and there exist relations

$$x_T(t) = \int_{-\infty}^{\infty} A_T(f)\exp(i2\pi ft)\, df,$$

and

$$A_T(f) = \int_{-\infty}^{\infty} x_T(t)\exp(-i2\pi ft)\, dt = \int_{-T}^{T} x(t)\exp(-i2\pi ft)\, dt.$$

Now we consider, instead of $A_T(f)$,

$$|A_T(f)|^2 \, df$$

which has the meaning of the contribution to the total energy of the wave $x(t)$ given by the frequency components in the interval $[f, f + df]$. We know that, for $T \to \infty$, the limit

$$\lim_{T \to \infty} |A_T(f)|^2$$

does not exist, but if we consider, instead of $|A_T(f)|^2$, $\dfrac{|A_T(f)|^2}{T}$, then the limit

$$\lim_{T \to \infty} \frac{|A_T(f)|^2}{2T}$$

does exist. Here we note that

$$\frac{|A_T(f)|^2}{T}$$

depends on each realization, $\{x(t), -T \le t \le T\}$, of the process $x(t)$. If we consider the average

$$E_x \left[\frac{|A_T(f)|^2}{2T} \right],$$

we can define the power spectrum of the process independent of the realization of the process.

Let us define $p(f)$ as a limit of the averaged value

$$E_x \left[\frac{|A_T(f)|^2}{2T} \right],$$

that is,

$$p(f) = \lim_{T \to \infty} E \left[\frac{|A_T(f)|^2}{2T} \right].$$

We call this $p(f)$ the "power spectrum" or "power spectral density function" of the nondeterministic process $x(t)$.

Let us define a sample auto covariance function $\hat{R}(\tau)$ by

$$\hat{R}(\tau) = \frac{1}{2T} \int_{-\infty}^{\infty} x_T(u) x_T(u - \tau) du,$$

where

$$x_T(u) = \begin{cases} x(u) & -T \le u \le T \\ 0 & \text{otherwise.} \end{cases}$$

Then we have

$$\frac{|A_T(f)|^2}{2T} = \frac{1}{2T} \int\limits_{-\infty}^{\infty} \exp(-i2\pi f\tau)\left\{ \int\limits_{-\infty}^{\infty} x_T(u)x_T(u-\tau)du \right\} d\tau$$

$$= \int\limits_{-\infty}^{\infty} \exp(-i2\pi f\tau)\hat{R}_T(\tau)d\tau.$$

Here we set

$$\hat{R}_T(\tau) = \frac{1}{2T} \int\limits_{-\infty}^{\infty} x_T(u)x_T(u-\tau)du.$$

Using this relation, the power spectrum $p(f)$ can be written as

$$p(f) = E\left[\lim_{T\to\infty} \frac{|A_T(f)|^2}{2T} \right]$$

$$= \lim_{T\to\infty} \left(E\left[\frac{|A_T(f)|^2}{2T} \right] \right)$$

$$= \lim_{T\to\infty} \left\{ \int\limits_{-\infty}^{\infty} \exp(-i2\pi f\tau)E[\hat{R}_T(\tau)]d\tau \right\}. \qquad (2.3)$$

By rewriting

$$\hat{R}_T(\tau) = \frac{1}{2T} \int\limits_{-\infty}^{\infty} x_T(u)x_T(u-\tau)du$$

as

$$\hat{R}(\tau) = \begin{cases} \dfrac{1}{2T} \int\limits_{-(T-|\tau|)}^{T} x(u)x(u-\tau)du & |\tau| \le T \\ 0 & |\tau| > T, \end{cases}$$

we have

$$E[\hat{R}_T(\tau)] = \begin{cases} \dfrac{1}{2T} \displaystyle\int_{-(T-|\tau|)}^{T} E[x(u)x(u-\tau)]du & |\tau| \le T \\[2ex] 0 & |\tau| > T. \end{cases}$$

For $|\tau| \le T$, we have

$$E[\hat{R}_T(\tau)] = \frac{1}{2T} \int_{-(T-|\tau|)}^{T} R(\tau)]du$$

$$= R(\tau)\left(\frac{2T-|\tau|}{2T}\right) = \left(1 - \frac{|\tau|}{2T}\right)R(\tau). \tag{2.4}$$

Inserting (2.4) into (2.3), we obtain a relation between the power spectrum and the auto covariance function $R(\tau)$:

$$p(f) = \lim_{T \to \infty} \left\{ \int_{-\infty}^{\infty} \exp(-i2\pi f\tau)\left(1 - \frac{|\tau|}{2T}\right)R(\tau)d\tau \right\}$$

$$= \int_{-\infty}^{\infty} \exp(-i2\pi f\tau)R(\tau)d\tau. \tag{2.5}$$

The relation (2.5) is sometimes called Wiener–Khinchin relation. This relation means that if $R(\tau)$ converges fast enough to 0, for $\tau \to \infty$, so that its Fourier integral exists, then the power spectrum of the process exists and is obtained as a Fourier transform of the auto covariance function $R(\tau)$ of the process $x(t)$.

   In general, we cannot expect a nondeterministic stochastic process $x(t) \to 0$ for $t \to \pm\infty$, but it is natural that we expect its auto covariance function $R(\tau) \to 0$ for $\tau \to \pm\infty$, so that

$$\int_{-\infty}^{\infty} |R(\tau)| d\tau < \infty.$$

Then the power spectrum of the process exists and is given by (2.5). When the power spectrum $p(f)$ of the process is given, then the auto covariance function $R(\tau)$ is given by the inverse Fourier transform of the $p(f)$ as

$$R(\tau) = \int_{-\infty}^{\infty} \exp(i2\pi f\tau)p(f)df. \tag{2.6}$$

Thus, the power spectrum and the auto covariance function are related by Fourier and inverse-Fourier transforms. Both the quantities carry the same information of the process $x(t)$. In applied time series analysis, the estimation of the power spectrum from finite length time series data still represents one of the most useful sources of information about the stochastic dynamics behind the data. In the succeeding chapters we will discuss how the power spectrum can be estimated through various types of linear models, both in continuous time and in discrete time. For the nonparametric approach to the estimation of the power spectrum, we refer to Priestley (1981).

2.2.2 Power Spectrum of Nondeterministic Processes in Discrete Time

So far we have seen the definition of power spectrum of continuous-time nondeterministic processes. When we apply this definition to time series process in discrete time with a sampling interval Δt, we need to pay attention to the so-called aliasing problem. Note that for time series with sampling interval Δt, that is, for $x(-l\Delta t)$, ..., $x(-\Delta t)$, $x(0)$, $x(\Delta t)$, $x(2\Delta t)$, ..., $x(s\Delta t)$, ..., we cannot distinguish

$$\exp\left\{i2\pi\left(f+\frac{k}{\Delta t}\right)s\Delta t\right\}$$

from

$$\exp\{i2\pi fs\Delta t\}$$

for $k = 1, 2, ...$ Naturally, this affects the Fourier and inverse-Fourier relations between the power spectrum and the auto covariance function of the discrete-time process. Although we do not have any difficulty in defining the discrete-time auto covariance function $R(\tau)$ for $\tau = m\Delta t$, $m = 0, 1, 2, 3, ...$ However, the power spectrum of the discrete-time process can be defined only for the limited frequency interval, that is, for

$$-\frac{1}{2\Delta t} \le f \le \frac{1}{2\Delta t}.$$

When the lag τ of the auto covariance function $R(\tau)$ is written as $\tau = m\Delta t$, from (2.6), we have

$$R(m\Delta t) = \int_{-\infty}^{\infty} \exp(i2\pi fm\Delta t)p(f)df$$

$$= \int_{-1/2\Delta t}^{1/2\Delta t} \exp(i2\pi fm\Delta t)\left\{\sum_{k=-\infty}^{\infty} p\left(f+\frac{k}{\Delta t}\right)\right\}df.$$

It will be reasonable to define the power spectrum $p(f; \Delta t)$ of the discrete-time process, $x(m\Delta t)$, where $m = \ldots, -2, -1, 0, 1, 2, \ldots$ by

$$p(f; \Delta t) = \sum_{k=-\infty}^{\infty} \left\{ p\left(f + \frac{k}{\Delta t} \right) \right\} \left(-\frac{1}{2\Delta t} \leq f \leq \frac{1}{2\Delta t} \right)$$

and then we have a discrete-time version of (2.6),

$$R(m\Delta t) = \int_{-1/2\Delta t}^{1/2\Delta t} \exp(i2\pi f m \Delta t) p(f; \Delta t) df. \qquad (2.7)$$

As for the Wiener–Khinchin relation (2.5), we have a discrete-time version as follows:

$$p(f; \Delta t) = \sum_{k=-\infty}^{\infty} \left\{ p\left(f + \frac{k}{\Delta t} \right) \right\}$$

$$= \int_{-\infty}^{\infty} \exp(-i2\pi f \tau) R(\tau) d\tau$$

$$= \sum_{m=-\infty}^{\infty} \exp(-i2\pi f m \Delta t) R(m\Delta t) \Delta t. \qquad (2.8)$$

For a time series with a fixed Δt, we sometimes write, for simplicity, f_d for $f\Delta t$, $R_d(m)$ or $R(m)$ for $R(m\Delta t)$, and $p(f)$ or $p_d(f)$ for $p(f; \Delta t)$. Then we may sometimes write the relations (2.7) and (2.8) between the power spectrum and the auto covariance function of time series as

$$p_d(f) = \sum_{m=-\infty}^{\infty} \exp(-i2\pi f m) R_d(m)$$

and

$$R_d(m) = \int_{-\frac{1}{2}}^{\frac{1}{2}} \exp(i2\pi f m) p_d(f) df$$

for the discrete-time process x_t ($t = \ldots, -2, -1, 0, 1, 2, \ldots$).

2.2.3 Aliasing Effect

Note that, when the continuous-time process is observed in discrete time with a small sampling interval Δt, then for any frequency f with $-1/(2\Delta t) \leq f \leq 1/(2\Delta t)$, we cannot distinguish between $\exp(i2\pi f s\Delta t)$ and $\exp\left\{i2\pi\left(f + \dfrac{k}{\Delta t}\right)s\Delta t\right\}$ $(k = 1, 2, \ldots)$. The power spectrum, $p(f; \Delta t)$, at the frequency f, such that $-1/(2\Delta t) \leq f \leq 1/(2\Delta t)$, of the discrete-time process, is a sum of all the powers at frequencies, $f, f + 1/(\Delta t), f + 2/(\Delta t), f + 3/(\Delta t), \ldots$ of the continuous-time process (see (2.8)). Note also that the power spectrum is a symmetric function such that $p(f) = p(-f)$. That means the power at $k/(2\Delta t) - f$ and the power at $k/(2\Delta t) + f$ are not distinguishable (see Figure 2.1a) since we have

$$\exp\left\{i2\pi\left(\frac{k}{2\Delta t} - f\right)s\Delta t\right\} = \exp\left\{i2\pi\left(\frac{-k}{2\Delta t} + f\right)s\Delta t\right\}$$

$$= \exp\left\{i2\pi\left(\frac{k}{2\Delta t} + f\right)s\Delta t\right\}.$$

As a result all the power spectrum information in the continuous-time process is compressed into the discrete-time power spectrum function,

$$p(f; \Delta t) = \sum_{k=-\infty}^{\infty}\left\{p\left(f + \frac{k}{\Delta t}\right)\right\}\left(-\frac{1}{2\Delta t} \leq f \leq \frac{1}{2\Delta t}\right)$$

where the original continuous-time power spectrum $p(f)$ is folded (see Figure 2.1b at $\pm k/(2\Delta t)$ $(k = 1, 2, \ldots)$ and at $\pm k/(\Delta t)(k = 1, 2, \ldots)$. This effect obviously comes from the discretization with the sampling interval Δt. In order to avoid the misinterpretation of the power spectrum in the time series analysis, we need to make sure that the sampling interval is small enough so that the main frequency, f, of the dynamic phenomena of interest is within the range of $-1/(2\Delta t) \leq f \leq 1/(2\Delta t)$.

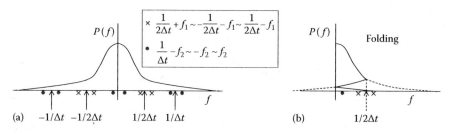

FIGURE 2.1
(a) Aliasing effect. (b) Folding effect.

2.2.4 From Discrete Time to Continuous Time

For the k-variate stationary discrete-time process, we have the relationships

$$R_d(m) = \int_{-1/2}^{1/2} \exp(i2\pi f_d m) p_d(f) df_d \tag{2.9}$$

and

$$p_d(f_d) = \sum_{m=-\infty}^{\infty} \exp(-i2\pi f_d m) R_d(m), \tag{2.10}$$

between the auto covariance function $R_d(m)$ and the power spectrum $p_d(f_d)$. If the parameter ϕ of the stationary autoregressive model,

$$x_t = \phi x_{t-1} + w_t,$$

is positive, that is, $1 > \phi > 0$, so that the trajectory of the discrete-time process x_t is sufficiently smooth, we can think of the case where the discrete-time process x_t comes originally from a continuous-time process $x(t)$ defined by a k-variate stochastic dynamical system model

$$\dot{x} = Ax + w(t)$$

with a sampling interval Δt. Here it makes sense to consider the auto covariance function $R_c(\tau)$ and the power spectrum $p_c(f)$ of the continuous-time process $x(t)$. Then we can rewrite (2.9), for the discrete-time process, as

$$R(m\Delta t) = \int_{-1/2}^{1/2} \exp\left(i2\pi \frac{f_d}{\Delta t} m\Delta t\right) \Delta t p_d\left(\frac{f_d}{\Delta t}\right) \frac{df_d}{\Delta t}$$

$$= \int_{-1/2\Delta t}^{1/2\Delta t} \exp(i2\pi f_{\Delta t} m\Delta t) p_c(f_{\Delta t}) df_{\Delta t}. \tag{2.11}$$

Here we have defined

$$f_{\Delta t} = f_d/\Delta t, \quad -1/2\Delta t < f_{\Delta t} < 1/2\Delta t,$$

and

$$\Delta t p_d(f_{\Delta t}) = p_c(f_{\Delta t}).$$

From (2.11), together with

$$f_{\Delta t} = f_d/\Delta t \xrightarrow[\Delta t \to 0]{} f$$

and

$$m\Delta t \xrightarrow[\Delta t \to 0]{} \tau$$

we obtain for the continuous-time process $x(t)$

$$R_c(\tau) = \int_{-\infty}^{\infty} \exp(i2\pi f \tau) p_c'(f) df. \qquad (2.12)$$

By rewriting (2.10), we obtain

$$\Delta t p_d\left(\frac{f_d}{\Delta t}\right) = \Delta t \sum_{m=-\infty}^{\infty} \exp\left(-i2\pi \frac{f_d}{\Delta t} m\Delta t\right) R'(m\Delta t). \qquad (2.13)$$

With $\Delta t \to 0$, this yields the continuous-time version of the relation (2.10) as

$$p_c(f) = \int_{-\infty}^{\infty} \exp(-i2\pi f \tau) R_c(\tau) d\tau. \qquad (2.14)$$

In summary, we obtain discrete-time relations of the power spectrum and the auto covariance function from the continuous-time relations as

$$\left. \begin{array}{l} R(\tau) = \int_{-\infty}^{\infty} \exp(i2\pi f \tau) p(f) df \\[2mm] p(f) = \int_{-\infty}^{\infty} \exp(-i2\pi f \tau) R(\tau) d\tau \end{array} \right) \xrightarrow{\Delta t} \left(\begin{array}{l} R_d(m) = \int_{-1/2}^{1/2} \exp(i2\pi f_d m) p_d(f) df \\[2mm] p_d(f) = \sum_{m=-\infty}^{\infty} \exp(-i2\pi f_d m) R_d(m). \end{array} \right.$$

Conversely, we can go from the discrete time to continuous time as

$$
\left.
\begin{aligned}
R_d(m) &= \int_{-1/2}^{1/2} \exp(i2\pi f_d m) p_d(f_d) df \\[2mm]
p(f) &= \sum_{m=-\infty}^{\infty} \exp(-i2\pi f_d m) R(m)
\end{aligned}
\right\}
\quad
\xrightarrow[\substack{m\Delta t \to \tau \\ f_{\Delta t} = \frac{f_d}{\Delta t} \to f}]{\Delta t \to 0}
\quad
\left(
\begin{aligned}
R(\tau) &= \int_{-\infty}^{\infty} \exp(i2\pi f \tau) p(f) df \\[2mm]
p(f) &= \int_{-\infty}^{\infty} \exp(-i2\pi f \tau) R(\tau) d\tau,
\end{aligned}
\right.
$$

where the Δt needs to tend to zero together with the extra conditions for τ and f $\left(\text{i.e., } m\Delta t \to \text{ and } \dfrac{f_d}{\Delta t} \to f\right)$. These relations are useful in deriving the parametric power spectrum representation for the continuous-time process from the parametric power spectrum representation for the discrete-time model such as AR model; they are also useful for the opposite direction, from continuous time to discrete time (see Section 5.2.1).

3

Discrete-Time Dynamic Models

We first look into discrete-time models since they are most useful in applications where the observed data are recorded at discrete-time points as time series.

3.1 Linear Time Series Models

3.1.1 AR, MA, and ARMA Models

The following three types of parametric time series models may be regarded as the most useful linear dynamical system models:

1. Autoregressive (AR) model of order p:

$$x_t = \phi_1 x_{t-1} + \cdots + \phi_p x_{t-p} + w_t$$

2. Moving average (MA) model of order q:

$$x_t = \theta_1 w_{t-1} + \cdots + \theta_q w_{t-q} + w_t$$

3. Autoregressive moving average (ARMA) model of orders p and q:

$$x_t = \phi_1 x_{t-1} + \cdots + \phi_p x_{t-p} + \theta_1 w_{t-1} + \cdots + \theta_q w_{t-q} + w_t$$

Here w_t is a Gaussian white noise with mean equal to zero and variance given by σ_w^2. In the AR model, the current value of the process, x_t, is expressed as a finite linear aggregate of previous values of the process and a white noise w_t. In the MA model, the current value of the process, x_t, is a finite linear aggregate of the previous values of the white noise, $w_{t-1}, w_{t-2}, \ldots, w_{t-q}$, plus the current value of the white noise w_t. The name "moving average" is somewhat misleading because the weights 1, θ_1, θ_2, ..., θ_q, which multiply the w's, need not total unity nor need they be positive. However, we use the name since this nomenclature is in common use. In ARMA model, x_t is expressed as a sum of finite linear aggregate of past values of the process and the past white noise inputs. This provides greater flexibility in fitting of time series data.

The coefficients of these models are free independent parameters. However, for these linear models to be useful in time series analysis, the coefficients of these models are expected to satisfy some conditions. For example, for an AR(1) model, $x_t = \phi x_{t-1} + w_t$, to be a reasonable model for a nonexplosive stationary time series, the coefficient ϕ should obey the condition $|\phi| < 1$. Otherwise, the process will diverge for $t \to \infty$, which is not very appropriate for most time series data. In time series analysis, people are usually interested in some dynamic and stochastic properties of the phenomena behind the data with a stationarity assumption.

For the convenience of algebraic manipulation, we shall introduce hereafter the so-called "backward shift operator," B, which is defined by $Bx_t = x_{t-1}$; hence, $B^m x_t = x_{t-m}$. The inverse operation is performed by the "forward shift operator," $F = B^{-1}$, defined by $Fx_t = x_{t+1}$; hence, $F^m x_t = x_{t+m}$. Then the AR(1) model, $x_t = \phi x_{t-1} + w_t$, can be expressed as $(1 - \phi B)x_t = w_t$ and also as $x_t = (1 - \phi B)^{-1} w_t$. Note that $(1 - \phi B)^{-1} = 1 + \phi B + \phi^2 B^2 + \cdots$. This means that the AR(1) model can also be expressed as an infinite order MA model, to be denoted as MA(∞):

$$x_t = w_t + \phi w_{t-1} + \phi^2 w_{t-2} + \phi^3 w_{t-3} + \cdots.$$

For an MA(1) model, $x_t = \theta w_{t-1} + w_t$, to be a reasonable model for a stationary time series, the coefficient θ should obey the condition $|\theta| < 1$. Otherwise, the prediction error will diverge for $t \to \infty$. This is because the prediction error is represented by a weighted sum of the present and the past values of time series as

$$w_t = (1 + \theta B)^{-1} x_t$$

$$= x_t - \theta x_{t-1} + \theta^2 x_{t-2} - \theta^3 x_{t-3} + \cdots,$$

where the weights, $-\theta$, $(-\theta)^2$, $(-\theta)^3$, ..., are explosive for $t \to \infty$ when $|\theta| > 1$.

In general, such constraining conditions for general AR(p) models, MA(q) models, and ARMA(p,q) models are interpreted as stationarity conditions and invertibility conditions.

3.1.2 Stationarity and Invertibility

1. Stationarity:
   Stationarity of a zero mean Gaussian process is equivalent to the existence of the finite variance of the process. Therefore, any MA(q) process is always stationary. For time series defined by the AR-MA model,

$$x_t = \phi_1 x_{t-1} + \cdots + \phi_p x_{t-p} + \theta_1 w_{t-1} + \cdots + \theta_q w_{t-q} + w_t,$$

   or AR model,

$$x_t = \phi_1 x_{t-1} + \cdots + \phi_p x_{t-p} + w_t,$$

to be stationary, the characteristic roots, $\lambda_1, \ldots, \lambda_p$, defined as the roots of the characteristic equation,

$$\Lambda^p - \phi_1 \Lambda^{p-1} - \cdots - \phi_p = 0,$$

need to stay inside the unit circle. This condition is called the "stationarity condition" of the AR(p) and ARMA(p,q) processes. For a given set of coefficients, we can test for stationarity by numerically solving the aforementioned characteristic equation. Also we can easily see in simulations with the Gaussian white noise input whether the process diverges or not. If the characteristic roots stay outside the unit circle, the process will diverge with geometrical speed.

2. Invertibility:

When we calculate the prediction errors from the ARMA model or MA model, we need to express w_t in terms of the present and past observations, $x_t, x_{t-1}, \ldots$. Since w_t is expressed with the backward shift operator as

$$w_t = \frac{(1 - \phi_1 B - \cdots - \phi_p B^p)}{(1 + \theta_1 B + \cdots + \theta_q B^q)} x_t,$$

for w_t to be finite (nonexplosive), the roots, $\mu_1, \ldots, \mu_q$ of the characteristic equation,

$$M^p + \theta_1 M^{p-1} + \cdots + \theta_q = 0,$$

need to stay inside the unit circle. Otherwise, the prediction error w_t will diverge. This condition is called the "invertibility condition." However, in the stage of actual model estimation, this condition is always automatically satisfied because when we are estimating the model parameters, we are numerically searching parameters yielding the smallest possible sum of squares of w_t. Therefore, those MA parameters that do not satisfy the invertibility condition are naturally disregarded in the stage of numerical optimization employed for the parameter estimation.

3.2 Parametric Characterization of Power Spectra

3.2.1 General Linear Process Representation

In order to discuss stochastic properties (such as the power spectrum and the auto covariances) of the AR, MA, and ARMA models, it is convenient to introduce the following infinite order MA representation,

$$x_t = w_t + \psi_1 w_{t-1} + \psi_2 w_{t-2} + \psi_3 w_{t-3} + \cdots,$$

which is called the general linear process representation. All the AR, MA, and ARMA models have general linear process representations with some $\psi_1, \psi_2, \psi_3, \ldots$.

Some simple examples are shown as follows:

1. Example 1: MA(1) model, $x_t = w_t + \theta w_{t-1}$:
   We can rewrite the MA(1) model as

   $$x_t = (1+\theta B)w_t.$$

   Thus, the model has a general linear process representation with

   $$\psi_1 = \theta \quad \text{and} \quad \psi_k = 0 \quad \text{for } k \geq 2.$$

2. Example 2: AR(1) model,

   $$x_t = \phi x_{t-1} + w_t: \tag{3.1}$$

   The model can be rewritten as

   $$x_t = (1-\phi B)^{-1} w_t.$$

   Since

   $$(1-\phi B)^{-1} = (1+\phi B + \phi^2 B^2 + \phi^3 B^3 + \cdots),$$

   we have

   $$x_t = (1+\phi B + \phi^2 B^2 + \phi^3 B^3 + \cdots)w_t$$

   $$= w_t + \phi w_{t-1} + \phi^2 w_{t-2} + \phi^3 w_{t-3} + \cdots$$

   Thus, for AR(1) model (3.1), we have a general linear process representation with

   $$\psi_1 = \phi, \quad \psi_2 = \phi^2, \quad \psi_3 = \phi^3, \cdots.$$

3. Example 3: ARMA(1,1) model: $x_t = \phi x_{t-1} + \theta w_{t-1} + w_t$:
   In the same way as for AR and MA models, we can rewrite the model as

   $$x_t = \phi x_{t-1} + \theta w_{t-1} + w_t$$

   $$= ((1+\theta B)/(1-\phi B))w_t$$

   $$= \left\{(1+\theta B)(1+\phi B + \phi^2 B^2 + \phi^3 B^3 + \cdots)\right\}w_t$$

   $$= w_t + (\phi+\theta)w_{t-1} + (\phi^2 + \theta\phi)w_{t-2} + (\phi^3 + \theta\phi^2)w_{t-3} + \cdots$$

Then we have a general linear process representation with

$$\psi_1 = (\phi + \theta), \quad \psi_2 = (\phi^2 + \theta\phi), \quad \psi_3 = (\phi^3 + \theta\phi^2), \ldots.$$

In general, the MA model of order q, that is, MA(q), the AR model of order p, that is, AR(p), the ARMA model of AR order p, and MA order q, that is, ARMA(p,q) have the following general linear process representations:

1. MA(q) model: $x_t = \theta_1 w_{t-1} + \cdots + \theta_q w_{t-q} + w_t$.
   A corresponding general linear process representation is

   $$x_t = w_t + \psi_1 w_{t-1} + \psi_2 w_{t-2} + \cdots.$$

   Thus, we have $\psi_1 = \theta_1$, $\psi_2 = \theta_2$, ..., $\psi_q = \theta_q$, $\psi_{q+1} = 0$, $\psi_{q+2} = 0$, ...

2. AR(p) model: $x_t = \phi_1 x_{t-1} + \phi_2 x_{t-2} + \cdots + \phi_p x_{t-p} + w_t$.
   A corresponding general linear process representation is

   $$x_t = w_t + \psi_1 w_{t-1} + \psi_2 w_{t-2} + \cdots$$

   where $\psi_1, \psi_2, \psi_3, \ldots$ are given from

   $$\phi(B)\psi(B) = \left(1 + \sum_{j=1}^{p} \phi_j B^j\right)\left(1 + \sum_{j=1}^{\infty} \psi_j B^j\right) = 1.$$

   Note that, for the AR(p) model, we have $\phi(B)x_t = w_t$ and

   $$x_t = \left(1/\phi(B)\right)w_t$$
   $$= \psi(B)w_t.$$

   Computationally, $\psi_1, \psi_2, \psi_3, \ldots$ are obtained by calculating the output of the AR(p) process driven by a unit impulse. The function h_t ($t = 0, 1, 2, 3, \ldots$) defined by $h_0 = 1$, $h_t = \psi_t (t = 1, 2, 3, \ldots)$ is called the impulse response function.

3. ARMA(p,q) model: $x_t = \phi_1 x_{t-1} + \cdots + \phi_p x_{t-p} + \theta_1 w_{t-1} + \cdots + \theta_q w_{t-q} + w_t$.
   A corresponding general linear process representation is

   $$x_t = w_t + \psi_1 w_{t-1} + \psi_2 w_{t-2} + \cdots,$$

   where $\psi_1, \psi_2, \psi_3, \ldots$ are given from

   $$\left(1 + \sum_{j=1}^{p} \phi_j B^j\right)\left(1 + \sum_{j=1}^{\infty} \psi_j B^j\right) = \left(1 + \sum_{j=1}^{q} \theta_j B^j\right).$$

Note that for the ARMA(p,q) model, we have $\phi(B)\, x_t = \theta(B)\, w_t$ and

$$x_t = \frac{\theta(B)}{\phi(B)} w_t = \psi(B)w_t,$$

and the coefficients are obtained from $\phi_1, \ldots, \phi_p$ and $\theta_1, \ldots, \theta_q$ by calculating the impulse response function of the ARMA(p,q) model.

3.2.2 Parametric Auto Covariance Function

The general linear process representation, $x_t = \psi(B)w_t$, of the process x_t is useful for the parametric characterization of the auto covariance function and the power spectrum of AR, MA, and ARMA models. For example, the auto covariance function of the process x_t is given by

$$\gamma_k = E[x_t x_{t+k}]$$

$$= E\left[\sum_{j=0}^{\infty}\sum_{h=0}^{\infty} \psi_j \psi_h w_{t-j} w_{t+k-h}\right]$$

$$= \sigma_w^2 \sum_{j=0}^{\infty} \psi_j \psi_{j+k}.$$

Here, x_t has a general linear process representation:

$$x_t = w_t + \psi_1 w_{t-1} + \psi_2 w_{t-2} + \cdots$$

$$= \left(1 + \sum_{j=1}^{\infty} \psi_j B^j\right) w_t$$

$$= \psi(B)w_t.$$

Let the generating function $\gamma(B)$ of the auto covariance function, $\gamma_k(-\infty < k < \infty)$, be defined as

$$\gamma(B) = \sum_{k=-\infty}^{\infty} \gamma_k B^k$$

$$= \sigma_w^2 \sum_{k=-\infty}^{\infty}\sum_{j=0}^{\infty} \psi_j \psi_{j+k} B^k$$

$$= \sigma_w^2 \sum_{j=0}^{\infty}\sum_{k=-j}^{\infty} \psi_j \psi_{j+k} B^k.$$

σ_w^2 is the variance of the white noise w_t. Let $j + k = h$, $k = h - j$. Then with the use of $\psi_1, \psi_2, \psi_3, \ldots$ of the given model (whether they are AR, MA, or ARMA), $\gamma(B)$ of the process defined by the model can be rewritten as

$$\gamma(B) = \sigma_w^2 \sum_{j=0}^{\infty} \sum_{h=0}^{\infty} \psi_j \psi_h B^{h-j}$$

$$= \sigma_w^2 \sum_{j=0}^{\infty} \psi_h B^h \sum_{h=0}^{\infty} \psi_j B^{-j}$$

$$= \sigma_w^2 \psi(B)\psi(B^{-1})$$

$$= \sigma_w^2 \psi(B)\psi(F). \tag{3.2}$$

From the previous discussion, we know that it holds, for AR(p) model,

$$\psi(B) = \frac{1}{\phi(B)},$$

and for MA(q) model,

$$\psi(B) = \theta(B),$$

and for AR-MA(p,q) model,

$$\psi(B) = \frac{\theta(B)}{\phi(B)}.$$

Then the parametric representation of the generating function of the auto covariance functions of these models are given as follows:

$$\begin{cases} \gamma(B) = \sigma_w^2 \dfrac{1}{\phi(B)\phi(F)} & \text{for AR}(p) \text{ model} \\[2mm] \gamma(B) = \sigma_w^2 \theta(B)\theta(F) & \text{for MA}(q) \text{ model} \\[2mm] \gamma(B) = \sigma_w^2 \dfrac{\theta(B)\theta(F)}{\phi(B)\phi(F)} & \text{for ARMA}(p,q) \text{ model.} \end{cases}$$

3.2.3 Parametric Power Spectrum Representation

Parametric power spectrum function of the process defined by the model (AR, MA or ARMA) can be also specified easily by the use of $\psi_1, \psi_2, \psi_3, \ldots$ Here, the power spectrum function $p(f)$ ($-1/2 \leq f \leq 1/2$) is defined by

$$p(f) = \sum_{k=-\infty}^{\infty} \gamma_k e^{-i2\pi fk}, \quad (-1/2 \leq f \leq 1/2)$$

with the auto covariances, γ_k ($k = 0, \pm1, \pm2, \pm3, \ldots$) Then using (3.2), we have

$$p(f) = \sigma_w^2 \psi(e^{-i2\pi fk}) \psi(e^{i2\pi fk})$$

$$= \sigma_w^2 \, | \, \psi(e^{-i2\pi fk}) \, |^2 \, .$$

From the previous discussion, we know that we have, for AR(p) model,

$$\psi(e^{-i2\pi fk}) = \frac{1}{\phi(e^{-i2\pi fk})} \, ,$$

for MA(q) model,

$$\psi(e^{-i2\pi fk}) = \theta(e^{-i2\pi fk}),$$

and for ARMA(p,q) model,

$$\psi(e^{-i2\pi fk}) = \frac{\theta(e^{-i2\pi fk})}{\phi(e^{-i2\pi fk})}.$$

Then the parametric representations of the power spectrum function of these models are given as follows:

$$p(f) = \frac{1}{\left| 1 - \phi_1 e^{-i2\pi f} - \cdots - \phi_p e^{-i2\pi fp} \right|^2} \sigma_w^2 \quad : \text{for AR}(p) \text{ model.}$$

$$p(f) = \left| 1 + \theta_1 e^{-i2\pi f} + \cdots + \theta_q e^{-i2\pi fq} \right|^2 \sigma_w^2 \quad : \text{for MA}(q) \text{ model.} \qquad (3.3)$$

$$p(f) = \frac{\left| 1 + \theta_1 e^{-i2\pi f} + \cdots + \theta_q e^{-i2\pi fq} \right|^2}{\left| 1 - \phi_1 e^{-i2\pi f} - \cdots - \phi_p e^{-i2\pi fp} \right|^2} \sigma_w^2 \quad : \text{for ARMA}(p,q) \text{ model.}$$

3.2.3.1 Stationarity Conditions for the Parameters

One of the main merits of the use of AR, MA, and ARMA models in time series analysis is that it provides us with a tool for parametric estimation of power spectral density. Here, the coefficients and the noise variance of the best model in the sense of the prediction error (i.e., giving the smallest prediction error variance) also give us a method for a characterization of the power spectral properties of the time series as well.

Note that these parametric power spectrum representations are valid only for the process defined by the coefficients satisfying the "stationary" conditions. "Stationary" AR process means that the stochastic properties of the model do not change over time. This implies that the model parameters are constant over time and that the process does not explode in time. To satisfy this nonexplosive property of the dynamics, we see in Section 3.1.2 that the coefficients need to satisfy the stationarity condition, that is,

"The roots $\lambda_1, \lambda_2, ..., \lambda_p$ of the characteristic equation,

$$(\Lambda^p - \phi_1 \Lambda^{p-1} - \cdots - \phi_p) = 0,$$

lie inside the unit circle, that is, $|\lambda_i| < 1$ $(i = 1, 2, ..., p)$."

Note that any MA process is stationary as long as the model is of finite order q.

3.2.3.2 Peak and Trough Frequencies

We can learn useful properties such as the peaks and the troughs of the power spectrum of the AR, MA, and ARMA processes, from their parametric power spectrum forms. For example, consider the case where the characteristic roots, $\lambda_1 (= a_1 e^{i2\pi f_1}), ..., \lambda_p (= a_p e^{i2\pi f_p})$, of an AR model are within the unit circle but they are close to the unit circle, that is, $a_1 \approx 1, ..., a_p \approx 1$. Since the process is stationary, we have

$$\left| \log p(f) \right| = \log \sigma_w^2 + \log \frac{1}{\left| 1 - \phi_1 e^{-i2\pi f} - \cdots - \phi_p e^{-i2\pi f p} \right|^2} < \infty.$$

Here the denominator is almost zero at the frequencies f_k for $k = 1, 2, ..., p$ so that the power spectrum is close to infinity when f is close to these frequencies. This implies that the power spectrum has very sharp peaks at these frequencies.

Similarly, for the MA models and for ARMA models, if they satisfy the invertibility condition, that is, "the roots, $\mu_1 (= b_1 e^{i2\pi g_1}), ..., \mu_q (= b_p e^{i2\pi g_p})$ of the characteristic equation,

$$M^q + \theta_1 M^{q-1} + \cdots + \theta_q = 0,$$

stay inside the unit circle," but if they are close to the unit circle, the numerator of the parametric power spectrum is almost zero at these frequencies, $g_1, ..., g_q$.

Remember that from the Kolmogorov theory (2.1), we have, for the nondeterministic MA(q) or ARMA(p,q) process,

$$\int_{-1/2}^{1/2} \log p(f) df = \int_{-1/2}^{1/2} \left\{ \log \sigma_w^2 + \left| 1 + \theta_1 e^{-i2\pi f} + \cdots + \theta_q e^{-i2\pi f q} \right|^2 \right\} df > -\infty.$$

This implies that the power spectrum of these MA models or ARMA models have very steep troughs at these frequencies.

3.2.3.3 Efficient Parametric Characterization of Power Spectrum

The earlier discussion shows that

1. AR models

$$x_t = \phi_1 x_{t-1} + \cdots + \phi_p x_{t-p} + w_t$$

   are useful for approximating the power spectrum of Gaussian time series with sharp peaks. When the absolute value of one pair of complex characteristic conjugate roots,

$$\lambda_k = ae^{\pm i2\pi f_k},$$

   of the model is very close to 1, the numerator is close to zero, which means $p(f)$ becomes very large when $f \to \pm f_k$.

2. MA models

$$x_t = \theta_1 w_{t-1} + \cdots + \theta_q w_{t-q} + w_t$$

   are useful for approximating the power spectrum of Gaussian time series with steep troughs. When the absolute value of one pair of complex conjugate roots, $be^{\pm i2\pi g_k}$, of the model is very close to 1, $p(f)$ becomes very close to zero when $f \to \pm g_k$.

3. ARMA models

$$x_t = \phi_1 x_{t-1} + \cdots + \phi_p x_{t-p} + \theta_1 w_{t-1} + \cdots + \theta_q w_{t-q} + w_t$$

   are the most general model for a stationary Gaussian process in the sense that they can approximate the power spectrum of a Gaussian process in an efficient and parsimonious way, because of their flexible functional form, when the power spectrum has either steep troughs or sharp peaks.

3.2.3.4 Whitening and Power Spectrum

Obviously these three parametric models give rise to three different ways of whitening the time series data, $x_1, x_2, \ldots, x_N$, into residuals, $w_1, w_2, \ldots, w_N$. For example,

1. AR model: With the AR(p) model, $x_{p+1}, x_{p+2}, \ldots, x_N$ is transformed into $w_{p+1}, w_{p+2}, \ldots, w_N$ by

$$w_t = x_t - (\phi_1 x_{t-1} + \cdots + \phi_p x_{t-p})$$

   for $t = p + 1, p + 2, \ldots, N$.

2. MA model: With the MA(q) model, $x_1, x_2, \ldots, x_N$ is transformed into $w_1, w_2, \ldots, w_N$ by

$$w_t = x_t - (\theta_1 w_{t-1} + \cdots + \theta_q w_{t-q})$$

with the initial conditions, $w_1 = 0, w_{-1} = 0, \ldots, w_{1-q} = 0$.

3. ARMA model: With the ARMA(p,q) model, $x_{p+1}, x_{p+2}, \ldots, x_N$ is transformed into $w_{p+1}, w_{p+2}, \ldots, w_N$ by

$$w_t = x_t - (\phi_1 x_{t-1} + \cdots + \phi_p x_{t-p} + \theta_1 w_{t-1} + \cdots + \theta_q w_{t-q})$$

with the initial conditions, $w_1 = 0, w_{-1} = 0, \ldots, w_{1-q} = 0$. This means that there are at least three different ways of whitening the time series. Intuitively, we expect that the whitening method that leads to the least sum of squares of the residuals will be the most preferable one. This topic is discussed in Chapter 10.

3.2.3.5 Parsimonious Parameterization

We have seen the multiplicity of the linear prediction model in the previous section. For example, the MA(1) Model $x_t = -\theta w_{t-1} + w_t$ has an equivalent AR(∞) model representation,

$$x_t = -\theta x_{t-1} - \theta^2 x_{t-2} - \theta^3 x_{t-3} - \theta^4 x_{t-4} - \cdots + w_t.$$

This is because the MA(1) model is written as $x_t = (1 - \theta B)w_t$ which is converted to

$$\{1/(1-\theta B)\}x_t = (1 + \theta B + \theta^2 B^2 + \theta^3 B^3 + \cdots)x_t$$

$$= w_t.$$

Similarly, from the ARMA(1,1) model

$$x_t = \phi x_{t-1} + \theta w_{t-1} + w_t$$

we have an equivalent AR(∞) representation,

$$x_t = (\phi + \theta)x_{t-1} - (\theta^2 + \phi\theta)x_{t-2} + (\theta^3 + \phi\theta^2)x_{t-3} - (\theta^4 + \phi\theta^3)x_{t-4} + \cdots + w_t,$$

and an equivalent MA(∞) representation,

$$x_t = (\phi + \theta)w_{t-1} + (\phi^2 + \theta\phi)w_{t-2} + (\phi^3 + \theta\phi^2)w_{t-3} + (\phi^4 + \theta\phi^3)w_{t-4} + \cdots + w_t.$$

From the inferential point of view, this multiplicity of the parametric model is quite an important problem. When the coefficients of the models are unknown, and if scientists want to estimate them from the time series data of finite length, they may naturally choose the model with the smaller number of parameters. If the number of the parameters to estimate is too large, the reliability of the estimates is lost. This principle is called "Occam's razor" or "principle of parsimony." In most scientific experiments, the amount of measured data is finite and limited. Here, the model with the most parsimonious parameterization is what scientists need. It is easy to see that between the three types of linear prediction models, that is, AR, MA, and ARMA models, the ARMA model is the ideal and the most useful one.

3.2.3.6 Causal Meaning of Model Coefficients

The multiplicity of the time series representation implies the danger of some type of causality argument, which results from putting too much emphasis on the significance of the nonzero coefficients of the AR model (Granger 1969). The AR(∞) model with AR parameters, $\phi_1 = -\theta$, $\phi_2 = -\theta^2$, $\phi^3 = -\theta^3$, ..., that is,

$$x_t = -\theta x_{t-1} - \theta^2 x_{t-2} - \theta^3 x_{t-3} - \cdots + w_t,$$

does not necessarily imply that x_t was casually influenced by its infinite past. The model is equivalent to $x_t = w_t - \theta w_{t-1}$, where the influence is only from one step behind.

In the situation of the multiplicity of the parameterization of the time series, more attention needs to be paid to the fact that the power spectrum function and the auto covariance function are unique. This can be easily seen by the use of the general linear process representation in the previous section. A detailed discussion of the causality based on the power spectrum of the multivariate time series models is given in Chapter 14.

3.2.4 Five Different Ways of Characterizing an AR(p) Model

We have seen that AR models, MA models, and ARMA models are useful for the prediction and characterization of the power spectrum of a stationary process. Among these three models, AR models have special advantage in applications because they are computationally much easier to identify from the time series data (a detailed discussion of computational problems is given in Chapter 10). Here we would like to point out that there are many ways of characterizing a Gaussian AR(p) process. At least the following five different ways of characterizing the AR(p) process are worth remembering for a better understanding of the statistical and computational problems associated with the AR models:

1. Characterization by the AR coefficients, ϕ_1, ϕ_2, ..., ϕ_p, and the noise variance σ_w^2
2. Characterization by the auto covariance function, γ_k, ($k = 0, 1, 2, ...$)
3. Characterization by the power spectrum function, $p(f)$, ($-0.5 < f < 0.5$)

4. Characterization by characteristic roots, $\lambda_1, \lambda_2, \ldots, \lambda_p$, and σ_w^2
5. Characterization by the partial auto correlation function, $\phi_{k,k}$, $(k = 1, 2, \ldots, p)$, (specified by (3.9) through (3.12)) and the residual variances, $\sigma^2(k)$, $(k = 0, 1, 2, \ldots, p)$

A brief sketch of the proof of the aforementioned equivalences is given in the following:

Here we start from (1) where an AR(p) model is defined by autoregressive coefficients, $\phi_1, \phi_2, \ldots, \phi_p$, and the noise variance σ_w^2 of Gaussian white noise w_t as

$$x_t = \phi_1 x_{t-1} + \phi_2 x_{t-2} +, \ldots, + \phi_p x_{t-p} + w_t.$$

Since the characteristic roots, $\lambda_1, \lambda_2, \ldots, \lambda_p$, are defined as roots of the characteristic polynomial,

$$\Lambda^p - \phi_1 \Lambda^{p-1} - \phi_2 \Lambda^{p-2} -, \ldots, -\phi_p = 0,$$

the equivalence of (1) and (4) is obvious. Also the equivalence of (2) and (3) is obvious by the definition of the power spectrum function and the auto covariance function, where the two are related by the Fourier and inverse-Fourier relations (see Section 2.2).

The equivalence of (1) and (2) is seen from the following:

The AR coefficients, $\phi_1, \phi_2, \ldots, \phi_p$, and the noise variance, σ_w^2, are derived from the auto covariance function, $\gamma_0, \gamma_1, \gamma_2, \ldots, \gamma_p$, using the relations

$$\gamma_0 = \phi_1 \gamma_1 + \phi_2 \gamma_2 + \cdots + \phi_p \gamma_p + \sigma_w^2$$

and

$$\begin{pmatrix} \gamma_0 & \gamma_1 & \cdots & \gamma_{p-1} \\ \gamma_1 & \gamma_0 & \cdots & \gamma_{p-2} \\ \cdots & \cdots & \cdots & \cdots \\ \gamma_{p-1} & \gamma_{p-2} & \cdots & \gamma_0 \end{pmatrix} \begin{pmatrix} \phi_1 \\ \phi_2 \\ \cdots \\ \phi_p \end{pmatrix} = \begin{pmatrix} \gamma_1 \\ \gamma_2 \\ \cdots \\ \gamma_p \end{pmatrix} \tag{3.4}$$

which follows from

$$E[x_t x_{t-k}] = E[(\phi_1 x_{t-1} + \phi_2 x_{t-2} +, \ldots, + \phi_p x_{t-p}) x_{t-k}] \quad (k = 0, 1, 2, \ldots, p).$$

Equation 3.4 is called Yule–Walker equation. If the auto covariance function γ_k ($k = 0, 1, 2, \ldots$) is given, we can obtain $\phi_1, \phi_2, \ldots, \phi_p$ and σ_w^2 from the aforementioned relations. On the other hand, if we are given $\phi_1, \phi_2, \ldots, \phi_p$

and σ_w^2, the auto covariance function, γ_k ($k = 0, 1, 2, ...$) is given by solving the following linear equation:

$$
\begin{pmatrix}
1 & -\phi_1 & -\phi_2 & \cdots & -\phi_{p-1} & -\phi_p \\
-\phi_1 & 1-\phi_2 & -\phi_3 & \cdots & -\phi_p & 0 \\
-\phi_2 & -\phi_1-\phi_3 & 1-\phi_4 & \cdots & 0 & 0 \\
\cdots & \cdots & \cdots & \cdots & \cdots & \cdots \\
-\phi_{p-1} & -\phi_{p-2}-\phi_p & -\phi_{p-3} & \cdots & 1 & 0 \\
-\phi_p & -\phi_{p-1} & -\phi_{p-2} & \cdots & -\phi_1 & 1
\end{pmatrix}
\begin{pmatrix}
\gamma_0 \\ \gamma_1 \\ \gamma_2 \\ \cdots \\ \gamma_{p-1} \\ \gamma_p
\end{pmatrix}
=
\begin{pmatrix}
\sigma_w^2 \\ 0 \\ 0 \\ \cdots \\ 0 \\ 0
\end{pmatrix},
$$

which comes from the relations

$$
\begin{cases}
\gamma_0 - \phi_1\gamma_1 - \cdots - \phi_p\gamma_p = \sigma_w^2 \\
\gamma_1 - \phi_1\gamma_0 - \cdots - \phi_p\gamma_{p-1} = 0 \\
\gamma_2 - \phi_1\gamma_1 - \cdots - \phi_p\gamma_{p-2} = 0 \\
\qquad \cdots \\
\gamma_p - \phi_1\gamma_{p-1} - \cdots - \phi_p\gamma_0 = 0
\end{cases}
\tag{3.5}
$$

which is obtained by considering

$$
E[x_t x_{t-k}] = E[(\phi_1 x_{k-1} + \cdots + \phi_p x_{t-p})x_{t-k}] \tag{3.6}
$$

for $k = 0, 1, ..., p$.

To understand the equivalence of (5) and (2), it is often useful to understand the symmetric nature of the scalar AR process, where the temporal dependence of the process is characterized by the ordinary forward prediction model

$$
x_t = \phi_1 x_{t-1} + \phi_2 x_{t-2} + \cdots + \phi_p x_{t-p} + w_t
$$

and the backward prediction model

$$
x_t = \pi_1 x_{t+1} + \pi_2 x_{t+2} + \cdots + \pi_p x_{t+p} + v_t.
$$

By comparing the Yule–Walker Equation 3.4 for the forward model and the linear relation,

$$
\begin{pmatrix}
\gamma_0 & \gamma_1 & \cdots & \gamma_{p-1} \\
\gamma_1 & \gamma_0 & \cdots & \gamma_{p-2} \\
\cdots & \cdots & \cdots & \cdots \\
\gamma_{p-1} & \gamma_{p-2} & \cdots & \gamma_0
\end{pmatrix}
\begin{pmatrix}
\pi_1 \\ \pi_2 \\ \cdots \\ \pi_p
\end{pmatrix}
=
\begin{pmatrix}
\gamma_1 \\ \gamma_2 \\ \cdots \\ \gamma_p
\end{pmatrix},
$$

which is obtained from the backward prediction model, as

$$E[x_t x_{t+k}] = E[(\pi_1 x_{t+1} + \pi_2 x_{t+2} + \cdots + \pi_p x_{t+p} + v_t) x_{t+k}]$$

$$= \pi_1 E[x_{t+1} x_{t+k}] + \pi_2 E[x_{t+2} x_{t+k}] + \cdots + \pi_p E[x_{t+p} x_{t+k}] + E[v_t x_{t+k}]$$

$$= \pi_1 \gamma(k-1) + \pi_2 \gamma(k-2) + \cdots + \pi_p \gamma(k-p)$$

(for $k = 0, 1, \ldots, p$), we can see that the coefficients $\pi_1, \pi_2, \ldots, \pi_p$ and $\phi_1, \phi_2, \ldots, \phi_p$ are equivalent. However, we distinguish them by using different symbols because they have different meaning. In addition, the equivalence of the forward coefficients and the backward coefficients holds only for scalar AR processes. For the multivariate AR process, they are no longer equal (see Section 4.4), while still playing an important role in the stage of model identifications by using Levinson's scheme (Levinson 1947) (see Section 10.4.2.2).

The partial autocorrelations, $\phi_{k,k}$, ($k = 1, 2, \ldots, p$), and the residual variances, $\sigma^2(k)$, ($k = 0, 1, 2, \ldots, p$), are defined by the solutions of

$$E\big[(x_t - \phi_{k,1} x_{t-1} - \cdots - \phi_{k,k} x_{t-k}) x_{t-l}\big] = 0 \quad \text{for} \quad l = 1, 2, \ldots, k,$$

and

$$\sigma^2(k) = E\big[(x_t - \phi_{k,1} x_{t-1} - \cdots - \phi_{k,k} x_{t-k}) x_t\big].$$

This means that they are obtained by solving the Yule–Walker equation (3.4) and

$$\sigma^2(k) = \gamma_0 - \phi_{k,1} \gamma_1 - \cdots - \phi_{k,k} \gamma_k$$

for $k = 0, 1, 2, \ldots, p$. For $k = 0$, we have $\sigma^2(0) = \gamma_0$. This means that they are specified if the auto covariance function, $\{\gamma_0, \gamma_1, \gamma_2, \ldots\}$, of the AR process is known.

On the other hand if we start from $\{\sigma^2(0), \phi_{1,1}, \sigma^2(1), \phi_{2,2}, \ldots, \sigma^2(p), \phi_{p,p}\}$, we can obtain the auto covariance function, $\{\gamma_0, \gamma_1, \gamma_2, \ldots, \gamma_p, \ldots\}$, of the process as the following:

$$\begin{cases} \gamma_0 = \sigma^2(0) \\ \gamma_1 = \gamma_0 \phi_{1,1} \\ \gamma_2 = \phi_{2,2} \sigma^2(1) + \phi_{1,1} \gamma_1 \\ \gamma_3 = \phi_{3,3} \sigma^2(2) + \phi_{2,1} \gamma_2 + \phi_{2,2} \gamma_1 \\ \quad \cdots \\ \gamma_p = \phi_{p,p} \sigma^2(p) + \phi_{p-1,1} \gamma_{p-1} + \cdots + \phi_{p-1,p-1} \gamma_1. \end{cases} \tag{3.7}$$

The higher auto covariances, γ_k ($k = p + 1, p + 2, \ldots$), are obtained by applying the highest order AR coefficients, $\phi_{p,1}, \phi_{p,2}, \ldots, \phi_{p,p}$, as

$$\gamma_k = \phi_{p,1}\gamma_{k-1} + \phi_{p,2}\gamma_{k-2} + \cdots + \phi_{p,p}\gamma_{k-p}.$$

Thus, we saw the equivalence of (5) and (2).

We note that relations used in (3.7) are obtained from the following recursive scheme originally proposed by Levinson (1947):

1. For $j = 0$ $\sigma^2(0) = \gamma_0$ $\qquad\qquad\qquad\qquad\qquad\qquad\qquad$ (3.8)

2. From $j = 0$ to $j = 1$ $\left(\begin{aligned} &\phi_{1,1} = \gamma(1)/\gamma_0 \\ &\sigma^2(1) = (1 - \phi_{1,1}^2)\sigma^2(0) \end{aligned} \right.$ $\qquad\qquad$ (3.9)

3. From $j = 1$ to $j = 2$ $\left(\begin{aligned} &\phi_{2,2} = \{\gamma_2 - \phi_{1,1}\gamma_1\}/\sigma^2(1) \\ &\phi_{2,1} = \phi_{1,1} - \phi_{2,2}\phi_{1,1} \\ &\sigma^2(2) = (1 - \phi_{2,2}^2)\sigma^2(1) \end{aligned} \right.$ $\qquad$ (3.10)

4. From $j = 2$ to $j = 3$ $\left(\begin{aligned} &\phi_{3,3} = \{\gamma_3 - \phi_{2,1}\gamma_2 - \phi_{2,2}\gamma_1\}/\sigma^2(2) \\ &\phi_{3,i} = \phi_{2,i} - \phi_{3,3}\phi_{2,3-i} \quad (i = 2, 1) \\ &\sigma^2(3) = (1 - \phi_{3,3}^2)\sigma^2(2) \end{aligned} \right.$ $\qquad$ (3.11)

. . . .

5. From $j = k$ to $j = k + 1$ $\left(\begin{aligned} &\phi_{k+1,k+1} = \{\gamma_{k+1} - \phi_{k,1}\gamma_k - \cdots - \phi_{k,k}\gamma_1\}/\sigma^2(k) \\ &\phi_{k+1,i} = \phi_{k,i} - \phi_{k+1,k+1}\phi_{k,k+1-i}. \quad (i = k-1, \ldots, 1) \\ &\sigma^2(k+1) = (1 - \phi_{k+1,k+1}^2)\sigma^2(k) \end{aligned} \right.$

$$\qquad\qquad\qquad\qquad\qquad\qquad\qquad\qquad\qquad\qquad\qquad\qquad (3.12)$$

The proof is omitted since readers can find it in most introductory time series textbooks. The Levinson–Durbin procedure is a special case of the Levinson scheme for the multivariate AR models, and the proof for the multivariate Levinson procedure is given in Section 10.4.2.2 (the proof for the scalar case is derived from the multivariate case by using the equivalence of the forward model coefficients and the backward model coefficients)

Incidentally, Levinson (1947) introduced recursive schemes for both scalar AR processes and multivariate AR processes. However, he modestly wrote that one might regard them as mathematically trivial. The same procedure was rediscovered later for a scalar case by Durbin (1960), and statisticians started attributing the procedure to Durbin and calling the procedure Durbin's scheme. The same procedure for the multivariate AR processes are developed by Whittle (1963a). Even though Levinson himself considered it mathematically trivial, however, we should not ignore his original contribution, which predates the work of Durbin and Whittle.

3.3 Tank Model and Introduction of Structural State Space Representation

3.3.1 State Space Representation and AR-MA Models

When Box and Jenkins (1970) promoted ARMA models as the most parsimonious parametric models for stationary Gaussian time series, they did so without much reference to state space models. Most time series textbooks (Harvey 1989, Durbin and Koopman 2001, Schumway and Stoffer 2000, West and Harrison 1997, Brockwell and Davis 2006, Lutkepohl 2006) introduce state space models as tools for designing a specific structural model for time series, without reference to its role as a general model for stationary Gaussian processes.

In this section, we focus our attention to the role of the model family of state space models as the most general and flexible models for stationary Gaussian Markov processes. It will be shown that for any stationary Gaussian time series a state space model can be identified, and the family of ARMA models is reintroduced as a family of models equivalent to state space models.

The advantage of the state space modeling approach will be understood well when we step forward from the scalar time series modeling to the multivariate time series modeling in Chapter 4; for multivariate time series, designing a specific structure of the state space model becomes too complicated because of higher dimensionality and complex feedback relationships between variables.

Besides being useful for the characterization of Gaussian processes, state space models turn out to be also useful, and in a sense most suitable, for the approximation and characterization of any nonlinear and non-Gaussian Markov diffusion-type processes, where the trajectories of the processes are continuous in time, and observed data are time series sampled in discrete time. The important role played by state space models in the characterization of nonlinear and non-Gaussian time series is described in Chapters 6 and 11 through 14.

3.3.2 Characteristic Roots and State Space Models

If the p characteristic roots of an AR(p) model are composed of p_1 pairs of complex conjugate roots and p_2 real roots (so that $p = 2p_1 + p_2$), we can regard the AR(p) process as an output of a consecutively chained $(p_1 + p_2)$ system, where the output of one system is the input to another system. Here each component system is either AR(1) or AR(2) model and is specified by either one of p_2 real roots or one of p_1 pairs of complex conjugate roots.

For example, if the fifth order AR model driven by a white noise n_t

$$x_t = \phi_1 x_{t-1} + \phi_2 x_{t-2} + \phi_3 x_{t-3} + \phi_4 x_{t-4} + \phi_5 x_{t-5} + n_t \tag{3.13}$$

has two pairs of complex conjugate roots, $\lambda_1, \bar{\lambda}_1$ and $\lambda_2, \bar{\lambda}_2,$ and one real root $\lambda_3,$ the characteristic polynomial of the AR(5) model (3.13) is factorized into 3 factors as

$$(\Lambda^5 - \phi_1 \Lambda^4 - \phi_2 \Lambda^3 - \phi_3 \Lambda^2 - \phi_4 \Lambda - \phi_5)$$

$$= (\Lambda - \lambda_1)(\Lambda - \bar{\lambda}_1)(\Lambda - \lambda_2)(\Lambda - \bar{\lambda}_2)(\Lambda - \lambda_3)$$

$$= (\Lambda^2 - \phi_1^{(1)} \Lambda - \phi_2^{(1)})(\Lambda^2 - \phi_1^{(2)} \Lambda - \phi_2^{(2)})(\Lambda - \phi_1^{(3)}).$$

Here we have

$$\phi_1^{(1)} = \lambda_1 + \bar{\lambda}_1, \quad \phi_2^{(1)} = -|\lambda_1|^2,$$

$$\phi_1^{(2)} = \lambda_2 + \bar{\lambda}_2, \quad \phi_2^{(2)} = -|\lambda_2|^2,$$

and

$$\phi_1^{(3)} = \lambda_3.$$

This implies that the present AR(5) process x_t is generated from an AR(2) process driven by $x_t^{(1)}$ as

$$x_t = \phi_1^{(1)} x_{t-1} + \phi_2^{(1)} x_{t-2} + x_t^{(1)} \tag{3.14}$$

and $x_t^{(1)}$ is generated from an AR(2) process driven by $x_t^{(2)}$ as

$$x_t^{(1)} = \phi_1^{(2)} x_{t-1}^{(1)} + \phi_2^{(2)} x_{t-2}^{(1)} + x_t^{(2)}. \tag{3.15}$$

Here $x_t^{(2)}$ is generated from an AR(1) process driven by a white noise n_t as

$$x_t^{(2)} = \phi_1^{(3)} x_{t-1}^{(2)} + n_t \tag{3.16}$$

Using the backward shift operator B, we can rewrite (3.14), (3.15), and (3.16) as

$$\phi^{(1)}(B)x_t = x_t^{(1)}, \tag{3.17}$$

$$\phi^{(2)}(B)x_t^{(1)} = x_t^{(2)}, \tag{3.18}$$

$$\phi^{(3)}(B)x_t^{(2)} = n_t \tag{3.19}$$

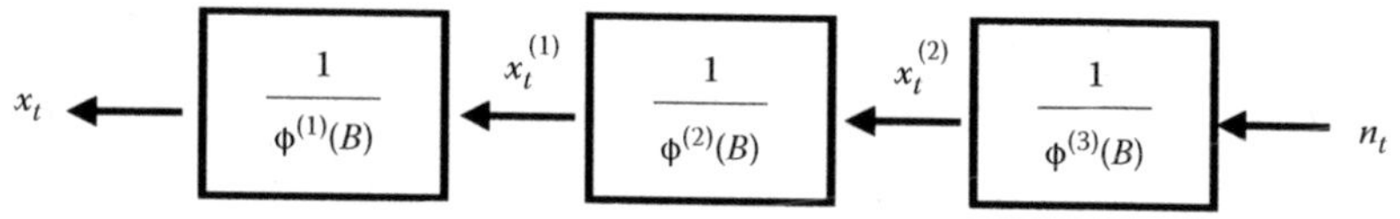

$$\left(1 - \phi_1^{(1)}B - \phi_2^{(1)}B^2\right)\left(1 - \phi_1^{(2)}B - \phi_2^{(2)}B^2\right)\left(1 - \phi_1^{(3)}B\right)x_t = n_t$$

FIGURE 3.1
Three subsystems in sequence.

and

$$\phi^{(3)}(B)\phi^{(2)}(B)\phi^{(1)}(B)x_t = n_t$$

where $\phi^{(1)}(B) = \left(1 - \phi_1^{(1)}B - \phi_2^{(1)}B^2\right)$, $\phi^{(2)}(B) = \left(1 - \phi_1^{(2)}B - \phi_2^{(2)}B^2\right)$ and $\phi^{(3)}(B) = \left(1 - \phi_1^{(3)}B\right)$. This means that the AR(5) model (3.13) can be regarded as a dynamic system composed of three subsystems chained in a sequential way as in Figure 3.1.

3.3.2.1 Sequential-Type Structural Model

Note that the model (3.13) can be considered as one of the many possible structural models derived from the set of characteristic roots, $\lambda_1, \bar{\lambda}_1$ and $\lambda_2, \bar{\lambda}_2$, and λ_3. In Figure 3.1, the three subsystems are coupling in a sequential way. The structural model has a state space representation as

$$\begin{cases} z_t = A^{(seq)}z_{t-1} + \eta_t^{(seq)} \\ x_t = C^{(seq)}z_t \end{cases} \tag{3.20}$$

where $z_t = (x_t, x_{t-1}, x_{t-2}, x_{t-3}, x_{t-4})'$, $\eta_t^{(seq)} = (n_t, 0, 0, 0, 0)'$, $C^{(seq)} = (1, 0, 0, 0, 0)$, and

$$A^{(seq)} = \begin{pmatrix} \phi_1 & \phi_2 & \phi_3 & \phi_4 & \phi_5 \\ 1 & 0 & 0 & 0 & 0 \\ 0 & 1 & 0 & 0 & 0 \\ 0 & 0 & 1 & 0 & 0 \\ 0 & 0 & 0 & 1 & 0 \end{pmatrix}.$$

3.3.2.2 Parallel-Type Structural Model

We note that, out of the same set of characteristic roots, $\lambda_1, \bar{\lambda}_1$ and $\lambda_2, \bar{\lambda}_2$, and λ_3, we can think of another different structural model as in Figure 3.2, where the same three subsystems are arranged in a parallel way, where each subsystem

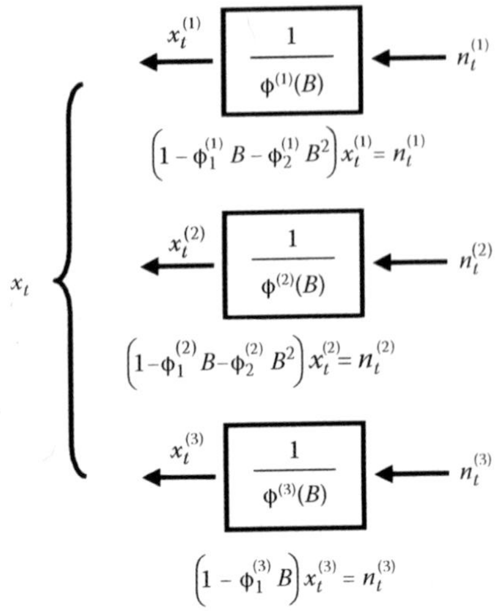

FIGURE 3.2
Three subsystems in parallel.

The diagram shows:

$$\dfrac{1}{\phi^{(1)}(B)}$$ with input $x_t^{(1)}$ and $n_t^{(1)}$

$$\left(1 - \phi_1^{(1)} B - \phi_2^{(1)} B^2\right) x_t^{(1)} = n_t^{(1)}$$

$$\dfrac{1}{\phi^{(2)}(B)}$$ with input $x_t^{(2)}$ and $n_t^{(2)}$

$$\left(1 - \phi_1^{(2)} B - \phi_2^{(2)} B^2\right) x_t^{(2)} = n_t^{(2)}$$

$$\dfrac{1}{\phi^{(3)}(B)}$$ with input $x_t^{(3)}$ and $n_t^{(3)}$

$$\left(1 - \phi_1^{(3)} B\right) x_t^{(3)} = n_t^{(3)}$$

is driven by its own noise source independently, and the time series x_t results as a sum of the outputs of these three subsystems.

The model has a state space representation as

$$\begin{cases} z_t = A^{(para)} z_{t-1} + \eta_t^{(para)} \\ x_t = C^{(para)} z_t \end{cases} \tag{3.21}$$

where $z_t = \left(x_t^{(1)}, x_{t-1}^{(1)}, x_t^{(2)}, x_{t-1}^{(2)}, x_t^{(3)}\right)'$, $\eta_t^{(para)} = \left(\varepsilon_t^{(1)}, 0, \varepsilon_t^{(2)}, 0, \varepsilon_t^{(3)}\right)'$, $C^{(para)} = (1, 0, 1, 0, 1)$ and

$$A^{(seq)} = \begin{pmatrix} \phi_1^{(1)} & \phi_2^{(1)} & 0 & 0 & 0 \\ 1 & 0 & 0 & 0 & 0 \\ 0 & 0 & \phi_1^{(2)} & \phi_2^{(2)} & 0 \\ 0 & 0 & 1 & 0 & 0 \\ 0 & 0 & 0 & 0 & \phi_1^{(3)} \end{pmatrix}.$$

3.3.2.3 Mixed-Type Structural Model

Out of the same set of characteristic roots, λ_1, $\bar{\lambda}_1$ and λ_2, $\bar{\lambda}_2$, and λ_3, the following two structures of the mixed sequential and parallel types are also possible (see Figures 3.3 and 3.4). The structural model of Figure 3.3 can be described as a state space model,

$$\begin{cases} z_t = A^{(mix1)} z_{t-1} + \eta_t^{(mix1)} \\ x_t = C^{(mix1)} z_t \end{cases} \tag{3.22}$$

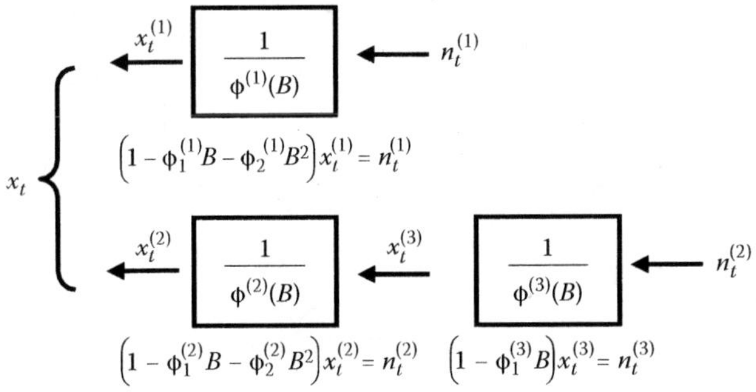

FIGURE 3.3
Three subsystems in mixed-type 1.

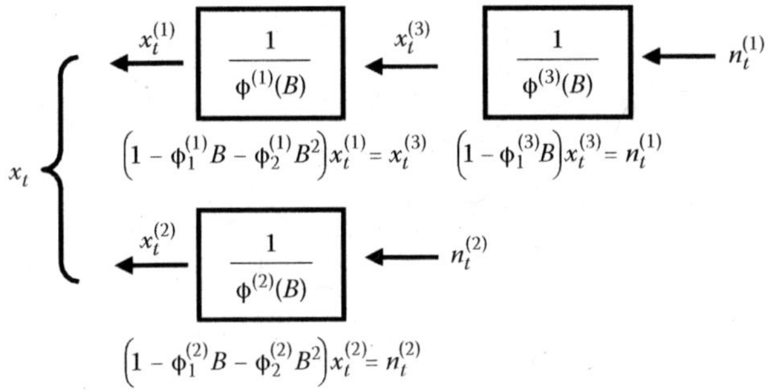

FIGURE 3.4
Three subsystems in mixed-type 2.

where $z_t = \left(x_t^{(1)}, x_{t-1}^{(1)}, x_{t-2}^{(1)}, x_t^{(2)}, x_{t-1}^{(2)} \right)$, $\eta_t^{(mix1)} = \left(\varepsilon_t^{(1)}, 0, 0, \varepsilon_t^{(2)}, 0 \right)$, $C^{(mix1)} = (1, 0, 0, 1, 0)$, and

$$
A^{(mix1)} = \begin{pmatrix}
\phi_1^{(1)} & \phi_2^{(1)} & \phi_3^{(1)} & 0 & 0 \\
1 & 0 & 0 & 0 & 0 \\
0 & 1 & 0 & 0 & 0 \\
0 & 0 & 0 & \phi_1^{(2)} & \phi_2^{(2)} \\
0 & 0 & 0 & 1 & 0
\end{pmatrix}.
$$

Here $\phi_1^{(1)}$, $\phi_2^{(1)}$ and $\phi_3^{(1)}$ come from $\phi^{(1)}(B)\,\phi^{(3)}(B) = 1 - \phi_1^{(1)}B - \phi_2^{(1)}B^2 - \phi_3^{(1)}B^3$.

The structural model of Figure 3.4 can be described as a state space model as

$$
\begin{cases}
z_t = A^{(mix2)} z_{t-1} + \eta_t^{(mix2)} \\
x_t = C^{(mix2)} z_t
\end{cases}
\tag{3.23}
$$

where $\quad z_t = \left(x_t^{(1)}, x_{t-1}^{(1)}, x_t^{(2)}, x_{t-1}^{(2)}, x_{t-2}^{(2)}\right)'$, $\quad \eta_t^{(mix2)} = \left(\varepsilon_t^{(1)}, 0, \varepsilon_t^{(2)}, 0, 0\right)'$, $\quad C^{(mix2)} \quad =$
$(1, 0, 1, 0, 0)$, and

$$
A^{(mix2)} = \begin{pmatrix} \phi_1^{(1)} & \phi_2^{(1)} & 0 & 0 & 0 \\ 1 & 0 & 0 & 0 & 0 \\ 0 & 0 & \phi_1^{(2)} & \phi_2^{(2)} & \phi_3^{(2)} \\ 0 & 0 & 1 & 0 & 0 \\ 0 & 0 & 0 & 1 & 0 \end{pmatrix}
$$

$\phi_1^{(2)}$, $\phi_2^{(2)}$ and $\phi_3^{(2)}$ come from $\phi^{(2)}(B)\,\phi^{(3)}(B) = 1 - \phi_1^{(2)}B - \phi_2^{(2)}B^2 - \phi_3^{(2)}B^3$.

3.3.3 Parametric Power Spectrum of the Structural Models

From the definition of the AR power spectrum, we can have parametric power spectrum representations of the structural models as well.

For x_t of the sequential-type structural model (3.20), we have

$$
p(f) = \frac{1}{2\pi}\left\{ \frac{\sigma_\varepsilon^2}{\left|1 - \phi_1 e^{-i2\pi f} - \phi_2 e^{-i2\pi f \times 2} - \phi_3 e^{-i2\pi f \times 3} - \phi_4 e^{-i2\pi f \times 4} - \phi_5 e^{-i2\pi f \times 5}\right|^2} \right\}. \quad (3.24)
$$

For x_t of the parallel-type structural model (3.21), we have

$$
p(f) = \frac{1}{2\pi}\left\{ \frac{\sigma_{\varepsilon^{(1)}}^2}{\left|1 - \phi_1^{(1)} e^{-i2\pi f} - \phi_2^{(1)} e^{-i2\pi f \times 2}\right|^2} + \frac{\sigma_{\varepsilon^{(2)}}^2}{\left|1 - \phi_1^{(2)} e^{-i2\pi f} - \phi_2^{(2)} e^{-i2\pi f \times 2}\right|^2} + \frac{\sigma_{\varepsilon^{(3)}}^2}{\left\|1 - \phi_1^{(3)} e^{-i2\pi f}\right\|} \right\}
$$

$$(3.25)$$

For x_t of the mixed-type structural model (3.22), we have

$$
p(f) = \frac{1}{2\pi}\left\{ \frac{\sigma_{\varepsilon^{(1)}}^2}{\left|1 - \phi_1^{(1)} e^{-i2\pi f} - \phi_2^{(1)} e^{-i2\pi f \times 2} - \phi_3^{(1)} e^{-i2\pi f \times 3}\right|^2} + \frac{\sigma_{\varepsilon^{(2)}}^2}{\left|1 - \phi_1^{(2)} e^{-i2\pi f} - \phi_2^{(2)} e^{-i2\pi f \times 2}\right|^2} \right\}
$$

$$(3.26)$$

For x_t of the mixed-type structural model (3.23), we have

$$
p(f) = \frac{1}{2\pi}\left\{ \frac{\sigma_{\varepsilon^{(1)}}^2}{\left|1 - \phi_1^{(1)} e^{-i2\pi f} - \phi_2^{(1)} e^{-i2\pi f \times 2}\right|^2} + \frac{\sigma_{\varepsilon^{(2)}}^2}{\left|1 - \phi_1^{(2)} e^{-i2\pi f} - \phi_2^{(2)} e^{-i2\pi f \times 2} - \phi_3^{(2)} e^{-i2\pi f \times 3}\right|^2} \right\}
$$

$$(3.27)$$

We note that each state space model, (3.20), (3.21), (3.22), and (3.23), is equivalent to an ARMA(5,4) model. This is because the p-dimensional state space model,

$$
z_t = Az_{t-1} + Bw_t
$$

$$
x_t = Cz_t
$$

is, in general, transformed to an ARMA($p,p-1$) model as follows,

$$x_t - \phi_1 x_{t-1} - \cdots - \phi_p x_{t-p}$$

$$= CBw_t + C(A - \phi_1 I)Bw_{t-1}$$

$$+ C(A^2 - \phi_1 A - \phi_2 I)Bw_{t-2}$$

$$+ \cdots$$

$$+ C(A^{p-1} - \phi_1 A^{k-2} - \cdots - \phi_{p-1} I)Bw_{t-p+1},$$

where $\phi_1, \phi_2, \ldots, \phi_p$ are characteristic coefficients of the transition matrix A, that is,

$$A^p - \phi_1 A^{p-1} - \cdots - \phi_{p-1} A - \phi_p I = 0.$$

An implication of the four power spectra and the four state space models presented earlier is worth noting. We had four parametric power spectra, (3.24), (3.25), (3.26), (3.27), corresponding to these four structural models, (3.20) through (3.23), where the models share the same set of eigenvalues, $\lambda_1, \bar{\lambda}_1$ and $\lambda_2, \bar{\lambda}_2$, and λ_3. Here the autoregressive parts of the four ARMA models are equivalent since they share the identical set of eigen–values, whereas the moving average parts of the four ARMA models are different. This means that the structural information is in the moving average part of the ARMA models.

Since the AR parts of the four models are the same, their spectra have peaks at the same frequencies. The differences of the spectra of the four models are in the shape of the dips, which are described by the MA part of the model. This means that in many applications of "structural modeling" to time series, we may start from AR modeling with some eigen-values in mind and then introduce an AR model with a constraint for the MA part, based on their subjective preference. Whether these subjective choices are justified or not should be checked by some objective statistical criterion. This is a structural inference problem for dynamical systems.

Since different moving average terms yield different predictions and different prediction errors, they yield different log-likelihoods and different AICs. Here, it may be reasonable to choose a model yielding the least prediction errors among the candidate structural models.

In ordinary time series modeling, based on AR, MA, and ARMA models, all variables are observed, whereas in state space modeling, not necessarily all state variables are observed. A discussion of the estimation of the unobserved state variables and the maximum likelihood method for state space models is given in Chapters 11 through 13.

3.3.4 Applications of Structural Models

3.3.4.1 EEG Power Spectrum Estimation

The power spectrum of EEG has been shown to be useful as a tool for monitoring and diagnosis in certain clinical applications such as the following:

1. Distinguishing epileptic seizures from other types of spells
2. Monitoring the depth of anesthesia
3. Serving as an adjunct test of brain death

EEG is conveniently divided into frequency bands, approximately defined as δ (0.5–4 Hz), θ (4–8 Hz), low α (8–10 Hz), high α (10–12 Hz), β (12–25 Hz), and γ (25–50 Hz).

A commonly used method for estimating the EEG spectrum is fitting an AR model. It must be remembered, however, that the spectrum of EEG time series is always changing in time, as is always the case for time series in neuroscience. Clinically useful information comes from the empirical specification of the dynamics of δ, θ, α, β, and γ oscillations in the individual EEG time series of the patient. Here, the change, that is, rise and fall of rhythms, is judged by doctors and neurophysiologists with their expertise and experience of reading the time plots of EEG recording.

If this rise and fall of the wave components in the EEG plots were automatized in an objective manner by a statistical method, such as the system identification of a parametric model for the dynamic change of δ, θ, α, β, and γ oscillations in the patients' EEG time series, this would help and reduce the heavy load of doctors and anesthesiologists in their judgment.

For the statistical estimation of a time–varying spectrum, a nonstationary generalization of some of the aforementioned structural models plays an important role. Applications of these state space models for the estimation of time-varying power spectra of various types of EEG data are given in Wong et al. (2006) and Galka et al. (2010).

3.3.4.2 Trend-Cycle Decomposition

The structural modeling has been used since the 1960s (Shyskin et al. 1967) in seasonal adjustment of official and economic time series, where people are interested in decomposing time series x_t ($t = 1, 2, ..., N$) into a sum of trend component T_t and cyclic(seasonal) component S_t as $x_t = T_t + S_t$. Here the trend component T_t ($t = 1, 2, ..., N$) and cyclic component s_t ($t = 1, 2, ..., N$) are assumed to be generated from stochastic processes that are independent of each other.

For example, for a monthly sampled seasonal economic time series data, the trend model and the seasonal model can be modeled as follows:

$$(1 - B)T_t = w_t^{(T)} \text{ (for the trend)}$$

$$(1 + B + \cdots + B^{11})S_t = w_t^{(s)} \text{ (for the seasonal)}$$

Note that we have

$$(1-B)(1+B+\cdots+B^{11}) = (1-B^{12})$$

By arranging the trend model and the seasonal model to be a sequential type, we have

$$(1-B^{12})x_t = w_t,$$

which is a kind of autoregressive model, that is, $x_t = x_{t-12} + w_t$. By arranging the trend model and the seasonal model in parallel, driven by the two independent driving noises $w_t^{(T)}$ and $w_t^{(s)}$, we have a special state space model,

$$\begin{cases} x_t = A^{(Para1)}x_{t-1} + \eta_t^{(para1)} \\ y_t = C^{(Para1)}x_t. \end{cases}$$

Here the state vector is given by $x_t = (T_t, S_t, S_{t-1}, \ldots, S_{t-9}, S_{t-10})'$,

$$A^{(Para1)} = \begin{pmatrix} 1 & 0 & 0 & \cdots & 0 & 0 \\ 0 & -1 & -1 & \cdots & -1 & -1 \\ 0 & 1 & 0 & \cdots & 0 & 0 \\ 0 & 0 & 1 & \cdots & 0 & 0 \\ \cdots & \cdots & \cdots & \cdots & \cdots & \cdots \\ 0 & 0 & 0 & \cdots & 1 & 0 \end{pmatrix}, \quad \eta_t^{(para1)} = \begin{pmatrix} w_t^{(T)} \\ w_t^{(s)} \\ 0 \\ \cdots \\ 0 \\ 0 \end{pmatrix},$$

and

$$C^{(Para1)} = (1, 1, 0, \ldots, 0, 0).$$

Note that the structural modeling is valid also for the non-stationary case where the characteristic roots, $\lambda_1, \ldots, \lambda_p$, are on the unit circle, that is, $|\lambda_1|^2 = 1$. We also note that the polynomial of the seasonal part can be further split into six factors as

$$(1+B+\cdots+B^{11}) = (1-\sqrt{3}\,B+B^2)(1+\sqrt{3}\,B+B^2)(1-B+B^2)(1+B+B^2)(1+B^2)(1+B).$$

This means that the 11 characteristic roots, $\dfrac{\sqrt{3}}{2} \pm \dfrac{1}{2}i, \dfrac{1}{2} \pm \dfrac{\sqrt{3}}{2}i, \pm i, -\dfrac{1}{2} \pm \dfrac{\sqrt{3}}{2}i,$

$-\dfrac{\sqrt{3}}{2} \pm \dfrac{1}{2}i$, and -1, corresponding to the 5 factors, define the harmonic oscillations of the 12 steps (months) cyclic oscillations,

$$(1 - 2\cos\omega_i B + B^2)S_{i,t} = u_{i,t} \quad (i = 1, 2, \ldots, 5),$$

where $\omega_1 = 2\pi/12$, $\omega_2 = 2\pi/6$, $\omega_3 = 2\pi/4$, $\omega_4 = 2\pi/3$, $\omega_5 = 2\pi/2.4$, and $\omega_6 = 2\pi/2$ as the following:

$(1-\sqrt{3}\,B+B^2)S_t^{(12)} = u_t^{(12)}$ defines 12 month cyclic oscillations by

$$S_t^{(12)} = \sqrt{3}\,S_{t-1}^{(12)} - S_{t-2}^{(12)} + u_t^{(12)}.$$

$(1-B+B^2)S_t^{(6)} = u_t^{(6)}$ defines 6 month cyclic oscillations by $S_t^{(6)} = S_{t-1}^{(6)} - S_{t-2}^{(6)} + u_t^{(6)}$.
$(1+B^2)S_t^{(4)} = u_t^{(4)}$ defines 4 month cyclic oscillations by $S_t^{(4)} = -S_{t-2}^{(4)} + u_t^{(4)}$.
$(1+B+B^2)S_t^{(3)} = u_t^{(3)}$ defines 3 month cyclic oscillations by $S_t^{(3)} = -S_{t-1}^{(3)} - S_{t-2}^{(3)} + u_t^{(3)}$.
$(1+\sqrt{3}\,B+B^2)S_t^{(2.4)} = u_t^{(2.4)}$ defines 2.4 month cyclic oscillations by

$$S_t^{(2.4)} = -\sqrt{3}\,S_{t-1}^{(2.4)} - S_{t-2}^{(2.4)} + u_t^{(2.4)}.$$

$(1+B)S_t^{(2)} = u_t^{(2)}$ defines 2 month cyclic oscillations by $S_t^{(2)} = -S_{t-1}^{(2)} + u_t^{(2)}$.

If we arrange all these submodels into a parallel-type state space model, we have another state space model,

$$\begin{cases} z_t = A^{(para2)}z_{t-1} + \eta_t^{(para2)} \\ x_t = C^{(para2)}z_t \end{cases} \tag{3.28}$$

where

$$A^{(para2)} = \begin{pmatrix} 1 & O & O & \cdots & O & O \\ O & A^{(12)} & 0 & \cdots & O & O \\ O & O & A^{(6)} & \cdots & O & O \\ \cdots & \cdots & \cdots & \cdots & \cdots & \cdots \\ O & O & O & \cdots & A^{(2.4)} & O \\ O & O & O & \cdots & O & A^{(2)} \end{pmatrix}$$

$C^{(para2)} = (1,1,0,1,0,1,0,1,0,1,0,1)$, $\eta_t^{(para2)} = \left(w_t^{(T)}, u_t^{(12)}, 0, u_t^{(6)}, 0, \ldots, u_t^{(2.4)}, 0, u_t^{(2)}\right)'$

$$Z_t = (T_t, S_t^{(12)}, S_{t-1}^{(12)}, S_t^{(6)}, S_{t-1}^{(6)}, \ldots, S_t^{(2.4)}, S_{t-1}^{(2.4)}, S_t^{(2)})',$$

$$A^{(12)} = \begin{pmatrix} \sqrt{3} & -1 \\ 1 & 0 \end{pmatrix}, \quad A^{(6)} = \begin{pmatrix} 1 & -1 \\ 1 & 0 \end{pmatrix}, \quad A^{(4)} = \begin{pmatrix} 0 & -1 \\ 1 & 0 \end{pmatrix}, \quad A^{(3)} = \begin{pmatrix} -1 & -1 \\ 1 & 0 \end{pmatrix},$$

$$A^{(2.4)} = \begin{pmatrix} -\sqrt{3} & -1 \\ 1 & 0 \end{pmatrix}, \text{ and } A^{(2)} = (-1).$$

3.4 Akaike's Theory of Predictor Space

So far we have seen the close relation between the state space model and AR (and ARMA) models in a heuristic way through structural modeling. A more theoretical approach to the analysis of the relations between the state space models and AR models for stationary Gaussian time series processes was introduced in the 1970s by Akaike (1974a,b, 1975) through the use of his theory of the predictor space. In order to see Akaike's predictor space approach to state space modeling and AR modeling, let us start from the discussion of a dynamic model of a simple scalar input–output system as in Figure 3.5.

3.4.1 Input–Output System and the Predictor Space

Consider a stochastic system with an input u_t and an output x_t. It is assumed that u_t and x_t are scalar time series. It is further assumed that both u_t and x_t have zero mean and finite second-order moments and that they are stationarily correlated.

Hereafter, we will be concerned with various linear spaces spanned by the components of the u_t and x_t. It is assumed throughout the current section that the metric in these spaces is given by the root mean square and the spaces are closed within this metric.

If we adopt the definition of the state space of a system at time t as the totality of the information from the present and past input to be transmitted to the present and future output, a natural mathematical representation of the state space in this case will be given by the linear space spanned by the components of the projections $x_{t+k|t}$ ($k = 0, 1, \ldots$), where $x_{t+k|t}$ denotes the projection of x_{t+k} onto the linear space $R_u(t-)$ spanned by the components of the present and past inputs $u_t, u_{t-1}, \ldots$. The projection $x_{t+k|t}$ is composed of the elements of $R_u(t-)$ and is characterized by the relation

$$E[x_{t+k|t}u_{t-m}] = E[x_{t+k}u_{t-m}] \quad \text{for } m = 0,1,\ldots \tag{3.29}$$

The very basic assumption to be made in this setting is that the dimension of the linear space spanned by the components of $x_{t+k|t}$ ($k = 0, 1, \ldots$) is finite. Let the dimension of the space be p. This space that is spanned by the components of the predictors is called the "predictor space" and is denoted by $R(t+|t-)$. Obviously, $R(t+|t-) \subset R_u(t-)$.

The relation (3.29) shows that the information of x_{t+k} within $R_u(t-)$ is condensed in $x_{t+k|t}$, in the sense that the dependence of x_{t+k} on $R_u(t-)$ can be explained by $x_{t+k|t}$. Take an arbitrary basis $z_t^{(1)}, z_t^{(2)}, \ldots, z_t^{(p)}$ of the predictor

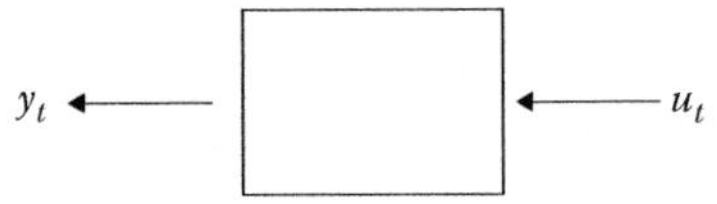

$y_t \longleftarrow \boxed{} \longleftarrow u_t$

FIGURE 3.5
Simple input–output system with a box and arrows.

space $R(t+|t-)$ and denote by z_t the vector $\left(z_t^{(1)}, z_t^{(2)}, ..., z_t^{(p)}\right)'$. Since $x_{t|t}$ is composed of the elements of $R(t+|t-)$, there is a $1 \times p$ matrix C such that

$$x_{t|t} = Cz_t.$$

Thus, x_t has a representation

$$x_t = Cz_t + \varepsilon_t \tag{3.30}$$

where ε_t is independent of the elements of $R_u(t-)$, that is, ε_t is independent of the present and past inputs. Denote by z_{t+1} the random vector obtained by replacing $u_t, u_{t-1}, ...$ in the definition of z_t by $u_{t+1}, u_t, ...$ From the assumption of stationarity, the components of z_{t+1} form a basis of the space $R((t+1)+|(t+1)-)$.

By an analogous reasoning as in the case of x_t, it can be seen that z_{t+1} admits a representation

$$z_{t+1} = Az_t + \eta_{t+1}, \tag{3.31}$$

with some A, where η_{t+1} is independent of $u_t, u_{t-1}, u_{t-2}, ...$ From (3.31), since the components of z_{t+1} and z_t are elements of $R_u((t+1)-)$, η_{t+1} is composed of the elements of $R_u((t+1)-)$, but its projection onto $R_u(t-)$ is equal to zero. Thus, η_{t+1} can be represented in the form

$$\eta_{t+1} = Bw_{t+1}, \tag{3.32}$$

where B is a $p{\times}1$ matrix and

$$w_{t+1} = u_{t+1} - u_{t+1|t}. \tag{3.33}$$

$u_{t+1|t}$ denotes the projection of u_{t+1} onto $R_u(t-)$ and w_{t+1} is the innovation of the input u_{t+1} at time $t+1$. The variance of the innovation w_t is denoted by σ_w^2, that is, $\sigma_w^2 = E\left[w_t^2\right]$.

Equations 3.30 through 3.33 provide a representation of a general stationary linear stochastic system. In this representation, ε_t denotes the part of x_t that is independent of the present and past inputs, $u_t, u_{t-1}, u_{t-2}, ...$ The innovations $w_t, w_{t+1}, ...$ are mutually independent for different values of t, whereas this does not necessarily hold for ε_t's. z_t provides a specification of the predictor space $R(t+|t-)$, and this may be considered to be a representation of the state of the stochastic system at time t. z_t is called the state vector.

3.4.2 Feedback System and the Predictor Space

When there are feedbacks from the outputs to the inputs, the distinction of the variables in terms of the inputs and outputs becomes meaningless. The current input is affected by an output from the past, and also the current output will affect the system as an input to the future output. To get a representation of this most general input–output system, we have only to redefine new multivariate variables $x_t^{(new)} = \left(x_t^{(old)}, u_t^{(old)}\right)'$ and $u_t^{(new)} = x_t^{(new)}$. If we redefine and reinterpret x_t and u_t in (3.30)–(3.33) with these new variables, $x_t\left(=x_t^{(new)}\right)$ and $u_t\left(=u_t^{(new)}\right)$, we have $x_t = u_t$, and since x_t is composed of the elements of $R_u(t-)$, we have $x_{t|t} = x_t$ and ε_t in (3.30) vanishes. Thus, we get a representation

$$z_{t+1} = Az_t + Bw_{t+1}$$
$$x_t = Cz_t$$

(3.34)

where the innovation w_{t+1} of x_t at time point $t + 1$ is defined by

$$w_{t+1} = x_{t+1} - x_{t+1|t}.$$

(3.35)

This representation (3.34) is called a Markovian representation of x_t.

Since any basis of the predictor space $R(t+|t-)$ can play the role of the state vector z_t, the matrices A, B and C of the aforementioned representation are not unique. A proper specification of the structure of z_t defines a canonical representation of a stochastic linear system.

Our canonical representation is obtained by choosing the maximum set of linearly independent elements within the sequence of the predictors $x_{t|t}$, $x_{t+1|t}$, ..., $x_{t+k|t}$ ... as the state vector. Conceptually, the state vector z_t can be constructed by the following procedure:

Step (1) Define the vector s by

$$s = \begin{pmatrix} x_{t|t} \\ x_{t+1|t} \\ \cdots \end{pmatrix},$$

(3.36)

and the ith component s_i of s is given by

$$s_i = x_{t+i-1|t}.$$

Step (2) Successively test the linear dependence of s_i on its antecedents,

$$s_{i-1}, s_{i-2}, \ldots, s_1 \quad \text{for} \quad i = 1, 2, \ldots$$

Step (3) z_t is defined as the vector obtained by discarding from s those components that are dependent on its antecedents.

Once the specification of z_t is given, the matrices A, B, and C of the corresponding canonical representation can be obtained very easily.

3.4.3 How to Construct A, B, and C

Now for $z_t^{(i)} = x_{t+i-1|t}$ we have $z_{t+1}^{(i)} = x_{t+i|t+1}$, and its projection on $R_z(t-)$, denoted by $z_{t+1|t}^{(i)}$, is equal to $x_{t+i|t}$. If $x_{t+i|t}$ is independent of its antecedents, $x_{t+i-1|t}$, $x_{t+i-2|t}$, ..., $x_{t|t}$ are augmented into the state vector z_t, and we have $z_{t+1|t}^{(i)} = z_t^{(i+1)}$. This can be expressed as

$$z_{t+1|t}^{(i)} = \sum_{m=1}^{p} A(i,m) z_t^{(m)}. \tag{3.37}$$

Here $A(i, m)$ $(m = 1, ..., p)$ is the ith row of the transition matrix A, where for $1 \le i \le p - 1$, $A(i, m) = 0$ for $m \ne i + 1$ and $A(i, i + 1) = 1$.

If $z_{t+1|t}^{(i)}$ $(=x_{t+i|t})$ is dependent, at $i = p$, on the antecedents, all the $x_{t+k|t}$ $(k > p)$ are dependent on the antecedents, $x_{t+p-1|t}$, $x_{t+p-2|t}$, ..., $x_{t|t}$. Then discard from s all the elements $x_{t+p|t}$, $x_{t+p+1|t}$ ..., and $s = (x_{t|t}, x_{t+1|t}, ..., x_{t+p-1|t})'$ makes the state vector z_t. Then p defines the dimension of the state vector z_t, and $z_{t+1|t}^{(p)}$ $(= x_{t+p|t})$ can be expressed as a linear combination of its antecedents, $z_t^{(1)} = x_{t|t}, z_t^{(2)} = x_{t+1|t}, ..., z_t^{(p)} = x_{t+p-1|t}$. Then $A(p, m)$, $(m = 1, ..., p)$ specifies how $x_{t+p|t}$ depends on $(x_{t|t}, x_{t+1|t}, ..., x_{t+p-1|t})$. Thus, for this case there is a unique representation

$$x_{t+p|t} = \sum_{m=1}^{p} A(p,m) x_{t+m-1|t}$$

which is equivalent to

$$z_{t+1|t}^{(p)} = \sum_{m=1}^{p} A(p,m) z_t^{(m)}, \tag{3.38}$$

where A has the form

$$A = \begin{pmatrix} 0 & 1 & 0 & \dots & 0 \\ 0 & 0 & 1 & \dots & 0 \\ \dots & \dots & \dots & \dots & \dots \\ 0 & 0 & 0 & \dots & 1 \\ * & * & * & \dots & * \end{pmatrix}.$$

Here * means a nonzero value.

The ith element $B(i, 1)$ of the $p \times 1$ input matrix B, is characterized as the impulse response of $z_t^{(i)}$ to the input w_t, which is the innovation. Note that if x_t has the following impulse response representation,

$$x_t = h_0 w_t + h_1 w_{t-1} + \dots + h_m w_{t-m} + \dots$$

We obtain

$$x_{t+i+1|t+1} = x_{t+i+1|t} + h_i w_{t+1}$$

from

$$x_{t+i+1} = w_{t+i+1} + h_1 w_{t+i} + \cdots + h_m w_{t+i+1-m} + \cdots$$

Then for $z_t^{(i)} = x_{t+i-1|t}$, $B(i, 1)$ can be obtained as the impulse response of x_t at time lag i to the input w_t. Thus, we have

$$B = \begin{pmatrix} 1 \\ h_1 \\ \cdots \\ h_p \end{pmatrix}.$$

The output matrix C, which is $1 \times p$, can be obtained as the matrix of regression coefficients of x_t on z_t. C takes the form

$$C = [1, O],$$

where O is a $1 \times (p - 1)$ zero matrix. This is the case when $u_t = x_t$ and the variance σ_w^2 of the innovation w_t is not zero. When $u_t = x_t$, under the same assumption as discussed earlier, the input matrix B is an identity matrix. Hereafter we consider the case where $u_t = x_t$ and $\sigma_w^2 \neq 0$.

It is obvious that in the aforementioned canonical Markovian representation of x_t the matrices A, B, C, and σ_w^2 are uniquely determined. Here A, B, C, and σ_w^2 are such that they are in the forms given by the canonical representation of x_t and define a covariance structure of the process x_t. This shows that under the assumption of $\sigma_w^2 \neq 0$ and the finiteness of the dimension of the predictor space, the correspondence between the covariance structure $R(\tau)$ of x_t and the set of A, B, C, and σ_w^2 is one to one. The totality of the sets $(A, B, C, \text{ and } \sigma_w^2)$ provides a basic framework for the identification of a stationary time series.

3.4.4 Canonical State Space Model and ARMA Representation

The relation between the Markovian representation model and the autoregressive moving average models of a stationary time series can be analyzed in the framework of predictor space as follows.

From Equation 3.38, we get

$$x_{t+p|t} = \sum_{m=1}^{p} \phi_m x_{t+p-m|t}. \qquad (3.39)$$

Note that x_{t+p-m}'s ($m = 1, \ldots, p$) are composed of the elements of $R((t + p - 1)-)$. From these relations (3.39), we have the representation

$$x_{t+p} - \sum_{m=1}^{p} \phi_m x_{t+p-m} = w_{t+p} + \sum_{l=1}^{p-1} \theta_l w_{t+p-l},$$

for some $\theta_1, \ldots, \theta_{p-1}$. Here we use the impulse response representation of x_t,

$$x_t = w_t + h_1 w_{t-1} + \cdots + h_m w_{t-m} + \cdots$$

and the relations of the conditional expectations,

$$x_{t+m} = x_{t+m|t+m-1} + w_{t+m}$$

$$x_{t+m|t} = x_{t+m|t-1} + h_m w_t$$

$$x_{t+m|t+k} = x_{t+m|t+k-1} + h_{m-k} w_{t+k},$$

in terms of the impulse responses, h_k ($k = 1, 2, \ldots$).

We have seen, in Section 3.3.2, that four different structural state space models are converted to four different ARMA models, where the AR parts are equivalent and the differences of the structures are in MA parts. Next it will be shown that any ARMA model can be converted into a state space model.

When the process is generated from an ARMA model,

$$x_{t+p} - \sum_{m=1}^{p} \phi_m x_{t+p-m} = w_{t+p} + \sum_{l=1}^{p-1} \theta_l w_{t+p-l},$$

we get (3.39) from the relation (3.38). This means that the dimension of the predictor space is p, where the state vector is

$$z_t = \left(z_t^{(1)}, z_t^{(2)}, \ldots, z_t^{(p)} \right)' = \left(x_{t|t}, x_{t+1|t}, \ldots, x_{t+p-1|t} \right)',$$

$$z_{t+1} = \left(x_{t+1|t+1}, x_{t+2|t+1}, \ldots, x_{t+p|t+1} \right)',$$

and $A(i, m) = 0$, for $m \neq i + 1$, and $A(i, i + 1) = 1$, for $i = 1, 2, \ldots, p - 1$. For the pth row of the matrix A, we have

$$A(p, p+1-m) = \phi_m \quad (m = 1, \ldots, p).$$

Thus, we have the unique (so-called canonical) state space representation,

$$\begin{cases} z_t = A z_{t-1} + B w_t \\ x_t = C z_t \end{cases}$$

where

$$A = \begin{pmatrix} 0 & 1 & 0 & \cdots & 0 \\ 0 & 0 & 1 & \cdots & 0 \\ \cdots & \cdots & \cdots & \cdots & \cdots \\ 0 & 0 & 0 & \cdots & 1 \\ \phi_p & \phi_{p-1} & \phi_{p-2} & \cdots & \phi_1 \end{pmatrix}, \quad B = \begin{pmatrix} 1 \\ h_1 \\ \cdots \\ h_p \end{pmatrix}$$

and $C = (1, 0, \ldots, 0)$. The impulse response function, h_i $(i = 1, 2, \ldots)$, is generated by the recursive relations,

$$\begin{cases} h_1 = 1 \\ h_j = \theta_{j-1} + \sum_{k=1}^{p-1} \phi_k h_{j-k} \end{cases}$$

We note that the ARMA representation for the given time series is not unique. Let us assume that we have an ARMA representation,

$$\phi_p(B)x_t = \theta_q(B)w_t, \tag{3.40}$$

where

$$\phi_p(B) = 1 - \phi_1 B - \cdots - \phi_p B^p,$$

and

$$\theta_q(B) = 1 - \theta_1 B - \cdots - \theta_q B^q.$$

Now let us multiply by ψ, where $-1 < \psi < 1$, on both sides of (3.40) and shift the time two steps backward, that is, we consider the model

$$\phi_p(B)\psi x_{t-2} = \theta_q(B)\psi w_{t-2},$$

which is equivalent to

$$\phi_p(B)\psi B^2 x_t = \theta_q(B)\psi B^2 w_t. \tag{3.41}$$

By adding (3.41) to (3.40), we obtain

$$\phi_p(B)(1 + \psi B^2)x_t = \theta_q(B)(1 + \psi B^2)w_t. \tag{3.42}$$

Both models, (3.40) and (3.42), have the same impulse response function, and they are not distinguishable. These models are called observationally equivalent. In the scalar time series case, we can avoid the nonunique parameterization problem by choosing the minimum order model between the observationally equivalent models. However, we must be careful to note that, in multivariate time series cases, this does not solve the problem (see Section 4.3.2).

In practice, however, at the stage of maximum likelihood estimation of the two models for a scalar time series, this is not a problem. ARMA($p + 2, q + 2$), when it is fitted to the data, tries to pick up the line spectrum in the periodogram of the white noise. Even though this is a numerically harder situation, the fit will always converge to a proper numerical solution, that is, a local maximum, of the likelihood maximization problem. Here the problem to be solved is the model identification problem (where overfitting needs to be avoided) rather than the problem of finding conditions for the uniqueness of the model representation. We have seen, however, in the discussion of Akaike's theory of predictor space that, apart from the ordinary parameterization of the ARMA model,

$$x_t = \phi_1 x_{t-1} + \cdots + \phi_p x_{t-p} + \theta_1 w_{t-1} + \theta_2 w_{t-2} + \cdots + \theta_q w_{t-q} + w_t,$$

an alternative parameterization of an ARMA model through the state space modeling is possible. Here the moving average parameters are dependent on the autoregressive parameters, which may help in obviating the aforementioned numerical difficulties (local maximum problem) commonly experienced and recognized for high-order ARMA model fitting.

3.5 Dynamic Models with Exogenous Input Variables

We have been studying, so far, time series models that are driven by Gaussian white noise. This is partly because we are considering predictors composed of the information of past time series observations, of x_{t-1}, x_{t-2}, of the concerned time series x_t. In some situations past observations of some other time series variables, $u_{t-1}, u_{t-2}, \ldots$, may carry useful information for the prediction of x_t. In such situations, it will be reasonable to consider including a linear combination of $u_{t-1}, u_{t-2}, \ldots, u_{t-r}$, in the predictor of x_t. For example, if we add extra terms such as $\psi_1 u_{t-1} + \cdots + \psi_r u_{t-r}$ to the AR(p) model with a constant term ϕ_0,

$$x_t = \phi_0 + \sum_{i=1}^{p} \phi_i x_{t-i} + w_t,$$

FIGURE 3.6
Input–output model, a box with 2 input arrows and one output arrow.

we can think of a prediction model

$$x_t = \phi_0 + \sum_{i=1}^{p} \phi_i x_{t-i} + \sum_{i=1}^{r} \psi_i u_{t-i} + w_t \tag{3.43}$$

where the parameters ϕ_0, ϕ_i ($i = 1, ..., p$) and, ψ_i ($i = 1, ..., r$) are to be estimated from the observed time series data. The model is called "AR model
with eXogenous variables" and is often abbreviated as ARX Model. With this
model, the time series x_t is regarded as an output of a system driven by both
Gaussian white noise and the exogenous variable u_t (see Figure 3.6).
In the same way as the ARX model, we can think of an ARMAX model,

$$x_t = \phi_0 + \sum_{i=1}^{p} \phi_i x_{t-i} + \sum_{i=1}^{r} \psi_i u_{t-i} + \sum_{i=1}^{q} \theta_i w_{t-i} + w_t.$$

Parameters of these models are estimated using the least squares method or
the maximum likelihood method (see Chapter 10). The ARX model is especially useful because the parameter estimation is as efficient as the AR models, while the ARMAX model needs a rather large computational load for the
estimation of the parameters as ARMA models do.
The stationarity conditions for ARX models and ARMAX models can be
given in the same way as for the original AR models and ARMA models,
respectively (Hannan and Deistler 1988). Also, the invertibility conditions
for ARMAX models can be given in the same way as for ARMA models.
Pure MAX models would also be possible, but so far they have not found any
useful applications. On the other hand, nonlinear versions of ARX models
have turned out to be very useful in engineering applications, both for the
scalar and the multivariate case. Some nonlinear multivariate ARX models
are discussed in Chapter 6.

4

Multivariate Dynamic Models

So far, we have discussed only univariate linear time series models and related problems. Before going to the more general case of nonlinear time series models, let us see how these "univariate" linear models can be extended to the case of "multivariate" linear time series models.

Theoretically speaking, all the univariate linear time series models, that is, AR, MA, ARMA, and state space models, can be naturally extended to the multivariate cases. Several methods and techniques have been developed for multivariate time series analysis since the 1960s. However, because of associated computational advantages, only multivariate AR models have been used and applied in most multivariate time series analyses in sciences and engineering. In the 1970s, when the available computational power was not as strong as it is now, methods for modeling multivariate (e.g., 5- or 7-dimensional) time series by AR models were implemented, for modeling and controlling complex systems in engineering (Akaike 1971, Akaike and Nakagawa 1972, Otomo et al. 1972, Akaike 1978, Fukunishi 1977, Ohtsu et al. 1979, Nakamura and Akaike 1981).

Although the system given by the dynamic functioning of the brain in neuroscience is much different from the industrial system, both represent feedback systems and, therefore, share many mathematical similarities. In this chapter, we will see how multivariate AR models are used for the analysis of feedback systems, which will have a bearing on the later discussion of the causality studies of brain functioning in Chapter 14.

As could be anticipated from the previous chapter, however, Akaike's method for the modeling of canonical state space representation and the AR modeling method can be easily extended to multivariate situations. In Section 4.4, we review Akaike's method, which provides a canonical multivariate ARMA model identifiable without any artificial conditions. Although in the 1970s, when it was first introduced, Akaike's method was felt to be extremely demanding in terms of computational cost, this no longer poses a problem in the current computational environment, making the method a very promising approach for developing the idea of causality in brain science data.

4.1 Multivariate AR Models

4.1.1 Power Spectrum

Nowadays many people may think that linear dynamic models are too simple to be useful for solving real-world time series problems. However, this is not true. It seems that in many important application problems, the strength and the power of modeling by multivariate AR models as predictive models is not widely recognized. This class of models is important not only because of its theoretical role in the Kolmogorov–Whittle relation (Kolmogorov 1939, Whittle 1953),

$$|\Sigma_w| = \exp\left\{ \int_{-1/2}^{1/2} \log p(f)df \right\},$$

which we saw in Sections 2.1 and 2.2, but also because it provides us with a powerful solution, without much computational cost, in many real-world problems, such as causality analysis, or variable selection problems in designing an optimal control system for complex high-dimensional feedback systems, occurring in engineering and science.

We will first look into the spectral theory of multivariate AR models of order 1 (note that a multivariate AR model of order $p > 1$ can be reformulated as an AR(1) model with a state variable of higher dimension). The auto covariance function $\Gamma(k)$, $-\infty < k < \infty$, and the power spectrum function $p(f)$, $-\frac{1}{2} < f < \frac{1}{2}$, of the stationary multivariate AR process x_t are related by the Fourier and inverse Fourier relations

$$p_{j,k}(f) = \sum_{s=-\infty}^{\infty} e^{-i2\pi fk} \Gamma_{j,k}(s)$$

and

$$\Gamma_{j,k}(s) = \int_{-1/2}^{1/2} e^{i2\pi fs} p_{j,k}(f)df.$$

Here $\Gamma_{j,k}(s)$ is (j, k)th element of $\Gamma(s) = E[x_t x'_{t-s}]$. We may obtain a parametric representation of the power spectrum for stationary multivariate AR processes in the same way as in the scalar case.

For example, the pth order AR model

$$x_t = \sum_{k=1}^{p} A_k x_{t-k} + w_t$$

can be rewritten using the backward shift operator B as

$$x_t = \left(I - \sum_{k=1}^{p} A_k B^k \right)^{-1} w_t.$$

Let us define $w(f)$ and $A(f)$, respectively, by

$$w(f) = \sum_{k=-\infty}^{\infty} e^{-i2\pi f k}(w_k)$$

and

$$A(f) = \sum_{k=-\infty}^{\infty} e^{-i2\pi f k}(-A_k), \quad (A_0 = -I), \quad 1/2 \le f \le 1/2.$$

Note that for the pth order AR model, $A_k = 0$ for $k = p + 1, p + 2, \ldots$ and for $k < 0$. Then, we have

$$x(f) = A(f)^{-1}w(f).$$

Since it holds that

$$p_{xx}(f) = E[x(f)\overline{x(f)'}],$$

$$p_{ww}(f) = E[w(f)\overline{w(f)'}] = \Sigma_w,$$

and

$$E\left[\left\{ A(f)^{-1}w(f) \right\} \left\{ \overline{A(f)}^{-1}\overline{w(f)} \right\}' \right] = \left\{ A(f)^{-1} \right\} p_{ww}(f) \left\{ \overline{A(f)'}^{-1} \right\},$$

we have

$$p_{xx}(f) = \{A(f)^{-1}\}\Sigma_w\{\overline{A(f)'}\}^{-1}, \quad -1/2 \le f \le 1/2. \tag{4.1}$$

For example, $x_t = Ax_{t-1} + w_t.$
For this multivariate AR(1) model, we have the following parametric power spectrum representation:

$$p_{xx}(f) = \{I - \exp(-i2\pi f)A\}^{-1}\Sigma_w\overline{\{I - \exp(-i2\pi f)A'\}}^{-1}.$$

4.1.2 Multivariate Yule–Walker Equations

Many of the theoretical properties that we saw for the scalar AR models still hold for the multivariate AR model:

$$x_t = A_1 x_{t-1} + \cdots + A_p x_{t-p} + w_t. \tag{4.2}$$

For example, the auto covariance function $\Gamma(k)$, $-\infty < k < \infty$, of the multivariate AR process and the power spectral function $p(f)$, $-1/2 < f < 1/2$, are related by the Fourier and inverse Fourier relations

$$p_{j,k}(f) = \sum_{s=-\infty}^{\infty} e^{-i2\pi f k} \Gamma_{j,k}(s)$$

and

$$\Gamma_{j,k}(s) = \int_{-1/2}^{1/2} e^{i2\pi f s} p_{j,k}(f) df,$$

and the AR coefficients, $A_1, \ldots, A_p$, and auto covariance function, $\Gamma(k)$, $-\infty < k < \infty$, are related by the Yule–Walker equations (4.3) and (4.4), which are derived from

$$E[x_t x'_{t-k}] = E[(A_1 x_{t-1} + \cdots + A_p x_{t-p} + w_t) x'_{t-k}] \quad (\text{for } k = 0, 1, 2, \ldots)$$

$$= A_1 E[x_{t-1} x'_{t-k}] + \cdots + A_p E[x_{t-p} x'_{t-k}] + E[w_t x'_{t-k}].$$

Here, since $E[w_t x'_{t-k}] = 0$ for $k \geq 1$ and $E[w_t x'_{t-k}] = \Sigma_w$ for $k = 0$, we have the multivariate version of the Yule–Walker equation,

$$(\Gamma_1 \quad \cdots \quad \Gamma_p) = (A_1 \quad \cdots \quad A_p) \begin{pmatrix} \Gamma_0 & \cdots & \Gamma_{p-1} \\ \cdots & \cdots & \cdots \\ \Gamma_{-p+1} & \cdots & \Gamma_0 \end{pmatrix}, \tag{4.3}$$

and

$$\Gamma_0 = \sum_{j=1}^{p} A_j \Gamma_{-j} + \Sigma_w = \sum_{j=1}^{p} A_j \Gamma'_j + \Sigma_w. \tag{4.4}$$

These Yule–Walker equations give the linear relations between the model parameters, $A_1, \ldots, A_p$, and the auto covariance function, $\Gamma(k)$, $-\infty < k < \infty$, of the process, and they are useful when we try to identify the model (4.2) from the observation data, $x_1, x_2, \ldots, x_N$. This topic is discussed in Chapter 10.

4.1.3 Eigenvalues

We can write down the k-variate AR(p) model,

$$x_t = A_1 x_{t-1} + \cdots + A_p x_{t-p} + w_t,$$

by a $(k \times p)$-variate AR(1) model,

$$
\begin{pmatrix} x_t \\ x_{t-1} \\ \cdots \\ x_{t-p+1} \\ x_{t-p} \end{pmatrix} =
\begin{pmatrix}
A_1 & A_2 & \cdots & A_{p-1} & A_p \\
I_k & O_k & \cdots & O_k & O_k \\
\cdots & \cdots & \cdots & \cdots & \cdots \\
O_k & O_k & \cdots & O_k & O_k \\
O_k & O_k & \cdots & I_k & O_k
\end{pmatrix}
\begin{pmatrix} x_{t-1} \\ x_{t-2} \\ \cdots \\ x_{t-p} \\ x_{t-p-1} \end{pmatrix} +
\begin{pmatrix} w_t \\ 0 \\ \cdots \\ 0 \\ 0 \end{pmatrix},
$$

where I_k is a k-dimensional identity matrix, O_k, is a $k \times k$ zero matrix, and

$$
x_t = \begin{pmatrix} x_t^{(1)} \\ x_t^{(2)} \\ \cdots \\ x_t^{(k)} \end{pmatrix}, \quad
w_t = \begin{pmatrix} w_t^{(1)} \\ w_t^{(2)} \\ \cdots \\ w_t^{(k)} \end{pmatrix},
$$

$$
A_1 = \begin{pmatrix}
a_{1,1,1} & a_{1,2,1} & \cdots & a_{1,k,1} \\
a_{2,1,1} & a_{2,2,1} & \cdots & a_{2,k,1} \\
\cdots & \cdots & \cdots & \cdots \\
a_{k,1,1} & a_{k,2,1} & \cdots & a_{k,k,1}
\end{pmatrix}, \ldots, A_p = \begin{pmatrix}
a_{1,1,p} & a_{1,2,p} & \cdots & a_{1,k,p} \\
a_{2,1,p} & a_{2,2,p} & \cdots & a_{2,k,p} \\
\cdots & \cdots & \cdots & \cdots \\
a_{k,1,p} & a_{k,2,p} & \cdots & a_{k,k,p}
\end{pmatrix}.
$$

Another useful tool to characterize the properties of the multivariate AR process is given by computing the eigenvalues of the transition matrix:

$$
A = \begin{pmatrix}
A_1 & A_2 & \cdots & A_{p-1} & A_p \\
I_k & O_k & \cdots & O_k & O_k \\
\cdots & \cdots & \cdots & \cdots & \cdots \\
O_k & O_k & \cdots & O_k & O_k \\
O_k & O_k & \cdots & I_k & O_k
\end{pmatrix}.
$$

For stationary k-variate AR models, the eigenvalues of the transition matrix A are supposed to stay inside the unit circle. The arguments of the complex conjugate pairs of eigenvalues characterize the peaks of the power spectrum, as in the case of scalar AR models, and the absolute value of the eigenvalues characterizes the damping characteristics of the oscillatory behaviors arising from the complex eigenvalues.

4.1.4 Forward and Backward Models

There are, however, some essential differences between the cases of scalar and multivariate AR models. The most important difference arises with respect to the symmetry of temporal direction. As we saw, in Section 3.2.4, in the scalar case, the parameters of the forward prediction model,

$$x_t = \phi_1 x_{t-1} + \phi_2 x_{t-2} + \cdots + \phi_p x_{t-p} + w_t,$$

and the parameters of the backward prediction model,

$$x_t = \pi_1 x_{t+1} + \pi_2 x_{t+2} + \cdots + \pi_p x_{t+p} + v_t,$$

are the same, that is, $(\phi_1, \phi_2, ..., \phi_p) = (\pi_1, \pi_2, ..., \pi_p)$. However, for the multivariate case, this is not true. The parameters of the forward model,

$$x_t = \sum_{i=1}^{p} \Phi_i x_{t-i} + w_t, \tag{4.5}$$

and the backward model,

$$x_t = \sum_{i=1}^{p} \Pi_i x_{t+i} + v_t, \tag{4.6}$$

are different, that is, $\Phi_i \neq \Pi_i$ ($i = 1, 2, ..., p$).

4.1.5 Multivariate Levinson Procedure

Even though for multivariate AR processes there is no symmetry with respect to temporal direction, there are recursive relations between the $(n + 1)$th-order AR coefficients and the nth-order AR coefficients, similar to the Levinson procedure for the scalar case. Furthermore, these recursions provide us with the transformation from the auto covariance function, $\Gamma(s)$, $-\infty < s < \infty$, to the AR model coefficients, $\{A_1, ..., A_n, \Sigma_{w_n}\}$, $n = 1, ..., p$, recursively from $n = 1$ to $n = p$. Here, $\{A_1, ..., A_n, \Sigma_{w_n}\}$ is the solution of the Yule–Walker equations (4.3) and (4.4) with $p = n$.

For notational convenience, let us introduce $F_i = -\Phi_i'$ and $B_i = -\Pi_i'$ and rewrite the nth order forward model (4.5) by the forward filter,

$$x_t + \sum_{i=1}^{n} F_i' x_{t-i} = w_t^{(n)},$$

and the nth order backward model (4.5) by the backward filter,

$$x_t + \sum_{i=1}^{n} B'_i y_{t+i} = v_t^{(n)},$$

where

$$E\left[\left\{w_t^{(n)}\right\}\left\{w_t^{(n)}\right\}'\right] = \Sigma_{w_n}$$

and

$$E\left[\left\{v_t^{(n)}\right\}\left\{v_t^{(n)}\right\}'\right] = \Sigma_{v_n}.$$

Then, the nth order forward filter and the nth order backward filter yield the equations

$$\begin{pmatrix} R(0) & R(1) & \cdots & R(n) \\ R(-1) & R(0) & \cdots & R(n-1) \\ \cdots & \cdots & \cdots & \cdots \\ R(-n) & R(-n+1) & \cdots & R(0) \end{pmatrix} \begin{pmatrix} I \\ F_1^{(n)} \\ \cdots \\ F_n^{(n)} \end{pmatrix} = \begin{pmatrix} \Sigma_{w_n} \\ O \\ \cdots \\ O \end{pmatrix} \quad (4.7)$$

and

$$\begin{pmatrix} R(0) & R(1) & \cdots & R(n) \\ R(-1) & R(0) & \cdots & R(n-1) \\ \cdots & \cdots & \cdots & \cdots \\ R(-n) & R(-n+1) & \cdots & R(0) \end{pmatrix} \begin{pmatrix} B_n^{(n)} \\ B_{n-1}^{(n)} \\ \cdots \\ I \end{pmatrix} = \begin{pmatrix} O \\ O \\ \cdots \\ \Sigma_{v_n} \end{pmatrix}, \quad (4.8)$$

which can be rewritten compactly in a single equation as

$$\begin{pmatrix} R(0) & R(1) & \cdots & R(n) \\ R(-1) & R(0) & \cdots & R(n-1) \\ \cdots & \cdots & \cdots & \cdots \\ R(-n) & R(-n+1) & \cdots & R(0) \end{pmatrix} \left(\begin{pmatrix} I \\ F_1^{(n)} \\ \cdots \\ F_n^{(n)} \end{pmatrix}, \begin{pmatrix} B_n^{(n)} \\ B_{n-1}^{(n)} \\ \cdots \\ I \end{pmatrix} \right) = \left(\begin{pmatrix} \Sigma_{w_n} \\ O \\ \cdots \\ O \end{pmatrix}, \begin{pmatrix} O \\ O \\ \cdots \\ \Sigma_{v_n} \end{pmatrix} \right).$$

$$(4.9)$$

The $(n + 1)$th order relations of the forward filter and the backward filter can also be written in a single equation as

$$
\begin{pmatrix}
R(0) & R(1) & \cdots & R(n) & R(n+1) \\
R(-1) & R(0) & \cdots & R(n-1) & R(n) \\
\cdots & \cdots & \cdots & \cdots & \cdots \\
R(-n) & R(-n+1) & \cdots & R(0) & R(1) \\
R(-n-1) & R(-n) & \cdots & R(-1) & R(0)
\end{pmatrix}
\left(
\begin{pmatrix}
I \\
F_1^{(n+1)} \\
\cdots \\
F_n^{(n+1)} \\
F_{n+1}^{(n+1)}
\end{pmatrix},
\begin{pmatrix}
B_{n+1}^{(n+1)} \\
B_n^{(n+1)} \\
\cdots \\
B_1^{(n+1)} \\
I
\end{pmatrix}
\right)
$$
$$
=
\left(
\begin{pmatrix}
\Sigma_{w_{n+1}} \\
O \\
\cdots \\
O \\
O
\end{pmatrix},
\begin{pmatrix}
O \\
O \\
\cdots \\
O \\
\Sigma_{v_{n+1}}
\end{pmatrix}
\right).
\tag{4.10}
$$

Now let us try to find the $(n + 1)$th order coefficient vectors, $F_k^{(n+1)}$ ($k = 1$, 2, ..., $n + 1$) and $B_k^{(n+1)}$ ($k = 1, 2, ..., n + 1$) in (4.10) in terms of the (n)th order coefficients, $F_k^{(n)}$ ($k = 1, 2, ..., n$) and $B_k^{(n)}$ ($k = 1, 2, ..., n$). Let the constants C_n and D_n be such that

$$
\left(
\begin{pmatrix}
I \\
F_1^{(n+1)} \\
\cdots \\
F_n^{(n+1)} \\
F_{n+1}^{(n+1)}
\end{pmatrix},
\begin{pmatrix}
B_{n+1}^{(n+1)} \\
B_n^{(n+1)} \\
\cdots \\
B_1^{(n+1)} \\
I
\end{pmatrix}
\right)
=
\left(
\begin{pmatrix}
I \\
F_1^{(n)} \\
\cdots \\
F_n^{(n)} \\
O
\end{pmatrix}
+ C_n
\begin{pmatrix}
O \\
B_n^{(n)} \\
\cdots \\
B_1^{(n)} \\
I
\end{pmatrix},
D_n
\begin{pmatrix}
I \\
F_1^{(n)} \\
\cdots \\
F_n^{(n)} \\
O
\end{pmatrix}
+
\begin{pmatrix}
O \\
B_n^{(n)} \\
\cdots \\
B_1^{(n)} \\
I
\end{pmatrix}
\right).
\tag{4.11}
$$

Then, from Equations 4.10 and 4.11, we have

$$
\Sigma_{w_n} + H_n C_n = \Sigma_{w_{n+1}},
$$
$$
G_n + \Sigma_{v_n} C_n = O,
\tag{4.12}
$$

and

$$
\Sigma_{v_n} + G_n D_n = \Sigma_{v_{n+1}},
$$
$$
H_n + \Sigma_{w_n} D_n = O,
\tag{4.13}
$$

where

$$G_n = R(-n-1) + R(-n)F_1^{(n)} + \cdots + R(-1)F_n^{(n)}$$

$$H_n = R(n+1) + R(n)B_1^{(n)} + \cdots + R(1)B_n^{(n)},$$

$$\Sigma_{w_n} = R(0) + R(1)F_1^{(n)} + \cdots + R(n)F_n^{(n)}, \qquad (4.14)$$

$$\Sigma_{v_n} = R(0) + R(1)B_1^{(n)} + \cdots + R(n)B_n^{(n)}.$$

From Equations (4.12) through (4.14) together with (4.11), we have the $(n + 1)$th order coefficients, $F_k^{(n+1)}$ ($k = 1, 2, \ldots, n + 1$) and $B_k^{(n+1)}$, from the (n)th coefficients, $F_k^{(n)}$ ($k = 1, 2, \ldots, n$) and $B_k^{(n)}$ ($k = 1, 2, \ldots, n$) as follows:

$$F_k^{(n+1)} = F_k^{(n)} - \Sigma_{v_n}^{-1}G_n B_{n+1-k}^{(n)} \quad k = 1, \ldots, n$$

$$F_{n+1}^{(n+1)} = -\Sigma_{v_n}^{-1}G_n$$

$$B_k^{(n+1)} = B_k^{(n)} - \Sigma_{w_n}^{-1}H_n F_{n+1-k}^{(n)} \quad k = 1, \ldots, n \qquad (4.15)$$

$$B_{n+1}^{(n+1)} = -\Sigma_{w_n}^{-1}H_n.$$

The recursion starts from $n = 0$, where we have

$$F^{(0)} = B^{(0)} = I$$

$$\Sigma_{w_0} = \Sigma_{v_0} = R(0).$$

The recursive relations (4.15), representing a multivariate version of the Levinson scheme (Levinson 1947), were derived by Whittle (1963a). These relations are useful for the estimation of the parameters of the multivariate AR models through the sample auto covariances, as discussed in Section 10.4.2.2.

4.1.6 Advantages and Disadvantages

As we saw in the scalar case, the AR model has some disadvantages. From a point of view of parsimonious parameterization of a general stationary Gaussian time series, the AR model class is inferior to the ARMA model class. In general, for modeling of stationary Gaussian time series, ARMA models are the ideal choice. However, ARMA models do not have the advantage of computational efficiency, whereas AR models do. In order to estimate the parameters of an ARMA model, we need to employ a time-consuming nonlinear optimization method for the maximization of the log-likelihood or the minimization of the prediction error variance. The same thing can be said about the multivariate case. The multivariate ARMA model class is a more general and parsimonious parametric model class than the model class of multivariate AR models, but the multivariate AR model is computationally more efficient in estimating its parameters with efficient recursive methods such as the Levinson–Whittle procedure.

So far, in the previous sections, we have seen the theoretical structure behind multivariate AR models. We note that these properties are useful in "whitening" the multivariate time series data. For example, since the time series, $x_1, x_2, ..., x_N$, is whitened into multivariate white noise, $w_1, w_2, ..., w_N$, by the multivariate AR model,

$$x_t = \sum_{k=1}^{p} A_k x_{t-k} + w_t,$$

the multivariate Levinson–Whittle procedure provides us with a whitening method,

$$w_t = x_t - \sum_{k=1}^{p} \hat{A}_k x_{t-k},$$

where $\hat{A}_k$ ($k = 1, ..., p$) are given from the data through the estimated auto covariance function, $\hat{\Gamma}_0, \hat{\Gamma}_1, \hat{\Gamma}_2, ..., \hat{\Gamma}_{N-1}$. The validity of these estimation methods for the "whitening" is discussed in Chapter 10.

4.2 Multivariate AR Models and Feedback Systems

4.2.1 Feedback System and the Difficulty of Parameter Estimation

The strength of multivariate AR modeling can be shown exemplarily in the modeling of complex feedback systems. Let us start from a simple example of a feedback system (see Figure 4.1) with a bivariate time series, $x_t, x_{t+1}, ...$ and $y_t, y_{t+1}, ...$, such that each variable results as output of a linear dynamic

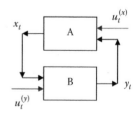

FIGURE 4.1
A simple feedback system (2 box) with one input and one output.

Here we note that the transformed dynamic models (4.18) and (4.19) define a special type of bivariate AR model,

$$
\begin{pmatrix} x_t \\ y_t \end{pmatrix} = \begin{pmatrix} d_1 & B_1 \\ A_1 & c_1 \end{pmatrix} \begin{pmatrix} x_{t-1} \\ y_{t-1} \end{pmatrix} + \cdots + \begin{pmatrix} d_L & B_L \\ A_L & c_L \end{pmatrix} \begin{pmatrix} x_{t-L} \\ y_{t-L} \end{pmatrix}
$$

$$
+ \begin{pmatrix} 0 & B_{L+1} \\ A_{L+1} & 0 \end{pmatrix} \begin{pmatrix} x_{t-L-1} \\ y_{t-L-1} \end{pmatrix} + \cdots + \begin{pmatrix} 0 & B_{L+M} \\ A_{L+M} & 0 \end{pmatrix} \begin{pmatrix} x_{t-L-M} \\ y_{t-L-M} \end{pmatrix} + \begin{pmatrix} \xi_t \\ \eta_t \end{pmatrix},
$$

$$(4.20)$$

where the diagonal elements of the coefficient matrices of lag order $k > L$ are zero. Since $(\xi_t, \eta_t)'$ is bivariate white noise, the model parameters can be estimated consistently from the bivariate time series data, and the original coefficients c_k, d_k ($k = 1, \ldots, L$) and a_m, b_m ($m = 1, \ldots, M$) can be derived from the AR coefficients of equation (4.20). For example, the a_m ($m = 1, \ldots, M$) can be obtained as follows:

$$
a_1 = A_1
$$

$$
a_m = A_m + \sum_{l=1}^{m-1} \tilde{c}_l a_{m-l} \quad (\text{for } m = 2, 3, \ldots, M).
$$

4.2.3 Multivariate Feedback Systems

The preceding idea of bivariate feedback system modeling can be easily extended to k-variate feedback systems (see Figure 4.2).

Suppose we have a k-variate feedback model,

$$
x_t^{(i)} = \sum_{\substack{j=1 \\ j \neq i}}^{k} \sum_{m=1}^{M} a_m^{(i,j)} x_{t-m}^{(j)} + u_t^{(i)} \quad (i = 1, 2, \ldots, k),
$$

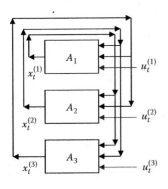

FIGURE 4.2
Multivariate feedback system.

driven by a colored noise $u_t^{(i)}$. By rewriting this with the whitening filters,

$$u_t^{(i)} - \sum_{l=1}^{L} c_l^{(i)} u_{t-l}^{(i)} = \varepsilon_t^{(i)} \quad (i = 1, 2, \ldots, k),$$

we obtain, for each variable $x_t^{(i)}$, a dynamic model,

$$x_t^{(i)} = \sum_{l=1}^{L} c_l^{(i)} x_{t-l}^{(i)} + \sum_{\substack{j=1 \\ j \neq i}}^{k} \sum_{m=1}^{M+L} A_{i,j,m} x_{t-m}^{(j)} + \varepsilon_t^{(i)}. \quad (i = 1, 2, \ldots, k).$$

This is a special case of a general k-variate AR model driven by a white noise, that is,

$$x_t^{(i)} = \sum_{m=1}^{M} \sum_{j=1}^{k} A_m^{(i,j)} x_{t-m}^{(j)} + \varepsilon_t^{(i)}, \quad (i = 1, 2, \ldots, k),$$

and we can estimate coefficients $A_m^{(i,j)}$'s consistently. From the estimated $A_m^{(i,j)}$'s, we can obtain estimates of coefficients $a_m^{(i,j)}$'s of the original feedback model consistently in the same way as for the bivariate case.

4.3 Multivariate ARMA Models

As we saw for the scalar ARMA models in Sections 3.1 and 3.2, ARMA models offer a more general and parsimonious representation for Gaussian processes for the multivariate case as well. The parsimonious parameterization is especially important when the number of data points available for parameter estimation is limited.

4.3.1 Multivariate ARMA Processes and the Power Spectrum

The power spectrum of a multivariate ARMA model has a more general representation form (i.e., rational functional form) than the multivariate AR model of Equation 4.1 in Section 4.1.1. For example, the r-dimensional ARMA(p,q) model,

$$x_t = \sum_{k=1}^{p} \Phi_k x_{t-k} + \sum_{k=1}^{q} \Theta_k w_{t-k} + w_t,$$

is rewritten using the backward shift operator B as

$$x_t = \left(I - \sum_{k=1}^{p} \Phi_k B^k \right)^{-1} \left(I + \sum_{k=1}^{p} \Theta_k B^k \right) w_t.$$

Let

$$w(f) = \sum_{k=-\infty}^{\infty} e^{-i2\pi f k}(w_k),$$

$$\Phi(f) = \sum_{k=-\infty}^{\infty} e^{-i2\pi f k}(-\Phi_k) \quad \text{where } \Phi_0 = -I, \quad -1/2 \le f \le 1/2$$

and

$$\Theta(f) = \sum_{k=-\infty}^{\infty} e^{-i2\pi f k}(\Theta_k)\,(\Theta_0 = I), \quad -1/2 \le f \le 1/2.$$

Then, we have

$$x(f) = \Phi(f)^{-1}\Theta(f)w(f).$$

Since

$$p_{xx}(f) = E\left[x(f)\overline{x(f)'} \right],$$

$$p_{ww}(f) = E\left[w(f)\overline{w(f)'} \right] = \Sigma_w,$$

and

$$E\left[\left\{ \Phi(f)^{-1}w(f)\Theta(f) \right\}\left\{ \overline{\Phi(f)^{-1}\Theta(f)w(f)} \right\}' \right] = \left\{ \Phi(f)^{-1}\Theta(f) \right\} p_{ww}(f) \left\{ \overline{\Phi(f)^{-1}\Theta(f)} \right\}',$$

we have

$$p_{yy}(f) = \left\{ \Phi(f)^{-1}\Theta(f) \right\}\Sigma_W \left\{ \overline{\Phi(f)^{-1}\Theta(f)'} \right\}^{-1}. \quad -1/2 \le f \le 1/2. \quad (4.21)$$

Example: $x_t = \Phi_1 x_{t-1} + \Theta_1 w_{t-1} + w_t.$

For this simple r-dimensional ARMA(1,1) model, we can write down the power spectral function explicitly as

$$p_{xx}(f) = \left\{ I - \exp(-i2\pi f)\Phi_1 \right\}^{-1} \left\{ I + \exp(-i2\pi f)\Theta_1 \right\} \Sigma_w$$

$$\times \left\{ \overline{I + \exp(-i2\pi f)\Theta_1{}'} \right\}^{-1} \left\{ \overline{I - \exp(-i2\pi f)\Phi_1{}'} \right\}.$$

Also similar to the scalar case, the advantage of the parsimonious and general parameterization of ARMA models is gained on the cost of computational load. For the computation of multivariate ARMA models, we cannot employ such a computationally efficient algorithm as the Levinson procedure for the computation of the multivariate AR models, discussed in the previous section.

The difficulty associated with the use of multivariate ARMA models in applications is not limited to the problem of high computational load. A more serious problem originates from an intrinsic theoretical difficulty, that is, the nonuniqueness of the model representation, a problem that does not arise for multivariate AR models.

4.3.2 Nonuniqueness of the Multivariate ARMA Representation

Although the ARMA representation of a stationary stochastic process has been used as one of the most general models in time series analysis, there is a serious conceptual difficulty inherent in this model, that is, the nonuniqueness of the representation.

When the following ARMA representation is given,

$$y_t + \Phi_1 y_{t-1} + \cdots + \Phi_p y_{t-p} = z_t + \Theta_1 z_{t-1} + \cdots + \Theta_q z_{t-q}, \tag{4.22}$$

there are infinitely many other representations of the same process. A simple example is shown here.

Let a bivariate ARMA(2,2) model be given by

$$\begin{pmatrix} y_t^{(1)} \\ y_t^{(2)} \end{pmatrix} = \begin{pmatrix} \phi_{11}^{(1)} & \phi_{12}^{(1)} \\ 0 & \phi_{22}^{(1)} \end{pmatrix} \begin{pmatrix} y_{t-1}^{(1)} \\ y_{t-1}^{(2)} \end{pmatrix} + \begin{pmatrix} \phi_{11}^{(2)} & \phi_{12}^{(2)} \\ 0 & 0 \end{pmatrix} \begin{pmatrix} y_{t-2}^{(1)} \\ y_{t-2}^{(2)} \end{pmatrix} + \begin{pmatrix} \theta_{11}^{(1)} & \theta_{12}^{(1)} \\ 0 & \theta_{22}^{(1)} \end{pmatrix} \begin{pmatrix} \varepsilon_{t-1}^{(1)} \\ \varepsilon_{t-1}^{(2)} \end{pmatrix}$$

$$+ \begin{pmatrix} \theta_{11}^{(2)} & \theta_{12}^{(2)} \\ 0 & 0 \end{pmatrix} \begin{pmatrix} \varepsilon_{t-2}^{(1)} \\ \varepsilon_{t-2}^{(2)} \end{pmatrix} + \begin{pmatrix} \varepsilon_t^{(1)} \\ \varepsilon_t^{(2)} \end{pmatrix}. \tag{4.23}$$

Then, the following ARMA(2,2) model defines the same process as (4.23) for any $c > 0$:

$$\begin{pmatrix} y_t^{(1)} \\ y_t^{(2)} \end{pmatrix} = \begin{pmatrix} \phi_{11}^{(1)} & \phi_{12}^{(1)} + c \\ 0 & \phi_{22}^{(1)} \end{pmatrix} \begin{pmatrix} y_{t-1}^{(1)} \\ y_{t-1}^{(2)} \end{pmatrix} + \begin{pmatrix} \phi_{11}^{(2)} & \phi_{12}^{(2)} - c\phi_{22}^{(1)} \\ 0 & 0 \end{pmatrix} \begin{pmatrix} y_{t-2}^{(1)} \\ y_{t-2}^{(2)} \end{pmatrix}$$

$$+ \begin{pmatrix} \theta_{11}^{(1)} & \theta_{12}^{(1)} - c \\ 0 & \theta_{22}^{(1)} \end{pmatrix} \begin{pmatrix} \varepsilon_{t-1}^{(1)} \\ \varepsilon_{t-1}^{(2)} \end{pmatrix} + \begin{pmatrix} \theta_{11}^{(2)} & \theta_{12}^{(2)} - c\theta_{22}^{(1)} \\ 0 & 0 \end{pmatrix} \begin{pmatrix} \varepsilon_{t-2}^{(1)} \\ \varepsilon_{t-2}^{(2)} \end{pmatrix} + \begin{pmatrix} \varepsilon_t^{(1)} \\ \varepsilon_t^{(2)} \end{pmatrix}.$$

$$(4.24)$$

The second model (4.24) is obtained by adding

$$c(y_{t-1}^{(2)} - \phi_{22}^{(1)} y_{t-2}^{(2)} - \theta_{22}^{(1)} \varepsilon_{t-2}^{(2)} - \varepsilon_{t-1}^{(2)}) = 0 \qquad (4.25)$$

to the first row of model (4.23). Relation (4.25) is derived from the second row of model (4.23) by multiplying $c > 0$ and shifting the time from t to $t - 1$, that is,

$$y_t^{(2)} = \phi_{22}^{(1)} y_{t-1}^{(2)} + \theta_{22}^{(1)} \varepsilon_{t-1}^{(2)} + \varepsilon_t^{(2)} \rightarrow cy_t^{(2)} = c\phi_{22}^{(1)} y_{t-1}^{(2)} + c\theta_{22}^{(1)} \varepsilon_{t-1}^{(2)} + c\varepsilon_t^{(2)}$$

$$\rightarrow c(y_{t-1}^{(2)} - \phi_{22}^{(1)} y_{t-2}^{(2)} - \theta_{22}^{(1)} \varepsilon_{t-2}^{(2)} - \varepsilon_{t-1}^{(2)}) = 0.$$

In general an equivalent process with a different representation is obtained from (4.22) by the transformations, which replace t of (4.22) by some $t - k$ ($k \geq 1$), then premultiply by an $r \times r$ matrix D_k and add the results to the original model (4.22). When y_t is a univariate process, the representation can be made unique by requiring the orders p and q of (4.22) to be minimal. When y_t is a multivariate process, this requirement of minimal order is not necessarily sufficient to make the representation unique, as we saw in the previous example of the bivariate ARMA(2,2) model (4.23).

It may appear that this lack of uniqueness of the ARMA model representation poses a problem for the derivation of methods for estimating the model parameters. Conditions for the uniqueness of ARMA representations and for identifiability of ARMA models have been derived (Hannan 1969a). To the surprise of those who were aware of the earlier work of Hannan (1969a), Akaike (1974a) showed that there is a uniquely determined standard ARMA representation, the parameters of which can be estimated from time series data without any conditions. Akaike (1974a, p. 369) says, "It is trivially true that the uniqueness of a representation is a prerequisite for the development of consistent estimation procedures. But this fact should not be considered as meaning the impracticability of developing a general estimation procedure of ARMA models without identifiability conditions. If the purpose of fitting an ARMA model is only to get an estimate of the covariance structure of the process under observation, any one of the possible equivalent representations can serve for the purpose, if only it can be specified properly."

Before we can discuss the estimation problem and the dynamic causality analysis for an r-dimensional time series, $x_1, x_2, ..., x_N$, two important problems have to be addressed:

1. How to choose a standard (canonical) ARMA model
2. How to parameterize the coefficient matrices, $\Phi_1, ..., \Phi_p, \Theta_1, ..., \Theta_q$, of the standard ARMA($p,q$)

Estimating all elements of the coefficient matrices, that is, $r \times r \times p + r \times r \times q$ parameters, certainly sounds very impractical. The same can be said for the parameterization of the transition matrix of state space models for multivariate time series.

A very pragmatic and practical solution for multivariate ARMA modeling was proposed by Akaike (1974a) without the need for any specific conditions for the uniqueness of the ARMA representation.

4.3.3 State Space Model for Multivariate ARMA Processes

As we have seen in the previous section for the scalar time series case, ARMA models and state space models are closely related. Since Akaike's method for multivariate ARMA modeling exploits these close relations between state space models and ARMA models, we briefly review here the relations between the two model classes for the multivariate time series case.

4.3.3.1 Nonuniqueness of the State Space Representation

Before discussing Akaike's method, we emphasize that for the class of state space models the nonuniqueness of the model representation is even more obvious than for multivariate ARMA models. This is because the state variables are not directly observed as time series. For example, the following two state space models are equivalent for any nonsingular matrix R:

$$\begin{cases} y_t = A_1 y_{t-1} + B_1 \varepsilon_t \\ x_t = C_1 y_t \end{cases} \tag{4.26}$$

$$\begin{cases} z_t = A_2 z_{t-1} + B_2 \varepsilon_t \\ x_t = C_2 z_t \end{cases}, \tag{4.27}$$

where $A_2 = R^{-1} A_1 R$, $B_2 = R^{-1} B_1$ and $C_2 = C_1 R$. The state space model (4.26) has a problem of efficient parameterization of the matrices A_1, B_1 and C_1.

Obviously, treating all the elements of these matrices as independent parameters would lead us to numerical difficulties in the actual estimation and model identification steps.

4.3.3.2 Stationary Multivariate ARMA Model and the Impulse Response Function

Let an r-dimensional ARMA process x_t be defined by

$$x_t - \Phi_1 x_{t-1} - \Phi_2 x_{t-2} - \cdots - \Phi_p x_{t-p}$$
$$= w_t + \Theta_1 w_{t-1} + \Theta_2 w_{t-2} + \cdots + \Theta_q w_{t-q}, \tag{4.28}$$

where Φ_i and Θ_i are matrices of coefficients, x_t is an $r \times 1$ vector, and w_t is an r-variate white noise with $E[w_t] = 0$, that is, zero vector, for $i \neq 0$, $E[w_t w'_{t+i}] = 0$, zero matrix, $E[w_t w'_t] = \Sigma_w$, and $E[x_t w'_{t+i}] = 0$ for $i = 1, 2, \ldots$. We note that the stationarity condition of the ARMA model (4.28), that is, "The roots of the characteristic equation, $|\Lambda^p - \Phi_1 \Lambda p - 1 - \cdots - \Phi_p| = 0$, lie inside the unit circle," guarantees that x_t has the general linear process (or an impulse response) representation,

$$x_t = \sum_{m=0}^{\infty} h_m w_{t-m}, \tag{4.29}$$

where $h_0 = I$. For $h_1, h_2, \ldots$ to be unique, a sufficient condition is "Σ_w is nonsingular."

If an r-dimensional ARMA model is stationary and has a general linear process representation, it can be transformed into a state space model (in the same manner as for the scalar time series case in Section 3.4). We see this in the next section, where the concept of predictor space and the impulse response function representation play important roles.

4.3.3.3 From ARMA Model to State Space Model

To obtain a Markovian representation of x_t, it is only necessary to analyze the structure of the predictors $x_{t|t}, x_{t+1|t}, \ldots$, where $x_{|t}$ denotes the projection of x on $R(t-)$, that is, the linear space spanned by the components of the present and the past values of the r-dimensional time series. Then, $x_{t+i|t}$ satisfies the relation

$$x_{t+i|t} - \Phi_1 x_{t+i-1|t} - \Phi_2 x_{t+i-2|t} - \cdots - \Phi_p x_{t+i-p|t}$$
$$= w_{t+i|t} + \Theta_1 w_{t+i-1|t} + \Theta_2 w_{t+i-2|t} + \cdots + \Theta_q w_{t+i-q|t}, \tag{4.30}$$

where $x_{t+d|t} = x_{t+d}$ for $d = 0, -1, \ldots$, and $w_{t+d|t} = 0$ for $d = 0, 1, \ldots$. For $i \geq q + 1$, the right-hand side of (4.30) vanishes. Thus, $x_{t+i|t}$ ($i = 0, 1, \ldots$) can be expressed as linear transforms of $x_{t|t}, x_{t+1|t}, \ldots, x_{t+K-1|t}$, where $K = \max(p, q + 1)$, and the components of these vectors form a system of generators of the linear space spanned by the components of $x_{t+i|t}$ ($i = 0, 1, \ldots$). In particular, it holds that

$$x_{t+K|t} = \Phi_1 x_{t+K-1|t} + \Phi_2 x_{t+K-2|t} + \cdots + \Phi_K x_{t|t}, \tag{4.31}$$

where by definition $\Phi_m = 0$ for $m = p + 1, p + 2, \ldots, K$. From (4.29), one can obtain

$$x_{t+i+1|t+1} = x_{t+i+1|t} + h_i w_{t+1}. \tag{4.32}$$

From (4.31) and (4.32), it can be seen that the vector,

$$z_t = (x_{t|t}, x_{t+1|t}, \ldots, x_{t+K-1|t})',$$

provides a Markovian representation,

$$z_{t+1} = \begin{pmatrix} 0 & I & 0 & \cdots & 0 \\ 0 & 0 & I & \cdots & 0 \\ \cdots & \cdots & \cdots & \cdots & \cdots \\ 0 & 0 & 0 & \cdots & I \\ \Phi_K & \Phi_{K-1} & \Phi_{K-2} & \cdots & \Phi_1 \end{pmatrix} z_t + \begin{pmatrix} h_0 \\ h_1 \\ \cdots \\ h_{K-1} \\ h_K \end{pmatrix} w_{t+1}, \tag{4.33}$$

$$x_t = \begin{pmatrix} I & 0 & 0 & \cdots & 0 \end{pmatrix} z_t$$

where $K = \max\{p, q + 1\}$. By definition, $\Phi_m = 0$ for $m = p + 1, p + 2, \ldots, K$. The impulse response matrices h_m's are easily calculated with the following scheme, which comes from the ARMA model (4.28),

$$h_m(.j) - \Phi_1 h_{m-1}(.j) - \cdots - \Phi_M h_{m-M}(.j)$$
$$= D_t(.j) + \Theta_1 D_{t-1}(.j) + \cdots + \Theta_L D_{t-L}(.j). \tag{4.34}$$

Here $h_m = 0$, zero vector, for $m < 0$ and $D_1(.j)$ denotes the jth column of a matrix D, which is by definition equal to I, identity matrix, for $t = 0$ and 0, zero matrix, for $t \neq 0$.

4.3.3.4 Block-Wise State Space Model

The aforementioned result shows that a multivariate ARMA process (4.28) always has a state space representation (4.33). However, we note that there is a limitation in the type of state space representations given by Equation 4.33.

Note that the dimension of the representation (4.33) is always an integral multiple of the dimension r of the time series x_t. There is no reason why the dimension of the background system should be an integral multiple of the dimension of the observed time series x_t. Thus, the aforementioned state space representation forms a rather limited class of state space representations for the time series x_t. We call this type of state space representations (4.33) a "block-wise" state space representation. The derivation of a more general class of state space representations, proposed by Akaike (1974a), is discussed next in Section 4.4.

Note that in the preceding derivation, the state space model (4.33) is constructed in such a way that both the controllability condition and the observability condition are satisfied, that is,

$$\text{Rank}(O_s) = s: \text{(Observability condition)}$$

and

$$\text{Rank}(C_s) = s: \text{(Controllability condition)}$$

where

$$C_s = [B,\ AB,\ A^2B,\ ...,\ A^{s-1}B]$$

and

$$O_s = \begin{pmatrix} C \\ CA \\ CA^2 \\ ... \\ CA^{s-1} \end{pmatrix},$$

and $s = r \times K$, that is, the dimension of the state. In other words, the state space model constructed by this method always satisfies the two conditions: the observability condition and the controllability condition.

Here the only necessary condition for the state space model (4.33) to be identifiable is the nonsingularity of the innovation variance matrix $\Sigma\varepsilon$. Since w_t is uniquely determined (if Σ_w is nonsingular) by (4.29) and satisfies

$$\Theta_i = w_i + \Phi_1 w_{i-1} + \cdots + \phi_M w_{i-M},$$

where $w_m = 0$ for $m < 0$, the ARMA representation is also uniquely determined from the state space representation (4.33).

4.3.3.5 *From State Space Model to ARMA Model*

Any state space model can be rewritten as an ARMA model. Suppose an r-dimensional process x_t has a state space representation,

$$z_{t+1} = Az_t + B\varepsilon_{t+1}$$
$$x_t = Cz_t,$$

(4.35)

where it is assumed that z_t is a $p \times 1$ vector of the state and ε_t is the innovation of x_t. If the characteristic polynomial of A is given by

$$|\lambda I - A| = \lambda^p + \sum_{m=1}^{p} a_m \lambda^{p-m},$$

then by the Cayley–Hamilton theorem, we have

$$A^p + \sum_{m=1}^{p} a_m A^{p-m} = 0.$$

From (4.35),

$$z_{t+i} = A^i z_t + A^{i-1} B\varepsilon_{t+1} + \cdots + B\varepsilon_{t+i},$$

and it follows that x_t has an ARMA representation,

$$x_{t+p} + a_1 x_{t+p-1} + a_2 x_{t+p-2} + \cdots + a_p x_t$$
$$= \varepsilon_{t+p} + C_1 \varepsilon_{t+p-1} + C_2 \varepsilon_{t+p-2} + \cdots + C_{p-1} \varepsilon_{t+1},$$

(4.36)

where

$$C_i = C(A^i + a_1 A^{i-1} + \cdots + a_i I)B.$$

This shows that any r-dimensional stationary stochastic process x_t with the state space representation (4.35) also has an ARMA representation (4.36).

4.4 Multivariate State Space Models and Akaike's Canonical Realization

4.4.1 Input–Output System and the Predictor Space

In this section, we look into Akaike's predictor space approach to the multivariate time series. Let us consider a stochastic system with the q-dimensional input u_t and r-dimensional output x_t, where it is assumed that $(u_t', x_t')'$ is

$(r + q)$-dimensional stationarity correlated zero mean Gaussian time series. Hereafter, we are concerned with various linear spaces spanned by the components of u_t and x_t, where the metric of these spaces is given by the root mean square.

If we adopt the definition of the state space of a system at time t as the totality of the information from the present and past input to be transmitted to the present and future output, a natural mathematical representation of the state space in this case will be given, as shown for the scalar case in Section 3.4, by the linear space spanned by the components of the projections $x_{t+k|t}$ ($k = 0, 1, \ldots$), where $x_{t+k|t}$ denotes the projection of x_{t+k} onto the linear space $R_u(t-)$ spanned by the components of the present and past inputs $u_t, u_{t-1}, \ldots$.

The projection $x_{t+k|t}$ is composed of the elements of $R_u(t -)$ and is characterized by the relation

$$Ex_{t+k|t}u'_{t-m} = Ex_{t+k}u'_{t-m} \quad \text{for } m = 0, 1, \ldots. \tag{4.37}$$

The very basic assumption to be made in this section is that the dimension of the linear space spanned by the components of $x_{t+k|t}$ ($k = 0, 1, \ldots$) is finite. This space that is spanned by the components of the predictors is called "predictor space" and is denoted by $R(t+|t-)$. The dimension of the space is denoted by p. Obviously, $R(t+|t-) \subset R_u(t-)$.

Equation 4.37 shows that the whole information of x_{t+k} within $R_u(t-)$ is condensed in $x_{t+k|t}$, in the sense that the dependence of x_{t+k} on $R_u(t-)$ can be explained by $x_{t+k|t}$. Take an arbitrary basis $z_t^{(1)}, z_t^{(2)}, \ldots, z_t^{(p)}$ of the predictor space $R(t+|t-)$ and denote by z_t the vector $(z_t^{(1)}, z_t^{(2)}, \ldots, z_t^{(p)})'$. Since $x_{t|t}$ is composed of the elements of $R(t+|t-)$, there is an $r \times p$ matrix C such that

$$x_{t|t} = Cz_t.$$

Thus, x_t has a representation,

$$x_t = Cz_t + \varepsilon_t, \tag{4.38}$$

where ε_t is independent of the elements of $R_u(t-)$, that is, ε_t is independent of the present and past inputs. Denote by z_{t-1} the random vector obtained by replacing $u_t, u_{t-1}, \ldots$, in the definition of z_t by $u_{t+1}, u_t, \ldots$. From the stationarity, the components of z_{t+1} form a basis of the space $R((t + 1)+|(t + 1)-)$.

By an analogous reasoning as in the case of x_t, it can be seen that z_{t+1} admits a representation

$$z_{t+1} = Az_t + \eta_{t+1}, \tag{4.39}$$

where η_{t+1} is independent of $u_t, u_{t-1}, \ldots$. From (4.39), since the components of z_{t-1}, and z_t are elements of $R_u((t + 1)-)$, η_{t+1} is composed of the elements of

$R_u((t + 1)-)$, but its projection onto $R_u(t-)$ is equal to zero. Thus, η_{t+1} can be represented in the form,

$$\eta_{t+1} = Bw_{t+1}, \tag{4.40}$$

where B is a $p \times q$ matrix and

$$w_{t+1} = u_{t+1} - u_{t+1|t}. \tag{4.41}$$

$u_{t+1|t}$ denotes the projection of u_{t+1} onto $R_u(t-)$, and w_{t+1} is the innovation of the input u_{t+1} at time $t + 1$. The covariance matrix of the innovation w_t is denoted by Σ_w, that is, $\Sigma_w = E[w_t w_t']$.

Equations 4.38 through 4.41 yield to a state space representation of a general stationary linear stochastic input–output system,

$$\begin{cases} z_{t+1} = Az_t + Bw_{t+1} \\ x_t = Cz_t + \varepsilon_t \end{cases}. \tag{4.42}$$

In this representation, ε_t denotes the part of x_t that is independent of the present and past inputs. w_t with different values of t is mutually independent, but ε_t's are not necessarily so. z_t provides a specification of the predictor space $R(t+|t-)$, and this may be considered to be a representation of the state of the stochastic system at time t. z_t is called the state vector.

4.4.2 Feedback System and the Predictor Space

When we are not sure about the clear separation of the input u_t and output x_t, or if there is a possibility of feedback from x_t to u_t, the state space representation (4.42) is not suitable for this. When there is a feedback from x_t to u_t, the output x_t could be an input to the same system at the same time, and the input u_t could be an output of the same system. Here, distinguishing input and output of the system becomes meaningless. Then, x_t and u_t are indistinguishable, and they are put into one joined vector time series $x_t^{(new)} = (x_t^{(old)}, u_t^{(old)})'$, which we rewrite as $x_t = x_t^{(new)}$. Here we have a higher dimensional newly observed multivariate time series data, $x_1, x_2, x_3, \ldots$. For this case, since $x_t = u_t$ in the previous notation, and since x_t is composed of the elements of $R_u(t-)$, we have $x_{t|t} = x_t$ and ε_t in (4.38) vanishes. Thus, we get a state space representation,

$$\begin{cases} z_{t+1} = Az_t + Bw_{t+1} \\ x_t = Cz_t, \end{cases} \tag{4.43}$$

where the innovation w_{t+1} of x_t at time $t + 1$ is defined by

$$w_{t+1} = x_{t+1} - x_{t+1|t}. \tag{4.44}$$

This state space representation is sometimes called a Markovian representation of x_t.

Mathematically the model (4.43) may appear less general, in its form, than the model (4.42), but actually it is more general than the nonfeedback model (4.42), in the sense that the feedback is taken care of in (4.43). In other words, the feedback from the variable $x_t^{(old)}$, the old output, to the variable $u_t^{(old)}$, the old input, can be accepted, if there is any, in (4.43) but not in (4.42).

4.4.3 How Are A, B, and C Determined?

Our canonical state space representation for the r-dimensional time series $x_t = \left(x_t^{(1)}, x_t^{(2)}, \ldots, x_t^{(r)}\right)'$ $(t = 1, 2, \ldots, N)$ is obtained by choosing the first maximum set of linearly independent elements within the sequence of the predictors,

$$x_{t|t}^{(1)}, x_{t|t}^{(2)}, \ldots, x_{t|t}^{(r)}, x_{t+1|t}^{(1)}, x_{t+1|t}^{(2)}, \ldots, x_{t+1|t}^{(r)}, \ldots, x_{t+k|t}^{(1)}, x_{t+k|t}^{(2)}, \ldots,$$

as the state vector, where $x_{t+k|t}^{(i)}$ denotes the ith component of $x_{t+k|t}$. In principle, the state vector z_t can be constructed by the following procedure.

1. Define the vector Y_t by

$$Y_t = \begin{pmatrix} x_{t|t} \\ x_{t+1|t} \\ \ldots \end{pmatrix}. \tag{4.45}$$

The ith component of $Y_t^{(i)}$ of Y_t is given by $Y_t^{(i)} = x_{t+k|t}^{(j)}$, where $i = kr + j$ and r is the dimension of x_t.

2. Successively test the linear dependence of $Y_t^{(i)}$ on its antecedents

$$Y_t^{(i-1)}, Y_t^{(i-2)}, \ldots, Y_t^{(1)} \quad \text{for } i = 1, 2, \ldots$$

3. z_t is simply defined as the vector obtained by discarding from Y_t those components that are dependent on its antecedents. The dimension of z_t is denoted by p.

The state vector z_t is completely characterized by a vector $H = (H^{(1)}, H^{(2)}, \ldots, H^{(p)})'$, which is defined by the relation,

$$H^{(j)} = i \quad \text{when} \quad z_t^{(j)} = Y_t^{(i)}.$$

$H^{(j)}$ gives the specification of the jth component of the state z_t as an element within the sequence of the predictors. This vector of H is called "structural characteristic vector" of z_t.

When there is a representation of $Y_t^{(i)} = x_{t+k|t}^{(j)}$ as a linear combination of its antecedents, by replacing t by $t + m$ ($m = 1, 2, \ldots$) in the definition of elements in the representation, one can see that $Y_t^{(i+mr)} = x_{t+m+k|t}^{(j)}$ can also be expressed as a linear combination of its antecedents. Thus, for each component $x_t^{(j)}$ of x_t there is a smallest value k_j of k such that $x_{t+k|t}^{(j)}$ is linearly dependent on its antecedents in Y_t. The r-dimensional vector $k = (k_1, k_2, \ldots, k_r)'$ can also specify the structure of the state vector. The state vector z_t is obtained by arranging $x_{t+k|t}^{(j)}$'s with $k = 0, 1, \ldots, k_j - 1$ and $j = 1, 2, \ldots, r$ in the increasing order of $j + kr$.

Those $z_t^{(i)}$'s that correspond to $x_{t+k_j-1|t}^{(j)}$'s are characterized by the fact that $(H^{(i)} + r)$'s do not belong to the set of the components of the structural characteristic vector H. Now for $z_t^{(i)} = x_{t+k_j-1|t}^{(j)}$, we have $z_{t+1}^{(i)} = x_{t+k_j|t+1}^{(j)}$, and its projection on $R(t-)$, denoted by $z_{t+1|t}^{(i)}$, is equal to $x_{t+k_j|t}^{(j)}$, which can be expressed as a linear combination of its antecedents in Y_t of (4.45). Thus, for this case there is a unique representation,

$$z_{t+1|t}^{(i)} = \sum_{m=1}^{p} A_{i,m} z_t^{(m)}, \tag{4.46}$$

where $A_{i,m} = 0$ for $m = m(i) + 1, m(i) + 2, \ldots, p$. $m(i)$ is the number of $H^{(k)}$'s that are smaller than $H^{(i)} + r$. $A_{i,m}$ defines the (i, m)th element of the matrix of A.

When $z_t^{(i)}$ is defined to be equal to $x_{t+k-1|t}^{(j)}$ with $k < k_j$, $x_{t+k|t}^{(j)}$ is another component of z_t that will be denoted by $z_t^{(m(i))}$. The corresponding component of the structural characteristic vector H is given by

$$H^{m(i)} = H^{(i)} + r.$$

$m(i)$ is equal to the number of $H^{(m)}$'s that are smaller than or equal to $H^{(i)} + r$. For this case, since $z_{t+1}^{(i)}$ is given by $x_{t+k|t+1}^{(j)}$, $z_{t+1|t}^{(i)}$ is equal to $x_{t+k|t}^{(j)}$, which is the $m(i)$th component of z_t. Thus, we get a unique representation

$$z_{t+1|t}^{(i)} = \sum_{m=1}^{p} A_{i,m} z_t^{(m)}, \tag{4.47}$$

where $A_{i,m} = 1$ for $m = m(i)$ and $A_{i,m} = 0$ otherwise. Equations 4.46 and 4.47 give a complete specification of the transition matrix A.

The (i,s)th element $B_{i,s}$ of the input matrix B, which is a $p \times r$ matrix, is characterized as the impulse response of $z_t^{(i)}$ to the input $w_t^{(s)}$, the sth component of the innovation. If $z_t^{(i)}$ is identical to $x_{t+k|t}^{(j)}$, then $B_{i,s}$ can be obtained as the impulse response of $x_t^{(j)}$ at time lag k to the input $w_t^{(s)}$.

The observation matrix C, which is $r \times p$, can be obtained as the matrix of regression coefficients of x_t on z_t. Under the assumption of nonsingularity of the covariance matrix of $x_{t|t}$, C takes the form

$$C = [I, O],$$

where I is an $r \times r$ identity matrix and O is an $r \times (p - r)$ zero matrix. This is for the cases where we cannot separate the input variables and the output variables because of the presence of the feedbacks between them (then we have $u_t = x_t$), and the covariance matrix Σ_w of the innovation w_t is nonsingular. For $u_t = x_t$, under the same assumption, the upper-most, $r \times r$ submatrix of the input matrix B is an identity matrix. In general multivariate time series modeling, unless otherwise stated, we will only be concerned with the case where $u_t = x_t$ and the covariance matrix Σ_w of w_t is nonsingular.

4.4.4 Uniqueness

It is obvious that in the aforementioned canonical state space representation of x_t the matrices A, B, C, and Σ_w are uniquely determined. Any modification of the elements of these matrices, which are not explicitly specified as equal to 1 or 0, produces a representation of a time series that is different from the original one. This shows that under the assumption of nonsingularity of Σ_w and the finiteness of the dimension of the predictor space, the correspondence between the covariance structure of x_t and the set of H, A, B, C, and Σ_w is one to one. Here A, B, C, and Σ_w are such that they are in the forms given by the canonical representation of x_t and define a covariance structure with the linear dependence characteristics of the predictors specified by H.

The totality of the sets (H, A, B, C, Σ_w) provides a basic framework for the identification of a stationary time series. The fundamental difficulty in identifying a multivariate stochastic linear system is caused by the complex structure of H, which is also reflected in the fact that k is a vector with the same dimension as that of x_t. We will call k_i the order of the ith component of x_t. It should be remembered that the value of k_i is independent on the ordering of the components within the vector x_t. The quantity $k = \max(k_1, k_2, \ldots, k_r)$ is independent of this ordering and will be called the order of x_t or of the linear stochastic system.

We note that the relation (4.47) shows, with some matrices, C_0, C_1, ... and C_p, the relation,

$$x_{t+p|t} = \sum_{m=0}^{p} C_m x_{t+p-m|t},$$ (4.48)

between the projections of the r-dimensional process, x_t, since the state variables are constructed from the process variables projected into the predictor space. Equation 4.48 is reorganized, with $k = \max(k_1, k_2, ..., k_r)$, into

$$(I - C_0)x_{t+k|t} - \sum_{m=1}^{k} C_m x_{t+k-m|t} = 0.$$

From this relation we can have an r-dimensional ARMA(k, $k - 1$) representation with the innovations, w_{t+k}, w_{t+k-1}, ..., w_{t+1}, and with some matrices D_0, D_1, ... and D_{k-1}, as

$$(I - C_0)x_{t+k} - \sum_{m=1}^{k} C_m x_{t+k-m} = D_0 w_{t+k} + D_1 w_{t+k-1} + \cdots + D_{k-1} w_{t+1}.$$ (4.49)

Here we use the innovation representation of x_{t+i} and $x_{t+i|t}$ such that

$$x_{t+i} = h_0 w_{t+i} + h_1 w_{t+i-1} + h_2 w_{t+i-2} + \cdots + h_{i-1} w_{t+1} + h_i w_t + \cdots$$

$$x_{t+i|t} = h_i w_t + h_{i+1} w_{t-1} + \cdots$$

$$x_{t+i} = x_{t+i|t} + h_0 w_{t+i} + h_1 w_{t+i-1} + h_2 w_{t+i-2} + \cdots + h_{i-1} w_{t+1}.$$

This is written as an ordinary ARMA model form,

$$x_{t+k} = \Phi_1 x_{t+k-1} + \cdots + \Phi_k x_t + \Theta_0 w_{t+k} + \Theta_1 w_{t+k-1} + \cdots + \Theta_{k-1} w_{t+1},$$

where

$$\Theta_i = (I - C_0)^{-1} D_i,$$

and

$$\Phi_i = -(I - C_0)^{-1} C_i.$$

Note that from the same canonical state space representation we can have another ARMA model with AR order p and MA order $p - 1$, where p is the dimension of the state. However, the ARMA model derived earlier is much more compact since the order $k = \max(k_1, k_2, \ldots, k_r)$ is usually much smaller than the dimension p of the state of the canonical state space model.

4.4.5 How It Works

In this section, we see how an ARMA process is characterized by a canonical state space model.

4.4.5.1 Example 1

Suppose we have a 2-dimensional ARMA(3,1) model,

$$x_t + \begin{pmatrix} -0.9 & 0 \\ 0 & -1.5 \end{pmatrix} x_{t-1} + \begin{pmatrix} 0.4 & 0 \\ 0 & 1.2 \end{pmatrix} x_{t-2} + \begin{pmatrix} 0 & 0 \\ 0 & -0.448 \end{pmatrix} x_{t-3}$$

$$= w_t + \begin{pmatrix} 0.8 & 0 \\ 0 & 0 \end{pmatrix} w_{t-1}, \tag{4.50}$$

where $\sum_w = I$. From this model, we can construct a state space vector z_t as follows:

1. Define the vector Y_t by

$$Y_t = \begin{pmatrix} x_{t|t} \\ x_{t+1|t} \\ \cdots \end{pmatrix}$$

$$= \left(x_{t|t}^{(1)}, x_{t|t}^{(2)}, x_{t+1|t}^{(1)}, x_{t+1|t}^{(2)}, x_{t+2|t}^{(1)}, x_{t+2|t}^{(2)}, x_{t+3|t}^{(1)}, x_{t+3|t}^{(2)}, \ldots \right)'.$$

Here, the ith component $Y_t^{(i)}$ of Y_t is given by

$$Y_t^{(i)} = x_{t+k|t}^{(j)},$$

the dimension of x_t is 2, and $i = 2k + j$.

2. Successively test the linear dependence of $Y_t^{(i)}$ on its antecedents,

$$Y_t^{(i-1)}, Y_t^{(i-2)}, \ldots, Y_t^{(1)} \quad \text{for } i = 1, 2, \ldots.$$

2-1. We first check the linear dependence of $Y_t^{(3)} = x_{t+1|t}^{(1)}$, assuming the linear independence of the first two elements, $Y_t^{(1)} = x_{t|t}^{(1)}$ and $Y_t^{(2)} = x_{t|t}^{(2)}$. Thus, we start from the structural vector $H = (1,2)'$.

2-2. Since we have

$$\begin{pmatrix} x_{t+1}^{(1)} \\ x_{t+1}^{(2)} \end{pmatrix}_{|t} = \begin{pmatrix} 0.9 & 0 \\ 0 & 1.5 \end{pmatrix}\begin{pmatrix} x_t^{(1)} \\ x_t^{(2)} \end{pmatrix}_{|t} + \begin{pmatrix} -0.4 & 0 \\ 0 & -1.2 \end{pmatrix}\begin{pmatrix} x_{t-1}^{(1)} \\ x_{t-1}^{(2)} \end{pmatrix}_{|t} + \begin{pmatrix} 0 & 0 \\ 0 & 0.448 \end{pmatrix}\begin{pmatrix} x_{t-2}^{(1)} \\ x_{t-2}^{(2)} \end{pmatrix}_{|t},$$

$Y_t^{(3)} = x_{t+1|t}^{(1)}$ cannot be dependent on $Y_t^{(1)} = x_{t|t}^{(1)}$ and $Y_t^{(2)} = x_{t|t}^{(2)}$, and it is included in the state space vector, and H is made into $H = (1, 2, 3)'$. Also from the aforementioned relation, $Y_t^{(4)} = x_{t+1|t}^{(2)}$ cannot be dependent on its antecedents, $x_{t|t}^{(1)}, x_{t|t}^{(2)}, x_{t+1|t}^{(1)}$, and it is included in the state vector; H is converted to $H = (1, 2, 3, 4)'$.

2-3. However, $Y_t^{(5)} = x_{t+2|t}^{(1)}$ is judged to be dependent on its antecedents, $x_{t|t}^{(1)}, x_{t|t}^{(2)}, x_{t+1|t}^{(1)}, x_{t+1|t}^{(2)}$, because it holds

$$x_{t+2|t}^{(1)} = 0.9x_{t+1|t}^{(1)} - 0.4x_{t|t}^{(1)},$$

because of the relation

$$\begin{pmatrix} x_{t+2}^{(1)} \\ x_{t+2}^{(2)} \end{pmatrix}_{|t} = \begin{pmatrix} 0.9 & 0 \\ 0 & 1.5 \end{pmatrix}\begin{pmatrix} x_{t+1}^{(1)} \\ x_{t+1}^{(2)} \end{pmatrix}_{|t} + \begin{pmatrix} -0.4 & 0 \\ 0 & -1.2 \end{pmatrix}\begin{pmatrix} x_t^{(1)} \\ x_t^{(2)} \end{pmatrix}_{|t} + \begin{pmatrix} 0 & 0 \\ 0 & 0.448 \end{pmatrix}\begin{pmatrix} x_{t-1}^{(1)} \\ x_{t-1}^{(2)} \end{pmatrix}_{|t}.$$

Then, $Y_t^{(5)} = x_{t+2|t}^{(1)}$ is discarded from the vector, and all the $Y_t^{(5+2m)} = x_{t+2+m|t}^{(1)}$ $(m = 1, 2, \dots)$ are discarded from the vector Y_t. Here, the order k_1 of the first element $x_t^{(1)}$ of the bivariate time series x_t is judged to be $k_1 = 2$.

2-4. $Y_t^{(6)} = x_{t+2|t}^{(2)}$ is still judged to be independent of its antecedents because of this relation for $x_{t+2|t}^{(2)}$. Then H is made into $H = (1, 2, 3, 4, 6)'$.

2-5. Since $Y_t^{(5+2m)} = x_{t+2+m|t}^{(1)}$ $(m = 1, 2, \dots)$ are already discarded from Y_t, we check the dependence of $Y_t^{(8)} = x_{t+3|t}^{(2)}$ on its antecedents, $x_{t|t}^{(1)}, x_{t|t}^{(2)}, x_{t+1|t}^{(1)}, x_{t+1|t}^{(2)}, x_{t+2|t}^{(2)}$. Because of the relation

$$\begin{pmatrix} x_{t+3}^{(1)} \\ x_{t+3}^{(2)} \end{pmatrix}_{|t} = \begin{pmatrix} 0.9 & 0 \\ 0 & 1.5 \end{pmatrix}\begin{pmatrix} x_{t+2}^{(1)} \\ x_{t+2}^{(2)} \end{pmatrix}_{|t} + \begin{pmatrix} -0.4 & 0 \\ 0 & -1.2 \end{pmatrix}\begin{pmatrix} x_{t+1}^{(1)} \\ x_{t+1}^{(2)} \end{pmatrix}_{|t} + \begin{pmatrix} 0 & 0 \\ 0 & 0.448 \end{pmatrix}\begin{pmatrix} x_t^{(1)} \\ x_t^{(2)} \end{pmatrix}_{|t},$$

it is judged to be dependent on its antecedents and is given by

$$x_{t+3|t}^{(2)} = 1.5x_{t+2|t}^{(2)} - 1.2x_{t+1|t}^{(2)} + 0.448x_{t|t}^{(2)}.$$

All the $Y_t^{(8+2m)} = x_{t+3+m|t}^{(2)}$ $(m = 1, 2, \dots)$ are discarded from the vector Y_t, and the checking procedure of the dependence terminates. The order of the second elements $x_t^{(2)}$ of the bivariate time series x_t is judged to be $k_2 = 3$. The structural characteristic vector H is given by $H = (1, 2, 3, 4, 6)'$. The order of the process is characterized by $(k_1, k_2) = (2,3)$.

Finally, the state vector z_t is defined by the remaining elements of Y_t as

$$z_t = \left(x_{t|t}^{(1)}, x_{t|t}^{(2)}, x_{t+1|t}^{(1)}, x_{t+1|t}^{(2)}, x_{t+2|t}^{(2)} \right)',$$

where the dimension of z_t is 5. The state space representation is finally given, with z_t by

$$z_{t+1} = \begin{pmatrix} 0 & 0 & 1 & 0 & 0 \\ 0 & 0 & 0 & 1 & 0 \\ -0.4 & 0 & 0.9 & 0 & 0 \\ 0 & 0 & 0 & 0 & 1 \\ 0 & 0.448 & 0 & -1.2 & 1.5 \end{pmatrix} z_t + \begin{pmatrix} 1 & 0 \\ 0 & 1 \\ 1.7 & 0 \\ 0 & 1.5 \\ 0 & 1.05 \end{pmatrix} w_{t+1}. \tag{4.51}$$

$$x_t = \begin{pmatrix} 1 & 0 & 0 & 0 & 0 \\ 0 & 1 & 0 & 0 & 0 \end{pmatrix} z_t.$$

Note that if we follow the procedure of Section 4.3.3, we get from the same ARMA(3,1) model (4.50) the following block-wise state space representation,

$$z_{t+1} = \begin{pmatrix} 0 & 0 & 1 & 0 & 0 & 0 \\ 0 & 0 & 0 & 1 & 0 & 0 \\ 0 & 0 & 0 & 0 & 1 & 0 \\ 0 & 0 & 0 & 0 & 0 & 1 \\ 0 & 0 & -0.4 & 0 & 0.9 & 0 \\ 0 & 0.448 & 0 & -1.2 & 0 & 1.5 \end{pmatrix} z_t + \begin{pmatrix} I_2 \\ h_1 \\ h_2 \end{pmatrix} w_{t+1}.$$

$$x_t = \begin{pmatrix} 1 & 0 & 0 & 0 & 0 & 0 \\ 0 & 1 & 0 & 0 & 0 & 0 \end{pmatrix} z_t.$$

Here, the state is 6-dimensional and is defined by

$$z_t = \left(x_{t|t}^{(1)} \quad x_{t|t}^{(2)} \quad x_{t+1|t}^{(1)} \quad x_{t+1|t}^{(2)} \quad x_{t+2|t}^{(1)} \quad x_{t+2|t}^{(2)} \right)'.$$

I_2, h_1, and h_2 are defined as

$$I_2 = \begin{pmatrix} 1 & 0 \\ 0 & 1 \end{pmatrix}, \quad h_1 = \begin{pmatrix} h_1^{(1,1)} & h_1^{(1,2)} \\ h_1^{(2,1)} & h_1^{(2,2)} \end{pmatrix}, \quad h_2 = \begin{pmatrix} h_2^{(1,1)} & h_2^{(1,2)} \\ h_2^{(2,1)} & h_2^{(2,2)} \end{pmatrix}.$$

Here h_1 and h_2 are the impulse response matrices of the ARMA model of (4.50), and their elements, $h_1^{(1,1)}, \ldots, h_1^{(2,2)}, h_2^{(1,1)}, \ldots, h_2^{(2,2)}$ are calculated from (4.50) using the scheme of (4.34) of Section 4.3.3.

4.4.5.2 Example 2

Suppose a 3-dimensional stationary Gaussian process is found to be of order $p = 4$ and with the structural characteristic vector, $H = \{1, 2, 3, 5\}$, with a nonsingular variance matrix for the innovations. Then we obtain, from $Y_t = (x_{t|t}', x_{t+1|t}' \ldots)'$, the 4-dimensional state vector,

$$z_t = (Y_t^{(1)}, Y_t^{(2)}, Y_t^{(3)}, Y_t^{(5)})' = (x_{t|t}^{(1)}, x_{t|t}^{(2)}, x_{t|t}^{(3)}, x_{t+1|t}^{(2)})'.$$

The structural vector $H = \{1, 2, 3, 5\}$ tells us how $Y_{t+1|t}$ is dependent on $Y_t^{(k_j)}$ for $k_j < 1 + r$. Here $r = 3$ is the dimension of the process. For example, $Y_{t+1|t}^{(1)}$ is dependent on the predecessors $Y_t^{(1)}, Y_t^{(2)}, Y_t^{(3)}$, and we have

$$Y_{t+1|t}^{(1)} = A_{11} Y_t^{(1)} + A_{12} Y_t^{(2)} + A_{13} Y_t^{(3)}. \tag{4.52}$$

Here A_{11}, A_{12}, A_{13} make the row elements of the transition matrix A of the state. Since $Y_{t+1|t}^{(1)}$ is independent of $Y_t^{(5)}$ of $z_t = (Y_t^{(1)}, Y_t^{(2)}, Y_t^{(3)}, Y_t^{(5)})'$, we have $A_{14} = 0$. Interpreting the dependence of $Y_{t+1|t}^{(2)}$, $Y_{t+1|t}^{(3)}$, and $Y_{t+1|t}^{(5)}$, we have $A_{21} = 0$, $A_{22} = 0, A_{23} = 0, A_{24} = 1$, which come from $Y_{t+1|t}^{(2)} = Y_t^{(5)} = x_{t+1|t}^{(2)}$,

$$Y_{t+1|t}^{(3)} = A_{31} Y_t^{(1)} + A_{32} Y_t^{(2)} + A_{33} Y_t^{(3)} + A_{34} Y_t^{(5)}, \tag{4.53}$$

and

$$Y_{t+1|t}^{(5)} = A_{41} Y_t^{(1)} + A_{42} Y_t^{(2)} + A_{43} Y_t^{(3)} + A_{44} Y_t^{(5)}. \tag{4.54}$$

Finally, for the given 3-dimensional process with $H = (1, 2, 3, 5)$, we have the following transition matrix,

$$A = \begin{pmatrix} A_{11} & A_{12} & A_{13} & 0 \\ 0 & 0 & 0 & 1 \\ A_{31} & A_{32} & A_{33} & A_{34} \\ A_{41} & A_{42} & A_{43} & A_{44} \end{pmatrix},$$

for the canonical state space model,

$$z_{t+1} = Az_t + Bw_{t+1}$$
$$x_t = Cz_t \tag{4.55}$$

with

$$z_t = (Y_t^{(1)}, Y_t^{(2)}, Y_t^{(3)}, Y_t^{(5)})' = (x_{t|t}^{(1)}, x_{t|t}^{(2)}, x_{t|t}^{(3)}, x_{t+1|t}^{(2)})'.$$

If we rewrite the three dependent relations of the canonical state space model (4.52) through (4.54), we have

$$x_{t+1|t}^{(1)} = A_{11}x_{t|t}^{(1)} + A_{12}x_{t|t}^{(2)} + A_{13}x_{t|t}^{(3)}$$

$$x_{t+1|t}^{(3)} = A_{31}x_{t|t}^{(1)} + A_{32}x_{t|t}^{(2)} + A_{33}x_{t|t}^{(3)} + A_{34}x_{t+1|t}^{(2)}$$

$$x_{t+2|t}^{(2)} = A_{41}x_{t|t}^{(1)} + A_{42}x_{t|t}^{(2)} + A_{43}x_{t|t}^{(3)} + A_{44}x_{t+1|t}^{(2)}.$$

Here,

$$Y_t^{(k_1)} = x_{t|t}^{(1)}, \quad Y_t^{(k_2)} = x_{t|t}^{(2)}, \quad Y_t^{(k_3)} = x_{t|t}^{(3)}, \quad Y_t^{(k_4)} = x_{t+1|t}^{(2)},$$

$$Y_{t+1|t}^{(k_1)} = x_{t+1|t}^{(1)}, \quad Y_{t+1|t}^{(k_2)} = x_{t+1|t}^{(2)}, \quad Y_{t+1|t}^{(k_3)} = x_{t+1|t}^{(3)}, \quad Y_{t+1|t}^{(k_4)} = x_{t+2|t}^{(2)},$$

Rewriting this again, we have

$$x_{t+2|t}^{(1)} = A_{11}x_{t+1|t}^{(1)} + A_{12}x_{t+1|t}^{(2)} + A_{13}x_{t+1|t}^{(3)}$$

$$x_{t+2|t}^{(3)} = A_{31}x_{t+1|t}^{(1)} + A_{32}x_{t+1|t}^{(2)} + A_{33}x_{t+1|t}^{(3)} + A_{34}x_{t+2|t}^{(2)},$$

$$x_{t+2|t}^{(2)} = A_{41}x_{t|t}^{(1)} + A_{42}x_{t|t}^{(2)} + A_{43}x_{t|t}^{(3)} + A_{44}x_{t+1|t}^{(2)}$$

which is written as

$$
\begin{pmatrix} x_{t+2|t}^{(1)} \\ x_{t+2|t}^{(2)} \\ x_{t+2|t}^{(3)} \end{pmatrix} = \begin{pmatrix} 0 & 0 & 0 \\ 0 & 0 & 0 \\ 0 & A_{34} & 0 \end{pmatrix} \begin{pmatrix} x_{t+2|t}^{(1)} \\ x_{t+2|t}^{(2)} \\ x_{t+2|t}^{(3)} \end{pmatrix} + \begin{pmatrix} A_{11} & A_{12} & A_{13} \\ 0 & A_{44} & 0 \\ A_{31} & A_{32} & A_{33} \end{pmatrix} \begin{pmatrix} x_{t+1|t}^{(1)} \\ x_{t+1|t}^{(2)} \\ x_{t+1|t}^{(3)} \end{pmatrix}
$$

$$
+ \begin{pmatrix} 0 & 0 & 0 \\ A_{41} & A_{42} & A_{43} \\ 0 & 0 & 0 \end{pmatrix} \begin{pmatrix} x_{t|t}^{(1)} \\ x_{t|t}^{(2)} \\ x_{t|t}^{(3)} \end{pmatrix} \tag{4.56}
$$

By putting $C_0 = \begin{pmatrix} 0 & 0 & 0 \\ 0 & 0 & 0 \\ 0 & A_{34} & 0 \end{pmatrix}$, $C_1 = \begin{pmatrix} A_{11} & A_{12} & A_{13} \\ 0 & A_{44} & 0 \\ A_{31} & A_{32} & A_{33} \end{pmatrix}$, $C_2 = \begin{pmatrix} 0 & 0 & 0 \\ A_{41} & A_{42} & A_{43} \\ 0 & 0 & 0 \end{pmatrix}$,

we can rewrite (4.56) in a vector form as

$$x_{t+2|t} = C_0 x_{t+2|t} + C_1 x_{t+1|t} + C_2 x_{t|t}.$$

This model yields to

$$(I - C_0)x_{t+2} - C_1 x_{t+1} - C_2 x_t = D_0 w_{t+2} + D_1 w_{t+1},$$

which is written finally as a 3-dimensional ARMA(2,1) representation,

$$x_{t+2} = \Phi_1 x_{t+1} + \Phi_2 x_t + \Theta_1 w_{t+1} + \Theta_0 w_{t+2},$$

where

$$\Phi_1 = (I - C_0)^{-1} C_1, \quad \Phi_2 = (I - C_0)^{-1} C_2, \quad \Theta_0 = (I - C_0)^{-1} D_0 \text{ and, } \Theta_1 = (I - C_0)^{-1} D_1.$$

We note that from the same canonical state space representation (4.55) we have another ARMA representation, that is, a block-wise 3-dimensional ARMA(4,3) representation, by using the Cayley–Hamilton theorem, where the AR order comes from the dimension of the state.

4.4.5.3 Example 3

Suppose a 2-dimensional stationary Gaussian process x_t is found to be on order $p = 4$, with the structural characteristic vector, $H = (1, 2, 3, 5)$, and with nonsingular variance matrix for the innovations. Then we obtain, from $Y_t = (x'_{t|t}, x'_{t+1|t}, \ldots)'$, the 4-dimensional state vector,

$$z_t = (Y_t^{(1)}, Y_t^{(2)}, Y_t^{(3)}, Y_t^{(5)})' = (x_{t|t}^{(1)}, x_{t|t}^{(2)}, x_{t+1|t}^{(1)}, x_{t+2|t}^{(1)})'.$$

The structural vector $H = (1, 2, 3, 5)$ means how $Y_{t+1|t}^{(1)}$, the projected process one-step ahead, is dependent on $Y_t^{(k_j)}$ for $k_j < 1 + r$. Here $r = 2$ is the dimension of the process. For example, since $Y_{t+1|t}^{(1)} = x_{t+1|t}^{(1)} = Y_t^{(3)}$ we have $A_{11} = 0$, $A_{12} = 0$, $A_{13} = 1$, and $A_{14} = 0$, which make the first row elements of the transition matrix A. Since $Y_{t+1|t}^{(2)} = x_{t+1|t}^{(2)}$ is dependent on the predecessors $Y_t^{(1)}, Y_t^{(2)}, Y_t^{(3)}$, we have, with some nonzero elements, A_{21}, A_{22}, and A_{23}, a linear relation,

$$Y_{t+1|t}^{(2)} = A_{21} Y_t^{(1)} + A_{22} Y_t^{(2)} + A_{23} Y_t^{(3)}. \tag{4.57}$$

Since $Y_{t+1|t}^{(2)} = x_{t+1|t}^{(2)}$ is not dependent on $Y_t^{(5)} = x_{t+2|t}^{(1)}$, we have $A_{24} = 0$. Since $Y_{t+1|t}^{(3)} = x_{t+2|t}^{(1)} = Y_t^{(5)}$, we have $A_{31} = 0$, $A_{32} = 0$, $A_{33} = 0$ and $A_{34} = 1$, which make the fourth row of the transition matrix A. For the fourth element of $Y_{t+1|t}$, we have $Y_{t+1|t}^{(4)} = x_{t+3|t}^{(1)}$, which is dependent on the predecessors, and we have

$$Y_{t+1|t}^{(4)} = A_{41}Y_t^{(1)} + A_{42}Y_t^{(2)} + A_{43}Y_t^{(3)} + A_{44}Y_t^{(5)}. \tag{4.58}$$

Finally, for the given 2-dimensional process x_t with $H = \{1, 2, 3, 5\}$, we have the following transition matrix,

$$A = \begin{pmatrix} 0 & 0 & 1 & 0 \\ A_{21} & A_{22} & A_{23} & 0 \\ 0 & 0 & 0 & 1 \\ A_{41} & A_{42} & A_{43} & A_{44} \end{pmatrix}.$$

for the canonical state space model,

$$\begin{aligned} z_{t+1} &= Az_t + Bw_{t+1} \\ x_t &= Cz_t \end{aligned} \tag{4.59}$$

with $z_t = (Y_t^{(1)}, Y_t^{(2)}, Y_t^{(3)}, Y_t^{(5)})' = (x_{t|t}^{(1)}, x_{t|t}^{(2)}, x_{t+1|t}^{(1)}, x_{t+2|t}^{(1)})'$.

If we rewrite the two dependent relations of the canonical state space model (4.57) and (4.58), we have

$$Y_{t+1|t}^{(k_2)} = A_{21}Y_t^{(k_1)} + A_{22}Y_t^{(k_2)} + A_{23}Y_t^{(k_3)},$$

$$Y_{t+1|t}^{(k_4)} = A_{41}Y_t^{(k_1)} + A_{42}Y_t^{(k_2)} + A_{43}Y_t^{(k_3)} + A_{44}Y_t^{(k_4)}.$$

Since it holds that

$$Y_t^{(k_1)} = x_{t|t}^{(1)}, \quad Y_t^{(k_2)} = x_{t|t}^{(2)}, \quad Y_t^{(k_3)} = x_{t+1|t}^{(1)}, \quad Y_t^{(k_4)} = x_{t+2|t}^{(1)},$$

$$Y_{t+1}^{(k_2)} = x_{t+1|t+1}^{(2)}, \quad Y_{t+1}^{(k_4)} = x_{t+3|t+1}^{(1)},$$

we obtain

$$x_{t+1|t+1}^{(2)} = x_{t+1|t}^{(2)} = A_{21}x_{t|t}^{(1)} + A_{22}x_{t|t}^{(2)} + A_{23}x_{t+1|t}^{(1)},$$

$$x_{t+3|t+1}^{(1)} = x_{t+3|t}^{(1)} = A_{41}x_{t|t}^{(1)} + A_{42}x_{t|t}^{(2)} + A_{43}x_{t+1|t}^{(1)} + A_{44}x_{t+2|t}^{(1)},$$

and

$$x_{t+3|t}^{(2)} = A_{21}x_{t+2|t}^{(1)} + A_{22}x_{t+2|t}^{(2)} + A_{23}x_{t+3|t}^{(1)},$$

$$x_{t+3|t}^{(1)} = A_{41}x_{t|t}^{(1)} + A_{42}x_{t|t}^{(2)} + A_{43}x_{t+1|t}^{(1)} + A_{44}x_{t+2|t}^{(1)}$$

which is written as

$$\begin{pmatrix} 1 & 0 \\ 0 & 1 \end{pmatrix} \begin{pmatrix} x_{t+3|t}^{(1)} \\ x_{t+3|t}^{(2)} \end{pmatrix} = \begin{pmatrix} 0 & 0 \\ A_{23} & 0 \end{pmatrix} \begin{pmatrix} x_{t+3|t}^{(1)} \\ x_{t+3|t}^{(2)} \end{pmatrix} + \begin{pmatrix} A_{44} & 0 \\ A_{21} & A_{22} \end{pmatrix} \begin{pmatrix} x_{t+2|t}^{(1)} \\ x_{t+2|t}^{(2)} \end{pmatrix}$$

$$+ \begin{pmatrix} A_{43} & 0 \\ 0 & 0 \end{pmatrix} \begin{pmatrix} x_{t+1|t}^{(1)} \\ x_{t+1|t}^{(2)} \end{pmatrix} + \begin{pmatrix} A_{41} & A_{42} \\ 0 & 0 \end{pmatrix} \begin{pmatrix} x_{t|t}^{(1)} \\ x_{t|t}^{(2)} \end{pmatrix}. \qquad (4.60)$$

By putting

$$C_0 = \begin{pmatrix} 0 & 0 \\ A_{23} & 0 \end{pmatrix}, \quad C_1 = \begin{pmatrix} A_{44} & 0 \\ A_{21} & A_{22} \end{pmatrix}, \quad C_2 = \begin{pmatrix} A_{43} & 0 \\ 0 & 0 \end{pmatrix}, \quad C_3 = \begin{pmatrix} A_{41} & A_{42} \\ 0 & 0 \end{pmatrix},$$

we can write (4.60) in a vector form as

$$x_{t+3|t} - (C_0 x_{t+3|t} + C_1 x_{t+2|t} + C_2 x_{t+1|t} + C_3 x_{t|t}) = 0.$$

By using the innovation representation of the projected variables, we obtain, with some matrices, D_0, D_1, and D_2,

$$(I - C_0)x_{t+3} - C_1 x_{t+2} - C_2 x_{t+1} - C_3 x_t = D_0 w_{t+3} + D_1 w_{t+2} + D_2 w_{t+1}, \qquad (4.61)$$

which has a 2-dimensional ARMA(3,2) representation,

$$x_{t+3} = \Phi_1 x_{t+2} + \Phi_2 x_{t+1} + \Phi_3 x_t + \Theta_0 w_{t+3} + \Theta_1 w_{t+2} + \Theta_2 w_{t+1},$$

where $\Phi_1 = (I - C_0)^{-1}C_1$, $\Phi_2 = (I - C_0)^{-1}C_2$, $\Phi_3 = (I - C_0)^{-1}C_3$, $\Theta_0 = (I - C_0)^{-1}D_0$, $\Theta_1 = (I - C_0)^{-1}D_1$, and $\Theta_2 = (I - C_0)^{-1}D_2$.

On the other hand, from the same canonical state space representation (4.59), we also have a block-wise 2-dimensional ARMA(4,3) representation, by using the Cayley–Hamilton theorem, where the AR order comes from the dimension of the state.

4.4.6 Discussion

We already saw that causality between variables should be discussed based on the correlations and co-variance structure (which contains the same information as the power spectrum) of the multivariate time series rather than the significance of the nonzero coefficients of the fitted multivariate AR model such as in the method of Granger causality (Granger 1969).

We know, from examples of scalar ARMA models in Sections 3.1 and 3.2, and also from the discussion, that the AR models are less efficient (than ARMA models) in parameterizing the power spectrum of the process, leading to smaller (−2)log-likelihood and AIC. The fact that the AR model set is a special restricted subset of a more general ARMA model set, together with the fact that multivariate AR models do not have the serious nonuniqueness problem of the multivariate ARMA representation, delivers to us a strong warning message toward the easygoing use of the method of the causality analysis of the Granger type. This point is explained further in Chapter 14.

On the other hand, the following two facts are useful and promising for the causality study:

1. A canonical multivariate ARMA model is identifiable from a given time series without any special artificial conditions.
2. The multivariate ARMA model uniquely determines the impulse response function, the power spectrum, and the auto covariance function of the process, even though the time domain model representation is nonunique.

It means that the Akaike causality method based on the estimated power spectrum through the multivariate AR modeling can be realized in a more refined way, that is, the estimated power spectrum through the multivariate ARMA modeling, on the cost of more heavy computational load. The topic is further discussed in Chapter 14.

Nonuniqueness of the parameterization of the multivariate ARMA model and the state space model implicitly delivers an essential message and lessons in the study of general state space modeling in recent Monte Carlo simulation-based maximum likelihood estimation method. Even for linear, Gaussian state space model estimation, we have great difficulties in estimating parameters, where we need to elucidate minimum necessary and sufficient coefficients out of all the linear coefficients. Without proper constraints in the searching space of parameters or without proper parameterizations of the state equation and the observation equation, those general state space model identification methods certainly end up in some numerical problems, such as slow convergence problem and the singularity of the Hessian matrix in the numerical optimization, which makes the results unreliable and doubtful in actual applications. We discuss this problem in Chapter 13.

4.5 Multivariate and Spatial Dynamic Models with Inputs

4.5.1 Multivariate ARX Model

In feedback system modeling, some exogenous variable is included only for the objective of improvement of the prediction and control of some important variables, and the dynamics of the exogenous variable are not of interest. Multivariate ARX models are more suitable for the modeling of such systems than the multivariate AR models. For example, if the two exogenous variables, u_t and v_t, are known to be useful for the prediction of the three main variables, x_t, y_t, and z_t, we have the 3-dimensional ARX model with two exogenous variables, u_t and v_t, as follows:

$$
\begin{pmatrix} x_t \\ y_t \\ z_t \end{pmatrix} = \begin{pmatrix} \phi_{1,1}^{(1)} & \phi_{1,2}^{(1)} & \phi_{1,3}^{(1)} \\ \phi_{2,1}^{(1)} & \phi_{2,2}^{(1)} & \phi_{2,3}^{(1)} \\ \phi_{3,1}^{(1)} & \phi_{3,2}^{(1)} & \phi_{3,3}^{(1)} \end{pmatrix} \begin{pmatrix} x_{t-1} \\ y_{t-1} \\ z_{t-1} \end{pmatrix} + \begin{pmatrix} \phi_{1,1}^{(2)} & \phi_{1,2}^{(2)} & \phi_{1,3}^{(2)} \\ \phi_{2,1}^{(2)} & \phi_{2,2}^{(2)} & \phi_{2,3}^{(2)} \\ \phi_{3,1}^{(2)} & \phi_{3,2}^{(2)} & \phi_{3,3}^{(2)} \end{pmatrix} \begin{pmatrix} x_{t-2} \\ y_{t-2} \\ z_{t-2} \end{pmatrix}
$$

$$
+ \begin{pmatrix} \phi_{1,1}^{(3)} & \phi_{1,2}^{(3)} & \phi_{1,3}^{(3)} \\ \phi_{2,1}^{(3)} & \phi_{2,2}^{(3)} & \phi_{2,3}^{(3)} \\ \phi_{3,1}^{(3)} & \phi_{3,2}^{(3)} & \phi_{3,3}^{(3)} \end{pmatrix} \begin{pmatrix} x_{t-3} \\ y_{t-3} \\ z_{t-3} \end{pmatrix} + \begin{pmatrix} \psi_{1,1}^{(1)} & \psi_{1,2}^{(1)} \\ \psi_{2,1}^{(1)} & \psi_{2,2}^{(1)} \\ \psi_{3,1}^{(1)} & \psi_{3,2}^{(1)} \end{pmatrix} \begin{pmatrix} u_{t-1} \\ v_{t-1} \end{pmatrix}
$$

$$
+ \begin{pmatrix} \psi_{1,1}^{(2)} & \psi_{1,2}^{(2)} \\ \psi_{2,1}^{(2)} & \psi_{2,2}^{(2)} \\ \psi_{3,1}^{(2)} & \psi_{3,2}^{(2)} \end{pmatrix} \begin{pmatrix} u_{t-2} \\ v_{t-2} \end{pmatrix} + \begin{pmatrix} \xi_t \\ \eta_t \\ \varsigma_t \end{pmatrix}, \tag{4.62}
$$

where $w_t = (\xi_t, \eta_t, \varsigma_t)'$ is a Gaussian white noise whose variance matrix could be nondiagonal. We introduce an efficient way of calculating least squares estimate of this type of linear multivariate ARX model in Section 10.4.3.

4.5.2 Spatial Time Series Models for fMRI Data

In this section, we will see that the idea of the innovation approach, developed for finite dimensional time series analysis, is also useful for the characterization of the spatial and temporal correlation structure of spatial time series data in neuroscience such as fMRI data, where the dimension of the time series is "very high."

4.5.2.1 fMRI Time Series

Functional magnetic resonance imaging (fMRI) is the measurement of the Blood-Oxygen-Level-Dependence (BOLD) signal, which represents the hemodynamic response related to neural activity in the brain or spinal cord of humans or other animals. fMRI BOLD signal data are obtained as a time series at each voxel inside the brain; thus, the fMRI data are typically a high-dimensional ($64 \times 64 \times 36$) time series observed from one experiment on a subject usually under some properly designed stimulus. The association between the spatially remote neurons in distinct brain regions gives us the key to understand higher human brain functions, and fMRI data are becoming a more and more common tool for brain connectivity studies. According to Friston (1994, 2003), the definition of functional connectivity is "correlation between spatially remote neurophysiological events." fMRI BOLD signal time series data are a kind of spatial temporal data, and what is needed here is an efficient statistical method for estimation of the "spatial temporal correlation structure" from the fMRI time series data, which is usually about 147,000 channel time series measured at $64 \times 64 \times 36$ grid points in the brain.

In time series analysis, a natural way of characterizing the temporal correlation structure of a stationary time series is to use a linear dynamic model such as an AR model. By identifying a suitable multivariate AR model from the observed data, we can characterize the multivariate auto covariance functions of the process behind the data. The innovation approach suggests finding a dynamic model yielding the smallest prediction errors from the time series data. In the case of the fMRI BOLD signal data of 147,000 channels, it suggests finding 147,000 dimensional prediction errors from a 147,000 dimensional time series. The optimal prediction of $x_t^{(i,j,k)}$ at time point $t - 1$ will be

$$x_{t|t-1}^{(i,j,k)} = E[x_t^{(i,j,k)} \mid x_{t-1}^{(*)}],$$

under the local Gaussian assumption. Here, $x_{t-1}^{(*)}$ denotes all the BOLD signal information available at all the grid points at the time point $t - 1$.

4.5.2.2 NN–ARX Model

A natural way of characterizing the temporal correlation structure of a stationary time series is to use linear models, such as an AR model. In theory, if the whole brain is in a stationary state, the multivariate 147,000 dimensional AR model of the BOLD signal determines the 147,000 dimensional auto covariance functions, where we can extract autocorrelations at each voxel and cross correlations between all the pair voxels. However, this is easy to say but difficult to do in practice where we need to handle 147,000 or more dimensional transition matrices of the model. Further, when the process is stimulated from outside the brain and is driven by an exogenous variable, the definition of the correlation structure of the process is not as straightforward as the stationary time series case.

Since the neighboring voxels contain the most useful information for the one-step ahead prediction of $x_t^{(i,j,k)}$, the conditional expectation of $x_t^{(i,j,k)}$ given all the observations up to time point $t - 1$ will be approximately equal to the conditional expectation of $x_t^{(i,j,k)}$ given all the neighboring observations at time point $t - 1$, that is,

$$x_{t|t-1}^{(v)} = E[x_t^{(v)} \mid x_{t-1}^{(*)}, x_{t-2}^{(*)}, \dots, x_1^{(*)}] \approx E[x_t^{(v)} \mid x_{t-1}^v, v \in N(v)],$$

where $x_t^{(*)}$ denotes all the BOLD signal information at time point t, and $N(v)$ denotes the set of all the neighboring voxels of the voxel v. Then, a natural approximate linear predictor of $x_t^{(v)}$ at the voxel $v = (i, j, k)$ will be a linear combination of the neighboring voxels $x_{t-1}^{(i-1,j,k)}, x_{t-1}^{(i+1,j,k)}, \dots, x_{t-1}^{(i,j,k+1)}$ and itself, $x_{t-1}^{(i,j,k)}$, that is,

$$E[x_t^{(i,j,k)} \mid x_{t-1}^{(v)}, v \in N(v)] \approx a_1^{(i,j,k)} x_{t-1}^{(i,j,k)} + b_1^{(i,j,k)} x_{t-1}^{(i-1,j,k)} + b_2^{(i,j,k)} x_{t-1}^{(i+1,j,k)}$$

$$+ \cdots + b_6^{(i,j,k)} x_{t-1}^{(i,j,k+1)}.$$

Then, the prediction error may be written as

$$\varepsilon_t^{(i,j,k)} = x_t^{(i,j,k)} - \left\{ a_1^{(i,j,k)} x_{t-1}^{(i,j,k)} + b_1^{(i,j,k)} x_{t-1}^{(i-1,j,k)} + b_2^{(i,j,k)} x_{t-1}^{(i+1,j,k)} + \cdots + b_6^{(i,j,k)} x_{t-1}^{(i,j,k+1)} \right\}.$$

This implies that the following spatial ARX-type model could be a reasonable starting point as an approximation of the dynamic model for fMRI time series:

$$x_t^{(i,j,k)} = a_1^{(i,j,k)} x_{t-1}^{(i,j,k)} + b_1^{(i,j,k)} x_{t-1}^{(i-1,j,k)} + b_2^{(i,j,k)} x_{t-1}^{(i+1,j,k)} + \cdots + b_6^{(i,j,k)} x_{t-1}^{(i,j,k+1)}$$

$$+ \theta_1^{(i,j,k)} s_{t-1} + \varepsilon_t^{(i,j,k)}.$$

The original idea of the model was introduced by the present author at the Workshop on Mathematical Methods in Brain Mapping at CRM, University of Montreal, in 2000, and has been further developed by Riera et al. (2004b), where the model is called NN–ARX model (Nearest Neighbor AutoRegressive model with eXogenous variable). A more general NN–ARX model with higher lag orders may be written as

$$x_t^{(v)} = a_1^{(v)} x_{t-1}^{(v)} + \cdots + a_p^{(v)} x_{t-p}^{(v)} + \frac{1}{6} \left(\sum_{v' \in N(v)} b_{v'}^{(v)} x_{t-1}^{(v')} \right) + \theta_1^{(v)} s_{t-1} + \cdots + \theta_r^{(v)} s_{t-r} + \varepsilon_t^{(v)}.$$

$$(4.63)$$

Here, $(v) = (i, j, k)$, $N(v) = \{(i - 1, j, k), (i + 1, j, k), (i, j - 1, k), (i, j + 1, k), (i, j, k - 1), (i, j, k + 1)\}$. The coefficients, $a_1^{(v)}, \dots, a_p^{(v)}, b_{v'}^{(v)}$ $(v' \in N(v))$, $\theta_1^{(v)}, \dots, \theta_r^{(v)}$, are calculated by solving the linear equation for each voxel v. Whether the system noise $\varepsilon_t^{(v)}$ is zero needs to be checked by a statistical method. Incidentally, we note that the discrete time model (4.63), with $p = 1$ and $r = 1$, can be obtained by discretizing the following partial differential equation model of a spatial stochastic process with an external input $s(t)$.

$$\frac{\partial x(\xi, \eta, \varsigma, t)}{\partial t} = a(\xi, \eta, \varsigma)x + b(\xi, \eta, \varsigma)\left(\frac{\partial^2 x}{\partial \xi^2} + \frac{\partial^2 x}{\partial \eta^2} + \frac{\partial^2 x}{\partial \varsigma^2}\right) + \theta(\xi, \eta, \varsigma)s(t) + \delta W(\xi, \eta, \varsigma, t).$$

In other words, we are interpreting the fMRI data as a realization of a spatial blurring process driven by stimulus $s(t)$ and a spatial Gaussian white noise process $\delta W(\xi, \eta, \varsigma, t)$ (Brown et al. 2000). Here the innovation approach with the NN–ARX model performs some kind of a deblurring procedure in order to improve the resolution so that we may discover important temporal spatial information in the data. The theoretical foundation of the innovation approach to spatial stochastic processes was given by Ito (1984).

It must be noted that we have so far ignored possible instantaneous correlations between the noise of neighboring voxels, and the noise covariance of the 147,000 dimensional AR model is assumed to be diagonal. This may not be an appropriate assumption for the NN–ARX model to be a general spatial time series model. One simple way of removing the instantaneous correlations between the neighboring voxels is to apply an instantaneous Laplacian operator L, which operates as

$$Lx_t^{(i,j,k)} = x_t^{(i,j,k)} - \frac{1}{6}\left(x_t^{(i+1,j,k)} + x_t^{(i-1,j,k)} + x_t^{(i,j+1,k)} + x_t^{(i,j-1,k)} + x_t^{(i,j,k-1)} + x_t^{(i,j,k+1)}\right),$$

$$(4.64)$$

for the 3-dimensional case. If we apply the Laplacian operator L to the original data before fitting the aforementioned NN–ARX model, then we have

$$Lx_t^{(v)} = y_t^{(v)}$$

$$y_t^{(v)} = \mu_t^{(v)} + \sum_{k=1}^{r_1} \alpha_k^{(v)} y_{t-k}^{(v)} + \sum_{k=1}^{r_2} \beta_k^{(v)} \xi_{t-k}^{(v)} + \sum_{k=1}^{r_3} \gamma_k^{(v)} s_{t-k} + n_t^{(v)}.$$

Although the variance matrix of the noise n_t in the transformed space is diagonal, so that $E[n_t n_t'] = \sigma_n^2 I$, the variance matrix of the noise $\varepsilon_t = (L^{-1}n_t)$ in the original space is nondiagonal and is given by $\Sigma_\varepsilon = \sigma_n^2 (L'L)^{-1}$. This is a simple but useful way of characterizing a spatially homogeneous instantaneous dependency between the noises of neighboring voxels and was used in EEG dynamic inverse solutions by Galka et al. (2004) and Yamashita et al. (2004).

We must remember, however, that "spatial differencing" is not the only way of eliminating the dependence. If the data were manipulated too much in the stage of preprocessing, we may sometimes need a "spatial smoothing" of the data instead of the "spatial differencing" to restore the spatial independence. One simple example of the instantaneous spatial smoothing operator S could be the following:

$$Sx_t^{(i,j,k)} = x_t^{(i,j,k)} + \frac{1}{6}\left(x_t^{(i+1,j,k)} + x_t^{(i-1,j,k)} + x_t^{(i,j+1,k)} + x_t^{(i,j-1,k)} + x_t^{(i,j,k-1)} + x_t^{(i,j,k+1)}\right).$$

(4.65)

The Laplacian operator (4.64) and the smoothing operator (4.65) can be generalized into parametric variable transformations, with $a > 0$ and $b > 0$, respectively, as follows:

$$Lx_t^{(i,j,k)} = x_t^{(i,j,k)} - \frac{a}{6}\left(x_t^{(i+1,j,k)} + x_t^{(i-1,j,k)} + x_t^{(i,j+1,k)} + x_t^{(i,j-1,k)} + x_t^{(i,j,k-1)} + x_t^{(i,j,k+1)}\right)$$

(4.66)

$$Sx_t^{(i,j,k)} = x_t^{(i,j,k)} + \frac{b}{6}\left(x_t^{(i+1,j,k)} + x_t^{(i-1,j,k)} + x_t^{(i,j+1,k)} + x_t^{(i,j-1,k)} + x_t^{(i,j,k-1)} + x_t^{(i,j,k+1)}\right).$$

(4.67)

When a wider range of spatial smoothing and differencing are needed, we could use the product of the operators such as S^2, S^3, ..., and L^2, L^3, ... Then the whole picture of the model for the spatial data could be such that

$$S^{k_1} L^{k_2} x_t^{(v)} = y_t^{(v)}$$

$$y_t^{(v)} = \mu_t^{(v)} + \sum_{k=1}^{r_1} \alpha_k^{(v)} y_{t-k}^{(v)} + \sum_{k=1}^{r_2} \beta_k^{(v)} \xi_{t-k}^{(v)} + \sum_{k=1}^{r_3} \gamma_k^{(v)} s_{t-k} + n_t^{(v)}.$$

The covariance matrix of the system noise ε_t of the original space is specified by

$$\varepsilon_t = \left(L^{-k_2} S^{-k_1} n_t\right),$$

and its covariance matrix is written as

$$\Sigma_\varepsilon = \sigma_n^2 \left\{(S^{k_1})'(L^{k_2})' S^{k_1} L^{k_2}\right\}_t^{-1},$$

which is determined by the two parameters a and b in (4.66) and (4.67).

We note that since the likelihood of the model with differencing and smoothing depends on the parameters a and b, the Jacobian of the instantaneous variable transformation needs to be taken into account in the computation of the likelihood in terms of the original data. The superiority of the NN–ARX model with or without the spatial smoothing and differencing could be checked by comparing the AIC of the two models, with the transformation and without the transformation, fitted to the same fMRI data. Estimation method together with associated computational topic for the spatial time series model is discussed in Section 10.3.2 (see also Galka et al. 2006). For those who are interested in the relations of the NN–ARX model to the SPM approach by Friston (1995), we refer them to Galka et al. (2010) and Ozaki (2008, in press).

5

Continuous-Time Dynamic Models

The concept of a dynamical system has its origins in Newtonian mechanics. There, as in other natural sciences and engineering disciplines, the evolution rule of dynamical systems is given implicitly by a relation that gives the state of the system only a short time into the future. At this point, the discrete-time dynamical system and continuous-time dynamical system are essentially the same, although they are treated as though different in most scientific fields. Both models have advantages and disadvantages compared with the other model. Discrete-time dynamic models, for example, have some limitations such as the "aliasing effect" compared with the continuous-time model, as seen in Section 2.2.3. In this section, we consider, instead of discrete-time dynamic models, continuous-time dynamic models as prediction models of dynamic phenomena.

The evolution rule of continuous-time dynamical systems is given implicitly by a differential equation (deterministic or stochastic) that gives the state of the system only a short time into the future. To determine the state for all future times requires iterating the relation many times—each advancing time a small step. The iteration procedure is referred to as solving the system or integrating the system. Once the system can be solved, given an initial point, it is possible to determine all its future points, a collection known as a trajectory or orbit.

Before the advent of fast computing machines, solving a dynamical system required sophisticated mathematical techniques and could only be accomplished for a small class of dynamical systems. Numerical methods for solving the dynamical system with modern fast computers have significantly simplified the task of determining the orbits of a dynamical system.

Leaving these computational topics to Chapters 9, 10, and 13, in this chapter, we see some examples of continuous-time models for prediction.

5.1 Linear Oscillation Models

5.1.1 Pendulum and Damping Oscillation Systems

A typical example of a continuous-time dynamic system is the oscillation system of a pendulum. The dynamics of the angle of an oscillating pendulum is usually modeled by a second-order differential equation:

$$\ddot{x}(t) = -\omega^2 x(t). \tag{5.1}$$

Here, the oscillating dynamics are generated by the restoring force, which is assumed to be proportional, with a constant ω^2, to the angle $x(t)$ of the pendulum from the vertical line. The equation is rewritten as

$$\begin{pmatrix} \ddot{x} \\ \dot{x} \end{pmatrix} = \begin{pmatrix} 0 & -\omega^2 \\ 1 & 0 \end{pmatrix} \begin{pmatrix} \dot{x} \\ x \end{pmatrix},$$

which leads to a 2-dimensional dynamical system,

$$\frac{dX}{dt} = AX \tag{5.2}$$

with

$$X = \begin{pmatrix} \dot{x} \\ x \end{pmatrix} \quad \text{and} \quad A = \begin{pmatrix} 0 & -\omega^2 \\ 1 & 0 \end{pmatrix}.$$

The left-hand side of the equation is a time derivative of $X(t)$ and is often written as $\dot{X}$. If $X(0)$ is specified at $t = 0$, the solution of $X(t)$ is given analytically as $X(t) = X(0)e^{At}$ for $0 < t$, where e^{At} is an exponential of the matrix At. Here the future behavior of the angle $x(t)$ and its time derivative $\dot{x}(t)$ are completely specified by the eigenvalues of A and the initial point $x(0)$ and the initial velocity $\dot{x}(0)$. When $\omega \neq 0$, the eigenvalues are complex conjugate roots, which leads to harmonic oscillatory behavior of the angle $x(t)$ with a period $T = 2\pi/\omega$.

The oscillations of the pendulum defined by (5.1) are supposed to keep the same oscillation forever without damping down. Actual pendulum, unlike the harmonic oscillations, damps down as time goes if it is detached from any driving force. One of the causes for the damping may be the resistance from the air. If we assume that the approximate resisting force to the movement of the pendulum is proportional to $\dot{x}(t)$, the time derivative of the angle of the pendulum, we can have a more realistic model for the dynamic of the damping pendulum as follows:

$$\ddot{x}(t) = -c\dot{x}(t) - \omega^2 x(t)$$

or

$$\ddot{x}(t) + c\dot{x}(t) + \omega^2 x(t) = 0. \tag{5.3}$$

This can be rewritten in the same form as (5.2) with

$$X = \begin{pmatrix} \dot{x} \\ x \end{pmatrix} \quad \text{and} \quad A = \begin{pmatrix} -c & -\omega^2 \\ 1 & 0 \end{pmatrix}.$$

If we rewrite (5.3) using the differential operator D, that is, $Dx = dx/dt$ and $D^p = DD^{p-1}$, we notice that $\dfrac{-c}{2} \pm i\sqrt{\omega^2 - \dfrac{c^2}{4}}$, of the characteristic equation,

$$(D^2 + cD + \omega^2) = 0,$$

of Equation 5.3, are equivalent to the eigenvalues of the transition matrix

$$A = \begin{pmatrix} -c & -\omega^2 \\ 1 & 0 \end{pmatrix}$$

of (5.2). Mathematically, if $c = 0$ the characteristic roots are purely complex, and the solution exhibits an undamped harmonic oscillatory behavior whose periodicity is specified by ω. If $c < 0$, the system defines an explosive oscillating system. If $c > 0$, the system defines a damping oscillating system.

5.1.2 Damping Oscillations with Random Force

Note that real-world dynamical systems are exposed to all sorts of noise affecting the future state of the system. Such effects of fluctuations have been of interest for scientists over a century since the seminal work of Einstein (1905, 1956). A stochastic dynamical system has been used to characterize such a dynamical system subjected to the effects of noise. A simple example is the following damping oscillations with random force:

$$\ddot{x}(t) + c\dot{x}(t) + \omega^2 x(t) = w(t) \tag{5.4}$$

which has a 2-dimensional stochastic dynamical system representation,

$$\begin{pmatrix} \ddot{x} \\ \dot{x} \end{pmatrix} = \begin{pmatrix} -c & -\omega^2 \\ 1 & 0 \end{pmatrix} \begin{pmatrix} \dot{x} \\ x \end{pmatrix} + \begin{pmatrix} w(t) \\ 0 \end{pmatrix}. \tag{5.5}$$

When $c > 0$, the system has basically damping oscillation dynamics, but the system is exposed to a constant external driving force $w(t)$ and may create a stationary random oscillatory dynamics. Then it may make sense to consider the auto covariance function and the power spectrum of the random process of $x(t)$.

Here we notice the obvious similarity of the model (5.4) to the AR(2) model,

$$x_t = \phi_1 x_{t-1} + \phi_2 x_{t-2} + w_t,$$

where the characteristic behavior of the time series is specified by the complex conjugate roots, $\Lambda = \lambda, \bar{\lambda}$, of the characteristic equation,

$$\Lambda^2 - \phi_1\Lambda - \phi_2 = 0.$$

The similarities between the two different types of dynamic models, continuous-time models and discrete-time models, are discussed in Chapter 9.

The model (5.2) covers not only oscillation systems but also many important examples of dynamical systems such as tank models in Section 5.3. In general, an analytical solution of the continuous-time dynamical system is possible only for linear cases. As for the nonlinear dynamical systems, we need to rely on the numerical solutions, which are obtained by taking advantage of the modern fast computer with recently developed numerical mathematics techniques. The examples of nonlinear dynamical systems will be shown in Section 5.4.

5.2 Power Spectrum

5.2.1 Power Spectrum and Auto Covariance Function

The Fourier and inverse Fourier relations, between the power spectrum and the auto covariance function, which we saw for the discrete-time process in Sections 3.2 and 3.3, hold also for the continuous-time processes. One easy way to understand the relations for the continuous-time case is to start from the relations for the discrete-time case and let $\Delta t \to 0$.

Suppose we have a k-variate AR(1) process z_t of

$$x_t = \Phi x_{t-1} + w,$$

we have a power spectrum (density) representation,

$$p_{xx}^{(d)}(f) = \left\{ I - \exp(-i2\pi f)\Phi \right\}^{-1} \Sigma_w \left\{ \overline{I - \exp(-i2\pi f)\Phi'} \right\}^{-1}.$$

By discretizing a continuous-time linear p-dimensional dynamical system model,

$$\dot{x}(t) = Ax(t) + w(t), \tag{5.6}$$

with a small time interval, Δt, we have a consistent discretized model,

$$x_t = e^{A\Delta t} x_{t-\Delta t} + w_t, \tag{5.7}$$

where $w(t)$ is a continuous-time Gaussian white noise with a variance matrix Σ_w, and w_t is a discrete-time Gaussian white noise with the variance matrix $\Delta t \Sigma_w$. (see Section 9.1.3 for details) If the real part of the eigenvalues of A is negative, that is,

$$\text{Re}(\mu_j) < 0 \quad (j = 1, \ldots, k),$$

the eigenvalues, $\lambda_j = e^{\mu_j}$ $(j = 1, \ldots, k)$, of the transition matrix $e^A \Delta t$ stay inside the unit circle, and the AR process defined by (5.7) is stationary and has a power spectral representation,

$$p_{xx}^{(d)}(f) = \left\{ I - \exp(-i2\pi f)e^{A\Delta t} \right\}^{-1} \Sigma_w \left\{ \overline{I - \exp(-i2\pi f)e^{A\Delta t'}} \right\}^{-1}. \tag{5.8}$$

Now remember (Section 2.2.4) that we have the following relation,

$$p_{xx}^{(c)}(f_{\Delta t}) = \Delta t p_{xx}^{(d)}\left(\frac{f_d}{\Delta t} \right),$$

between the power spectrum density, $p_{xx}^{(d)}(f_d)$, of a discretized process of (5.7) and the power spectrum density, $p_{xx}^{(c)}(f)$, of the continuous-time process of (5.6). From the relation, by letting $\Delta t \to 0$, we have

$$f_{\Delta t} = \frac{f_d}{\Delta t} \xrightarrow[\Delta t \to 0]{} f$$

and

$$p_{xx}^{(c)}(f_{\Delta t}) = \Delta t p_{xx}^{(d)}\left(\frac{f_d}{\Delta t} \right)$$

$$= \Delta t \left\{ I - \exp(-i2\pi f_{\Delta t})e^{A\Delta t} \right\}^{-1} \Delta t \Sigma_w \left\{ \overline{I - \exp(-i2\pi f_{\Delta t})e^{A\Delta t'}} \right\}^{-1}$$

$$\xrightarrow[\Delta t \to 0]{} (i2\pi f - A)^{-1} \Sigma_w \overline{(i2\pi f - A)'}^{-1} = p_{xx}^{(c)}(f).$$

Then from (5.8), we have a power spectral density representation,

$$p_{xx}^{(c)}(f) = (i2\pi f - A)^{-1} \Sigma_w \overline{(i2\pi f - A)'}^{-1} \tag{5.9}$$

of a continuous-time dynamical system (5.6).

The inverse Fourier relation for the continuous-time process is deduced, from the relation for the discrete-time process, in a similar heuristic way. We start from the relation

$$R_{xx}^{(d)}(m) = \int_{-1/2}^{1/2} \exp\left(i2\pi f^{(d)}m\right) p_{xx}^{(d)}(f) df^{(d)} \tag{5.10}$$

for the discrete-time process. If the discrete-time process comes originally from a continuous-time process defined by a k-variate stochastic dynamical system model,

$$\dot{x} = Ax + w(t),$$

with sampling interval Δt, and if the real part of the eigenvalues of A are all negative, it makes sense to consider the auto covariance function and the power spectrum density of the continuous-time process. Here we can rewrite (5.10) as

$$R_{xx}^{(d)}(m\Delta t) = \int_{-1/2}^{1/2} \exp\left(i2\pi \frac{f_d}{\Delta t} m\Delta t\right) \Delta t p_{xx}^{(d)}\left(\frac{f_d}{\Delta t}\right) \frac{df_d}{\Delta t}$$

$$= \int_{-1/2\Delta t}^{1/2\Delta t} \exp\left(i2\pi f_{\Delta t} m\Delta t\right) p_{xx}^{(c)}(f_{\Delta t}) df_{\Delta t}. \tag{5.11}$$

Note that for the discrete-time case with a sampling interval Δt, we have

$$-1/2\Delta t < f_{\Delta t} < 1/2\Delta t,$$

where $f_{\Delta t} = f_d / \Delta t$ and $\Delta t p_{xx}^{(d)}(f_{\Delta t}) = p_{xx}^{(c)}(f_{\Delta t})$. Suppose f and τ are defined as a limit of $f_{\Delta t}$ and $m\Delta t$ for $\Delta t \to 0$, that is,

$$f_{\Delta t} = f_d / \Delta t \xrightarrow[\Delta t \to 0]{} f$$

and

$$m\Delta t' \xrightarrow[\Delta t \to 0]{} \tau.$$

Then we have

$$R_{xx}^{(c)}(\tau) = \int_{-\infty}^{\infty} \exp(i2\pi f \tau) p_{xx}^{(c)}(f) df. \tag{5.12}$$

Also from the inverse relation of the discrete-time model,

$$\Delta t p_{xx}^{(d)}\left(\frac{f_d}{\Delta t}\right) = \Delta t \sum_{m=-\infty}^{\infty} \exp\left(-i2\pi \frac{f_d}{\Delta t} m\Delta t\right) R_{xx}^{(d)}(m\Delta t), \tag{5.13}$$

we have a relation with the continuous-time version as

$$p_{xx}^{(c)}(f) = \int_{-\infty}^{\infty} \exp(-i2\pi f \tau) R_{xx}^{(c)}(\tau) d\tau. \tag{5.14}$$

5.2.2 Characteristic Roots and the Power Spectrum

Next let us see how the power spectrum of the continuous-time process $x(t)$ is characterized by the parameters of the differential equation models.

5.2.2.1 Example 1: Damping Oscillation Model

$$\ddot{x} + c\dot{x} + \omega^2 x = w(t) \tag{5.15}$$

Suppose $w(t)$ is a continuous-time Gaussian white noise. The model can be rewritten as a continuous-time state space model,

$$\dot{z} = Az + \eta(t)$$
$$x_t = Cz(t), \tag{5.16}$$

where $C = (0, 1)$ and

$$z = \begin{pmatrix} \ddot{x} \\ \dot{x} \end{pmatrix}, \quad A = \begin{pmatrix} -c & -\omega^2 \\ 1 & 0 \end{pmatrix}, \quad \eta(t) = \begin{pmatrix} w(t) \\ 0 \end{pmatrix}.$$

From (5.9), we have the power spectrum density representation for $z(t)$ of (5.16) as

$$p_{zz}^{(c)}(f) = \begin{pmatrix} p_{xx}^{(c)}(f) & p_{xx}^{(c)}(f) \\ p_{xx}^{(c)}(f) & p_{xx}^{(c)}(f) \end{pmatrix}$$

$$= \frac{1}{2\pi} (2\pi i f I - A)^{-1} \begin{pmatrix} \sigma_w^2 & 0 \\ 0 & 0 \end{pmatrix} \overline{(2\pi i f I - A')}^{-1}$$

$$= \frac{\sigma_w^2}{2\pi} \frac{1}{|\det(2\pi i f I - A)|^2} \begin{pmatrix} f^2 & -if \\ if & 1 \end{pmatrix}.$$

where σ_w^2 is the variance of the white noise $w(t)$. Then the power spectrum density of the process $x(t)$ is written as

$$p_{xx}(f) = \frac{1}{2\pi} \frac{\sigma_w^2}{|\det(2\pi i f I - A)|^2} = \frac{1}{2\pi} \frac{\sigma_w^2}{(f^2 - \omega^2)^2 + c^2 f^2}. \qquad (5.17)$$

Note that the spectral properties of the damping oscillations are explicitly specified in (5.17), where we can see that

1. The power spectrum density function $p(f)$ has its peak at the frequency $f_0 = \omega^2 - \dfrac{c^2}{2}$.

2. When $c \to 0$ the peak of the power spectrum goes to infinity, that is, the process approaches the undamped oscillations of the frequency $f = \omega$.

5.2.2.2 Example 2: pth-Order Linear Differential Equation System

$$x^{(p)}(t) + b_1 x^{(p-1)}(t) + \cdots + b_{p-1} x^{(1)}(t) + b_p x(t) = w(t). \qquad (5.18)$$

For the process defined with the pth order linear differential equation model (5.18), we have the following power spectrum representation,

$$p(f) = \frac{1}{2\pi} \frac{\sigma_w^2}{|(if)^p + b_1(if)^{p-1} + \cdots + b_p|^2},$$

where $w(t)$ is a continuous-time Gaussian white noise with the variance σ_w^2. Note that (5.18) is rewritten, using the differential operator D, as

$$(D^p + b_1 D^{p-1} + \cdots + b_{p-1} D + b_p) x(t) = w(t). \qquad (5.19)$$

If the characteristic equation

$$D^p + b_1 D^{p-1} + \cdots + b_{p-1} D + b_p = 0$$

has r complex roots and $(p - 2r)$ real roots, the differential operator is expressed as

$$(D^p + b_1 D^{p-1} + \cdots + b_{p-1} D + b_p)$$

$$= (D^2 + c_1 D + \omega_1^2) \cdots (D^2 + c_r D + \omega_r^2)(D + c_{2r+1}) \cdots (D + c_p), \qquad (5.20)$$

and the power spectrum density is expressed as

$$p(f) = \frac{1}{2\pi} \frac{\sigma_0^2}{\left|(if)^2 + c_1(if) + \omega_1^2\right|^2 \cdots \left|(if)^2 + c_r(if) + \omega_r^2\right|^2 \left|(if) + c_{2r+1}^2\right|^2 \cdots \left|(if) + c_p^2\right|^2}$$

$$= \frac{1}{2\pi} \frac{\sigma_0^2}{\left\{(f^2 - \omega_1^2)^2 + c_1^2 f^2\right\} \cdots \left\{(f^2 - \omega_r^2)^2 + c_r^2 f^2\right\}(f^2 + c_{2r+1}^2) \cdots (f^2 + c_p^2)}.$$

$$(5.21)$$

Here we can see that

1. The power spectrum function $p(f)$ has peaks at the frequencies $f_i^2 = \omega_i^2 - \dfrac{c_i^2}{2}$ $(i = 1, 2, \ldots, r)$.
2. When $c_i \to 0$, the ith peak of the power spectrum goes to infinity, that is, the process approaches the undamped oscillations of the frequency $f_i = \omega_i$.

5.3 Continuous-Time Structural Modeling

5.3.1 Parallel Structural Model in Continuous Time

Remember that in Section 3.3, we derived, from the set of eigenvalues, λ_1, $\bar{\lambda}_1, \ldots, \lambda_r, \bar{\lambda}_r, \lambda_{2r+1}, \ldots, \lambda_p$, of the AR($p$) model,

$$x_t = \phi_1 x_{t-1} + \cdots + \phi_1 x_{t-p} + w_t,$$

a parallel-type structural model in discrete time,

$$\begin{cases} z_t = A^{(para)} z_{t-1} + \eta_t^{(oara)} \\ x_t = C^{(para)} z_t \end{cases},$$

$$(5.22a)$$

where

$$z_t = \left(\xi_t^{(1)} \quad \xi_{t-1}^{(1)} \quad \cdots \quad \xi_t^{(r)} \quad \xi_{t-1}^{(r)} \quad \xi_t^{(2r+1)} \quad \cdots \quad \xi_t^{(p)}\right)',$$

$$\xi_t^{(j)} = 2\mathrm{Re}(\lambda_j)\xi_{t-1}^{(j)} - |\lambda_j|^2 \xi_{t-2}^{(j)} + w_t^{(j)}, \quad (j = 1, \ldots, r),$$

$$\xi_t^{(k)} = \lambda_k \xi_{t-1}^{(k)} + w_t^{(k)}, \quad (k = 2r+1, \ldots, p),$$

$$\eta_t^{(para)} = (w_t^{(1)} \quad 0 \quad \cdots \quad w_t^{(r)} \quad 0 \quad w_t^{(2r+1)} \quad \cdots \quad w_t^{(p)})',$$

$$C^{(para)} = (1, 0, \ldots, 1, 0, 1, \ldots, 1),$$

and

$$A^{(para)} = \begin{pmatrix} 2\operatorname{Re}(\lambda_1) & -|\lambda_1|^2 & \cdots & 0 & 0 & 0 & \cdots & 0 \\ 1 & 0 & \cdots & 0 & 0 & 0 & \cdots & 0 \\ \cdots & \cdots & \cdots & \cdots & \cdots & \cdots & \cdots & \cdots \\ 0 & 0 & \cdots & 2\operatorname{Re}(\lambda_r) & -|\lambda_r|^2 & 0 & \cdots & 0 \\ 0 & 0 & \cdots & 1 & 0 & 0 & \cdots & 0 \\ 0 & 0 & \cdots & 0 & 0 & \lambda_{2r+1} & \cdots & 0 \\ \cdots & \cdots & \cdots & \cdots & \cdots & \cdots & \cdots & \cdots \\ 0 & 0 & \cdots & 0 & 0 & 0 & \cdots & \lambda_p \end{pmatrix}.$$

In the same way, we can introduce a continuous-time parallel-type structural model,

$$\begin{cases} \dot{z}(t) = A^{(para)}z(t) + \eta^{(para)}(t) \\ x_t = C^{(para)}z_t \end{cases} \tag{5.22b}$$

using the same set of eigenvalues, $\mu_1, \bar{\mu}_1, \ldots, \mu_r, \bar{\mu}_r, \mu_{2r+1}, \ldots, \mu_p$, of the characteristic equation,

$$\mu^p + b_1\mu^{p-1} + \cdots + b_{p-1}\mu + b_p = 0$$

of the pth order stochastic linear differential equation system,

$$x^{(p)}(t) + b_1 x^{(p-1)}(t) + \cdots + b_{p-1}x^{(1)}(t) + b_p x(t) = w(t).$$

Here, the state vector is specified by

$$z(t) = \left(\dot{\xi}_{(1)}(t), \xi_{(1)}(t), \ldots, \dot{\xi}_{(r)}(t), \xi_{(r)}(t), \xi_{(2r+1)}(t), \ldots, \xi_{(p)}(t) \right)'.$$

The dynamics of the r pairs of complex conjugate eigenvalues yield the second-order stochastic differential equation models, that is,

$$\ddot{\xi}_{(j)}(t) = 2\operatorname{Re}(\mu_j)\dot{\xi}_{(j)}(t) - |\mu_j|^2 \xi_{(j)}(t) + w_{(j)}(t), \quad (j = 1, \ldots, r).$$

The dynamics of the $(p - 2r)$ real eigenvalues yield the first-order stochastic differential equation models, that is,

$$\dot{\xi}_{(k)}(t) = \mu_k \xi_{(k)}(t) + w_{(k)}(t), \quad (k = 2r+1, \ldots, p).$$

Then the transition matrix $A^{(para)}$ of (5.22) is given as

$$A^{(para)} = \begin{pmatrix} 2\,\mathrm{Re}(\mu_1) & -|\mu_1|^2 & \cdots & 0 & 0 & 0 & \cdots & 0 \\ 1 & 0 & \cdots & 0 & 0 & 0 & \cdots & 0 \\ \cdots & \cdots & \cdots & \cdots & \cdots & \cdots & \cdots & \cdots \\ 0 & 0 & \cdots & 2\,\mathrm{Re}(\mu_r) & -|\mu_r|^2 & 0 & \cdots & 0 \\ 0 & 0 & \cdots & 1 & 0 & 0 & \cdots & 0 \\ 0 & 0 & \cdots & 0 & 0 & \mu_{2r+1} & \cdots & 0 \\ \cdots & \cdots & \cdots & \cdots & \cdots & \cdots & \cdots & \cdots \\ 0 & 0 & \cdots & 0 & 0 & 0 & \cdots & \mu_p \end{pmatrix}.$$

The system noise $\eta^{(para)}(t)$ of (5.22) is

$$\eta^{(para)}(t) = \left(w_{(1)}(t), 0, w_{(2)}(t), 0, \ldots, w_{(r)}(t), 0, w_{(2r+1)}(t), \ldots, w_{(p)}(t)\right)',$$

and the observation matrix $C^{(para)}$ of (5.22) is

$$C^{(para)} = (0, 1, \ldots, 0, 1, 1, \ldots, 1).$$

$w_j(t)$, $(j = 1, 2, \ldots, r, 2r + 1, \ldots, p)$ are continuous-time Gaussian white noise with the variance $\sigma_{(j)}^2$ for $j = 1, \ldots, r$ and for $j = 2r + 1, \ldots, p$. The power spectrum density of the model is given by

$$p(f) = \frac{1}{2\pi}\left\{ \frac{\sigma_1^2}{\left|(if)^2 + c_1(if) + \omega_1^2\right|^2} + \cdots + \frac{\sigma_r^2}{\left|(if)^2 + c_r(if) + \omega_r^2\right|^2} + \frac{\sigma_{2r+1}^2}{(f^2 + c_{2r+1}^2)} + \cdots + \frac{\sigma_p^2}{(f^2 + c_p^2)} \right\}'$$

$$= \frac{1}{2\pi}\left\{ \frac{\sigma_1^2}{\left\{(f^2 - \omega_1^2)^2 + c_1^2 f^2\right\}} + \cdots + \frac{\sigma_r^2}{\left\{(f^2 - \omega_r^2)^2 + c_r^2 f^2\right\}} + \frac{\sigma_{2r+1}^2}{(f^2 + c_{2r+1}^2)} + \cdots + \frac{\sigma_p^2}{(f^2 + c_p^2)} \right\}.$$

$$(5.23)$$

Comparing (5.23) with (5.20), we notice that the power spectrum of the parallel model (5.22) and the power spectrum of the sequential model (5.18) have peaks at the same frequencies. The difference of the spectra is in their shape, which is determined by their different dynamic structure.

5.3.2 Advantage of Continuous-Time Structural Modeling

Continuous-time structural modeling is often useful when we need to deal with two types of dynamic time series with different frequency bands, such as very fast neural time series (EEG and MEG) and comparatively slow hemodynamic time series(fMRI or NIRS). To explain the dynamic interaction of the two series, we usually need to make the sampling interval small enough to capture the dynamics of the faster series (i.e., EEG or MEG). Then, the sampling interval is too small for the other time series (i.e., fMRI or NIRS), which leads to a model with extraordinary large lag order model in discrete-time parameterization. On the other hand, if a continuous-time model is used for the slower dynamic series, the number of parameters in the model stays reasonably small.

In the following sections, we see some typical examples of structural modeling: the first example is the dynamical system model for the hemodynamic response, which turns out to be equivalent to the tank modeling for riverflow prediction in hydrology by Sugawara (1962). Another example is the spectral decomposition model, which is derived from the trend–seasonal decomposition type modeling by Ozaki and Thomson (1992).

5.3.3 Tank Model for fMRI Hemodynamic Response

The problem of predicting riverflow from rainfall is one of the examples where the two time series of different speed of dynamics are involved. Here, daily riverflow is predicted using the past rainfall time series data. The two time series, rainfall time series and riverflow time series, are regarded as input and output of the riverflow system, where the response function depends on the topographical structure of the basin surrounding the river. Sugawara (1962) realized that discrete-time modeling is inefficient in the sense that the effect of the rainfall takes many time lags to come out and affects the riverflow increase, and he introduced the continuous-time tank modeling method, in which several tanks are combined in sequential structure, parallel structure, and mixed structure, considering the topographical structure of the river basin.

The view and approach of Sugawara (1962) is valid in the characterization of the hemodynamic response function of the fMRI against various kinds of stimulus and the characterization of evoked response function of EEG or MEG against the stimuli. For example, for the fMRI hemodynamic response function, h_k, a combination of Gamma functions (see Lange and Zeger [1997] and Worsley et al. [2002]) such as $t^r e^{-\lambda t}$ is assumed. The system implied by the response function $t^k e^{-\lambda t}$ is given by (5.24), with a state vector

$$Z(t) = \left(d^k z(t)/dt^k, \ldots, dz(t)/dt, z(t) \right)',$$

where the $(k + 1) \times (k + 1)$ matrix A, $k + 1$-dimensional vectors B and C are given by

$$\dot{Z}(t) = AZ(t) + Bs(t)$$
$$x_t = CZ(t) + \varepsilon_t,$$

(5.24)

where the $(k + 1) \times (k + 1)$ matrix A, $k + 1$-dimensional vectors B and C are given by

$$A = \begin{pmatrix} -\lambda & 0 & \cdots & 0 & 0 \\ 1 & -\lambda & \cdots & 0 & 0 \\ \cdots & \cdots & \cdots & \cdots & \cdots \\ 0 & 0 & \cdots & -\lambda & 0 \\ 0 & 0 & \cdots & 1 & -\lambda \end{pmatrix}, \quad B = \begin{pmatrix} b \\ 0 \\ \cdots \\ 0 \\ 0 \end{pmatrix}, \quad \text{and} \quad C = \begin{pmatrix} 0 & 0 & \cdots & 0 & 1 \end{pmatrix}.$$

A typical example of the structural tank model is shown in Figure 5.1.

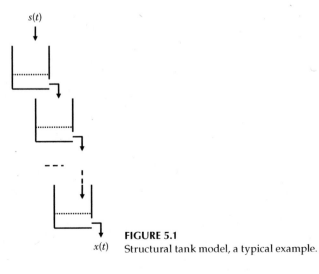

FIGURE 5.1
Structural tank model, a typical example.

5.3.4 More General Tank Models

If we need a more general response function, we need to use a more general state space model. For example, in order to have the following impulse response function,

$$h(t) = h_1 e^{-\lambda_1 t} + h_2 t e^{-\lambda_2 t} + \cdots + h_k t^{k-1} e^{\lambda_k t},$$

we need to have the following state space model:

$$
d\begin{pmatrix} z_1(t) \\ dz_2(t)/dt \\ z_2(t) \\ \cdots \\ d^{k-1}z_k(t)/dt^{k-1} \\ d^{k-2}z_k(t)/dt^{k-2} \\ \cdots \\ dz_k(t)/dt \\ z_k(t) \end{pmatrix} /dt =
\begin{pmatrix}
-\lambda_1 & 0 & 0 & \cdots & 0 & 0 & \cdots & 0 & 0 \\
1 & -\lambda_2 & 0 & \cdots & 0 & 0 & \cdots & 0 & 0 \\
0 & 1 & -\lambda_2 & \cdots & 0 & 0 & \cdots & 0 & 0 \\
\cdots & \cdots & \cdots & \cdots & \cdots & \cdots & \cdots & \cdots & \cdots \\
0 & 0 & 0 & \cdots & -\lambda_k & 0 & \cdots & 0 & 0 \\
0 & 0 & 0 & \cdots & 1 & -\lambda_k & \cdots & 0 & 0 \\
\cdots & \cdots & \cdots & \cdots & \cdots & \cdots & \cdots & \cdots & \cdots \\
0 & 0 & 0 & \cdots & 0 & 0 & \cdots & -\lambda_k & 0 \\
0 & 0 & 0 & \cdots & 0 & 0 & \cdots & 1 & -\lambda_k
\end{pmatrix}
$$

$$
\times \begin{pmatrix} z_1(t) \\ dz_2(t)/dt \\ z_2(t) \\ \cdots \\ d^{k-1}z_k(t)/dt^{k-1} \\ d^{k-2}z_k(t)/dt^{k-2} \\ \cdots \\ dz_k(t)/dt \\ z_k(t) \end{pmatrix} + \begin{pmatrix} b_1 \\ b_2 \\ 0 \\ \cdots \\ b_k \\ 0 \\ \cdots \\ 0 \\ 0 \end{pmatrix} s(t) \tag{5.25}
$$

$$x_t = z_1(t) + z_2(t) + \cdots + z_k(t) + \varepsilon_t.$$

The model (5.25) is written with a state vector,

$$Z(t) = \{z_1(t), dz_2(t)/dt, z_2(t), \ldots, d^{k-1}z_k(t)/dt^{k-1},$$

$$d^{k-2}z_k(t)/dt^{k-2}, \ldots, dz_k(t)/dt, z_k(t)\}',$$

and an observation matrix, $C = (1, 0, 1, 0, ..., 0, ..., 0, 1)$, as

$$\dot{Z} = AZ + Bs(t)$$

$$x_t = CZ_t + \varepsilon_t,$$

(5.26)

where A and B are

$$
A = \begin{pmatrix}
-\lambda_1 & 0 & 0 & \cdots & 0 & 0 & \cdots & 0 & 0 \\
1 & -\lambda_2 & 0 & \cdots & 0 & 0 & \cdots & 0 & 0 \\
0 & 1 & -\lambda_2 & \cdots & 0 & 0 & \cdots & 0 & 0 \\
\cdots & \cdots & \cdots & \cdots & \cdots & \cdots & \cdots & \cdots & \cdots \\
0 & 0 & 0 & \cdots & -\lambda_k & 0 & \cdots & 0 & 0 \\
0 & 0 & 0 & \cdots & 1 & -\lambda_k & \cdots & 0 & 0 \\
\cdots & \cdots & \cdots & \cdots & \cdots & \cdots & \cdots & \cdots & \cdots \\
0 & 0 & 0 & \cdots & 0 & 0 & \cdots & -\lambda_k & 0 \\
0 & 0 & 0 & \cdots & 0 & 0 & \cdots & 1 & -\lambda_k
\end{pmatrix}, \quad
B = \begin{pmatrix}
b_1 \\
b_2 \\
0 \\
\cdots \\
b_k \\
0 \\
\cdots \\
0 \\
0
\end{pmatrix},
$$

respectively. Here the dimension of the state Z_t is $K = k(k + 1)/2$.

What this deterministic model (5.25) or (5.26) implies is a very strong assumption that the future value of the signal $x_{t+\tau}$ ($\tau > 0$) is exactly predicted by the initial state $Z_0^{(v)}$ and the input $s(t)$ ($0 < t < T$). Here the noise ε_t is an observation error and never affects the future value of Z_t or x_t.

5.3.5 Structural Modeling for Spectral Decomposition

We have seen, in Section 3.3.4, that trend–seasonal decomposition is another typical example of discrete-time structural modeling applied in official and econometric time series data analysis. The same modeling is possible for continuous-time cases, which may be more suitable for neuroscience time series data analysis. Remember that in discrete-time annual trend–seasonal adjustment structural modeling, the basic annual seasonal structure is determined by the characteristic roots $\frac{\sqrt{3}}{2} \pm i \frac{1}{2}, \frac{1}{2} \pm i \frac{\sqrt{3}}{2}, \pm i, \frac{-1}{2} \pm i \frac{\sqrt{3}}{2}, \frac{-\sqrt{3}}{2} \pm i \frac{1}{2},$ and -1 which generate six cyclic submodels,

$$S_t^{(1)} = \sqrt{3} S_{t-1}^{(1)} - S_{t-2}^{(1)} + \varepsilon_t^{(1)},$$

$$S_t^{(2)} = -\sqrt{3} S_{t-1}^{(2)} - S_{t-2}^{(2)} + \varepsilon_t^{(2)},$$

$$S_t^{(3)} = S_{t-1}^{(3)} - S_{t-2}^{(3)} + \varepsilon_t^{(3)},$$

$$S_t^{(4)} = -S_{t-1}^{(4)} - S_{t-2}^{(4)} + \varepsilon_t^{(4)},$$

$$S_t^{(5)} = -S_{t-2}^{(5)} + \varepsilon_t^{(5)},$$

$$S_t^{(6)} = -S_{t-1}^{(6)} + \varepsilon_t^{(6)}.$$

The parallel-type seasonal model is made with these submodels as

$$\begin{cases} z_t = A^{(para2)} z_{t-1} + \eta_t^{(para2)} \\ x_t = C^{(para2)} z_t \end{cases}, \tag{5.27}$$

where

$$A^{(para2)} = \begin{pmatrix} A^{(1)} & 0 & \cdots & O & O \\ O & A^{(2)} & \cdots & O & O \\ \cdots & \cdots & \cdots & \cdots & \cdots \\ O & O & \cdots & A^{(5)} & O \\ O & O & \cdots & O & A^{(6)} \end{pmatrix},$$

$$C^{(para2)} = (1,0,1,0,1,0,1,0,1,0,1),$$

$$\eta_t^{(para2)} = \left(\varepsilon_t^{(1)}, 0, \varepsilon_t^{(2)}, 0, \ldots, \varepsilon_t^{(5)}, 0, \varepsilon_t^{(6)} \right)',$$

with

$$A^{(1)} = \begin{pmatrix} \sqrt{3} & -1 \\ 1 & 0 \end{pmatrix}, \quad A^{(2)} = \begin{pmatrix} 1 & -1 \\ 1 & 0 \end{pmatrix}, \quad A^{(3)} = \begin{pmatrix} 0 & -1 \\ 1 & 0 \end{pmatrix},$$

$$A^{(4)} = \begin{pmatrix} -1 & -1 \\ 1 & 0 \end{pmatrix}, \quad A^{(5)} = \begin{pmatrix} -\sqrt{3} & -1 \\ 1 & 0 \end{pmatrix}, \quad A^{(6)} = (-1).$$

If we assume the superposition principle, similar structural seasonal modeling is possible in EEG spectral decomposition, where the EEG signal $S(t)$ is considered as a sum of rhythms of different oscillations, such as delta, theta, alpha, beta, and gamma as

$$S(t) = S^{(1)}(t) + S^{(2)}(t) + S^{(3)}(t) + S^{(4)}(t) + S^{(5)}(t). \tag{5.28}$$

Cyclic components, $S^{(i)}(t)$ ($i = 1, ..., 5$), can be specified, respectively, by the dynamical system models

$$(D^2 + \omega_1^2)S^{(1)}(t) = w^{(1)}(t)\text{: delta oscillation}$$

$$(D^2 + \omega_2^2)S^{(2)}(t) = w^{(2)}(t)\text{: theta oscillation}$$

$$(D^2 + \omega_3^2)S^{(3)}(t) = w^{(3)}(t)\text{: alpha oscillation}$$

$$(D^2 + \omega_4^2)S^{(4)}(t) = w^{(4)}(t)\text{: beta oscillation}$$

$$(D^2 + \omega_5^2)S^{(5)}(t) = w^{(5)}(t)\text{: gamma oscillation}$$

where $w^{(i)}(t)$ ($i = 1, 2, ..., 5$) are continuous-time Gaussian white noise with variance $\sigma_{(i)}^2$ ($i = 1, 2, ..., 5$), respectively. A state space representation for $S(t)$ is

$$dz/dt = A^{(para)}z + w(t)$$

$$S(t) = Cz(t),$$

where $S(t) = (dS_1(t)/dt, S_1(t), dS_2(t)/dt, S_2(t), ..., dS_5(t)/dt, S_5(t))'$,

$$C = (0,1,0,1,0,1,0,1,0,1),$$

$$A^{(para)} = \begin{pmatrix} A_1 & O & \cdots & O \\ O & A_2 & \cdots & O \\ \cdots & \cdots & \cdots & \cdots \\ O & O & \cdots & A_5 \end{pmatrix}$$

$$A_i = \begin{pmatrix} 0 & -\omega_i^2 \\ 1 & 0 \end{pmatrix}, \quad (i = 1, 2, ..., 5)$$

$$w(t) = \left(w^{(1)}(t), 0, w^{(2)}(t), 0, ..., w^{(5)}(t), 0\right)'.$$

The variances of $w^{(i)}(t)$ ($i = 1, 2, ..., 5$) are $\sigma_{(i)}^2$ ($i = 1, 2, ..., 5$), respectively, and the variance matrix of $w(t)$ is

$$\Sigma_w = \begin{pmatrix} \sigma_{(1)}^2 & 0 & \cdots & 0 & 0 \\ 0 & 0 & \cdots & 0 & 0 \\ \cdots & \cdots & \cdots & \cdots & \cdots \\ 0 & 0 & \cdots & \sigma_{(5)}^2 & 0 \\ 0 & 0 & \cdots & 0 & 0 \end{pmatrix}.$$

The estimation problem of the model parameters such as ω_i ($i = 1, 2, ..., 5$) and $\sigma_{(i)}^2$ ($i = 1, 2, ..., 5$) is discussed in Chapters 10 and 13.

5.4 Nonlinear Differential Equation Models

Structural models, as seen so far, are composed of linear submodels. Dynamic models in real world are often much more complex than the examples of structural models we have seen. They are often nonlinear dynamic models, and the nonlinearity of the model depends on each phenomenon to be analyzed.

We discuss in the present section some typical nonlinear dynamic models seen in neuroscience data analysis. They are nonlinear oscillation models, Zetterberg model for EEG, Hodgkin–Huxley model for neural data, Balloon model for BOLD (Blood-Oxygen-Level Dependence) signal data, and DCM (Dynamic Causal Model) for neurovascular data.

5.4.1 Nonlinear Random Oscillations

We have seen that the first-order linear differential equation model

$$\dot{x}(t) = ax(t) + w(t) \tag{5.29}$$

and the second-order linear differential equation model

$$\ddot{x}(t) + c\dot{x}(t) + ax(t) = w(t) \tag{5.30}$$

constitute the basic model components for general multidimensional linear dynamical systems of more complex structural continuous-time models. This observation naturally comes from the fact that eigenvalues of any transition matrix of multidimensional linear dynamical systems are either complex conjugate eigenvalues or real eigenvalues. These two basic models, especially the second-order model (5.30), play an important role, not only in the introduction of linear structural models but also in nonlinear dynamical system models in continuous time.

Let us briefly review some typical phenomena of nonlinear vibrations, which are supposed to be fundamentally related to most random processes in science and engineering. The general nonlinear vibrations follow the nonlinear differential equation,

$$\ddot{x} + \phi(\dot{x}) + f(x) = 0, \tag{5.31}$$

where $\phi(\dot{x})$ is called the damping force and $f(x)$ is called the restoring force, both of which are nonlinear functions of $\dot{x}$ and x, respectively. The pendulum, which approximately follows the equation

$$ml^2\ddot{x} + mglx = 0, \tag{5.32}$$

is a special case of (5.31), where $f(x)$ is approximated by the linear function of x, that is, $mglx$, where l is the length of the pendulum, m is the attached mass, x is the angular displacement, and g is the acceleration due to gravity. Usually, the quantity $-f(x)$, which is the force exerted by the spring when it is subjected to a displacement x, is not necessarily a linear function. The nonlinear function $f(x)$ is sometimes approximated by a third-order polynomial,

$$f(x) = \alpha x + \beta x^3 \quad \alpha > 0.$$

In the nonlinear oscillations,

$$\ddot{x} + c\dot{x} + \alpha x + \beta x^3 = 0, \tag{5.33}$$

where $\beta \neq 0$, and the period T of the oscillations is not independent of the amplitude as the linear case. For a hard spring ($\beta > 0$), the period decreases as the amplitude x increases, and thus, the frequency increases when the amplitude increases, while just the opposite effect of increasing period and decreasing frequency occurs with the increase in the amplitude when the spring is soft ($\beta > 0$). This property, which is called amplitude-dependent frequency shift, is known to be one of the most typical nonlinear phenomena in engineering science.

It is often the case that this nonlinear oscillation system (5.33) is driven by an external force such as a sinusoidal wave, $F \cos \omega t$, or white noise $w(t)$. The nonlinear differential equation model,

$$\ddot{x} + c\dot{x} + \alpha x + \beta x^3 = F \cos \omega t,$$

was first studied by Duffing (1918) and is called Duffing's equation. The model has been used as a prototype of nonlinear oscillations such as ship rolling motion in the sea and EEG rhythms in the human brain (Zeeman 1976). Here, the restoring force of the ship rolling is not proportional to the rolling angle $x(t)$ if $\beta \neq 0$. Among the most interesting results of this system are jump phenomena (see Stoker [1950]). As is schematically shown in Figure 5.2a, if F is made constant and ω is decreased, the amplitude $|A|$ slowly increases through point 2 until point 3 is reached. Further decrease in ω causes the jump from point 3 to point 4 with accompanying increase in the amplitude $|A|$, after which $|A|$ decreases with ω. Upon performing the experiment in the other direction, that is, starting at point 5 and increasing ω, the amplitude follows the 5–4–6 portion of the curve, then jumps to point 2, and afterward slowly decreases. The circumstances are quite similar with a soft spring force ($\beta < 0$), but the jumps in amplitude take place in the reverse direction (see Figure 5.2b). These phenomena are a kind of hysteresis, typical for nonlinear oscillation systems.

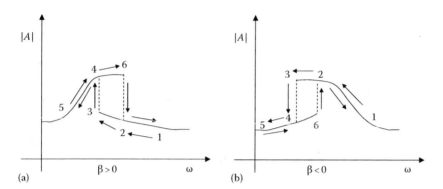

FIGURE 5.2
Hard spring jump phenomena for (a) $\beta > 0$ and (b) $\beta < 0$.

The oscillation whose damping force is nonlinear is called a self-sustained or self-excited oscillation and follows the differential equation,

$$\ddot{x} + f(\dot{x}) + \alpha x = 0. \tag{5.34}$$

This type of nonlinear oscillation that occurs in electric circuits (Stoker 1950), where the following condition is satisfied:

$$\begin{aligned}\dot{x}f(\dot{x}) &< 0 \quad \text{for small } |\dot{x}| \\ \dot{x}f(\dot{x}) &> 0 \quad \text{for large } |\dot{x}|.\end{aligned} \tag{5.35}$$

This condition means that the damping is negative for small $|\dot{x}|$, when the system absorbs energy, and one could expect the amplitude of x to increase. However, for large $|\dot{x}|$, the system dissipates energy and hence one could expect the amplitude of x to be limited finally from the one mentioned earlier. The differential equation (5.34) was studied by Rayleigh (1894) in connection with acoustical problems and by van der Pol (1927) in connection with electrical circuit problems. If we differentiate (5.34) with respect to t and put $y = \dot{x}$, then we have, with some nonlinear function $\varphi(y)$, a differential equation representation,

$$\ddot{y} + \varphi(y)\dot{y} + \alpha y = 0, \tag{5.36}$$

which many electrical engineers prefer to use. In this case, the condition (5.35) is equivalent to

$$\begin{aligned}\varphi(\dot{y}) &< 0 \quad \text{for small } |y| \\ \varphi(\dot{y}) &> 0 \quad \text{for large } |y|.\end{aligned}$$

A typical example is the following van der Pol equation model,

$$\ddot{y} + c(1 - y^2)\dot{y} + \alpha y = 0.$$

It can be shown that any solution of (5.34) or (5.36) tends to a periodic solution as $t \to \infty$. This is the salient fact about self-excited oscillations. Occurrences of this kind were first studied by Poincare (1881) and were given the name "limit cycles."

The oscillation of most physical systems, however, is excited by a more complicated force than the sinusoidal wave. For example, the dynamics of an excitatory submodel, which constitute a part of a neural mass model, may be specified by a stochastic nonlinear oscillation model, where its random driving force, coming from the other subcomponent of the mass-model, may have a continuous spectrum. The estimation and identification of these stochastic continuous-time dynamic models from time series data are discussed in Chapters 10 and 13.

5.4.2 Zetterberg Model

The nonlinear vibration models may provide us with basic models for the analysis of nonlinear dynamic phenomena, but the real world phenomena is much more complex to be explained by these second-order nonlinear differential equation models.

An interesting and more realistic example of nonlinear continuous-time dynamic models in neuroscience is the Zetterberg model, which is a kind of feedback system model, composed of five subsystems(see Figure 5.3).

Zetterberg model is an example of a neural mass model, introduced by Lopes da Silva et al. (1967) and Zetterberg et al. (1978), which is for describing the macroscopic dynamics of neural populations such as the alpha, beta rhythms or epileptic spike-like waves in EEG. It comprises two excitatory neural sets (KIe1 and KIe2) interconnected with one inhibitory neural set (KIi) (see Figure 5.3, which is derived from Figure 1b in Zetterberg et al. (1978) by setting $C_s = 0$ and $P_1(t) = 0$).

A general class of neural mass models was described earlier by Freeman (1975) in terms of a hierarchical classification of interacting sets of neurons. In his terminology, a KIe (KIi) set is a conglomerate of interconnected excitatory (inhibitory) neurons with a common input and output. The model is formulated in terms of the following variables:

1. $E_1(t)$, $E_2(t)$, and $I(t)$ are the proportions of cells firing per unit time in the neural populations Kle1, Kle2, and Kli at time t.
2. $V_{1e}(t)$, $V_{2e}(t)$, and $V_i(t)$ are the average membrane potentials of the neural populations Kle1, Kle2, and Kli at time t.
3. c_1, c_2, c_3, and c_4 are parameters that characterize the synaptic efficiency with which each neural populations influences the others.
4. $P(t)$ is an external input.
5. V_{1f} is the EEG signal.

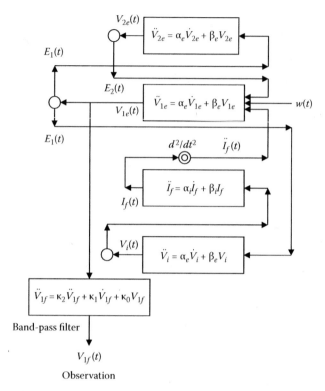

FIGURE 5.3
Zetterberg model structure.

The whole system is described by an 11-dimensional dynamical system model,

$$\frac{d^2V_{1e}}{dt^2} = \alpha_e \frac{dV_{1e}}{dt} + \beta_e V_{1e} + \varepsilon_e c_4 g(V_{2e}) + c_2 \frac{d^2I_f}{dt^2} + \varepsilon_e \Pi + \omega(t) \qquad (5.37)$$

$$\frac{d^2V_{2e}}{dt^2} = \alpha_e \frac{dV_{2e}}{dt} + \beta_e V_{2e} + \varepsilon_e c_3 g(V_{1e}) \qquad (5.38)$$

$$\frac{d^2V_i}{dt^2} = \alpha_e \frac{dV_i}{dt} + \beta_e V_i + \varepsilon_e c_1 g(V_{1e}) \qquad (5.39)$$

$$\frac{d^2I_f}{dt^2} = \alpha_i \frac{dI_f}{dt} + \beta_i I_f + \varepsilon_i g(V_i) \qquad (5.40)$$

$$\frac{d^3V_{1f}}{dt^3} = \kappa_2 \frac{d^2V_{1f}}{dt^2} + \kappa_1 \frac{dV_{1f}}{dt} + \kappa_0 V_{1f} + a\omega_n^2 \frac{dV_{1e}}{dt}. \qquad (5.41)$$

The time series data $x_1, x_2, ..., x_N$ are supposed to be from $V_{1f}(t)$ for $t = t_0 + \Delta t$, $t_0 + 2\Delta t, ..., t_0 + N\Delta t$ with some starting point t_0 and a fixed sampling interval Δt. The whole system is described as an 11-dimensional state space model with a scalar observation equation (5.42),

$$\frac{dz}{dt} = f(z) + w(t)$$

$$x_k = Cz_k + \varepsilon_k, \tag{5.42}$$

where $z = \left(\dfrac{dV_{1e}}{dt}, V_{1e}, \dfrac{dV_{2e}}{dt}, V_{2e}, \dfrac{dV_i}{dt}, V_i, \dfrac{dI_f}{dt}, I_f, \dfrac{d^2V_{1f}}{dt^2}, \dfrac{dV_{1f}}{dt}, V_{1f} \right)$ and $C =$

$(0, 0, ..., 0, 1)$. Note also that the 11-dimensional state dynamics is split into 5 subsystems, where each subsystem (5.37) through (5.41) is an oscillatory system specified by a second-order linear differential equation model, except the third-order linear subsystem (5.41) for the recording of EEG data. Nonlinearities are implemented by the instantaneous nonlinear sigmoid type transformation $g(.)$ of the output of each subsystem, which also act as inputs to other subsystems, as is typically seen in usual nonlinear feedback systems. The model can be considered as a kind of nonlinear mixed-type structural model, where the nonlinearity is in the instantaneous transformation $g(.)$, which is shown in Figure 5.3 by the symbol "O."

Estimation of the parameters of the model is made possible using the innovation approach, details of which are in Chapters 10 and 13.

5.4.3 Hodgkin–Huxley Model

The Hodgkin–Huxley model is a scientific model that describes how action potentials in neurons are initiated and propagated. It is a set of nonlinear ordinary differential equations that approximates the electrical characteristics of excitable cells such as neurons and cardiac myocytes. Hodgkin and Huxley described the model in 1952 to explain the ionic mechanisms underlying the initiation and propagation of action potentials in the squid giant axon.

The model is described with the Hodgkin–Huxley equations (5.43) through (5.46):

$$\frac{dV}{dt} = \frac{1}{C_M}\left\{ I - \bar{g}_{Na}m^3 h(V - E_{Na}) - \bar{g}_K n^4 (V - E_K) - \bar{g}_l(V - E_l) \right\} \tag{5.43}$$

$$\frac{dm}{dt} = \alpha_m(V)(1-m) + \beta_m(V)m, \tag{5.44}$$

$$\frac{dh}{dt} = \alpha_h(V)(1-h) + \beta_h(V)h, \tag{5.45}$$

$$\frac{dn}{dt} = \alpha_n(V)(1-n) + \beta_n(V)n, \tag{5.46}$$

which come in part from the statement concerning the equation

$$I = C_M \frac{dV}{dt} + I_i. \tag{5.47}$$

Which says that the total current is the sum of the capacitance current and the ionic currents. The capacitance of the membrane per unit area (i.e., C_M) has already been determined. There are three ionic currents: The two important are sodium current I_{Na} and potassium current I_K, and a third leakage current, denoted by I_l, consisting largely of chloride ions (details of the leakage current are discussed in Hodgkin and Huxley [1952]). Thus, we have

$$I_i = I_{Na} + I_K + I_l.$$

Using the independence principle, we regard I_{Na}, I_K, and I_l as having no influence on one another. Here we may deal with the sodium and potassium currents separately and write

$$I_{Na} = g_{Na}(V - E_{Na}),$$

$$I_K = g_K(V - E_K).$$

where
   g_{Na} is the sodium conductance
   E_{Na} is the equilibrium potential for the sodium ion
   g_k is the potassium conductance
   E_k is the equilibrium potential for the potassium

It turns out that the leakage current has an especially simple description,

$$I_l = g_l(V - E_l),$$

where
   g_l is a positive constant $g_l = \bar{g}_l$ over time
   E_l is the equilibrium potential for the ions (mostly chloride) that constitute
      the leakage current

Thus, (5.47) becomes

$$I = C_M \frac{dV}{dt} + g_{Na}(V - E_{Na}) + g_K(V - E_K) + \bar{g}_l(V - E_l)$$

or

$$C_M \frac{dV}{dt} = -g_{Na}(V - E_{Na}) - g_K(V - E_K) - \bar{g}_l(V - E_l) + I. \tag{5.48}$$

The remaining task is to characterize the dynamics of g_K and g_{Na}, by some differential equations such as

$$\frac{dg_{Na}}{dt} = G(t, V, g_{Na}, g_K),$$

$$\frac{dg_K}{dt} = H(t, V, g_{Na}, g_K).$$

A typical characteristic we see in the behavior of the potassium conductance g_K, for example, is that it rises with a marked inflection if V is increased, say from 0 to 25 mV. But if V is changed from 25 mV to 0, then $g_K(t)$ decreases in a simple exponential way. Finding a differential equation of $g_K(t)$ whose solution would fit the increasing and decreasing patterns is rather complicated.

In order to obtain a valid description of this kind of behavior of $g_K(t)$, Hodgkin and Huxley introduced a new dimensionless variable n. They proposed to express the function $g_K(t)$ as

$$g_K(t) = \bar{g}_K[n(t)]^4, \tag{5.49}$$

where $\bar{g}_K$ is a positive constant whose value will be obtained from the experimental data, and to require the function $n(t)$ to satisfy a differential equation of the form

$$dn/dt = \alpha_n(1-n) - \beta_n n, \tag{5.50}$$

where α_n and β_n are nonnegative functions of V. Here the dynamics of $g_K(t)$ is characterized by the dynamics of $n(t)$ of (5.50) together with the instantaneous relation of $n(t)$ and $g_K(t)$ by (5.49). Note that the idea of this particular nonlinear model structure is raised from the phenomenological concern. The best justification for this proposal and a similar proposal made for $g_{Na}(t)$ is that it works (Cronin 1987).

The study of sodium conductance, that is the function $g_{Na}(t)$, is analogous to the study of potassium conductance except that it is somewhat more complicated. In particular, if V is fixed at a value above the threshold value, then the sodium conductance first rises to a maximum value and then decreases. By considerations similar to those used to arrive at the description of potassium conductance given by (5.49) and (5.50), Hodgkin and Huxley decided to describe the dynamics of sodium conductance $g_{Na}(t)$ by means of the following equations:

$$g_{Na}(t) = m^3 h \bar{g}_{Na}, \tag{5.51}$$

$$dm/dt = \alpha_m(1-m) - \beta_m m, \tag{5.52}$$

$$dh/dt = \alpha_h(1-h) - \beta_h h, \tag{5.53}$$

where
$\bar{g}_{Na}$ is a positive constant
$\alpha_m, \beta_m, \alpha_h,$ and β_h are certain nonnegative functions of V

The dependent variables $m(t)$ and $h(t)$ are called the activation and inactivation variables, respectively. They can be regarded as measures of the way by which the membrane permits Na ions to pass through. In order to describe the dynamics of the variables in the dynamic model (5.48) for $V(t)$, we now have three extra dynamical systems, (5.50), (5.52) and (5.53), where the variables $g_K(t)$ and $g_{Na}(t)$ are replaced by $m(t)$, $h(t)$ and $n(t)$. These four equations form the celebrated Hodgkin–Huxley equations (5.43) through (5.46). It remains to choose functions $\alpha_m(V)$, $\beta_m(V)$, $\alpha_h(V)$, $\beta_h(V)$, $\alpha_n(V)$, and $\beta_n(V)$. The functions that Hodgkin and Huxley chose are

$$\alpha_m(V) = \frac{0.1(V + 25)}{\exp[(V + 25)/10] - 1},$$

$$\beta_m(V) = 4\exp(V/18),$$

$$\alpha_h(V) = 0.07\exp(V/20),$$

$$\beta_h(V) = \frac{1}{\exp[(V + 30)/10] + 1},$$

$$\alpha_n(V) = \frac{0.01(V + 10)}{\exp[(V + 10)/10] - 1},$$

$$\beta_n(V) = 0.125\exp(V/80).$$

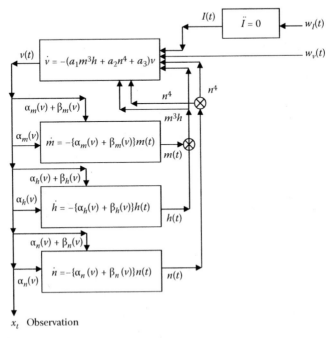

FIGURE 5.4
Hodgkin–Huxley model structure.

The Hodgkin–Huxley equations (5.43) through (5.46) can be regarded as a feedback system (see Figure 5.4) that is composed of four subsystems, each of which is a first-order nonlinear dynamical system,

$$\dot{V} = a_V V + b_V$$
$$\dot{m} = a_m m + b_m$$
$$\dot{h} = a_h h + b_h$$
$$\dot{n} = a_n n + b_n.$$

where the additive inputs to each subsystem are, respectively,

$$b_V = \frac{1}{C_m}\left\{ g_{Na} m^3 h E_{Na} + g_K n^4 E_K + \bar{g}_l E_l + I(t) \right\},$$
$$b_m = \alpha_m(V(t)),$$
$$b_h = \alpha_h(V(t)),$$
$$b_n = \alpha_n(V(t)).$$

The dynamic character of each subsystem is determined, respectively, by

$$a_V = -\frac{1}{C_m}\left\{ g_{Na} m(t)^3 h(t) + g_K n(t)^4 + \bar{g}_l \right\},$$
$$a_m = -\left\{ \alpha_m(V(t)) + \beta_m(V(t)) \right\},$$
$$a_h = -\left\{ \alpha_h(V(t)) + \beta_h(V(t)) \right\},$$
$$a_n = -\left\{ \alpha_n(V(t)) + \beta_n(V(t)) \right\}.$$

In time series analysis where the Hodgkin–Huxley model is considered, the datum x_t is usually regarded as a measurement of the variable $V(t)$, possibly with a Gaussian white observation noise ε_t. Then the whole system, including the measurement system, is regarded as a 4-dimensional nonlinear state space model,

$$C_m \frac{dV(t)}{dt} = -\left\{ \bar{g}_{Na} m(t)^3 h(t) + \bar{g}_K n(t)^4 + \bar{g}_l \right\} V(t)$$
$$+ \bar{g}_{Na} m^3 h E_{Na} + \bar{g}_K n^4 E_K + \bar{g}_l E_l + I(t)$$
$$\frac{dm(t)}{dt} = -\left\{ \alpha_m(V(t)) + \beta_m(V(t)) \right\} m(t) + \alpha_m(V(t))$$
$$\frac{dh(t)}{dt} = -\left\{ \alpha_h(V(t)) + \beta_h(V(t)) \right\} h(t) + \alpha_h(V(t))$$
$$\frac{dn(t)}{dt} = -\left\{ \alpha_n(V(t)) + \beta_n(V(t)) \right\} n(t) + \alpha_n(V(t))$$
$$x_t = V_t + \varepsilon_t.$$

Our remaining problem would be to determine the parameters of the state space model from the experimental data. Here we are required to estimate the parameters such as $\bar{g}_{Na}$, $\bar{g}_K$, $\bar{g}_l$, E_{Na}, E_K, and E_l from the observed time series data. The problem can be solved in the framework of nonlinear state space model identification, which we study in Chapters 11 through 13.

5.4.4 Balloon Model

The balloon model for the dynamics of the blood flow is another interesting example of nonlinear continuous-time dynamic models in neuroscience. Friston et al. (2000) introduced, following Buxton et al. (1998)'s work, a set of four nonlinear and nonautonomous ordinary differential equation models, which govern the dynamics of the intrinsic physiological variables: the blood flow-inducing signal, the cerebral blood flow (CBF), the cerebral blood volume (CBV), and the total de-oxyhemoglobin (dHb). Here the time varying vector, $z(t) = (z_1(t), z_2(t), z_3(t), z_4(t))'$, summarizes the dynamic of the system, where the normalized intrinsic variables $z_1(t)$, $z_2(t)$, $z_3(t)$, $z_4(t)$ are the flow-inducing signal, CBF, CBV, and dHb, respectively.

The model was generalized by Riera et al. (2004a) to include a Gaussian white noise $w(t)$ (i.e., an increment $dW(t)$ of a Wiener process $W(t)$), representing an additive physiological system noise, with a vector $g = (g_1, g_2, g_3, g_4)'$ defining the strength of randomness for each variable,

$$\dot{z} = f(z) + w(t). \tag{5.54}$$

The vector function $f(z,u)$ is defined by the following equations:

$$f_1(z,u) = \varepsilon u(t) - a_1 z_1(t) - a_2(z_2(t) - 1) \tag{5.55}$$

$$f_2(z,u) = z_1(t) \tag{5.56}$$

$$f_3(z,u) = \frac{1}{a_3}\left(z_2(t) - z_3(t)^{\frac{1}{a_4}} \right) \tag{5.57}$$

$$f_4(z,u) = \frac{1}{a_3}\left(\frac{z_2(t)}{a_5}\left[1 - (1 - a_5)^{\frac{1}{z_2}} \right] - z_4(t)z_3(t)^{\frac{1-a_4}{a_4}} \right). \tag{5.58}$$

The physiological meaning of the parameters $\{a_1, a_2, a_3, a_4, a_5\}$ is explained in detail in Riera et al. (2004a) and Friston et al. (2000). We note that the deterministic part of the stochastic dynamical system model (5.54), that is, $\dot{z} = f(z)$, is decomposed into the following three subsystems:

$$\ddot{z}_2(t) + a_1 \dot{z}_2(t) + a_2 z_2(t) = a_2 + \varepsilon s(t), \tag{5.59}$$

$$\dot{z}_3 = a_3(z_3)z_3 + u_3(z_2),$$ (5.60)

$$\dot{z}_4 = a_3(z_3)z_4 + u_4(z_2).$$ (5.61)

Here (5.59) comes from (5.55) and (5.56), and it yields to a linear oscillation system of the variable $z_2(t)$ driven by $a_2 + \varepsilon s(t)$. Equations 5.60 and 5.61 come from (5.57) and (5.58). respectively. They show that the variables $z_3(t)$ and $z_4(t)$ are both regarded as outputs of nonlinear first-order dynamical systems, $\dot{z}_3 = a_3(z_3)z_3$, and $\dot{z}_4 = a_3(z_3)z_4$, driven by nonlinear functions of $z_2(t)$, i.e.,

$$u_3(z_2) = \frac{1}{a_3} z_2(t)$$

and

$$u_4(z_2) = \frac{1}{a_3} \frac{z_2(t)}{a_5} \left[1 - (1 - a_5)^{\frac{1}{z_2}} \right],$$

respectively. Here $a_3(z_3)$ is specified from (5.57) and (5.58) as

$$a_3(z_3) = -\frac{1}{a_3} z_3(t)^{\frac{1-a_4}{a_4}}.$$

The balloon model is a kind of mixed-type nonlinear structural model (see Figure 5.5).

The nonlinearity of the dynamics of the system (5.54) is summarized in the z_3-dependent coefficient $a_3(z_3)$, while static nonlinearities are in the nonlinear

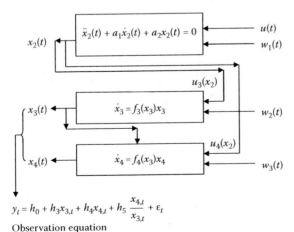

$$y_t = h_0 + h_3 x_{3,t} + h_4 x_{4,t} + h_5 \frac{x_{4,t}}{x_{3,t}} + \varepsilon_t$$

Observation equation

FIGURE 5.5
Balloon model structure.

driving forces $u_3(z_2)$ and $u_4(z_2)$ as well as in the fourth term of the nonlinear observation equation:

$$x_t = h_0 + h_3 z_{3,t} + h_4 z_4 + h_5 \frac{z_{4,t}}{z_{3,t}}. \tag{5.62}$$

Estimation of the parameters of the model is made possible using the innovation methods, details of which are in Chapters 10 and 13.

5.4.5 Dynamic Causal Model

The dynamic causal model (DCM) of Friston et al. (2003) is somehow different from the previous three nonlinear models in continuous time. It was introduced as a bilinear approximation of a nonlinear input–output system, $\dot{z} = F(z,u,\theta)$, as follows:

$$dz/dt = \left(A + \sum u_j B^{(j)}\right) z + Cu. \tag{5.63}$$

Here, $z = (z_1, \ldots, z_l)'$ is a l-dimensional output, and $u = (u_1, \ldots, u_m)'$ is an m-dimensional input. The matrices, A, $B^{(j)}$ ($j = 1, \ldots, m$), and C, are given by

$$A = \frac{\partial F}{\partial z} = \frac{\partial \dot{z}}{\partial z},$$

$$B^{(j)} = \frac{\partial^2 F}{\partial z \partial u_j} = \frac{\partial}{\partial u_j} \frac{\partial \dot{z}}{\partial z},$$

$$C = \frac{\partial F}{\partial u}.$$

Then DCM, for fMRI data, combines this bilinear model of neural dynamics with an empirically validated hemodynamic model that describes the transformation of neuronal activity into a BOLD signal response.

Although the model was introduced with the aim of describing the specific dynamics of the local hemodynamic responses under the neural stimulus, it turns out to be a kind of general input–output system model commonly seen in the field of control engineering. Here the system is a linear system (in terms of the state $z(t)$) driven by an input variable $u(t)$. However, the dynamic characteristics of the linear system for $z(t)$ are affected and regulated by the input $u(t)$. Here the inputs, linear or nonlinear functions of a stimulus, drive the system additively and also multiplicatively.

The measured BOLD signal response x_t is modeled as the predicted BOLD signal $h(z_t, u)$ (the generalized convolution of inputs) plus a linear mixture of confounded variables (e.g., signal drift X) and Gaussian observation error.

If we propose a parametric model, with parameter θ and β, for the observation x_t, we have an observation equation,

$$x_t = h(z_t, u_t, \theta) + X_t\beta + \varepsilon_t. \tag{5.64}$$

Thus, from (5.63) and (5.64), we have a state space model,

$$dz/dt = \left(A + \sum u_j B^{(j)} z\right) + Cu$$

$$x_t = h(z_t, u_t, \theta) + X_t\beta + \varepsilon_t, \tag{5.65}$$

for describing the dynamics of the BOLD signal time series data, $x_1, x_2, \ldots, x_N$. Note that $B^{(j)}$ $(j = 1, 2, \ldots, m)$ show the existence and strength of the connectivity between the l-dimensional state $z(t)$ and the jth input $u^{(j)}(t)$. With the state space model (5.65), we can, in principle, estimate these connectivity parameters from the observed BOLD signal time series data, $x_1, x_2, \ldots, x_N$ using the maximum likelihood method, which we see in Chapters 10 and 13. However, we must be careful when the speed of change (main frequency) of the two dynamics, dynamic of $z(t)$ and the dynamic of the BOLD signal, is very different. If the speed of $z(t)$ is much faster than x_t, we will not be able to extract useful information for the connectivity of $z(t)$ from x_t.

6

Some More Models

6.1 Nonlinear AR Models

The best possible predictor for Gaussian time series is the linear predictor (Whittle 1963b, p. 10), but not many articles dwell on the fact that the linear predictor can be outperformed by nonlinear predictors for the non-Gaussian case. Whittle (1963b) is one of the exceptional cases, where there is an example showing how a nonlinear predictor can outperform the best linear predictor in the sense of minimizing the sum of squares of the errors. Later experience in the 1970s and 1980s has proved that many non-Gaussian processes can actually be better predicted by nonlinear predictors (Ozaki and Oda 1978, Priestley 1988). Statisticians' effort on nonlinear time series modeling and engineers' effort on neural network modeling can be seen as being in line with this trend for obtaining better predictors in the sense of least squares criterion $E[.]^2$, directing to the strategy of minimizing the sum of squares of errors, that is,

$$\text{Min} \sum_{t=d+1}^{N} \left[x_t - f(x_{t-1}, x_{t-2}, \ldots, x_{t-d}) \right]^2,$$

where a nonlinear autoregressive model driven by a Gaussian white noise w_t,

$$x_t = f(x_{t-1}, x_{t-2}, \ldots, x_{t-d}) + w_t, \tag{6.1}$$

is implicitly assumed.

Finding an appropriate functional form $f(.)$ out of a huge number of possible candidates is very important but is obviously extremely difficult. In order to improve the prediction performance of the model (6.1), many different types of parametric models have been introduced in the late 1970s–1980s. Experience gained in earlier works by many applied time series analysts (including the present author) showed that representing (6.1) by a state-dependent AR model of the form

$$x_t = \phi_1(x_{t-1}, x_{t-2}, \ldots, x_{t-d})x_{t-1} + \cdots + \phi_d(x_{t-1}, x_{t-2}, \ldots, x_{t-d})x_{t-d} + w_t \tag{6.2}$$

141

and parameterizing the dynamics of its coefficients, $\phi_1(x_{t-1}, x_{t-2}, ..., x_{t-d})$, $\phi_d(x_{t-1}, x_{t-2}, ..., x_{t-d})$ by some simple smooth functions such as $\exp(-\gamma x_{t-1}^2)$ provided a steady step forward toward the goal of general parametric and nonlinear AR models.

Some of the examples of Gaussian white noise–driven nonlinear AR(2) models are shown here:

1. ExpAR(2) model (see Ozaki 1980, 1981b, Haggan and Ozaki 1981):

$$x_t = (\phi_1 + \pi_1 e^{-\gamma x_{t-1}^2})x_{t-1} + (\phi_2 + \pi_2 e^{-\gamma x_{t-1}^2})x_{t-2} + w_t.$$

2. RBF-AR(2) model (see Vesin 1993, Ozaki et al. 1997, 2004):

$$x_t = \left(\phi_1 + \pi_1^{(1)} e^{-\gamma_1(x_{t-1}-z_1)^2} + \pi_1^{(2)} e^{-\gamma_2(x_{t-1}-z_2)^2}\right)x_{t-1}$$
$$+ \left(\phi_2 + \pi_2^{(1)} e^{-\gamma_1(x_{t-1}-z_1)^2} + \pi_2^{(2)} e^{-\gamma_2(x_{t-1}-z_2)^2}\right)x_{t-2} + w_t.$$

3. STAR(2) model (see Terasvirta 1994):

$$x_t = \{\phi_1 x_{t-1} + \phi_2 x_{t-2}\}e^{-\gamma x_{t-1}^2} + \{\pi_1 x_{t-1} + \pi_2 x_{t-2}\}\left(1 - e^{-\gamma x_{t-1}^2}\right) + w_t.$$

Actually this model is essentially equivalent to ExpAR(2) model since

$$\{\phi_1 x_{t-1} + \phi_2 x_{t-2}\}e^{-\gamma x_{t-1}^2} + \{\pi_1 x_{t-1} + \pi_2 x_{t-2}\}\left(1 - e^{-\gamma x_{t-1}^2}\right)$$
$$= \left(\pi_1 + (\phi_1 - \pi_1)e^{-\gamma x_{t-1}^2}\right)x_{t-1} + \left(\pi_2 + (\phi_2 - \pi_2)e^{-\gamma x_{t-1}^2}\right)x_{t-2}.$$

4. Linear threshold AR(2) model (see Tong and Lim 1980):

$$x_t = \begin{cases} \phi_1^{(1)} x_{t-1} + \phi_2^{(1)} x_{t-2} + w_t & \text{for } x_{t-1} \geq \theta \\ \phi_1^{(2)} x_{t-1} + \phi_2^{(2)} x_{t-2} + w_t & \text{for } x_{t-1} < \theta \end{cases}.$$

5. Nonlinear threshold AR(2) model (see Ozaki 1981a):

$$x_t = \begin{cases} \{\phi_1^{(1)} + \pi_1^{(1)} x_{t-1}^2\}x_{t-1} + \{\phi_2^{(1)} + \pi_2^{(1)} x_{t-1}^2\}x_{t-2} + w_t & \text{for } |x_{t-1}|^2 < \theta^2 \\ \{\phi_1^{(2)} + \pi_1^{(2)} \theta^2\}x_{t-1} + \{\phi_2^{(2)} + \pi_2^{(2)} \theta^2\}x_{t-2} + w_t & \text{for } |x_{t-1}|^2 \geq \theta^2 \end{cases}.$$

Computationally, estimation of these models is quite straightforward and easy compared with the estimation of state space models in Chapter 13. We discuss the inferential and computational topic of the previous models in Chapter 10.

6.2 Neural Network Models

6.2.1 Multilayer Neural Network Models

Using neural network models is another approach to the prediction of time series. The reconstruction of a predictor of some complex system is one of the aims of neural network studies, where the nonlinear function,

$$f(x_{t-1}, \ldots, x_{t-d})$$

representing the predictor is approximated by a special type of nonlinear function constructed from the time series sample, $x_1, \ldots, x_N$. In fact, the multilayer neural network model, in particular the three-layer neural network model, seems to be the most popular and widely used neural network model for the model identification and prediction of the future, since there is a theory that shows that it is sufficient to use a maximum of three layers to solve an arbitrarily complex pattern classification problem (Rumelhart and McClelland 1986).

In the first layer of the multilayer neural network model, we can see that the h_1 hidden units (processing elements) perform a weighted summation of the inputs, which are then transformed by the nonlinear "activation" functions, ψ_{j_1}, $j_1 = 1, 2, \ldots, h_1$. One of the most common examples of the activation function is a sigmoid (S shaped) function of the form

$$\psi(x) = \tan h(-\gamma x) = \frac{1 - e^{-2\gamma x}}{1 + e^{-2\gamma x}},$$

where γ is a parameter controlling the "steepness" (slope) of the function. The relationship between inputs, $x_{t-1}, \ldots, x_{t-n_0}$, and the hidden outputs, $x_1^{(1)}, \ldots, x_{h_1}^{(1)}$, in the first layer may be expressed by

$$x_{j_1}^{(1)} = \psi_{j_1} \left[\sum_{i=1}^{n_0} a_{j,i} x_{t-i} + a_{j0} \right], \quad j_1 = 1, 2, \ldots, h_1,$$

where the synaptic weights $a_{j,i}$ are adjustable and can take positive, negative, or zero values.

The same kind of procedure is repeated in the second and later layers, using the output of the previous layer as inputs: for example,

$$\begin{pmatrix} x_1^{(1)} = \psi\left(\sum_k a_{1,k}^{(1)} x_{t-k}\right) \\ \cdots \\ x_{h_1}^{(1)} = \psi\left(\sum_k a_{h_1,k}^{(1)} x_{t-k}\right) \end{pmatrix} \rightarrow \begin{pmatrix} x_1^{(2)} = \psi\left(\sum_{k=1} a_{1,k}^{(2)} x_k^{(1)}\right) \\ \cdots \\ x_{h_2}^{(2)} = \psi\left(\sum_{k=1} a_{h_2,k}^{(2)} x_k^{(1)}\right) \end{pmatrix} \rightarrow \cdots$$

$$\cdots \rightarrow \begin{pmatrix} x_1^{(m)} = \psi\left(\sum_{k=1} a_{1,k}^{(m)} x_k^{(m-1)}\right) \\ \cdots \\ x_{h_m}^{(m)} = \psi\left(\sum_{k=1} a_{h_m,k}^{(m)} x_k^{(m-1)}\right) \end{pmatrix} \rightarrow x_t = \psi\left(\sum_{k=1} a_k^{(m+1)} x_k^{(m)}\right).$$

We could think of using this kind of model with samples of time series sequences $x_{t-1}, \ldots, x_{t-d}$, $(t = d + 1, d + 2, \ldots, N)$ as inputs, and one output $x_{t|t-1}$ at the final layer, where $x_{t|t-1}$ is the one-step-ahead prediction of x_t at the time point $(t-1)$, so that $x_{t|t-1} = \psi\left(\sum_{k=1} a_k^{(m+1)} x_k^{(m)}\right)$ could be written as $x_{t|t-1} = f(x_{t-1}, \ldots, x_{t-d})$ for some (very complicated) nonlinear function $f(.)$. The prediction of the time series x_t by $f(x_{t-1}, \ldots, x_{t-d})$ is not exact, of course, for the same reasons as those given for the nonlinear AR model case. The prediction error is defined as the difference between the actual value x_t and the value, $x_{t|t-1} = f(x_{t-1}, \ldots, x_{t-d})$, predicted by the neural network model, that is,

$$w_t = x_t - x_{t|t-1}$$

$$= x_t - f(x_{t-1}, \ldots, x_{t-d}).$$

We naturally expect that the predictor is unbiased, that is, that $E[w_t] = 0$, and we try to find the best model by minimizing the sum of squares of the errors, that is,

$$\sum_{i=d+1}^{N} [x_i - f(x_{t-1}, \ldots, x_{t-d})]^2,$$

which is asymptotically equivalent to maximizing the log-likelihood of the model, $x_t = f(x_{t-1}, \ldots, x_{t-d}) + w_t$, if w_t is a Gaussian white noise. Although most neural network researchers believe that the model is a deterministic approximation to the dynamics of a deterministic time series, what they are actually doing in reconstructing and approximating the dynamics is mathematically equivalent to the method for nonlinear AR models in Section 6.1. The special feature of the neural network model is its unique method of constructing the nonlinear function $f(.)$. Here the functional form of each layer is fixed and the weights are estimated for each layer using the sample data. There are several algorithms for the estimation of these coefficients. However, the computational speed of these estimation (or learning) procedures is extremely slow compared with ordinary least squares estimation of nonlinear models parameterized as linear functions of nonlinear functions, such as polynomial AR models. Although we fix the nonlinear functional form for each layer, the choice is rather arbitrary. There are many possible forms for the activation function (some examples are given in Cichoski and Unbehauen 1993), and the prediction results will depend on the initial choice of the functional form.

Although the multilayer neural network models provide us with a wide range of tractable nonlinear prediction models, the computational burden of estimating the weights of each layer in the model generally forces researchers to move from the general multilayer neural network model to a family of single layer networks with a general nonlinear function family. The shift is commonly seen in many scientific fields where scientists become more and more interested in a specific dynamic structure for their own problem, as their vision becomes clearer in the light of preliminary analysis with a general model such as a multilayer neural network model.

6.2.2 Single-Layer Neural Network Models and RBFs

An alternative model to the multilayer neural network model for the reconstruction of the dynamics and for prediction is provided by the single layer neural network model employing RBFs (radial basis functions). An RBF is a multidimensional function, which depends on the distance $r = \|X - \varsigma\|$ (where $\|\cdot\|$ denotes a vector norm) between a d-dimensional input vector X and a "center" ς. One of the simplest approaches to the approximation of a nonlinear function is to represent it by a linear combination of fixed nonlinear basis functions $B_i(X)$, that is,

$$f(X) = \sum_{i=1}^{m} c_i B_i(X).$$

Typical choices for radial basis functions $B(r) = B(\|X - \varsigma\|)$ are

1. $B(r) = c_r$: piecewise constant base function
2. $B(r) = r$: piecewise linear base function
3. $B(r) = r^3$: piecewise cubic base function
4. $B(r) = \exp(-r^2/h^2)$: piecewise Gaussian base function
5. $B(r) = r^2 \log r$: piecewise thin-plate spline base function
6. $B(r) = (r^2 + h^2)^{1/2}$: piecewise multiquadratic base function
7. $B(r) = (r^2 + h^2)^{-1/2}$: piecewise inverse multiquadratic base function

where h is a real coefficient called the width or scaling parameter. Among the functions described earlier, the most popular and widely used is the Gaussian function (4), which has a peak at the center ς and decreases monotonically as the distance from the center increases.

If the approximation is to remain valid for points of the embedded trajectory not used in estimating the model, that is, for predicting the future of the dynamics, the model should have well-behaved generalization capabilities. This requires that the reconstructed map $f(.)$ should be smooth and have good interpolation properties for scattered data points. $f(.)$ may be written in the form

$$f(x_{t-1}, \ldots, x_{t-d}) = c_0 + \sum_{i=1}^{m} c_i B_i(x_{t-1}, \ldots, x_{t-d})$$

with appropriate functional forms for the multivariate localized basis functions $B_i(x_{t-1}, \ldots, x_{t-d})$ ($i = 1, \ldots, m$). Since the class of RBFs has been recognized (Dyn 1989a,b) to be a class of localized basis functions possessing a high degree of smoothness or regularity, and can very easily incorporate multivariate data, this model should provide a good basis for prediction. RBF expansions have good interpolation properties in dealing with scattered data points and are endowed with the "universal approximation" and "best approximation" capabilities of any continuous function. Universal approximation implies the possibility of approximating a function to any required degree of accuracy. The stronger property of best approximation entails that the approximation error surface always has a unique global minimum for any approximation performance measure (Park and Sandberg 1991, 1993). As a consequence of these approximation capabilities, the RBF model has been recognized as an alternative to the multilayer neural network model (Lapedes and Farber 1987). However, the RBF model has a clear computational advantage over the multilayer neural network models, deriving from its "linear-in-the-parameter" formulation. The one-layer structure of the RBF model is a feature that can be exploited for parameter estimation and allows a faster learning scheme in comparison with the back-propagation techniques used for the multilayer neural network models.

6.3 RBF-AR Models

Instead of a single layer RBF network model,

$$x_t = c_0 + \sum_{i=1}^{m} c_i B_i \left(\left\| X_{t-1} - \varsigma_i \right\| \right) \} + w_t,$$

we can think of using the RBF function approximation for the nonlinear AR coefficients, $\phi_j(x_{t-1}, \ldots, x_{t-r})$, of the nonlinear AR model,

$$x_t = \phi_1(x_{t-1}, \ldots, x_{t-r})x_{t-1} + \cdots + \phi_p(x_{t-1}, \ldots, x_{t-r})x_{t-p} + w_t,$$

as

$$\phi_j(x_{t-1}, \ldots, x_{t-r}) = c_{j,0} + \sum_{l=1}^{m} c_{j,l} B_l \left(\left\| X_{t-1} - \varsigma_l \right\| \right) \} \quad (j = 1, \ldots, p).$$

This idea leads to an RBF-AR model, which was originally given by Vesin (1993), with properly chosen centers, $\varsigma_l = (\varsigma_l^{(1)}, \varsigma_l^{(2)}, \ldots, \varsigma_l^{(r)})'$ $(l = 1, 2, \ldots, m)$, and scaling parameters h_l $(l = 1, 2, \ldots, m)$, as

$$x_t = \sum_{j=1}^{p} \phi_j(X_{t-1}, \varsigma_1, \varsigma_2, \ldots, \varsigma_m) x_{t-j} + w_t,$$

$$\phi_j(X_{t-1}, \varsigma_1, \varsigma_2, \ldots, \varsigma_m) = c_{j,0} + \sum_{l=1}^{m} c_{j,l} \exp\left\{ -\left\| X_{t-1} - \varsigma_l \right\|_{H_l}^2 \right\}$$

$$= c_{j,0} + \sum_{l=1}^{m} c_{j,l} \prod_{k=1}^{r} \exp\left\{ -\frac{(x_{t-k} - \varsigma_l^{(k)})^2}{h_l} \right\},$$

where $X_{t-1} = (x_{t-1}, \ldots, x_{t-d})'$. Here we used the notation, $\left\| \cdot \right\|_{H_l}^2$, of the quadratic norm defined by

$$\left\| X_{t-1} - \varsigma_l \right\|_{H_l}^2 = \sum_{k=1}^{r} \frac{\left(x_{t-k} - \varsigma_l^{(k)} \right)^2}{h_l}.$$

Note that m and p should be sufficiently small compared with N to enable us to obtain a reasonable estimate of the coefficients $c_{j,l}$ $(j = 1, 2, \ldots, p, l = 1, 2, \ldots, m)$.

Vesin (1993) commented, in his introduction of RBF-AR models, that a possible defect of the ExpAR model is that its nonlinear AR coefficients depend only on x_{t-1}, and he suggested making them dependent on past values, x_{t-1}, $x_{t-2}, \ldots, x_{t-r}$. He also showed some interesting successful numerical results applied to real data using this idea (Vesin, 1993). The results seemed to suggest giving up the idea of the original ExpAR model, that is, looking for the dynamics of a continuous time model through ExpAR modeling. We think, however, that Vesin's idea could be implemented with regard to the two original ideas of the ExpAR model by using a generalized state X, where $X = (x, dx/dt, d^2x/dt^2, \ldots, d^{r-1}x/dt^{r-1})'$, so that the model can still maintain its "instantaneous" dependence on the state vector of the dynamics. If the differential operator d/dt is replaced by the difference operator, $\Delta = (1 - B)$, so that, for example, d/dt is replaced by $\Delta x_{t-1} = x_{t-1} - x_{t-2}$, the discrete time state at time point $t - 1$ is $X_{t-1} = (x_{t-1}, \Delta x_{t-1}, \Delta^2 x_{t-1}, \ldots, \Delta^{r-1}x_{t-1})'$, which is essentially equivalent to $(x_{t-1}, x_{t-2}, \ldots, x_{t-r})'$. Therefore, we could have another RBF-AR model,

$$x_t = \phi_0 + \sum_{j=1}^{d} \phi_j(X_{t-1}, \varsigma_1, \varsigma_2, \ldots, \varsigma_m) x_{t-j} + w_t,$$

but with alternative state vector and centers, $X_{t-1} = (x_{t-1}, \Delta x_{t-1}, \Delta^2 x_{t-1}, \ldots, \Delta^{r-1}x_{t-1})'$ and $\varsigma_l = (\varsigma_l^{(1)}, \varsigma_l^{(2)}, \ldots, \varsigma_l^{(r)})'$ $(l = 1, 2, \ldots, m)$, respectively. Then RBF-coefficients could be

$$\phi_j(X_{t-1}, \varsigma_1, \varsigma_2, \ldots, \varsigma_m) = c_{j,0} + \sum_{l}^{m} c_{j,l} \exp\left\{-\|X_{t-1} - \varsigma_l\|_{H_l}^2\right\}, \tag{6.3}$$

where we could have

$$\|X_{t-1} - \varsigma_l\|_{H_l}^2 = \sum_{k=1}^{r} \frac{\left(\Delta^{k-1}x_{t-j} - \varsigma_l^{(k)}\right)^2}{h_l^{(k)}}.$$

Figure 6.1a shows the plot of time series, $x_t, \Delta x_t, \Delta^2 x_t, \Delta^3 x_t$, of the EEG epilepsy data. Figure 6.1b shows the plot of $x_t, \Delta x_t, \Delta^2 x_t, \Delta^3 x_t$ of the same data for $100 \leq t \leq 140$ so that we can see more detail of each $\Delta^k x_t$ $(k = 0, 1, \ldots 3)$. Obviously, $\Delta^k x_t$'s $(k = 1, 2, 3)$ near the spike (around $t = 120$–130) carries much more information than in other periods, $100 \leq t < 120$ and $130 < t \leq 140$. However, a clear feature is that as k increases, the corresponding plots become flatter and are less informative.

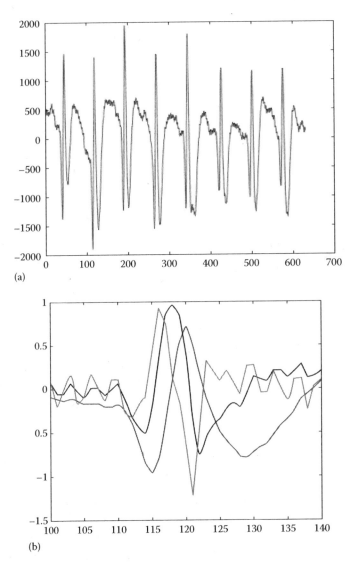

FIGURE 6.1
(a) Epilepsy EEG data. (b) $x(t)$, $\Delta x(t)$, $\Delta^2 x(t)$, $\Delta^3 x(t)$ of epilepsy data for $t = 100$–140.

The ExpAR model was originally introduced by Ozaki and Oda (1978) for the nonlinear modeling of ship-rolling angle time series data, where each AR coefficient is a special case of (6.3). Here the only center is the origin.

Note that most nonlinear AR models in Section 6.1 are regarded as simple examples of RBF-AR models. For example, the threshold AR model is a special form of RBF-AR model, where the base functions are piece-wise constant functions, and ExpAR model (and its copy, STAR model) is a special form of RBF-AR model, where the base functions are Gaussian functions.

6.4 Characterization of Nonlinearities

6.4.1 State-Dependent Eigenvalues

We can regard the nonlinear AR models of Section 6.1 as examples of multidimensional state space models for pragmatic reasons. Two main benefits of considering state space representations of nonlinear AR models are the following:

1. Application reasons (e.g., nonlinear control problems):
   Nonlinear AR models are generally described as p-lag order models, $p > 1$, but the dimension of the time series is mostly one dimension. However, for the nonlinear time series model to be useful in real-world problems, such as prediction and control of complex feedback systems, AR type model representation with high lag order, $p > 1$, is rather awkward and make the design of optimal control difficult. An alternative p-dimensional Markov state space representation of these nonlinear AR models is essential and indispensable for nonlinear extension of the modern control methods developed for linear state space models (see Haggan–Ozaki et al. (2009)).

2. Theoretical reasons (e.g., ergodicity of nonlinear AR processes):
   Time series processes defined by nonlinear AR models of lag order $p > 1$ are included in the family of Markov chain processes defined on the continuous Euclidian state space. Many advanced techniques have been developed in Markov chain on the continuous Euclidian state space. Since these techniques and mathematical theories are presented, in general, in the framework of multidimensional and Markov (i.e., lag order 1) form, they are easily used and applied to the theoretical analysis of many time series processes defined by a nonlinear autoregressive model of lag order p (see, e.g., Jones 1978, Ozaki 1985a).

Typical general pth order nonlinear AR models such as ExpAR(p), and Threshold AR(p) models, for example, are transformed into state space models as follows:

1. ExpAR(p) model:

$$x_t = \left(\phi_1 + \pi_1 e^{-\gamma x_{t-1}^2}\right)x_{t-1} + \cdots + \left(\phi_p + \pi_p e^{-\gamma x_{t-1}^2}\right)x_{t-p} + w_t$$

$$\rightarrow \begin{cases} z_t = A_{t-1}z_{t-1} + \eta_t \\ x_t = (1,0,\ldots,0)z_t \end{cases}$$

$$A_{t-1} = \begin{pmatrix} \phi_1 + \pi_1 e^{-\gamma x_{t-1}^2} & \phi_2 + \pi_2 e^{-\gamma x_{t-1}^2} & \cdots & \phi_p + \pi_p e^{-\gamma x_{t-1}^2} \\ 1 & 0 & \cdots & 0 \\ \cdots & \cdots & \cdots & \cdots \\ 0 & 0 & \cdots & 0 \end{pmatrix},$$

$$z_t = \begin{pmatrix} x_t \\ x_{t-1} \\ \cdots \\ x_{t-p+1} \end{pmatrix}, \quad \eta_t = \begin{pmatrix} w_t \\ 0 \\ \cdots \\ 0 \end{pmatrix}$$

2. Threshold AR(p) model:

$$x_t = \begin{cases} \phi_1^{(1)} x_{t-1} + \cdots + \phi_p^{(1)} x_{t-p} + w_t & \text{for } x_{t-1} \geq \theta \\ \phi_1^{(2)} x_{t-1} + \cdots + \phi_p^{(2)} x_{t-p} + w_t & \text{for } x_{t-1} < \theta \end{cases}$$

$$\rightarrow \begin{cases} z_t = A_{t-1}z_{t-1} + \eta_t \\ x_t = (1,0,\ldots,0)z_t \end{cases}$$

$$A_{t-1} = \begin{cases} \begin{pmatrix} \phi_1^{(1)} & \phi_2^{(1)} & \cdots & \phi_p^{(1)} \\ 1 & 0 & \cdots & 0 \\ \cdots & \cdots & \cdots & \cdots \\ 0 & 0 & \cdots & 0 \end{pmatrix} & \text{for } x_{t-1} \geq \theta \\[2em] \begin{pmatrix} \phi_1^{(2)} & \phi_2^{(2)} & \cdots & \phi_p^{(2)} \\ 1 & 0 & \cdots & 0 \\ \cdots & \cdots & \cdots & \cdots \\ 0 & 0 & \cdots & 0 \end{pmatrix} & \text{for } x_{t-1} < \theta \end{cases}$$

$$z_t = \begin{pmatrix} x_t \\ x_{t-1} \\ \cdots \\ x_{t-p+1} \end{pmatrix}, \quad \eta_t = \begin{pmatrix} w_t \\ 0 \\ \cdots \\ 0 \end{pmatrix}$$

Here the eigenvalues, $\lambda_1, \ldots, \lambda_p$, of the transition matrix, A_{t-1}, are explicit functions of x_{t-1}, and they provide us with useful information about the dynamics of the process, such as the stability, ergodicity, limit cycles, Lyapunov exponent, and chaos (see Haggan and Ozaki 1981, Ozaki 1982, 1985a, 1994b, Ozaki et al. 2000, Carbonell et al. 2002).

We note that each of the discrete-time nonlinear dynamic models discussed so far in the previous sections present a specific parametric nonlinear predictor function, $f(x_{t-1}, x_{t-2}, \ldots, x_{t-d})$, in the form of

$$x_t = f(x_{t-1}, x_{t-2}, \ldots, x_{t-d}) + w_t,$$

where w_t is a Gaussian white noise (examples of nonlinear AR models are summarized in Table 6.1).

Although these nonlinear AR models with nonlinear autoregressive predictors contain useful information about the underlying nonlinear dynamics, it is not very easy to get a clear picture of the dynamics of the model just by looking at a set of coefficients of the nonlinear regressors. We note that one useful way of characterizing the dynamics of the nonlinear AR models is plotting the state-dependent behavior of the eigenvalues of the transition matrix of the state space representation derived from the nonlinear AR models, where the elements of the transition matrix of each nonlinear AR model are state dependent.

TABLE 6.1

Examples of Nonlinear AR Models

1. Single-layer Gaussian RBF neural network model

$$x_t = c_0 + \pi_1 e^{-\|Y_{t-1} - \varsigma_1\|_{in1}^2} + \pi_2 e^{-\|Y_{t-1} - \varsigma_2\|_{in2}^2} + w_t$$

2. ExpAR(2) model

$$x_t = (\phi_1 + \pi_1 e^{-\gamma x_{t-1}^2}) x_{t-1} + (\phi_2 + \pi_2 e^{-\gamma x_{t-1}^2}) x_{t-2} + w_t$$

3. STAR(2) model

$$x_t = \{\phi_1 x_{t-1} + \phi_2 x_{t-2}\} e^{-\gamma x_{t-1}^2} + \{\pi_1 x_{t-1} + \pi_2 x_{t-2}\}(1 - e^{-\gamma x_{t-1}^2}) + w_t$$

4. RBF-AR(2) model of Vesin type

$$x_t = \left\{\phi_1 + \pi_1^{(1)} e^{-\|Y_{t-1} - \varsigma_1\|_{in1}^2}\right\} x_{t-1} + \left\{\phi_2 + \pi_2^{(1)} e^{-\|Y_{t-1} - \varsigma_2\|_{in2}^2}\right\} x_{t-2} + w_t$$

5. RBF-AR(2) model of Ozaki type

$$y_t = \left\{\phi_1 + \pi_1^{(1)} e^{-\|Y_{t-1} - \varsigma_1\|_{in1}^2}\right\} x_{t-1} + \left\{\phi_2 + \pi_2^{(1)} e^{-\|Y_{t-1} - \varsigma_2\|_{in2}^2}\right\} x_{t-2} + w_t$$

6. Linear threshold AR(2) model

$$x_t = \begin{cases} \phi_1^{(1)} x_{t-1} + \phi_2^{(1)} x_{t-2} + w_t & \text{for } x_{t-1} \geq \theta \\ \phi_1^{(2)} x_{t-1} + \phi_2^{(2)} x_{t-2} + w_t & \text{for } x_{t-1} < \theta \end{cases}$$

7. Nonlinear threshold AR(2) model

$$x_t = \begin{cases} \{\phi_1^{(1)} + \pi_1^{(1)} x_{t-1}^2\} x_{t-1} + \{\phi_2^{(1)} + \pi_2^{(1)} x_{t-1}^2\} x_{t-2} + w_t & \text{for } |x_{t-1}|^2 < \theta^2 \\ \{\phi_1^{(2)} + \pi_1^{(2)} \theta^2\} x_{t-1} + \{\phi_2^{(2)} + \pi_2^{(2)} \theta^2\} x_{t-2} + w_t & \text{for } |x_{t-1}|^2 \geq \theta^2 \end{cases}$$

Although the power spectrum (and the auto covariance function) is useful only for the characterization of "linear" dynamics with a constant transition matrix in a state space representation, the eigenvalues of the transition matrix are useful for characterizing both "linear and nonlinear" dynamics of the model.

Suppose we have a nonlinear AR model such as

$$x_t = f(x_{t-1}, x_{t-2}, \ldots, x_{t-d}) + w_t$$
$$= \phi_1(x_{t-1})x_{t-1} + \cdots + \phi_d(x_{t-1})x_{t-d} + w_t.$$

Then we have a simple state space representation,

$$\begin{cases} z_t = A_{t-1}z_{t-1} + \eta_t \\ x_t = (1, 0, \ldots, 0)z_t \end{cases}'$$

where

$$A_{t-1} = \begin{pmatrix} \phi_1(x_{t-1}) & \phi_2(x_{t-1}) & \cdots & \phi_{d-1}(x_{t-1}) & \phi_d(x_{t-1}) \\ 1 & 0 & \cdots & 0 & 0 \\ 0 & 1 & \cdots & 0 & 0 \\ \cdots & \cdots & \cdots & \cdots & \cdots \\ 0 & 0 & \cdots & 1 & 0 \end{pmatrix}, \quad z_t = \begin{pmatrix} x_t \\ x_t \\ \cdots \\ x_{t-d+1} \\ x_{t-d} \end{pmatrix}, \quad \eta_t = \begin{pmatrix} w_t \\ 0 \\ \cdots \\ 0 \\ 0 \end{pmatrix},$$

and so if we look into the dynamics of the x_{t-1}-dependent eigenvalues of the transition matrix A_{t-1}, that is, characteristic roots, $\lambda_1, \lambda_2, \ldots, \lambda_d$ of the following equation,

$$\Lambda^d - \sum_{k=1}^{d} \phi_k(x_{t-1})\Lambda^{d-k} = 0,$$

we can characterize the nonlinear behavior of the process, by looking at the behavior of the eigenvalues. They are supposed to move inside the unit circle, depending on x_{t-1}, if the nonlinear AR model defines a stable and nonexplosive stationary Markov process.

6.4.2 Examples of State-Dependent Eigenvalues

6.4.2.1 ExpAR(2) Model

The following nonlinear AR(2) model

$$x_t = a_1(x_{t-1})x_{t-1} + \phi_2 x_{t-2} + w_t \tag{6.4}$$

can be written in a 2-dimensional state space representation as

$$\begin{bmatrix} x_t \\ x_{t-1} \end{bmatrix} = \begin{bmatrix} a_1(x_{t-1}) & \phi_2 \\ 1 & 0 \end{bmatrix} \begin{bmatrix} x_{t-1} \\ x_{t-2} \end{bmatrix} + \begin{bmatrix} w_t \\ 0 \end{bmatrix}$$

$$x_t = (1,0) \begin{bmatrix} x_t \\ x_{t-1} \end{bmatrix}.$$

(6.5)

Here the transition matrix,

$$A = \begin{bmatrix} a_1(x_{t-1}) & \phi_2 \\ 1 & 0 \end{bmatrix},$$

has two eigenvalues, $\lambda_1(x_{t-1})$ and $\lambda_2(x_{t-1})$, which are the roots of the following characteristic equation:

$$\Lambda^2 - a_1(x_{t-1})\Lambda - \phi_2 = 0.$$

(6.6)

In the ExpAR(2) model, $a(x_{t-1})$ is usually parameterized with 2 coefficient parameters, ϕ_1 and π_1, and a scaling parameter γ as

$$a_1(x_{t-1}) = \phi_1 + \pi_1 \exp(-\gamma x_{t-1}^2).$$

If we assume that ϕ_1 and π_1 satisfy the condition that the roots $\lambda_1(x_{t-1})$ and $\lambda_2(x_{t-1})$ stay inside the unit circle, as in Figure 6.2, for $-\infty < x_{t-1} < \infty$, then we can say (see Ozaki 1985a) that the state space model (6.5) is stable and the nonlinear AR(2) model (6.4) is ergodic and stationary.

Let λ_0 and $\bar{\lambda}_0$ be complex conjugate roots of

$$\Lambda^2 - (\phi_1 + \pi_1)\Lambda - \phi_2 = 0.$$

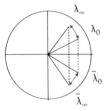

FIGURE 6.2
Two sets of eigenvalues of ExpAR(2)-Duffing-type.

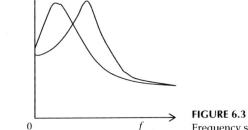

$p(f)$

0   f

FIGURE 6.3
Frequency shift in power spectrum by EXPAR(2).

Let λ_∞ and $\bar\lambda_\infty$ be complex conjugate roots of

$$\Lambda^2 - \phi_1\Lambda - \phi_2 = 0.$$

Then we can see that the roots $\lambda_1(x_{t-1})$ and $\lambda_2(x_{t-1})$ of (6.6) are complex conjugate roots and they move between $\lambda_0, \bar\lambda_0$ and $\lambda_\infty, \bar\lambda_\infty$ depending on x_{t-1} (see Figure 6.2). Remember that the argument of the complex roots of the AR(2) model specifies the frequency characteristic of the AR(2) process, giving the peak of the spectrum with the unimodal shape. Here in the nonlinear AR(2) model (6.4), we can interpret the movement of the roots as the nonlinearity characteristic of the amplitude-dependent frequency shift, (see Figure 6.3) which we saw for the stochastic Duffing model

$$\ddot{x} + c\dot{x} + \alpha x + \beta x^3 = w(t)$$

in the previous section (Section 5.4).

From this example, we see that typical nonlinear oscillations, such as the amplitude-dependent frequency shift and the limit cycle can be characterized by the following discrete-time state space model,

$$\begin{aligned} z_{t+1} &= A_t z_t + Bw_{t+1} \\ x_t &= Cz_t \end{aligned} \qquad (6.7)$$

where $z_t = (x_t, x_{t-1})'$, $B = (1, 0)'$, and

$$A_t = \begin{pmatrix} \phi_1(x_t) & \phi_2(x_t) \\ 1 & 0 \end{pmatrix}.$$

Here $\phi_1(x_t)$ and $\phi_2(x_t)$ are bounded functions of x_t and need to guarantee that the eigenvalues of the transition matrix A_t stay inside the unit circle for $|x_t| \to \infty$.

ExpAR models are also useful for the characterization of nonlinear dynamics of the process such as stable and unstable singular points and chaos (see Ozaki 1985a). The general order ExpAR(p) model is defined by

$$x_t = \left\{\phi_{1,0} + \phi_{1,1}\exp\left(-\gamma x_{t-1}^2\right)\right\} x_{t-1} + \cdots + \left\{\phi_{p,0} + \phi_{p,1}\exp\left(-\gamma x_{t-1}^2\right)\right\} x_{t-p} + w_t,$$

and it has been used in many fields for the characterization of nonlinear dynamic structure, such as limit cycles, singular points, and bifurcations (see, e.g., Shi et al. 2001).

6.4.2.2 RBF-AR(p) Model

A more general parametric model of nonlinear AR type is the general pth order RBF-AR model, which is given by

$$x_t = \phi_0 + \sum_{i=1}^{p} \phi_i(x_{t-1})x_{t-i} + w_t \tag{6.8}$$

or

$$x_t = \phi_0 + \sum_{i=1}^{p} \phi_i(x_{t-1}, \Delta x_{t-1})x_{t-i} + w_t. \tag{6.9}$$

$\phi_i(x)$ ($i = 1, \ldots, p$) are specified by a linear combination of the radial basis functions, $B_1(x), \ldots, B_m(x)$. A typical example of the radial basis function is the Gaussian function, such as

$$B_k(x_{t-1}) = \exp\left(-\gamma_k \left\| x_{t-1} - \varsigma_k \right\|^2\right) \quad (k = 1, \ldots, m)$$

Then the coefficients $\phi_i(x_{t-1})$ ($i = 1, \ldots, p$) of (6.8) or $\phi_i(x_{t-1}, \Delta x_{t-1})$ ($i = 1, \ldots, p$) of (6.9) are specified as

$$\phi_i(x_{t-1}) = c_{i,0} + \sum_{k=1}^{m} c_{i,k}\exp(-\gamma_k \left\| x_{t-1} - \varsigma_k \right\|^2) \quad (i = 1, \ldots, p)$$

or

$$\phi_i\left(x_{t-1}, \Delta x_{t-1}\right) = c_{i,0} + \sum_{k=1}^{m} c_{i,k} \exp\left(-\gamma_k \left\| \begin{pmatrix} x_{t-1} \\ \Delta x_{t-1} \end{pmatrix} - \begin{pmatrix} \varsigma_k^{(1)} \\ \varsigma_k^{(2)} \end{pmatrix} \right\|^2 \right) \quad (i = 1,\ldots,p)$$

The model (6.8) is reformulated in a p-dimensional Markov state space model as

$$
\begin{pmatrix} x_t \\ x_{t-1} \\ \ldots \\ x_{t-p+2} \\ x_{t-p+1} \end{pmatrix} =
\begin{pmatrix} \phi_1(x_{t-1}|\theta) & \phi_2(x_{t-1}|\theta) & \ldots & \phi_{p-1}(x_{t-1}|\theta) & \phi_p(x_{t-1}|\theta) \\ 1 & 0 & \ldots & 0 & 0 \\ \ldots & \ldots & \ldots & \ldots & \ldots \\ 0 & 0 & \ldots & 0 & 0 \\ 0 & 0 & \ldots & 1 & 0 \end{pmatrix}
\begin{pmatrix} x_{t-1} \\ x_{t-2} \\ \ldots \\ x_{t-p+1} \\ x_{t-p} \end{pmatrix}
$$

$$
+ \begin{pmatrix} \phi_0 \\ 0 \\ \ldots \\ 0 \\ 0 \end{pmatrix} + \begin{pmatrix} w_t \\ 0 \\ \ldots \\ 0 \\ 0 \end{pmatrix}. \tag{6.10}
$$

θ is the parameter vector, $\theta = (C_{1,0}, C_{1,1} \ldots, C_{1,m}, \ldots, C_{p,0}, C_{p,1}, \ldots, C_{p,m}, \gamma_1, \ldots, \gamma_m, \zeta_1, \ldots, \zeta_m)'$.

Here the eigenvalues of the transition matrix $A(x_{t-1}|\theta)$

$$
A(x_{t-1}|\theta) = \begin{pmatrix} \phi_1(x_{t-1}|\theta) & \phi_2(x_{t-1}|\theta) & \ldots & \phi_{p-1}(x_{t-1}|\theta) & \phi_p(x_{t-1}|\theta) \\ 1 & 0 & \ldots & 0 & 0 \\ \ldots & \ldots & \ldots & \ldots & \ldots \\ 0 & 0 & \ldots & 0 & 0 \\ 0 & 0 & \ldots & 1 & 0 \end{pmatrix}
$$

are more flexible functions of x_{t-1}, and they can characterize more complex patterns (see Figure 6.4) of the dynamics of the eigenvalues (see Haggan and Ozaki 1981, Ozaki et al. 1997, Ozaki et al. 2004).

6.4.3 Discrete-Time Nonlinear Model and Local Linearization

The method for the characterization of the nonlinear dynamics with the state-dependent eigenvalues is regarded as a method of local linearization

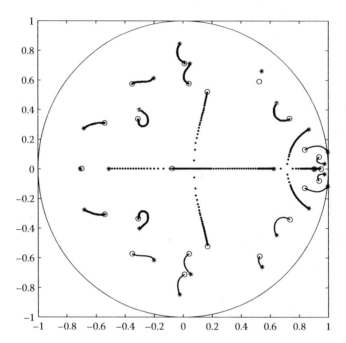

FIGURE 6.4
A complex pattern of RBF-AR eigenvalues.

for discrete time dynamic model. The method is different from the commonly used linearization method, which employs the first order Taylor approximation of the nonlinear dynamics.

The difference of the two linearizations can be typically shown in the following simple example of a nonlinear AR model:

$$x_{t+1} = f(x_t) + w_{t+1}.$$

The first order Taylor expansion method suggests approximating this nonlinear dynamic model with the Jacobian

$$J(x_t) = \frac{\partial f(x)}{\partial x}\bigg|_{x=x_t}$$

as

$$x_{t+1} = f(x_t) + w_{t+1}.$$
$$\approx f(0) + J(x_t)x_t + w_{t+1}.$$

On the other hand, the present approach uses the following exact relation:

$$x_{t+1} = f(x_t) + w_{t+1},$$
$$= f(0) + \phi(x_t)x_t + w_{t+1},$$

where $\phi(x_t)$ is defined by

$$\phi(x_t) = f'(0) + \sum_{k=1}^{\infty} \frac{1}{(k+1)!} \left. \frac{\partial^{k+1} f(x)}{\partial x^{k+1}} \right|_{x=0} x_t^k,$$

which comes from the Taylor expansion

$$f(x_t) = f(0) + \left. \frac{\partial f(x)}{\partial x} \right|_{x=0} x_t + \sum_{k=1}^{\infty} \frac{1}{(k+1)!} \left. \frac{\partial^{k+1} f(x)}{\partial x^{k+1}} \right|_{x=0} x_t^{k+1}$$
$$= \phi(x_t)x_t + f(0).$$

Note that, unlike the first Taylor expansion method, our method preserves the equality relation of the original dynamic model between x_{t+1} and $f(x_t) + w_{t+1}$.

For the 2-dimensional discrete-time nonlinear model,

$$\begin{pmatrix} x_{t+1} \\ y_{t+1} \end{pmatrix} = \begin{pmatrix} f^{(1)}(x_t, y_t) \\ f^{(2)}(x_t, y_t) \end{pmatrix} + \begin{pmatrix} w_{t+1}^{(1)} \\ w_{t+1}^{(2)} \end{pmatrix}, \tag{6.11}$$

we have

$$\begin{pmatrix} f^{(1)}(x_t, y_t) \\ f^{(2)}(x_t, y_t) \end{pmatrix} = \begin{pmatrix} f^{(1)}(0,0) \\ f^{(2)}(0,0) \end{pmatrix} + \begin{pmatrix} \dfrac{\partial f^{(1)}(x,y)}{\partial x} & \dfrac{\partial f^{(1)}(x,y)}{\partial y} \\[2mm] \dfrac{\partial f^{(2)}(x,y)}{\partial x} & \dfrac{\partial f^{(2)}(x,y)}{\partial y} \end{pmatrix}_{(x,y)'=(0,0)'} \begin{pmatrix} x_t \\ y_t \end{pmatrix}$$

$$+ \begin{pmatrix} \left. \displaystyle\sum_{i\geq 1, j\geq 1} a_{i,j}^{(1)}(x,y) \right|_{(x,y)'=(0,0)'} x_t^i y_t^j \\[4mm] \left. \displaystyle\sum_{i\geq 1, j\geq 1} a_{i,j}^{(2)}(x,y) \right|_{(x,y)'=(0,0)'} x_t^i y_t^j \end{pmatrix}.$$

Then we can rewrite (6.11) as

$$\begin{pmatrix} x_{t+1} \\ y_{t+1} \end{pmatrix} = \begin{pmatrix} f^{(1)}(0,0) \\ f^{(2)}(0,0) \end{pmatrix} + A(x_t, y_t) \begin{pmatrix} x_t \\ y_t \end{pmatrix} + \begin{pmatrix} w_{t+1}^{(1)} \\ w_{t+1}^{(2)} \end{pmatrix},$$

where

$$A(x_t, y_t) = \begin{pmatrix} \dfrac{\partial f^{(1)}(x,y)}{\partial x} + \displaystyle\sum_{i\geq 1, j\geq 1} a_{i,j}^{(1)}(x,y)x_t^{i-1}y_t^j & \dfrac{\partial f^{(1)}(x,y)}{\partial y} \\[20pt] \dfrac{\partial f^{(2)}(x,y)}{\partial x} & \dfrac{\partial f^{(2)}(x,y)}{\partial y} + \displaystyle\sum_{i\geq 1, j\geq 1} a_{i,j}^{(2)}(x,y)x_t^i y_t^{j-1} \end{pmatrix}_{(x,y)'=(0,0)'}$$

Here,

$$f^{(1)}(x_t, y_t) = f^{(1)}(0,0) + \frac{\partial f^{(1)}(x,y)}{\partial x}x_t + \frac{\partial f^{(1)}(x,y)}{\partial y}y_t + \frac{1}{2!}\frac{\partial^2 f^{(1)}(x,y)}{\partial x^2}\bigg|_{(x,y)'=(0,0)'} x_t^2$$

$$+ \frac{\partial^2 f^{(1)}(x,y)}{\partial x \partial y}\bigg|_{(x,y)'=(0,0)'} x_t y_t + \frac{1}{2!}\frac{\partial^2 f^{(1)}(x,y)}{\partial y^2}\bigg|_{(x,y)'=(0,0)'} y_t^2 + \cdots$$

$$f^{(2)}(x_t, y_t) = f^{(2)}(0,0) + \frac{\partial f^{(2)}(x,y)}{\partial x}x_t + \frac{\partial f^{(2)}(x,y)}{\partial y}y_t + \frac{1}{2!}\frac{\partial^2 f^{(2)}(x,y)}{\partial x^2}\bigg|_{(x,y)'=(0,0)'} x_t^2$$

$$+ \frac{\partial^2 f^{(2)}(x,y)}{\partial x \partial y}\bigg|_{(x,y)'=(0,0)'} x_t y_t + \frac{1}{2!}\frac{\partial^2 f^{(2)}(x,y)}{\partial y^2}\bigg|_{(x,y)'=(0,0)'} y_t^2 + \cdots$$

$a_{i,j}^{(1)}(x,y)\big|_{(x,y)'=(0,0)'}$ and $a_{i,j}^{(2)}(x,y)\big|_{(x,y)'=(0,0)'}$ are functions of $\dfrac{\partial^{i+j} f(x,y)}{\partial x^i \partial y^j}$ $(i \geq 1\ j \geq 1)$, evaluated at $(x, y)' = (0, 0)'$, that is,

$$a_{i,j}^{(1)}(x,y)\bigg|_{(x,y)'=(0,0)'} = g_{i,j}^{(1)}\left(\frac{\partial^2 f^{(1)}(x,y)}{\partial x^2}\bigg|_{(x,y)'=(0,0)'}, \frac{\partial^2 f^{(1)}(x,y)}{\partial x \partial y}\bigg|_{(x,y)'=(0,0)'} \right.$$

$$\left. \frac{\partial^2 f^{(1)}(x,y)}{\partial y^2}\bigg|_{(x,y)'=(0,0)'}, \cdots \right),$$

and

$$
a_{i,j}^{(2)}(x,y)\Big|_{(x,y)'=(0,0)'} = g_{i,j}^{(2)} \left(\frac{\partial^2 f^{(2)}(x,y)}{\partial x^2}\Big|_{(x,y)'=(0,0)'}, \frac{\partial^2 f^{(2)}(x,y)}{\partial x \partial y}\Big|_{(x,y)'=(0,0)'} \right.
$$

$$
\left. \frac{\partial^2 f^{(2)}(x,y)}{\partial y^2}\Big|_{(x,y)'=(0,0)'}, \cdots \right),
$$

with some functions $g_{i,j}^{(1)}(\cdot)$ and $g_{i,j}^{(2)}(\cdot)$.

The expression for $A(x_t, y_t)$ is not unique, whereas the Jacobian $J(x_t, y_t)$ is unique. However, the Jacobian yields an inconsistent approximate model in the sense that

$$
J(x_t, y_t)\begin{pmatrix} x_t \\ y_t \end{pmatrix} \neq \begin{pmatrix} f^{(1)}(x_t, y_t) \\ f^{(2)}(x_t, y_t) \end{pmatrix},
$$

while $A(x_t, y_t)$ is consistent and satisfies the equality

$$
A(x_t, y_t)\begin{pmatrix} x_t \\ y_t \end{pmatrix} = \begin{pmatrix} f^{(1)}(x_t, y_t) \\ f^{(2)}(x_t, y_t) \end{pmatrix}.
$$

Whether we should take the unique but inconsistent Jacobian or the nonunique but consistent $A(x_t, y_t)$ will depend on the situation. We can tell at least which is a more desirable expression of the transition matrix, when two candidates of the transition matrix are presented explicitly. Obviously, the one which gives the best value of the model fitting criterion (i.e., prediction performance) will be chosen in the data analysis.

6.4.4 History of ExpAR Model and RBF-AR Model

The ExpAR model is useful not only in analyzing random oscillatory time series data but also for the characterization of nonlinear dynamics of the process such as stable and unstable singular points and chaos (see Ozaki 1985a). A general order ExpAR(p) model, defined by

$$
x_t = \left\{\phi_{1,0} + \phi_{1,1}\exp(-\gamma x_{t-1}^2)\right\}x_{t-1} + \cdots + \left\{\phi_{p,0} + \phi_{p,1}\exp(-\gamma x_{t-1}^2)\right\}x_{t-p} + w_t,
$$

has been used in many fields for the characterization of nonlinear dynamic structure, such as checking the limit cycles of nuclear reactors (Shi et al. 2001). It turned out, however, that there are many different types of nonlinear dynamic phenomena in real data analysis in applications. ExpAR models are not really suitable for modeling some nonlinear dynamic processes where the amplitude is not the only variable affecting the nonlinear dynamics. Several generalizations have been proposed since (e.g. Kato and Ozaki 2002). Among several generalizations, the following Gaussian type RBF-AR model (Ozaki et al. 1997, Ozaki et al. 1999) is known to have a much stronger capability in prediction and simulation than the ExpAR model:

$$x_t = \phi_0(X_{t-1}) + \sum_{i=1}^{p} \phi_i(X_{t-1})x_{t-i} + w_t$$

$$\phi_i(X_{t-1}) = c_{i,0} + \sum_{k=1}^{m} c_{i,k} \exp\left\{-\gamma_k \left\|X_{t-1} - Z_k\right\|^2\right\}$$

$$X_{t-1} = \left\{x_{t-1}, \Delta x_{t-1}, \Delta^2 x_{t-1}, \ldots, \Delta^{d-1} x_{t-1}\right\}'$$

$$Z_k = \left\{\varsigma_{k,1}, \ldots, \varsigma_{k,d}\right\}'.$$

The instantaneous dynamics of the model is dependent not only on the present amplitude of the series but also its velocity Δx_t and accelerations $\Delta^2 x_t$, ... Therefore, it could produce, for example, asymmetric nonlinear wave patterns in time series, since the model dynamics could be different when the series is increasing ($\Delta x_t > 0$) or decreasing ($\Delta x_t < 0$).

RBF expansions have good interpolation properties in dealing with scattered data points and are endowed with the "universal approximation" and "best approximation" capabilities of any continuous function. Universal approximation implies the possibility of approximating a function to any required degree of accuracy. The stronger property of best approximation entails that the approximation error surface always has a unique global minimum for any approximation performance measure (Park and Sandberg 1991, 1993). Vesin (1993) did not try to optimize the parameter estimation based on the maximum likelihood method or least squares method. Shi et al. (1999) tried to use the maximum likelihood method but had to use a genetic algorithm because the conditions for the optimization of the likelihood were very bad. It turned out that the resulting estimate by the genetic algorithm was one of the local maximum values. Peng et al. (2002) found that this ill-conditioned convergence problem is easily solved by using an iterative method, in which the parameter space is divided into nonlinear parameters and linear parameters.

6.5 Hammerstein Model and RBF-ARX Model

6.5.1 Hammerstein Model

In control engineering, one of the general approaches to modeling a nonlinear input–output system is to use the Hammerstein model family (see Narendra and Gallman 1966). The model is typically written as

$$x_t = c_0 + \sum_{i=1}^{p} a_i x_{t-i} + \sum_{i=0}^{p-1} b_i^{(1)} u_{t-d-i}$$

$$+ \sum_{i=0}^{p-1} b_i^{(2)} u_{t-d-i}^2 + \sum_{i=1}^{p} b_i^{(3)} v_{t-i} + \sum_{i=1}^{p} b_i^{(4)} v_{t-i}^2 + w_t.$$

Here u_t and v_t are exogenous input variables possibly with the delay d, and w_t is a Gaussian white noise. Here the nonlinearity of the system is restricted to the static nonlinearity of the inputs. The model can be written in a state space form as

$$z_t = A z_{t-1} + \xi_{t-d} + \eta_t,$$

$$x_t = (1, 0, \ldots, 0) z_t$$

where

$$A = \begin{pmatrix} a_1 & a_2 & \cdots & a_p \\ 1 & 0 & \cdots & 0 \\ \cdots & \cdots & \cdots & \cdots \\ 0 & 0 & \cdots & 0 \end{pmatrix}, \quad \xi_{t-d} = \begin{pmatrix} b_{t-d} \\ 0 \\ \cdots \\ 0 \end{pmatrix}, \quad \eta_t = \begin{pmatrix} w_t \\ 0 \\ \cdots \\ 0 \end{pmatrix}$$

$$b_{t-d} = \sum_{i=0}^{p-1} b_i^{(1)} u_{t-d-i} + \sum_{i=0}^{p-1} b_i^{(2)} u_{t-d-i}^2 + \sum_{i=1}^{p} b_i^{(3)} v_{t-i} + \sum_{i=1}^{p} b_i^{(4)} v_{t-i}^2.$$

The dynamic part of the system is assumed to be linear so that the system is free from the danger of computational explosion. If the eigenvalues of the transition matrix A

$$A = \begin{pmatrix} a_1 & a_2 & \cdots & a_p \\ 1 & 0 & \cdots & 0 \\ \cdots & \cdots & \cdots & \cdots \\ 0 & 0 & \cdots & 0 \end{pmatrix}$$

are within the unit circle, the system is computationally stable.

6.5.2 RBF-ARX Model

The idea of RBF-AR modeling is naturally extended to the modeling of general nonlinear input–output systems, where the system is driven by an observable exogenous input variable u_{t-i} ($i = 1, 2, \ldots$) as well as an unobserved Gaussian white noise input w_t. The model is called RBF-ARX model and is specified, with $X_{t-1} = (x_{t-1}, \ldots, x_{t-r})'$, and $\varsigma_l = (\varsigma_l^{(0)}, \varsigma_l^{(1)}, \ldots, \varsigma_l^{(r)})'$ as

$$x_t = \phi_0 + \sum_{i=1}^{p} \phi_i(X_{t-1})x_{t-i} + \sum_{i=1}^{q} \theta_i u_{t-i} + w_t$$

where

$$\phi_i(X_{t-1}) = c_{i,0} + \sum_{l}^{m} c_{i,l} \exp\left\{ -\left\| X_{t-1} - \varsigma_l \right\|_{H_l}^2 \right\},$$

$$\left\| X_{t-1} - \varsigma_l \right\|_{H_l}^2 = \sum_{k=1}^{r} \frac{(\Delta^{k-1}x_{t-1} - \varsigma_l^{(k)})^2}{h_l^{(k)}}.$$

The idea of RBF-ARX modeling is in contrast to the Hammerstein modeling. The RBF-ARX modeling tries to characterize the nonlinear system with nonlinear dynamics together with linear input variables, while the Hammerstein modeling approach tries to use (stable) linear dynamics together with (static) nonlinear inputs. Here the most critical difficulty, that is, the stability of the system, is achieved, in the Hammerstein approach, by fixing the linear dynamics to be stable, while in the RBF-ARX modeling approach, the stability of the system is achieved by using the Gaussian radial basis functions, which guarantees the nonlinear AR coefficients of the system to stay bounded even if the state value goes to infinity. However, whether the actual concerned system is modeled better by linear dynamics with nonlinear inputs or by nonlinear dynamics with linear inputs depends on each problem. For those who are interested in comparative studies of the RBF-ARX modeling and the Hammerstein modeling, we refer Peng et al. (2002) and Ozaki et al. (2004).

6.5.3 Multivariate RBF-ARX Models

It is easy and straightforward to introduce a nonlinear generalization of multivariate ARX models. For example, we can generalize the 3-dimensional linear ARX model

$$\begin{pmatrix} x_t \\ y_t \\ z_t \end{pmatrix} = \sum_{k=1}^{3} \begin{pmatrix} \phi_{1,1}^{(k)} & \phi_{1,2}^{(k)} & \phi_{1,3}^{(k)} \\ \phi_{2,1}^{(k)} & \phi_{2,2}^{(k)} & \phi_{2,3}^{(k)} \\ \phi_{3,1}^{(k)} & \phi_{3,2}^{(k)} & \phi_{3,3}^{(k)} \end{pmatrix} \begin{pmatrix} x_{t-k} \\ y_{t-k} \\ z_{t-k} \end{pmatrix} + \sum_{k=1}^{2} \begin{pmatrix} \psi_{1,1}^{(k)} & \psi_{1,2}^{(k)} \\ \psi_{2,1}^{(k)} & \psi_{2,2}^{(k)} \\ \psi_{3,1}^{(k)} & \psi_{3,2}^{(k)} \end{pmatrix} \begin{pmatrix} u_{t-k} \\ v_{t-k} \end{pmatrix} + \begin{pmatrix} \xi_t \\ \eta_t \\ \zeta_t \end{pmatrix}$$

into the following 3-dimensional RBF-ARX model:

$$
\begin{pmatrix} x_t \\ y_t \\ z_t \end{pmatrix} = \sum_{k=1}^{3} \begin{pmatrix} \phi_{1,1}^{(k)}(RBF) & \phi_{1,2}^{(k)}(RBF) & \phi_{1,3}^{(k)}(RBF) \\ \phi_{2,1}^{(k)}(RBF) & \phi_{2,2}^{(k)}(RBF) & \phi_{2,3}^{(k)}(RBF) \\ \phi_{3,1}^{(k)}(RBF) & \phi_{3,2}^{(k)}(RBF) & \phi_{3,3}^{(k)}(RBF) \end{pmatrix} \begin{pmatrix} x_{t-k} \\ y_{t-k} \\ z_{t-k} \end{pmatrix}
$$

$$
+ \sum_{k=1}^{2} \begin{pmatrix} \psi_{1,1}^{(k)} & \psi_{1,2}^{(k)} \\ \psi_{2,1}^{(k)} & \psi_{2,2}^{(k)} \\ \psi_{3,1}^{(k)} & \psi_{3,2}^{(k)} \end{pmatrix} \begin{pmatrix} u_{t-k} \\ v_{t-k} \end{pmatrix} + \begin{pmatrix} \xi_t \\ \eta_t \\ \varsigma_t \end{pmatrix},
$$

where $\phi_{i,j}^{(k)}(RBF)$'s are given, for example, by

$$
\phi_{i,j}^{(k)}(RBF) = \phi_{i,j}^{(k)} + \sum_{m=1}^{M} \pi_{i,j,m}^{(k)} Exp \left\{ - \left\| \frac{R_{t-1} - c_m}{\Delta R_{t-1}} \right\|_Q^2 \Big/ h_m \right\}.
$$

R_{t-1} is a variable that is known to cause the change in the characteristics of the system dynamics, which could be one of the state variables, $x_{t-1}, y_{t-1}, z_{t-1}$, or one of the exogenous variables, u_{t-1}, v_{t-1}. The noise w_t is a 3-dimensional Gaussian white noise with a possibly nondiagonal variance matrix Σ_w.

It is always easier to discuss than to actually do the identification of the model from the actual time series data. The nonlinear generalization of multivariate AR models, in general, yields a model with too many parameters and also too many possible combinations of nonlinear radial basis functions. The computational problems of multivariate RBF-ARX models are discussed in Section 10.3.1.4.

6.6 Discussion on Nonlinear Predictors

In most nonlinear AR modeling, the parameters are estimated by least squares, which suggests minimizing the sum of squares of the prediction errors. The sum of squares of errors is employed as a criterion for model estimation because implicitly it is assumed that the error distribution is close to being Gaussian. Thus, the idea behind the procedure becomes the same as the maximum likelihood method for nonlinear AR modeling with Gaussian white noise w_t when $f(x_{t-1}, \ldots, x_{t-d})$ is specified parametrically with parameter vector θ, that is,

$$
x_t = f(x_{t-1}, \ldots, x_{t-d} \mid \theta) + w_t, \tag{6.12}
$$

where the nonlinear AR part specifies the conditional mean, and the variance of the noise w_t yields the conditional variance of the predictive distribution, $p(x_t | x_{t-1}, \ldots, x_{t-d})$. It is known, however, that even if the process is non-Gaussian, the best predictor $x_{t|t-1}$ in the sense of minimizing $E[x_t - x_{t|t-1}]^2$ is still given by the conditional mean, $x_{t|t-1} = E[x_t | x_{t-1}, x_{t-2}, \ldots]$. This is not trivial, although it may seem so to time series analysts (see, e.g., Kailath 1981 for a proof). We know that for Gaussian processes the conditional mean $x_{t|t-1}$ is given by a linear predictor. However, in the non-Gaussian case, there is no reason why a nonlinear predictor might not give a smaller least squares than the best linear predictor. Numerical experience with non-Gaussian processes tends to confirm the superiority (in the least squares sense) of nonlinear predictors over the best linear predictor.

The conditional mean and the conditional variance may be written mathematically using the conditional density function as

$$x_{t|t-1} = E\left[x_t | x_{t-1}, x_{t-2}, \ldots\right] = \int \xi_t p(\xi_t | x_{t-1}, \ldots, x_{t-d}) d\xi_t$$

and

$$\sigma^2_{t|t-1} = E[x_t - x_{t|t-1}]^2$$

$$= \int (\xi_t - x_{t|t-1})^2 p(\xi_t | x_{t-1}, \ldots, x_{t-d}) d\xi_t.$$

Masani and Wiener's (1959) mathematical work shows that $x_{t|t-1} = E[x_t | x_{t-1}, x_{t-2}, \ldots]$ may be approximated more and more closely with increasing numbers of higher order moments for a non-Gaussian process. Wiener (1958) suggested using a linear combination of nonlinear functions of past observations for the prediction of non-Gaussian processes. In general, it is necessary to use an infinite number of moments to characterize a non-Gaussian process. Hence, an infinitely large number of moments will also generally be needed to give the best predictor. There are non-Gaussian processes that can be characterized (at least approximately) with only a few parameters in a time domain model. However, in practice, explosion problems associated with polynomial models turned out to be serious for a truncated finite parameter model. For these reasons, Wiener's general approach with high order moments and high order polynomial models died away in the 1960s.

Nonlinear AR modeling in the 1970s, such as ExpAR modeling, aimed for parsimonious parameterization and for stability. In nonlinear AR modeling, the following two problems are critical for the model validity in practice:

1. Efficient parameterization
2. Computational stability

The number of possible nonlinear terms is huge in nonlinear AR modeling. Since there is no standard way to set an order for model complexity, modeling must proceed on a "case-by-case" basis. In simulating the estimated nonlinear AR model, computational stability is a well-known cumbersome problem. Estimated nonlinear AR models are often explosive in simulations, casting doubt on their validity in practical applications. Nonlinear AR models such as ExpAR models (Ozaki and Oda 1978, 1981a, Haggan and Ozaki 1981, Ozaki 1985a) or the nonlinear threshold AR model (Ozaki 1981a, 1985a) tried to tame computational instability by considering the dynamics of the instantaneous eigenvalues yielding coefficients of the AR models that were amplitude dependent. The use of the sigmoid function, by the neural network modeling in 1980s, was an ingenious way to perform the same function of taming the explosiveness of the discrete-time nonlinear dynamics. Neural network modeling avoids the problem of over-parameterization by shifting the stance from statistical parametric modeling to an algorithmic modeling problem.

Suppose we have a computationally stable nonlinear AR model (6.12) with zero mean white noise w_t and are able to find an estimate of the conditional mean $x_{t|t-1}$ given by

$$\hat{x}_{t|t-1} = f(x_{t-1}, \ldots, x_{t-d} \mid \hat{\theta})$$

with an estimate of the error variance $\sigma^2_{t|t-1}$ given by the estimated variance $\hat{\sigma}^2_w$ of the noise w_t. There could be many candidate nonlinear models f giving different unbiased estimates $\hat{x}_{t|t-1}$ of the conditional mean and different estimates $\hat{\sigma}^2_w$ of the error variance. The question of which model provides a better estimate is related to the statistical assessment of the identified model.

Finding an appropriate functional form $f(.)$ out of a huge number of possible candidates is very important but obviously extremely difficult. Nonlinear AR modeling is closely related to the function approximation problem, where an efficient parsimonious approximation is as important as the consistency. Some type of parameterization, such as a linear threshold AR model, seems to be popular among some statisticians' circle. However, there is not any mathematical reason why one type of approximation (e.g., a step function approximation) is always superior to another type of approximation (e.g., Gaussian basis function approximation). It depends on the problem that the model is aimed at and applied to. We discuss this problem again in Chapters 9 and 10, Part 2.

In the study of nonlinear predictors, either in neural network modeling or nonlinear AR prediction, there is little explanation of why $E[.]^2$ is used as a criterion for a good predictor. Most nonlinear predictors, such as the multilayer neural network, or RBF function approximations, $E[x(t) - x_{t|t-1}]^2$ is used as the objective function to minimize. However, from the previous discussion, it is clear that the use of the quadratic criterion $E[.]^2$ is closely connected to the concept of conditional densities and the conditional mean as the best predictor. Whether $E[.]^2$ is an appropriate criterion to be used, for example, when we have a multimodally distributed prediction error seems to be beyond the concern of most people. Another point is that some time series analysts also assess an m-step ahead prediction error using the $E[.]^2$ criterion. However when system dynamics are nonlinear, the several-step ahead predictive distribution could be asymmetric or multimodal even though the one-step ahead prediction error is exactly Gaussian (Ozaki 1993b).

Some Bayesian methods (see, e.g., Sage and Melsa 1971) consider a maximum *a posteriori* filter which takes the mode of the conditional density $p(x_t|z_\tau)$ as an estimate of the state. For the $E[.]^2$ criterion, a Bayes' modal predictor is not necessarily the best, since it does not coincide with the conditional mean unless the conditional density is unimodal and symmetric. If the criterion $E[.]^2$ is not used, however, the conditional mean no longer plays a central role. For a non-$E[.]^2$ criterion, a modal criterion may make more sense than the conditional mean. Therefore, a non-$E[.]^2$ criterion may be more appropriate, for example, for the assessment of m-step ahead predictors.

6.7 Heteroscedastic Time Series Models

6.7.1 Temporally Inhomogeneous Prediction Errors and the Innovation Approach

The innovation approach is a guideline principle for us in finding a suitable dynamic model for time series from an observed time series. It suggests finding a dynamic model that yields a prediction model, for example, $x_t = f(x_{t-1})$, with as much small prediction error variance as possible, so that the temporally dependent time series, $x_1, x_2, \ldots, x_N$, is transformed into a white noise sequence, $w_1, w_2, \ldots, w_N$. The Least Squares Method is one of the computational methods for obtaining such a dynamic model by minimizing the sum of squares of prediction errors, calculated by the model $f(.)$ as

$$w_t = x_t - \underset{f(.)}{E}\left[x_t|x_{t-1}, \ldots\right], \quad \text{for } t = 1, 2, \ldots, N.$$

An implicit assumption behind the Least Squares Method is the homoscedasticity of the prediction error, that is, the time series data, is whitened into temporally homogeneous Gaussian white noise w_t, that is, $w_t \sim N(0, \sigma^2)$ ($t = 1, \ldots, N$). Minimization of the sum of squares of the errors is justified only when the homoscedasticity is guaranteed.

What should we do if we can produce only heteroscedastic prediction errors with the dynamic models in our hand? Here the time series data may be whitened, with a model, into a Gaussian white noise, but the prediction error variance is not necessarily time homogeneous, that is, $w_t \sim N(0, \sigma_t^2)$ and $\sigma_t^2 \neq \sigma_{t'}^2$ for $t \neq t'$.

In such cases there are two reasons (one is practical and another is methodological) why heteroscedastic modeling should be considered instead of homoscedastic models, such as AR models and ARMA models:

1. Often the heteroscedasticity is a useful index for the characterization of the volatile phenomena, and characterization of the time-varying error variance from the data, in real time, yields useful information for the analysts.

2. The use of a crude estimate of the time-varying prediction error variance, instead of the estimated variance of a homoscedastic model, improves the likelihood, even when the prediction errors of a new model are equivalent to the errors of the homoscedastic model, so that the sum of squares of the errors does not make any difference.

To understand the meaning of the heteroscedastic modeling, let us compare the following two nonlinear AR models with some parameter ϕ:

$$x_t = f(x_{t-1}|\phi) + \sigma \varepsilon_t \tag{6.13}$$

$$x_t = f(x_{t-1}|\phi) + \sigma_t \varepsilon_t \tag{6.14}$$

where

$$\sigma_t^2 = \frac{1}{q}\left\{ \left(x_{t-1} - x_{t-1|t-2} \right)^2 + \cdots + \left(x_{t-q} - x_{t-q|t-q-1} \right)^2 \right\}. \tag{6.15}$$

Here ε_t is a white noise of variance 1, and the predictor $x_{t-j|t-j-1} = E[x_{t-j}|x_{t-j-1}]$ is given by $x_{t-j|t-j-1} = f(x_{t-j-1}|\phi)$.

When the noise variance of the time series data is changing slowly in time, the data-adaptive moving average of the past prediction errors given by (6.15) should be a much more suitable estimate for the time-varying error variance of the process than the constant error variance σ^2 of the model (6.13). This is actually confirmed in many heteroscedastic time series in financial engineering, where several parametric heteroscedastic models have been introduced for the financial risk analysis.

What we actually need to do to use the MLE method in data analysis for heteroscedastic time series is to specify (i.e., predict) the time-varying prediction error variance, $\sigma_{t|t-1}^2$, together with the prediction error, $w_t = x_t - E[x_t|x_{t-1}, \ldots]$, by some parametric model, at each time point. For the maximum likelihood estimation of the parameters of the heteroscedastic models, we can still take advantage of the Gaussian probability density representation of the innovations as

$$(-2)\log p(x_1, x_2, \ldots, x_N) = (-2)\log p(w_1, w_2, \ldots, w_N)$$

$$= \sum_{1}^{N}\left(\log \sigma_t^2 + \frac{w_t^2}{\sigma_t^2} \right) + N \log 2\pi, \qquad (6.16)$$

and the model parameters can be estimated by minimizing (6.16) as long as we can specify the time varying error variance, σ_t^2, by a parametric model at each time point.

6.7.2 ARCH Model and GARCH Model

In financial engineering, several parametric heteroscedastic models have been introduced. These models are useful not only in financial engineering but also in other areas where the dynamics of the phenomena are not precisely known and the prediction errors are temporally inhomogeneous. Typical examples of heteroscedastic time series models are the ARCH(p,q) model,

$$x_t = \sigma_t \varepsilon_t$$

$$\sigma_t^2 = \alpha_0 + \beta_1 x_{t-1}^2 + \cdots + \beta_q x_{t-q}^2, \qquad (6.17)$$

and the GARCH(p,q) model,

$$x_t = \sigma_t \varepsilon_t$$

$$\sigma_t^2 = \alpha_0 + \sum_{j=1}^{p} \alpha_j \sigma_{t-j}^2 + \sum_{i=1}^{q} \beta_i x_{t-i}^2. \qquad (6.18)$$

In both models, (6.17) and (6.18), the first equation has a simple form $x_t = \sigma_t \varepsilon_t$, that is, x_t is a white noise, but its variance is changing in time. This comes from the special feature of financial engineering, where most interesting time series, such as the return of stock indices or the return of the exchange rate, are more or less white. For heteroscedastic modeling other than the financial time series, this part could be replaced by some predictor such as $f(x_{t-1}|\phi)$, and the terms, $x_{t-1}^2, \ldots, x_{t-q}^2$, could be replaced by $(x_{t-1} - x_{t-1|t-2})^2, \ldots, (x_{t-q} - x_{t-q|t-q-1})^2$ as is in (6.15).

The ARCH model is regarded, in a sense, as a generalization of a simple moving average model of (6.15) type, where $f(x_{t-1}|\phi) = 0$ and the constant coefficient $(1/q)$, of x_{t-i}^2 ($i = 1, \ldots, q$) are replaced by an independent parameter

β_i ($i = 1, ..., q$), with a constant term α_0 added. Here the prediction error variance σ_t^2 is updated using the weighted average of the volatility information available from the past observed data.

The GARCH model comes out from the ARCH model with an idea of parsimonious parameterization. When the order q of the ARCH(q) model is high, the GARCH type parameterization is more parsimonious and efficient. This is similar to the situation where the ARMA model family is introduced as a parsimonious generalization of the AR or MA model family.

The idea of the characterization of time varying prediction errors is quite universal and the ARCH and GARCH are easily extended to more general situations (which we may see more often outside the financial applications) where not only the noise variance but also the process itself has some nontrivial dynamics. For example, we can think of a nonlinear AR(1) model with a heteroscedastic noise structure as

$$x_t = f(x_{t-1}|\phi) + \sigma_t \varepsilon_t$$
$$\sigma_t^2 = \sigma_0 + \alpha\sigma_{t-1}^2 + \beta\tilde{x}_{t-1|t-2}^2, \tag{6.19}$$

where $\tilde{x}_{t-1|t-2}^2 = (x_{t-1} - x_{t-1|t-2})^2$ and $x_{t-1|t-2} = E[x_{t-1}|x_{t-2}, ...] = f(x_{t-2}|\phi)$. Here we try to specify, with the model (6.19), the predictor $x_{t|t-1}$ of the series x_t as well as its prediction error variance $\tilde{x}_{t|t-1}^2 = E[(x_t - x_{t|t-1})^2]$ simultaneously.

Another essential point of volatility modeling is that σ_t^2 is supposed to be non-negative. The process of σ_t^2 defined by the estimated parameters of ARCH or GARCH may not necessarily be positive valued. In order to make sure that σ_t^2 always stays positive, several alternative parameterizations are available (see, e.g., Tsay 2002). One of the most useful models may be the GARCH model, where the dynamics of the error variance is specified with a linear dynamics in a log-transformed space, for example,

$$x_t = f(x_{t-1}|\phi) + \sigma_t \varepsilon_t$$
$$\log \sigma_t^2 = \alpha + \beta \log \sigma_{t-1}^2 + \gamma \log \tilde{x}_{t-1|t-2}^2.$$

So far we have considered only discrete-time heteroscedastic models and have not considered continuous-time heteroscedastic models. In fact these discrete-time heteroscedastic models were introduced as crude (but simple and easy-to-use) approximations to the original continuous-time heteroscedastic model, such as

$$dx(t) = \{a + bx(t)\}dt + \sigma(t)dW^{(1)}(t)$$
$$d\log \sigma^2(t) = \{\alpha + \beta \log \sigma^2(t)\}dt + \gamma dW^{(2)}(t). \tag{6.20}$$

The model (6.20) is called the stochastic volatility model. In practice the variable $x(t)$ is observed but the variable $\sigma(t)$ (which is called volatility) cannot be observed. Because of this, people have thought (and many people may be still thinking) that estimation of $\sigma(t)$ and identification of the model (6.20) from the time series $x_1, x_2, ..., x_N$, are impossible. Then people started bringing a method of estimating $\sigma^2(t-1)$ by using information coming from the squares of the past prediction errors of x_τ ($\tau = t - 1, t - 2, ...$) such as (6.15). This is partly because Kalman filter techniques (which we see in Chapters 11 through 13) for estimating continuous-time dynamic models were not well known in the financial engineering community. In fact $\sigma(t)$ and the model parameters, a, b, α, β and γ, can be easily estimated (see Jimenez et al. 2006) if we use the methods in Chapters 11 through 13. However, we do not discuss this model any further in this book since it requires rather heavy computation. Instead, we consider a state space generalization of the discrete-time heteroscedastic models in Chapter 13.

Conventional heteroscedastic models found in financial engineering applications are only those models where the dimension of the prediction error and the dimension of the state variable are equal. In neuroscience modeling and data analysis, this is a rather strong assumption and because of this assumption, the heteroscedastic models have strong limitations in their applicability outside financial engineering. It will be useful if the similar idea of ordinary heteroscedastic modeling is generalized to the state space situations, where the volatility of state variables can be estimated from the lower (than the dimension of the state variable) dimensional observed time series. This could be especially useful in neuroscience time series analysis, because the dynamics in neuroscience are quite complex, and observed time series are related to many other variables, some of which are not observable as exogenous variables. We see (in Section 13.6) that this idea of state space generalization of the GARCH modeling is possible and can be actually implemented. It turns out to work quite well, for solving dynamic EEG inverse problems, and for the characterization of the time-varying rhythm components in EEG signals. For the readers who are interested in the neuroscience applications of the present methods, we refer to Galka et al. (2004) and Wong et al. (2006).

Part II

Related Theories and Tools

So far we have seen in Chapters 3 through 6, that there are many types of dynamic models for prediction. These models have been developed and used not only in neuroscience but also in many scientific areas such as physics, electrical engineering, chemistry, biology, and economics. These models are classified into two groups: continuous-time models and discrete-time models. Both groups share similarities as a dynamic model if we look at them from two different viewpoints: one is the predictional viewpoint and another is the distributional viewpoint. In the present chapter we see how the viewpoint of prediction errors is useful for the classification of the nature of time series. The distributional viewpoint of time series is discussed in Chapter 8, where the close relations between the three, that is, (1) Pearson system of distributions, (2) stochastic differential equation model, and (3) Markov diffusion process, are discussed. In Chapter 9, we introduce a bridge connecting the field of stochastic differential equation model and the field of nonlinear time series models using the local linearization scheme, which transforms a continuous-time finite-variance process model to a discrete-time finite-variance process model. The problem of statistical identifications and estimations of the model in Part I are discussed in Chapter 10.

7

Prediction and Doob Decomposition

7.1 Looking at the Time Series from Prediction Errors

It is often useful and helpful in clarifying the nature of time series if we look at the time series as a sum of two components, that is, the predictor (predictable component) and the prediction error (unpredictable component). The Doob decomposition theorem, which we see in the following, provides us with a simple and useful base for our innovation approach to time series analysis.

Theorem (Doob Decomposition):

For any process x_t such that $E[|x_t|] < \infty$ for all $t = 0, 1, 2, \ldots$, there exists a decomposition

$$x_t = y_t + U_t \quad t = 0, 1, 2, \ldots,$$

where y_t is the predictable part of the time series, that is, $f(x_{t-1}, x_{t-2}, \ldots)$ of the past $x_{t-1}, x_{t-2}, \ldots$. U_t is the unpredictable part of the time series, that is, a process with the martingale property:

$$E[U_t | U_{t-1}, U_{t-2}, \ldots, U_0] = U_{t-1}.$$

For the prediction of a given time series data, so many dynamic models are possible and available for use. If the data are real-world data, such as experimental neuroscience or clinical data, we need to choose an appropriate model between them. The prediction errors of each dynamic model provide us with a natural guideline for the model selection (Kailath 1968, 1969), where people may simply choose the model attaining the smallest prediction error variance. Which model to choose is an important problem, and we will discuss this problem separately in Chapter 10.

7.2 Innovations and Doob Decompositions

The Doob decomposition is closely related to one of the most important concepts in the study of dynamical systems, that is, innovations (or equivalently prediction errors).

Let x_t, $t = 0, 1, \ldots$, be a process (not necessarily Gaussian) such that

$$E[\,|x_t|\,] < \infty \quad \text{for all } t = 0, 1, \ldots.$$

The "best" mean square prediction of $\Delta x_t = x_t - x_{t-1}$ is given by the conditional expectation,

$$E[\Delta x_t | x_{t-1}, x_{t-2}, \ldots, x_0].$$

This is because we have, from the definition of the conditional expectation,

$$E\left[\left\{ \Delta x_t - E[\Delta x_t | x_{t-1}, x_{t-2}, \ldots, x_0] \right\}^2 \right] \le E\left[\left\{ \Delta x_t - g(x_{t-1}, x_{t-2}, \ldots, x_0) \right\}^2 \right],$$

$$t = 0, 1, 2, \ldots$$

for any function $g(.)$ of $x_{t-1}, x_{t-2}, \ldots, x_0$.

Let us denote the prediction of Δx_t by $\Delta y_t = E[\Delta x_t | x_{t-1}, x_{t-2}, \ldots, x_0]$, $t = 0, 1, 2, \ldots$
Then, we have

$$\Delta x_t - \Delta y_t = x_t - E\left[x_t | x_{t-1}, x_{t-2}, \ldots, x_0 \right].$$

Let us denote $u_{x,t} = \Delta x_t - \Delta y_t$; then $u_{x,t}$ could be interpreted as the unpredictable part of the increment Δx_t. Here, $u_{x,t}$ is called innovation process. Using the predictable part Δy_t and the unpredictable innovation $u_{x,t}$, Δx_t can be expressed as

$$\Delta x_t = \Delta y_t + u_{x,t}.$$

Then, we have

$$x_t - x_0 = \sum_{k=1}^{t} \Delta y_k + \sum_{k=1}^{t} u_{x,k}.$$

We define new processes y_t and U_t by

$$y_0 = 0$$

$$y_t = \sum_{k=1}^{t} \Delta y_k$$

and

$$U_0 = x_0,$$

$$U_t = \sum_{k=1}^{t} u_{x,k} \quad t = 1, 2, \ldots$$

Then, we obtain the discrete-time version of Doob decomposition

$$x_t = y_t + U_t, \quad t = 0, 1, \ldots. \tag{7.1}$$

Note that y_t is a deterministic function of $x_{t-1}, x_{t-2}, \ldots, x_0$. Note also that we have the martingale property,

$$E[U_t | U_{t-1}, U_{t-2}, \ldots, U_0] = U_{t-1}.$$

Here, we have $u_{x,t} = U_t - U_{t-1}$ and $u_{x,t} = \Delta x_t - \Delta y_t = x_t - E[x_t | x_{t-1}, \ldots, x_0]$, so that $E[u_{x,t} | x_{t-1}, \ldots, x_0] = 0$.

7.2.1 Causal Relations and Causally Invertible Relations

We say the process x_t and the process z_t are related by causal and causally invertible transformations if they satisfy the relations with S and T such that

$$z_t = S(x_0, \ldots, x_{t-1}) + x_t,$$

$$x_t = T(z_0, \ldots, z_{t-1}) + z_t.$$

The processes $(x_0, x_1, \ldots, x_t)$ and $(z_0, z_1, \ldots, z_t)$ are essentially equivalent because they are obtained from the other process by the causal and causally inverse transformations S and T.

The importance of causal and causally invertible relations (between two processes) is understood with the innovation processes using the Doob decomposition. Let x_t, $t = 0, 1, \ldots$, be a discrete-time process (not necessarily Gaussian) such that

$$E[|x_t|] < \infty \quad \text{for all } t = 0, 1, \ldots,$$

we have a Doob decomposition

$$x_t = x_0 + y_t + U_t$$

$$= x_0 + \sum_{k=1}^{t} \Delta y_k + \sum_{k=1}^{t} u_{x,k}.$$

Here, $\Delta y_t = E[\Delta x_t | x_{t-1}, x_{t-2}, \ldots, x_0] = E[x_t | x_{t-1}, x_{t-2}, \ldots, x_0] - x_{t-1}$. Then the innovation process, $u_{x,t}, t = 0, 1, \ldots,$ is defined by

$$u_{x,0} = x_0$$

$$u_{x,t+1} = \Delta x_t - E[\Delta x_t | x_0, \ldots, x_t]$$

$$= x_{t+1} - E[x_{t+1} | x_0, \ldots, x_t].$$

This is a mean square unpredictable process, that is,

$$E[u_{x,t} | u_{x,0}, \ldots, u_{x,t-1}] = 0, \quad t = 1, 2, \ldots$$

and is related to the original process x_t by a causal and causally invertible (nonlinear in general) transformation, in symbols

$$u_{x,t} = G(x_0, \ldots, x_{t-1}) + x_t,$$

and

$$x_t = H(u_{x,0}, \ldots, u_{x,t-1}) + u_{x,t}.$$

7.2.2 Input–Output System Representations

It is often useful to regard a stationary time series as an output of a system driven by an innovation process. The Doob decomposition theorem allows us to consider an input–output system model for a stationary time series. The input–output representation of a process is derived from the causal and causally invertible relations. Let $x_t, t = 0, 1, \ldots$ be a finite-variance non-Gaussian process. From the earlier discussion, we have causal and causally invertible transformations $G(x_0, \ldots, x_{t-1})$ and $H(u_{x,0}, \ldots, u_{x,t-1})$ such that $u_{x,t} = G(x_0, \ldots, x_{t-1}) + x_t$, and $x_t = H(u_{x,0}, \ldots, u_{x,t-1}) + u_{x,t}$. If x_t is a Gaussian process and $E[x_t | x_{t-1}, \ldots, x_0]$ is a linear function of $x_{t-1}, \ldots, x_0$, then x_t and $u_{x,t}$ have "linear" causal and "linear" causally invertible relations:

$$u_{x,t} = \sum_{k=0}^{t} a_{t,k} x_k \tag{7.2}$$

and

$$x_t = \sum_{k=0}^{t} b_{t,k} u_{x,k}, \tag{7.3}$$

where $a_{t,k}$ and $b_{t,k}$ are constants, and $a_{t,t} = b_{t,t} = 1$. Here we say, "The process x_t and $u_{x,t}$ are linear causally equivalent." Note that (7.2) shows the whitening of time series x_t and (7.3) gives a representation of time series as a linear function of the past innovations.

If x_t is not only Gaussian but also time invariant, that is, for $t - k = t' - k'$, $a_{t,k} = a_{t',k'} = a_{t-k}$ and $b_{t,k} = b_{t',k'} = b_{t-k}$, then we have

$$u_{x,t} = \sum_{k=0}^{t} a_{t-k} x_k, \tag{7.4}$$

$$x_t = \sum_{k=0}^{t} b_{t-k} u_{x,k}. \tag{7.5}$$

Typical examples of (7.4) and (7.5) are those relations between the linear stationary AR(p) model, the MA(q) model, and the ARMA(p, q) model and their general linear process representations, which we saw in Chapter 3. These linear stationary (finite-variance) Gaussian processes can be regarded as outputs of input–output systems driven by innovations $w_t(=u_{x,t})$ (see Figure 7.1).

7.2.3 Markov Processes and Innovation Representations

If a finite-variance sample-continuous process is Markov, using the Doob decomposition, the process can be approximated by an output X_{t_n} of a discrete-time input–output representation model

$$X_{t_k} = X_{t_{k-1}} + a(X_{t_{k-1}})\Delta t_k + b(X_{t_{k-1}})w_{t_k},$$

where the input w_{t_k} is a Gaussian white noise and $\Delta t_k = t_k - t_{k-1} (=\Delta t)$.

From the Doob decomposition theorem, any finite-variance process X_t is decomposed into

$$X_t = X_0 + Y_t + U_t, \tag{7.6}$$

where $Y_t = \sum_{k=1}^{t} \Delta Y_k$, that is, a predictable part of X_t, and $U_t = \sum_{k=1}^{t} \Delta U_k$, that is, an unpredictable martingale with $U_0 = 0$. Since we can expect the predictable part of the process to be smooth, we have, for small $\Delta t_i = t_i - t_{i-1}$ ($i = 1, \ldots, k$), $\Delta Y_{t_k} = Y_{t_k} - Y_{t_{k-1}} = A_{t_{k-1}}\Delta t_k$, where $A_{t_{k-1}}$ is a

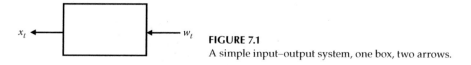

FIGURE 7.1
A simple input–output system, one box, two arrows.

function of the past $X_{t'}$ ($0 \leq t' \leq t_{k-1}$) written as $A_{t_{k-1}} = a(X_{t_0}, X_{t_1}, \ldots, X_{t_{k-1}})$. Then, we have $Y_{t_k} = Y_{t_{k-1}} + a(X_{t_0}, X_{t_1}, \ldots, X_{t_{k-1}})\Delta t_k$. For most physical systems, it is not unreasonable to assume that the average energy increment $E[dU_t^2 = u_t^2 | \cdots]$ or $E[\Delta U_t^2 = u_t^2 | \cdots]$ will either be constant or at most a smoothly varying function of past states. Then, we have, for $\Delta t_k = \Delta t$ ($k = 1$, 2, 3, ...).

$$E[\Delta U_{t_k}^2] \doteq B_{t_{k-1}}^2 \Delta t_k = b^2(X_{t_0}, X_{t_1}, \ldots, X_{t_{k-1}})\Delta t$$

This means we have $\Delta U_{t_k} \approx b(X_{t_0}, X_{t_1}, \ldots, X_{t_{k-1}})w_{t_k}$. Let $w_{t_k} \doteq \Delta U_{t_k}/b(X_{t_0}, X_{t_1}, \ldots, X_{t_{k-1}})$ be the redefined sample continuous martingale difference with $E[w_{t_k}^2] = \Delta t$. Then we have, from the Doob decomposition (7.6), a discrete-time approximate representation:

$$X_{t_n} = X_0 + \sum_{k=1}^{n} a(X_{t_0}, \ldots, X_{t_{k-1}})\Delta t + \sum_{k=1}^{n} b(X_{t_0}, \ldots, X_{t_{k-1}})w_k,$$

where w_{t_k} is a redefined sample continuous martingale difference with $E[w_{t_k}^2] = \Delta t$. If the process is Markov, the predictable process no longer depends on the entire past but only on the current state $X_{t_{k-1}}$. Then we have $a(X_{t_1}, X_{t_2}, \ldots, X_{t_{k-1}}) = a(X_{t_{k-1}})$ and $b(X_{t_1}, X_{t_2}, \ldots, X_{t_{k-1}}) = b(X_{t_{k-1}})$. Here the decomposition becomes

$$X_{t_n} = X_0 + \sum_{k=1}^{n} a(X_{t_{k-1}})\Delta t + \sum_{k=1}^{n} b(X_{t_{k-1}})w_{t_k}$$

or

$$X_{t_n} = X_{t_{n-1}} + a(X_{t_{n-1}})\Delta t_n + b(X_{t_{n-1}})w_{t_n}.$$

This shows that if we look at the time series sampled from a continuous-time Markov process, a model family of nonlinear AR-type models with Gaussian white noise with state-dependent variance, that is, of type $X_t = \phi(X_{t-1}) + \theta(X_{t-1})w_t$, is a reasonable approximate model family for the process. Examples of the nonlinear AR models that fall in this model family are those such as polynomial AR models, threshold AR models, exponential AR models, and RBF-AR models.

Remember that the Doob decomposition allows us to regard the finite-variance Markov process x_t as an output of an input–output system driven by the innovations $\Delta U_x t = u_x t$ which has a form

$$x_t = H(u_{x,0}, \ldots, u_{x,t-1}) + u_{x,t}.$$

We will see in Section 7.3 that the innovation is Gaussian white noise if the finite-variance (possibly non-Gaussian) process x_t is sample continuous.

7.2.4 Initial-Value Effect and Wold Decomposition

The initial-value problem of time series is not a problem only for deterministic chaos. It has also been studied in the classic Gaussian time series analysis. Let x_t, $t = 0, 1, 2, \ldots$ be the zero–mean Gaussian process. In the innovation representation of x_t,

$$x_t = \sum_{k=0}^{t} b_{t,k} u_{x,k},$$

the innovation at the starting point, $u_{x,0}$, was in fact specified by $u_{x,0} = x_0$. This means that the current value x_t of the process is decomposed into two components (called a singular process),

$$x_{t,s} = b_{t,0} x_0 \quad t = 0, 1, 2, \ldots,$$

which can be predicted with certainty from its initial value x_0 and an unpredictable part of the process (called a regular process),

$$x_{t,r} = \sum_{k=1}^{t} b_{t,k} u_{x,k} \quad t = 1, 2, \ldots$$

Thus, after time $t = 0$, this Gaussian process x_t has a decomposition

$$x_t = x_{t,s} + x_{t,r}$$

which is called Wold decomposition. Here the two component processes are both Gaussian and mutually independent.

If x_t, $t = \ldots, -1, 0, 1, 2, \ldots$ is a stationary Gaussian process from the infinitely remote past, the innovation is defined by

$$u_{x,t} = x_t - E\left[x_t \mid x_{t'}, t' < t \right]$$

instead of

$$u_{x,t} = x_t - E[x_t \mid x_0, \ldots, x_{t-1}].$$

Here, the singular process component $x_{t,s}$ should be interpreted as the best predictor of x_t based on the infinitely remote past of the process. Then, the Wold decomposition is given by the redefined singular process component and regular process component as

$$x_t = x_{t,s} + x_{t,r} \quad t = \ldots, -1, 0, 1, 2, \ldots,$$

where

$$x_{t,s} = \lim_{t_0 \to -\infty} E[x_t | x_{t'}, t' < t_0] \quad t = \ldots, -1, 0, 1, 2, \ldots$$

$$x_{t,r} = \sum_{k=-\infty}^{t} b_{t-k} u_{x,k} \quad t = \ldots, -1, 0, 1, 2, \ldots$$

Thus, we have the following Wold decomposition theorem (Wold 1938):

Theorem (Wold decomposition):

Any zero-mean covariance-stationary processes x_t can be written in the form

$$x_t = x_{t,s} + x_{t,r}, \quad t = 0, 1, \ldots,$$

where

$$x_{t,s} = \lim_{t_0 \to -\infty} E[x_t | x_{t'}, t' < t_0]$$

$$x_{t,r} = x_t - x_{t,s}.$$

The Wold decomposition theorem, sometimes called Wold–Cramer decomposition, was generalized to the multivariate cases as well as the continuous-time cases by Hanner (1949), Karhunen (1949), and Cramer (1960).

7.3 Innovations and Doob Decomposition in Continuous Time

We can also have a continuous-time version of the Doob Decomposition theorem. It provides us with useful ideas and tools for characterizing and modeling the process and time series. The Theorem is sometimes called Doob–Mayer–Fisk theorem with the names of those who contributed to the theorem and is summarized as follows:

Theorem (Doob–Meyer–Fisk):

Let $X(t)$ be a sample-continuous finite-variance process. Then, we have the following decomposition:

$$X(t) - X(0) = Y(t) + U(t), \quad t \geq 0,$$

where
  $Y(t)$ is a sample-continuous finite-variance process with respect to t
  $U(t)$ is a sample-continuous finite-variance martingale with $U(0) = 0$

A brief sketch of the proof is given in the following. Let us denote a forward increment of $X(t)$ by $dX(t) = X(t + dt) - X(t)$. Then the best prediction of $dX(t)$ is given by $E[dX(t)|X(t'), 0 \le t' \le t]$, which we write as $dY(t)$. Let us define a process $dU(t)$ as the difference of $dX(t)$ and $E[dX(t)|X(t'), 0 \le t' \le t]$, that is,

$$dU(t) = dX(t) - dY(t),$$

which represents the prediction error (or unpredictable part) of the increment $dX(t)$. For $dX(t)$, $dY(t)$, and $dU(t)$ with the partition, $t_0 = 0 < t_1 <, ..., < t_n < t_{n+1} = t$, we have

$$dX(t_k) = dY(t_k) + dU(t_k).$$

Adding these over, $k = 0, 1, ..., n$, we obtain

$$X(t) - X(0) = \sum_{k=0}^{n} dY(t_k) + \sum_{k=0}^{n} dU(t_k).$$

By making the partition finer with $n \to \infty$, we obtain the decomposition

$$X(t) - X(0) = Y(t) + U(t).$$

7.3.1 Martingale in Continuous Time

The process defined as a limit of $\sum_{k=0}^{n} dU(t_k)$ is a martingale, and so it has the following property:

$$E[U(t)|X(t'), 0 \le t' \le t_0] = E[U(t_0)|X(t'), 0 \le t' \le t_0]$$

$$= U(t_0).$$

$U(t)$ is called innovation martingale. A very important point with the innovation martingale $U(t)$ is that its increment $dU(t)$ is not only zero mean but also Gaussian (no matter what distribution the original process $X(t)$ has). The theorem is called Levy–Doob theorem and is stated as the following.

Theorem (Levy–Doob):

Let $U(t)$, $t \ge 0$ be a finite-variance sample-continuous martingale with $U(0) = 0$, such that for any $t \ge 0$ and $dt > 0$, it holds

$$E[dU(t)|U(t'), 0 \le t' \le t] = 0$$

and

$$E[(dU(t))^2 | U(t'), 0 \le t' \le t_0] \approx dt,$$

where $dU(t) = U(t + dt) - U(t)$ denotes a forward increment. Then, $U(t), t \ge 0$ is the standard Wiener process, that is, $dU(t)$ is a Gaussian white noise with unit variance.

It means that the prediction errors of the sample-continuous finite-variance process is Gaussian white even though the process itself is not Gaussian. This is a very important point when we go to the stage of estimating the dynamic models from time series data in later sections.

What can we say if prediction errors are not only zero-mean martingale but also "stationary"? Prediction error processes are not only zero-mean martingale but also often stationary and independent. A stochastic process that has a stationary independent increment is called Levy process. Levy process is a martingale processes, but a martingale process is not necessarily a Levy process. A detailed study of Levy process has been given by Levy since the 1950s (see Levy 1956, Protter 1990). A useful outcome of Levy's work, which is summarized in a theorem (called Levy–Khintchin theorem), is that a general Levy process can be seen as comprising three components (called Levy-Khintchin triplet): (1) a drift, (2) a diffusion component, and (3) a jump component. Therefore, if we model a time series of "continuous type" (i.e., no jump) with some dynamic model and if the prediction errors are unavoidable, a natural assumption for the error is Gaussian white noise.

7.3.2 Doob Decomposition and Stochastic Differential Equation Model

Since we expect the predictable part $Y(t)$ of the Doob decomposition

$$X(t) = X(0) + Y(t) + U(t)$$

to be smooth, we have, for small dt_k,

$$dY(t_k) = A(t_k)dt_k,$$

where $A(t_k) = a(X(t'), 0 \le t' \le t_k)$. Then, for a small time increment dt, we have

$$Y(t + dt) = Y(t) + A(t)dt,$$

where $A(t)$ is a function of the past $X(t)$.

By the definition, $dU(t) = U(t + dt) - U(t)$ is a zero-mean martingale difference, that is, $E[dU(t)] = 0$. For most physical systems, it is not unreasonable that we assume the average energy increment $E[dU^2(t)|...]$ will either be constant or at most a smoothly varying function of time. Then, we could have $E[dU^2(t)X(t'), 0 \le t' \le t] \doteq B^2(t)dt$ and $B(t) = b(X(t') (0 \le t' \le t)$.

With the $A(t)$ and $B(t)$ introduced earlier, we can write the process $X(t)$ as

$$X(t) = X(0) + Y(t) + U(t)$$

$$= X(0) + \int_0^t A(\tau)d\tau + \int_0^t B(\tau)dW(\tau),$$

where

$W(t)$ is a Wiener process
$A(t)$ and $B(t)$ are smooth functions of the past state

We can rewrite this in a differential form as

$$dX(t) = A(t)dt + B(t)dW(t).$$

Here $A(t)$ and $B(t)$ are smooth functions of the past state, $X(t')$, $0 \le t' \le t$. If the process is Markov, the predictable processes $A(t)$ and $B(t)$ can no longer depend on the entire past $X(t')$, $0 \le t' \le t$ but only on the current value $X(t)$ and possibly on the time t. Thus, $A(t) = a(X(t), t)$ and $B(t) = b(X(t), t)$. The Doob decomposition of $X(t)$ yields an integral representation of the process

$$X(t) = X(0) + \int_0^t a(X(\tau), \tau)d\tau + \int_0^t b(X(\tau), \tau)dW(\tau),$$

or a differential representation of the process

$$dX(t) = a(X(t), t)dt + b(X(t), t)dW(t).$$

With the informal notation, $\dot{W}(t) = dW(t)/dt$ and $\dot{X}(t) = dX(t)/dt$, this differential representation is often written as

$$\dot{X}(t) = a(X(t), t) + b(X(t), t)\dot{W}(t).$$

On the other hand, a sample-continuous finite-variance process $X(t)$ that satisfies this stochastic differential equation is known to be identical with the Markov process that satisfies the conditions (see Goel and Richter-Dyn 1974 and Larson and Schubert 1979)

$$\lim_{dt \to 0} \frac{1}{dt} P(|dX(t)| \ge \varepsilon \mid X(t) = x) = 0,$$

$$\lim_{dt \to 0} \frac{1}{dt} E[|dX(t)| X(t) = x] = a(x, t), \tag{7.7}$$

$$\lim_{dt \to 0} \frac{1}{dt} E[|(dX(t))^2| X(t) = x] = b^2(x, t).$$

This is summarized in the following Theorem (A).

Theorem (A):

A finite-variance sample-continuous process that satisfies

$$dX(t) = a(X(t),t)dt + b(X(t),t)dW(t)$$

is identical with the Markov process that satisfies condition (7.7).

7.3.3 Markov Process and Fokker–Planck Equation

The probability law of a Markov process $X(t)$ with a continuous-time domain $t \geq 0$ is completely specified by its initial density $p_0(x)$ and by the family of transition probability densities $p_{t_2|t_1}(x_2|x_1)$ defined for all $0 \leq t_1 \leq t_2$. It must satisfy the following condition, called Chapman–Kolmogorov equation:

$$p_{t_2|t_1}(x_2|x_1) = \int_{-\infty}^{\infty} p_{t_2|t}(x_2|x)p_{t|t_1}(x|x_1)dx$$

for all $t_1 < t < t_2$ (see Feller (1966)). This transition probability density $p_{t_2|t_1}(x_2|x_1)$ and the smooth functions $a(x, t)$ and $b(x, t)$ of Theorem (A) are related by the following Theorem (B).

Theorem (B):

Let $X(t)$ be a finite-variance sample-continuous Markov process of Theorem (A). Then, the transition probability densities $p_{t_2|t_1}(x_2|x_1)$ of $X(t)$ satisfy the following Kolmogorov forward equation:

$$\frac{\partial p_{t_2|t_1}(x_2|x_1)}{\partial t_2} = -\frac{\partial}{\partial x_2}[a(x_2,t_2)p_{t_2|t_1}(x_2|x_1)] + \frac{1}{2}\frac{\partial^2}{\partial x_2^2}[b^2(x_2,t_2)p_{t_2|t_1}(x_2|x_1)].$$

Also a dual equation, called Kolmogorov backward equation, holds:

$$\frac{\partial p_{t_2|t_1}(x_2|x_1)}{\partial t_1} = -a(x_1,t_1)\frac{\partial}{\partial x_1}p_{t_2|t_1}(x_2|x_1) - \frac{1}{2}b^2(x_1,t_1)\frac{\partial^2}{\partial x_1^2}p_{t_2|t_1}(x_2|x_1).$$

In applications, the Kolmogorov forward equation is rewritten in a simple form, known as Fokker–Planck equation:

$$\frac{\partial p(x,t)}{\partial t} = -\frac{\partial}{\partial x}[a(x,t)p(x,t)] + \frac{1}{2}\frac{\partial^2}{\partial x^2}[b^2(x,t)p(x,t)],$$

where we denote $p(x,t) = p_{t_1+t|t_1}(x|x_1)$. Here, $t_1 = 0$, $0 \leq t = t_2 - t_1$, and x_1 is fixed. These equations are useful when we need to characterize stationary distributions of some nonlinear dynamic models in continuous time, which we discuss in Chapter 8.

7.3.4 Frost–Kailath Theorem for the Processes with Observation Noise

So far, we have considered the continuous-time version of Doob decomposition and its applications without considering the presence of observation noise. An especially useful fact derived from the Levy–Doob theorem is that the innovation martingale $U(t)$ in continuous time is the standard Wiener process so that its increment $dU(t)$ is a Gaussian white noise. This theorem cannot be directly applied in the situation where the observation data of the original process is contaminated by an observation noise. Fortunately, a similar theorem about the Gaussianity of the innovation martingale is known to be obtained from the following theorem by Frost and Kailath (1971).

Theorem (Frost and Kailath):

Let a process $x(t)$ be observed with observation noise $\varepsilon(t)$,

$$x(t) = z(t) + \varepsilon(t), \quad t \geq 0, \tag{7.8}$$

where the observation noise $\varepsilon(t)$ is assumed to be white Gaussian noise possibly multiplied by a deterministic square integrable function, $b(t)$, of time t, that is, $\varepsilon(t) = b(t)\dot{W}(t)$, to count for variations in its intensity. $z(t)$, $t \geq 0$ is a finite-variance mean square integrable process, and the white Gaussian noise $\dot{W}(t)$ is for every $t > 0$ independent of the past $X(t')$, $0 \leq t' \leq t$. Then, there is an equivalent representation

$$x(t) = \tilde{z}(t) + u(t), \quad t \geq 0, \tag{7.9}$$

where

$$\tilde{z}(t) = E[z(t)|x(t'), 0 \leq t' \leq t],$$

and $u(t)$, $t \geq 0$ is again $b(t)$ times white Gaussian noise independent of the past $x(t')$, $0 \leq t' \leq t$, for every $t > 0$.

The process defined by

$$u(t) = x(t) - \tilde{z}(t) \tag{7.10}$$

is called an innovation process of the process $x(t)$. Since $u(t)$ is the formal derivative of an inhomogeneous Wiener process $U(t)$, it may be called an "innovation martingale" of the process $X(t) = \int_0^t x(\tau)d\tau$. Note that $u(t)$, $x(t)$, and $z(t)$ are formal derivatives of $U(t)$, $X(t)$, and $Z(t)$, respectively. Thus, $u(t)dt = dU(t)$, $x(t)dt = dX(t)$, and $\tilde{z}(t)dt = E[dZ(t)|X(t'), 0 \leq t' \leq t]$. Then, Equation 7.10 is also written as

$$dU(t) = dX(t) - E\big[dZ(t)|X(t'), 0 \leq t' \leq t\big].$$

The forward increment $dU(t)$ can be interpreted as a new information or a random innovation brought by the increment $dX(t)$ of the integrated observation process $X(t)$ during the time dt after all the information contained in the past $X(t')$, $0 \leq t' \leq t$ has been extracted. Thus, the formal derivative $u(t) = \dot{U}(t)$ can be interpreted as an instantaneous innovation brought by $dX(t)$ after extracting the past, hence the term "innovation process."

The innovation process $u(t)$, or equivalently, the innovation martingale $U(t)$, is completely determined by the observation process $x(t)$ or $X(t)$, according to its definition. The discussion also suggests that the converse statement may be true as well. That is, the observation process can be thought of as being built up from the infinitely dense stream of random innovations and hence completely determined by its innovation process. In other words, it seems that the observation process must be causally equivalent to its innovation process. Thus, for any $t \geq 0$, $u(t)$ can be obtained from $x(t')$, $0 \leq t' \leq t$, and $x(t)$ is obtainable from $u(t')$, $0 \leq t' \leq t$. Then one could construct a causal filter (input–output system) that will produce $u(t)$ as an output when $x(t)$ is applied as an input to the system. Then there should be another causal filter (input–output system) that will produce an output $x(t)$ when $u(t)$ is applied to the system as an input (see Figure 7.2).

The relations can be written, in the linear Gaussian case, by the pair of relations with the impulse response functions $h(.)$ and $g(.)$ as

$$u(t) = x(t) - \int_0^t h(t - \tau)x(\tau)d\tau$$

$$x(t) = u(t) + \int_0^t g(t - \tau)u(\tau)d\tau.$$

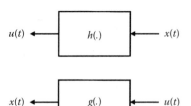

FIGURE 7.2
Causal and causally invertible relations, continuous-time version.

This means that the portion of the innovation process $u(t')$, $0 \leq t' \leq t$ contains all the information contained in the portion of the observation process $x(t)$ and vice versa. We refer to such a situation by saying that the processes $x(t)$ and $u(t)$ are causally linearly equivalent.

We see, in Chapter 11, that the idea of causal equivalence of a finite-variance process $x(t)$ and its innovations $u(t)$ is especially useful in deriving the linear (and local-linear) Kalman filter for the linear (and local-linear) state space models.

8

Dynamics and Stationary Distributions

8.1 Time Series and Stationary Distributions

Usually, a histogram of a time series provides us with useful information for data analysis. It shows a certain shape implying the stationarity with its stationary density distribution function. In the late 1970s, many stationary nonlinear time series models were introduced for the analysis of non-Gaussian time series, but it is not known how the parameters of these nonlinear time series models affect the shape of the non-Gaussian density distribution of the process. The ExpAR model

$$x_{t+1} = (\phi_1 + \phi_2 e^{-\gamma x_t^2}) x_t + w_{t+1} \tag{8.1}$$

is one of a few exceptional examples whose parameters clearly explain what kind of shape the histogram of the generated time series is going to be.

8.1.1 Exponential AR Models and Marginal Density Distributions

Figure 8.1a through c shows the histogram of three sets of time series, each with 80,000 observations, generated from the following three ExpAR models with the unit noise variance, $\sigma_w^2 = 1$.

Example D1:

$$x_{t+1} = (0.8 + 0.2e^{-x_t^2}) x_t + w_{t+1}. \tag{8.2}$$

Example D2:

$$x_{t+1} = (1 - 0.2e^{-x_t^2}) x_t + w_{t+1}. \tag{8.3}$$

Example D3:

$$x_{t+1} = (0.8 + 0.4e^{-x_t^2}) x_t + w_{t+1}. \tag{8.4}$$

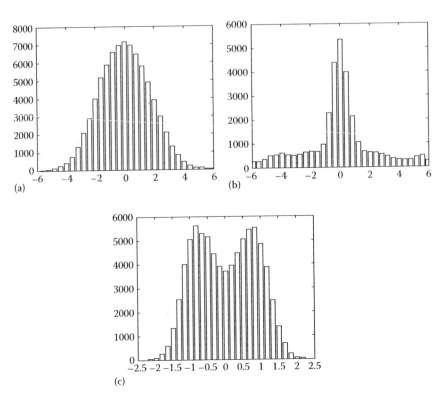

FIGURE 8.1
(a) ExpAR model histogram-1. (b) ExpAR model histogram-2. (c) ExpAR model histogram-3.

The three histograms show three different non-Gaussian characteristics of density distributions. The first histogram shows short light tails and a sharper (than Gaussian density distribution) peaked center, which are characteristic to leptokurtic distributions. This type of histogram is generated from any ExpAR model with $\phi_1 + \phi_2 = 1$, $0 < \phi_1 < 1$, and $0 < \phi_2 < 1$. The second histogram clearly shows a distribution with long heavy tails like a stable distribution. This type of histogram is generated from any ExpAR model (8.1) with $\phi_1 = 1$ and $-1 < \phi_2 < 0$. The third histogram shows a bimodal distribution with light and short tails, which is characteristic of distributions in the exponential family. This type of histogram is generated from any ExpAR model with $\phi_1 + \phi_2 > 1$, $0 < \phi_1 < 1$, and $0 < \phi_2 < 1$.

These examples of ExpAR models clearly show that we can generate time series with various types of non-Gaussian marginal distributions by controlling the two parameters, ϕ_1 and ϕ_2, of ExpAR model (8.1).

Many time series models, if they can generate nonexplosive trajectories, yield stationary processes with stationary marginal distributions. Sufficient conditions for the ergodicity and the stationarity of nonlinear non-Gaussian time series models are known for a certain class of models

(Tweedie 1975, Ozaki 1985a). However, the analytical derivation of stationary marginal distributions of each individual nonlinear time series model is difficult to obtain in most cases. Although we can roughly guess the shape of the density distribution by looking at some of the ExpAR coefficients, we do not have any convenient analytical tools for characterizing the non-Gaussian distributed process defined by a discrete-time model.

8.1.2 Continuous-Time Cases

Situations are different for continuous-time nonlinear dynamic model cases. Characteristic shapes similar to the three examples mentioned earlier, (8.2), (8.3), and (8.4), are observed in the following three examples of continuous-time dynamic models:

Example C1:

$$\dot{x}(t) = -x^3(t) + w(t) \tag{8.5}$$

Example C2:

$$\dot{x}(t) = -\sqrt{2}\tanh\sqrt{2}x(t) + w(t) \tag{8.6}$$

Example C3:

$$\dot{x}(t) = 2x(t) - x^3(t) + w(t) \tag{8.7}$$

The basic shape of the histogram of the simulated process of these models is determined by the nonlinear function $-f(x)$ of the model, $dx/dt = f(x) + w(t)$, that is, $-f(x) = x^3$, $-f(x) = \sqrt{2}\tanh\sqrt{2}x$, and $-f(x) = -2x + x^3$ (see Figure 8.2a through c).

By comparing the shapes of the nonlinear functions in Figure 8.2, we notice that the non-Gaussian character of each process is closely related to the nonlinear character of $-f(x)$ of the dynamical system. The dx/dt of the dynamical system of (8.5) is negative for $x(t) > 0$, and its absolute value increases, to ∞ as $x \rightarrow \infty$, much faster than the linear case. Then the trajectory is less likely to stay far away from the origin compared with the linear case. This means that the stationary density distribution of the process may have a much thinner tail than the Gaussian distribution. The opposite case to Example C1 is Example C2, where dx/dt is negative for $x > 0$ and its absolute value increases when $x \rightarrow \infty$, but the absolute value, that is, the strength of the force of pulling the trajectory back to the origin, is much less than the linear case. That means the process is likely to stay away from the origin for longer periods than the linear Gaussian case. This implies that the stationary density distribution of the process of Example C2 has a fatter tail than the Gaussian distribution.

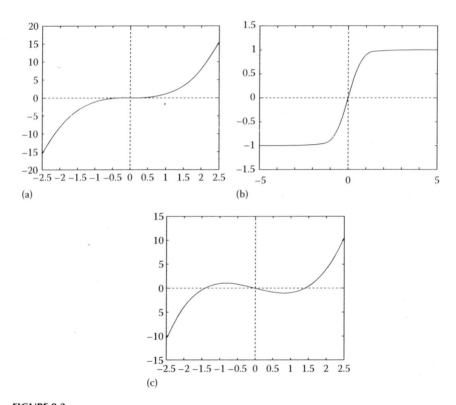

FIGURE 8.2
(a) Nonlinear function-1. (b) Nonlinear function-2. (c) Nonlinear function-3.

The close relation between a nonlinear dynamical system and its equilibrium distribution has been known in statistical physics. Here the stationary distribution $p(x)$ of a process $x(t)$ generated from a stochastic dynamical system, $\dot{x}(t) = f(x) + w(t)$, driven by a Gaussian white noise $w(t)$ is known to have a stationary density distribution,

$$p(x) = p_0 e^{2\int_{-\infty}^{x} f(\xi)d\xi},$$

where p_0 is a normalizing constant. The function $V(x)$ defined by

$$V(x) = -\int_{-\infty}^{x} f(\xi)d\xi$$

is sometimes called potential function of the dynamical system, $\dot{x}(t) = f(x)$. (This relation is reviewed in a more general setting of Markov diffusion processes in a later section.) The potential function $V(x)$ of the dynamical systems (8.5), (8.6), and (8.7) are given, respectively, by

$$V(x) = \frac{x^4}{4},$$

$$V(x) = \log(\cosh \sqrt{2}x),$$

$$V(x) = -x^2 + \frac{1}{4}x^4.$$

The shape of the density distributions, whether they are fat-tailed or thin-tailed and whether they are unimodal or multimodal, is determined by the shape of their potential functions, which originally comes from the nonlinear function $f(x)$ of the dynamical system $\dot{x}(t) = f(x)$.

8.2 Pearson System of Distributions and Stochastic Processes

8.2.1 Stationary Markov Diffusion Processes and Marginal Distributions

Let us assume that $x(t)$ is a Markov diffusion process determined by a drift $a(x)$ and a diffusion coefficient $b^2(x)$ so that the Fokker–Planck equation is

$$\frac{\partial p}{\partial t} = -\frac{\partial}{\partial x}[a(x)p] + \frac{1}{2}\frac{\partial^2}{\partial x^2}[b^2(x)p]. \tag{8.8}$$

Let the initial value of the process be $x(t) = y$; then its steady state density distribution is given by $p(x|y,\infty) = \lim_{t\to\infty} p(x|y,t)$, which is obtained by setting $\frac{\partial p}{\partial t} = 0$ with certain manipulations (see Goel and Richter-Dyn 1974) as

$$p(x|y,\infty) = \frac{C}{b^2(x)}\exp\left\{2\int^{x}\frac{a(x)}{b^2(x)}d\xi\right\}. \tag{8.9}$$

This relation between the non-Gaussian density distribution $p(x) = p(x|y, \infty)$ and the nonlinear dynamical system, $dx = a(x)dt + b(x)dW(t)(\Leftrightarrow \dot{x} = a(x) + b(x)w(t))$ corresponding to the Fokker–Planck equation (8.8), has been known since the era of Kolmogorov and Pearson (see Kolmogorov 1931, section 18, p. 105). Kolmogorov pointed out that the diffusion process defined by a drift,

$$a(x) = c_0 + c_1 x,$$

and a diffusion coefficient,

$$b(x) = d_0 + d_1 x + d_2 x^2,$$

gives rise to the Pearson system,

$$\frac{p'(x)}{p(x)} = \frac{x - p_0}{q_2 x^2 + q_1 x + q_0},$$

with

$$p_0 = \frac{d_1 - c_0}{c_1 - 2d_2}, \quad q_0 = \frac{d_0}{c_1 - 2d_2}, \quad q_1 = \frac{d_1}{c_1 - 2d_2}, \quad q_2 = \frac{d_0}{c_1 - 2d_2},$$

implying that any probability distribution of the Pearson system can be a steady stationary distribution of a Markov diffusion process.

The derivative $p' = \partial p / \partial x$ of (8.9) is

$$p' = C(-2)b'(x)b^{-3}(x)\exp\left\{2\int^x \frac{a(x)}{b^2(x)}\,d\xi\right\} + \frac{C}{b^2(x)}\left[2\frac{a(x)}{b^2(x)}\right]\exp\left\{2\int^x \frac{a(x)}{b^2(x)}\,d\xi\right\}.$$

Then we have

$$\frac{p'}{p} = \frac{1}{C}b^2(x)\exp-\left\{2\int^x \frac{a(x)}{b^2(x)}\,d\xi\right\}$$

$$\times \left[C(-2)b'(x)b^{-3}(x) + \frac{C}{b^2(x)}\left[2\frac{a(x)}{b^2(x)}\right]\right]\exp\left\{2\int^x \frac{a(x)}{b^2(x)}\,d\xi\right\}$$

$$= (-2)b'(x)b^{-1}(x) + 2\frac{a(x)}{b^2(x)}$$

$$= \frac{a(x) - b'(x)b(x)}{\frac{1}{2}b^2(x)} = \frac{c(x)}{d(x)}.$$

This means that we can construct a Markov diffusion process whose limiting state distribution is any of the distributions defined by the so-called generalized Pearson System:

$$\frac{p'(x)}{p(x)} = \frac{c(x)}{d(x)}. \tag{8.10}$$

Here the drift of the Markov diffusion process is specified as

$$a(x) = c(x) + b'(x)b(x) = c(x) + d'(x), \tag{8.11}$$

and the diffusion coefficient is specified by

$$b(x) = \sqrt{2d(x)}. \tag{8.12}$$

8.2.2 Pearson System and the Associated Dynamical Systems

Suppose the process $x(t)$ is generated from a stochastic differential equation model,

$$\dot{x} = a(x) + b(x)w(t),$$

or equivalently,

$$dx = a(x)dt + b(x)dW(t).$$

Then, for a transformed variable, $y = u(x)$, we have

$$dy = du(x) = \left\{ u_x(x)a(x) + \frac{1}{2}u_{xx}(x)b(x)^2 \right\} dt + u_x(x)b(x)dW(t),$$

with Ito's stochastic calculus. Here we rewrite the same equation in our convenient informal style, that is,

$$\dot{y} = \dot{u}(x) = \left\{ u_x(x)a(x) + \frac{1}{2}u_{xx}(x)b(x)^2 \right\} + u_x(x)b(x)w(t)$$

We take, hereafter, this informal style to elucidate the similarity between the deterministic dynamical system model and the stochastic dynamical system model. Then, if we choose such $u(x)$ that gives rise to $u_x(x)b(x) = 1$, we have, with $y = u(x)$,

$$\dot{y} = \left\{ u_x(x)a(x) + \frac{1}{2}u_{xx}(x)b(x)^2 \right\} + w(t)$$

$$= \left\{ u_x(x)a(x) + \frac{1}{2}\frac{\partial}{\partial x}\left[\frac{1}{b(x)}\right]b(x)^2 \right\} + w(t)$$

$$= \left\{ u_x(x)a(x) + \frac{1}{2}\left[\frac{-b'(x)}{b(x)^2}\right]b(x)^2 \right\} + w(t)$$

$$= \left\{ u_x(x)a(x) - \frac{1}{2}b'(x) \right\} + w(t)$$

$$= f(y) + w(t).$$

This means that for any Markov diffusion process $x(t)$ defined by

$$\dot{x} = a(x) + b(x)w(t),$$

we can always find a dynamical system,

$$dy/dt = f(y),$$

that generates the Markov diffusion process $x(t)$ with a Gaussian white noise $w(t)$ and the variable transformation, $x = u^{-1}(y) = h(y)$ as

$$\left(\begin{array}{l} \dot{y} = f(y) + w(t) \\ x(t) = h(y(t)). \end{array} \right.$$

Let $V(y)$ be defined with $f(y)$ as

$$V(y) = -\int_{-\infty}^{y} f(\eta)d\eta.$$

Since $dy/dt = f(y)$ and $V(y)$ are uniquely defined and attached to any stationary homogeneous Markov diffusion process, we call them "associated dynamical system" and "associated potential function" respectively.

From the earlier discussions, we have seen that many important concepts are involved in determining the dynamic and stochastic behavior of the process, and some of our time-series observation could be sampled from the Markov diffusion processes at discrete time points. The dynamic aspects of the characterization are summarized as the following:

(D1) A homogeneous Markov diffusion process $x(t)$ may be defined by a stochastic differential equation:

$$\dot{x} = a(x) + b(x)w(t).$$

(D2) The process $x(t)$ has an alternative representation through the transformed variable, $y(t) = h(x(t))$:

$$\left(\begin{array}{l} \dot{y} = f(y) + w(t) \\ x(t) = h(y(t)). \end{array} \right.$$

(D3) The process gives rise to a unique associated dynamical system:

$$dy/dt = f(y).$$

(D4) The process gives rise to a unique associated potential function:

$$V(y) = -\int_{-\infty}^{y} f(\eta)d\eta.$$

The probabilistic aspects of the characterization are summarized in the following:

(P1) The total stochastic behavior of the process $x(t)$ may be characterized by the Fokker–Planck equation:

$$\frac{\partial p}{\partial t} = \frac{\partial}{\partial x}[a(x)p] + \frac{1}{2}\frac{\partial^2}{\partial x^2}[b^2(x)p].$$

(P2) Let $Q(x)$ be the density function of the steady-state stationary distribution of the process $x(t)$, that is, $Q(x) = p(x|y, \infty)$; then, we have a generalized Pearson system representation:

$$\frac{dQ(x)}{dx} = \frac{a(x) - b'(x)b(x)}{\frac{1}{2}b^2(x)}Q(x).$$

(P3) With the transformed variable $y(t) = h(x(t))$, we have another associated system characterizing the probability density of the transformed variable,

$$\frac{dQ(y)}{dy} = 2f(y)Q(y),$$

where $f(y)$ comes from (D3).

(P4) The density function of the transformed variable, $y(t) = h(x(t))$, has the following special representation called the exponential family in statistics (Barndorff-Nielsen 1978):

$$Q(y) = Q_0 e^{-2V(y)},$$

where $y(t)$ comes from (D4). The relations (D1)–(D4) and (P1)–(P4) are schematically shown in Figure 8.3.

In order to have useful guidelines for finding a whitening operator for non-Gaussian distributed time series, we first investigate the nature and the dynamic mechanism of the nonlinear non-Gaussian processes, using the examples of continuous time homogeneous Markov diffusion process models in the next section.

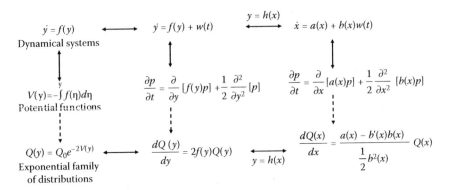

FIGURE 8.3
Schematic graph of potential-DS-SDE-Pearson system.

8.3 Examples

We would like to see some examples of the relations between typical non-linear dynamical systems and non-Gaussian density distributions, where the non-Gaussian stationary distributions are generated from the nonlinear dynamical systems driven by Gaussian white noise.

8.3.1 Leptokurtic Distributions and Nonlinear Dynamics

Example 1: $\dot{x} = -ax^3 + w(t)$.

The trajectory $x(t)$ of the deterministic nonlinear system $\dot{x} = -ax^3$ is, in a sense, similar to the linear dynamical system, $\dot{x} = -ax$, which is known to damp down to zero as $t \to \infty$. However, the difference is, because of the third-order nonlinearity, the decreasing speed is much faster for large x than $\dot{x} = -ax$. If the system is driven by a Gaussian white noise $w(t)$, the linear stochastic dynamical system, $\dot{x} = -ax + w(t)$, is known to have a Gaussian steady-state distribution. Then we may expect that the nonlinear stochastic dynamical system, $\dot{x} = -ax^3 + w(t)$, may have a steady-state distribution with a much thinner tail than the Gaussian distribution. This is in fact confirmed quantitatively by using the relations (D1)–(D4) and (P1)–(P4) in the previous discussions.

The Fokker–Planck equation of the Markov diffusion process, defined by

$$\dot{x} = -ax^3 + \sigma w(t),$$

is given using the drift $-ax^3$ and the diffusion coefficient σ^2 as

$$\frac{\partial p}{\partial t} = -\frac{\partial}{\partial x}[-ax^3 p] + \frac{1}{2}\frac{\partial^2}{\partial x^2}[\sigma^2 p].$$

By transforming x into $y = x/\sigma$, we have the normalized stochastic dynamical system representation,

$$\dot{y} = -\sigma^2 a y^3 + w(t),$$

with the associated dynamical system,

$$\dot{y} = -\sigma^2 a y^3.$$

Then, the associated potential function is

$$V(y) = \frac{\sigma^2}{4} a y^4.$$

The density function of the steady-state distribution of $y(t)$ is given, with a normalizing constant p_0, as

$$p(x) = p_0 e^{-\frac{a x^4}{2\sigma^2}}.$$

8.3.2 Trimodal Density Distributions and Nonlinear Dynamics

Example 2:

$$\dot{x} = -6x + 5.5x^3 - x^5. \tag{8.13}$$

The function $f(x) = -6x + 5.5x^3 - x^5$ has five zero points, $\xi_0 = 0$, $\xi_1^+ = \sqrt{\frac{3}{2}}$, $\xi_1^- = -\sqrt{\frac{3}{2}}$, $\xi_2^+ = 2$, $\xi_2^- = -2$. They are called singular points of the dynamical system (8.13). If the initial value x_0 of (8.13) is one of the five singular points, then $x(t)$ stays at x_0 for $t > 0$. If the dynamical system is driven by a Gaussian white noise $\sigma w(t)$, then the process $x(t)$, defined by

$$\dot{x} = -6x + 5.5x^3 - x^5 + \sigma w(t)$$

becomes a diffusion process and its Fokker–Planck equation is given, using the drift function, $-6x + 5.5x^3 - x^5$, and the diffusion coefficient σ^2, by

$$\frac{\partial p}{\partial t} = -\frac{\partial}{\partial x}[(-6x + 5.5x^3 - x^5)p] + \frac{1}{2}\frac{\partial^2}{\partial x^2}[\sigma^2 p].$$

By transforming x into $y = x/\sigma$, we have the normalized stochastic dynamical system representation,

$$\dot{y} = -6y + 5.5\sigma^2 y^3 - \sigma^4 y^5 + w(t),$$

with the associated dynamical system,

$$\dot{y} = -6y + 5.5\sigma^2 y^3 - \sigma^4 y^5.$$

Then, the associated potential function is

$$V(y) = 3y^2 - \frac{11\sigma^2}{8} y^4 + \frac{\sigma^4}{6} y^6.$$

The density function of the steady-state distribution of $y(t)$ is given, with a normalizing constant p_0, as

$$p(y) = p_0 e^{-2V(y)} = p_0 e^{-2\left(3y^2 - \frac{11\sigma^2}{8} y^4 + \frac{\sigma^4}{6} y^6\right)}.$$

Then, the density function of the distribution of x is written, with a normalizing constant p_0, as

$$p(x) = p_0 e^{\left(-6x^2 + \frac{11}{4} x^4 - \frac{1}{3} x^6\right)/\sigma^2}$$

8.3.3 Distributions of the Exponential Family and Nonlinear Dynamics

Example 3:

$$\dot{x} = a_1 x(t) + a_2 x(t)^3 + \cdots + a_r x(t)^{2r+1} + \sigma w(t). \tag{8.14}$$

Let us see the distributional properties of the process $x(t)$ defined by the earlier stochastic differential equation model, where the variance of the Gaussian white noise $w(t)$ is 1.

When the stochastic differential equation model (8.14) defines a finite-variance stationary process, the density function $Q(x)$ of the marginal steady-state distribution satisfies the differential equation,

$$\frac{dQ(x)}{dx} = \frac{a_1 x(t) + \cdots + a_r x(t)^{2r+1}}{\sigma^2/2} Q(x),$$

and the density function $Q(x)$ is given by

$$Q(x) = Q_0 \exp\left\{\frac{2}{\sigma^2}\left(\frac{1}{2} a_1 x^2 + \frac{a_2}{4} x^4 + \cdots + \frac{a_r}{(2r+2)} x^{2r+2}\right)\right\}, \tag{8.15}$$

where Q_0 is a normalizing constant. Since $Q(x)$ satisfies the following relation,

$$\frac{d}{dx}\left\{x^{2k+1} Q(x)\right\} = (2k+1) x^{2k} Q(x) + \left\{\frac{2}{\sigma^2}\left(a_1 x^{2k+2} + a_2 x^{2k+4} + \cdots + a_r x^{2k+2r+2}\right)\right\} Q(x),$$

$$\tag{8.16}$$

we have, for $k = 1, 2, 3, \ldots$, the relation,

$$(2k+1)\mu_{2k} + \frac{2}{\sigma^2}(a_1\mu_{2k+2} + a_2\mu_{2k+4} + \cdots + a_r\mu_{2k+2r+2}) = 0, \qquad (8.17)$$

between the moments, μ_{2k} of order $2k$, μ_{2k+2} of order $2k + 2$, ..., $\mu_{2k+2r+2}$ of order $2k + 2r + 2$.

The process $x(t)$ is symmetric about $x(t) = 0$, and so we have $\mu_0 = 0$, and $\mu_{2k+1} = 0$ for $k = 0, 1, 2, \ldots$. Then Equation 8.17 shows that $\mu_2, \mu_4, \ldots, \mu_{2r+2}$ are sufficient statistics since all the higher order moments, $\mu_{2r+4}, \mu_{2r+6}, \mu_{2r+8}, \ldots$, may be calculated recursively from $\mu_2, \mu_4, \ldots, \mu_{2r+2}$ and the relation (8.17). Process $x(t)$ of (8.14) is completely specified by the parameters $a_1, a_2, \ldots, a_r$ and σ^2. The marginal distribution, and so $\mu_2, \mu_4, \ldots, \mu_{2r+2}$, could also be explicitly specified using these parameters. This type of distribution belongs to the so-called exponential family of distributions. In spite of the importance of this type of distribution in nonlinear and non-Gaussian time-series analysis, not much work has been devoted to the study of the exponential family, except for a general treatment (Barndorff–Nielsen 1978), and even the moment generating function of the density function (8.15) does not seem to have been discussed in the statistics literature.

8.3.4 Fat-Tailed Distributions and Nonlinear Dynamics

We would like to see how these relations (D1)–(D4) and (P1)–(P4) actually work in some of the typical non-Gaussian distributions defined by the Pearson system. The Cauchy distribution is one of the typical non-Gaussian distributions where the tail of the distribution is fatter than the Gaussian distribution. The density of the Cauchy distribution is a special case ($\alpha = 1/2$) of the following density function, which is known to be a general stable distribution:

$$Q(x) = \frac{\Gamma(\alpha+1/2)}{\Gamma(1/2)\Gamma(\alpha)}(1+x^2)^{-(\alpha+1/2)}.$$

Its Pearson system representation is given by

$$\frac{dQ(x)}{dx} = \frac{-(2\alpha+1)x}{1+x^2}Q(x).$$

This yields the Fokker–Planck equation (Wong 1963) given by

$$\frac{\partial p}{\partial t} = -\frac{\partial}{\partial x}[-(2\alpha-1)x)p] + \frac{1}{2}\frac{\partial^2}{\partial x^2}[2(1+x^2)p].$$

The stochastic differential equation of this Markov diffusion process is given by

$$\dot{x} = -(2\alpha-1)x + \sqrt{2(1+x^2)}\,w(t).$$

By transforming the process $x(t)$ into $y(t) = \dfrac{1}{\sqrt{2}}\sinh^{-1}x(t)$, we have

$$\dot{y} = -\sqrt{2}\alpha\tanh(\sqrt{2}y) + w(t).$$

Here we have the associated dynamical system,

$$\dot{y} = -\sqrt{2}\alpha\tanh(\sqrt{2}y),$$

and the associated potential function,

$$V(y) = \alpha\log(\cosh\sqrt{2}y).$$

With this potential function, we have a density function for $y(t)$:

$$Q(y) = \frac{\sqrt{\pi}\,\Gamma(\alpha+1/2)}{\Gamma(1/2)\Gamma(\alpha)}(\cosh\sqrt{2}y)^{-2\alpha}.$$

This density function is known to be one of the typical fat-tailed stable distributions. The process $x(t)$ is obtained from $y(t)$ by a nonlinear transformation, $x(t) = \sinh(\sqrt{2}y(t))$, which transforms $y(t)$ in a distorting manner that the larger part of the positive $y(t)$ is transformed into a further larger value, and the smaller part of the negative $y(t)$ is transformed into further smaller value, so that the tail of the distribution of $x(t)$ is even fatter than the distribution of $y(t)$.

8.3.5 Beta Distributions and Nonlinear Dynamics

The Beta distribution, whose density function is given by

$$Q(x) = \frac{\Gamma(\alpha+\gamma+2)}{\Gamma(\alpha+1)\Gamma(\gamma+1)}\frac{(1+x)^{\alpha}(1-x)^{\gamma}}{2^{\alpha+\gamma+1}}, \quad \alpha,\gamma \geq -1$$

is used to model events that are constrained to take place within a finite interval defined by a minimum and a maximum value. The range of some of the time-series data are limited within an upper bound and a lower bound so that their steady-state distribution is defined on the finite interval. It will be intriguing to see what type of nonlinear dynamic model could be involved for defining a Beta-distributed Markov diffusion process. The Pearson system representation of the Beta distribution is given by

$$\frac{dQ(x)}{dx} = \frac{(\alpha-\gamma)-(\alpha+\gamma)x}{1-x^2}Q(x).$$

By the use of the relations (8.11) and (8.12), we have a Fokker–Planck equation

$$\frac{\partial p}{\partial t} = -\frac{\partial}{\partial x}[(\alpha - \gamma) - (\alpha + \gamma + 2)x)p] + \frac{1}{2}\frac{\partial^2}{\partial x^2}[2(1 - x^2)p]$$

of the process $x(t)$ whose stationary distribution is the Beta distribution. The stochastic differential equation model for the process $x(t)$ is given by

$$\dot{x} = \{(\alpha - \gamma) - (\alpha + \gamma + 2)x\} + \sqrt{2(1 - x^2)}\, w(t)$$

and the associated potential function,

$$V(y) = -\frac{\alpha - \gamma}{2}\log(1 + \sin\sqrt{2}y) - \frac{\alpha + \gamma + 1}{2}\log(\cos\sqrt{2}y). \tag{8.18}$$

With this potential function, we have a density function for $y(t)$ as

$$Q(y) = \frac{\Gamma(\alpha + \gamma + 2)}{\Gamma(\alpha + 1)\Gamma(\gamma + 1)2^{\alpha + \gamma + 0.5}}(1 + \sin\sqrt{2}y)^{\alpha - \gamma}(\cos\sqrt{2}y)^{\alpha + \gamma + 1}. \tag{8.19}$$

8.3.6 Gamma Distributions and Nonlinear Dynamics

It is often the case that the observed time-series data are positive-valued. A typical positive-valued distribution is the Gamma distribution, whose density function is specified with two parameters, α and β, as

$$Q(x) = \frac{x^{\alpha - 1}\exp(-x/\beta)}{\Gamma(\alpha)\beta^{\alpha}}.$$

It has the Pearson system representation:

$$\frac{dQ(x)}{dx} = \frac{(\alpha - 1)\beta - x}{\beta x}Q(x).$$

By the use of the relations (8.11) and (8.12), we have a Fokker–Planck equation,

$$\frac{\partial p}{\partial t} = -\frac{\partial}{\partial x}[(\alpha\beta - x)p] + \frac{1}{2}\frac{\partial^2}{\partial x^2}[2\beta x\, p],$$

of the process $x(t)$, whose stationary distribution is the Gamma distribution. The stochastic differential equation model for the process $x(t)$ is given by

$$\dot{x} = \{\alpha\beta - x\} + \sqrt{2\beta x}\, w(t).$$

By transforming the process into $y(t) = \sqrt{2\beta x(t)}/\beta$, we have the associated dynamical system,

$$\dot{y} = \frac{\alpha\beta - \dfrac{\beta}{2} - \dfrac{(\beta y)^2}{2\beta}}{\beta y}$$

$$= \left\{\left(\alpha - \frac{1}{2}\right)\frac{1}{y} - \frac{y}{2}\right\}, \tag{8.20}$$

and the associated potential function,

$$V(y) = \frac{y^2}{4} - \left(\alpha - \frac{1}{2}\right)\log y.$$

With this potential function, we have a density function for $y(t)$ as

$$Q(y) = \frac{1}{\Gamma(\alpha)2^{\alpha-1}} y^{(2\alpha-1)} \exp\left(-\frac{y^2}{2}\right).$$

8.4 Different Dynamics Can Arise from the Same Distribution

Next, we would like to point out some useful results arising out of Kolmogorov's discussion. Note that $c(x)$ and $d(x)$ in (8.10) do not need to be the first-order and the second-order polynomials as is the case for the Pearson system. They could be any analytic functions. In addition, they do not need to be mutually irreducible. For example,

$$\frac{Q'}{Q} = \frac{c(x)}{d(x)} \tag{8.21}$$

and

$$\frac{Q'}{Q} = \frac{xc(x)}{xd(x)} \tag{8.22}$$

have the same distribution $Q(x)$, but they yield different diffusion processes. From (8.21), we have

$$\frac{\partial p}{\partial t} = -\frac{\partial}{\partial x}[a_1(x)p] + \frac{1}{2}\frac{\partial^2}{\partial x^2}[b_1^2(x)p],$$

where $a_I(x) = c(x) + d'(x)$ and $b_I^2(x) = 2d(x)$. From (8.22), we have

$$\frac{\partial p}{\partial t} = -\frac{\partial}{\partial x}[a_{II}(x)p] + \frac{1}{2}\frac{\partial^2}{\partial x^2}\left[b_{II}^2(x)p\right],$$

where $a_{II}(x) = c(x)x + d(x) + d'(x)x$ and $b_{II}^2(x) = 2d(x)x$.

This means that for any distribution defined by the generalized Pearson system, we can have infinitely many different diffusion processes whose stationary distributions are the same. This may look surprising, but it is in a sense natural, because the static information such as the histogram of the time series shows only partial information of the data. In order to identify the model behind the observed time series, we ought to pay attention to the dynamic information, in the data, as well. We discuss this point in more detail in Section 8.4. Before that, we discuss the relations of distributions and the diffusion models in more detail with typical examples.

In Section 8.3.6, we had an example of a Markov diffusion process $x(t)$ defined by

$$\dot{x} = \{\alpha\beta - x\} + \sqrt{2\beta x}\, w(t), \tag{8.23}$$

whose stationary distribution is the Gamma distribution of the density function,

$$Q(x) = \frac{x^{\alpha-1}\exp(-x/\beta)}{\Gamma(\alpha)\beta^\alpha},$$

and its Pearson system representation is

$$\frac{dQ(x)}{dx} = \frac{(\alpha-1)\beta - x}{\beta x}Q(x). \tag{8.24}$$

If we multiply the numerator and the denominator of (8.24) by x, we have the following generalized Pearson system representation:

$$\frac{dQ(x)}{dx} = \frac{(\alpha-1)\beta x - x^2}{\beta x^2}Q(x). \tag{8.25}$$

From this modified system (8.25), we have another diffusion process defined by the Fokker–Planck equation:

$$\frac{\partial p}{\partial t} = -\frac{\partial}{\partial x}[(\alpha\beta - x)p] + \frac{1}{2}\frac{\partial^2}{\partial x^2}[2\beta xp].$$

Its stochastic differential equation representation is

$$\dot{x} = (\alpha\beta x - x^2) + \sqrt{2\beta x}\, w(t). \tag{8.26}$$

By transforming the process into $y(t) = \dfrac{1}{\sqrt{2\beta}} \log x(t)$, we have an associated dynamical system,

$$\dot{y} = \left(\frac{(\alpha-1)\beta}{\sqrt{2\beta}} - \frac{\exp(\sqrt{2\beta}y)}{\sqrt{2\beta}} \right). \qquad (8.27)$$

Thus, the process $x(t)$ defined either by (8.26) or by

$$x(t) = \exp\{\sqrt{2\beta}y\},$$

$$\dot{y} = \frac{(\alpha-1)\beta}{\sqrt{2\beta}} - \frac{\exp(\sqrt{2\beta}y)}{\sqrt{2\beta}} + w(t), \qquad (8.28)$$

has a Gamma-distributed stationary distribution. We call the Gamma-distributed process of (8.23) Type-I Gamma-distributed process and the process defined by (8.26) or (8.28), Type-II Gamma-distributed process. Some of the simulated data of Type-I and Type-II Gamma processes, with $\alpha = 4$, $\beta = 1$, and $\Delta t = 0.01$, are shown in Figure 8.4a and b respectively.

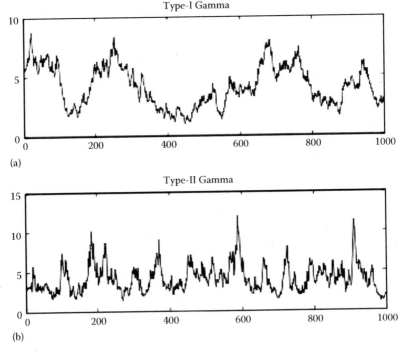

(a)

(b)

FIGURE 8.4
(a) Type-I Gamma simulation data. (b) Type-II Gamma simulation data.

Note that the model (8.26) is written as a first-order dynamical system,

$$\dot{x} = a(x, w)x,$$

whose coefficient $a(x, w)$ is a function of x and $w(t)$,

$$a(x, w) = \alpha\beta - x + \sqrt{2\beta}\, w(t).$$

This shows that bilinear type models with exogenous random inputs can generate various types of positive-valued processes by tuning the constant parameters α and β. This idea could be exploited well in the modeling of bilinear type dynamic models such as DCM (Friston et al. 2003), where the dynamics of some region in the brain is supposed to have an intrinsic influence from remote regions through neural connections.

8.4.1 Oscillatory Dynamics and Distributions

The two examples of Gamma-distributed processes in the previous section were generated by combinations of a memory-less variable transformation and a nonlinear dynamical system. In both cases, the dynamical systems used are one dimensional and cannot have an oscillating mechanism contributing to the existence of a peak in the spectrum of the process. A stochastic process having both an oscillating mechanism and a non-Gaussian marginal distribution character is easily obtained by combining a second-order stochastic differential equation model of type

$$\ddot{x}(t) + a\dot{x}(t) + bx = w(t) \tag{8.29}$$

and one of these models of type

$$\dot{y} = f(y) + w(t). \tag{8.30}$$

If a nonlinear random vibration model has a general nonlinear restoring function $b(x)x$, instead of the linear restoring force bx, we can write the model for nonlinear oscillation as

$$\ddot{x}(t) + a\dot{x}(t) + b(x(t))x(t) = w(t), \tag{8.31}$$

where the variance of Gaussian white noise $w(t)$ is σ^2. The marginal steady-state density distribution of (8.31) is given (see Caughey 1963) by

$$p(\dot{x}, x) = p_0 \exp\left(-\frac{a\dot{x}^2}{\sigma^2}\right)\exp\left(-\frac{2\int^x b(\xi)\xi d\xi}{\sigma^2}\right). \tag{8.32}$$

On the other hand, we know that the steady-state marginal density distribution of $y(t)$ of (8.30) is

$$p(y) = q_0 \exp\left(-\frac{2\int^y f(\eta)d\eta}{\sigma^2}\right),$$

where p_0 and q_0 are normalizing constants.

8.4.2 Gamma-Distributed Oscillation Model

If we replace $b(x)$ in model (8.31) by $f(.)$ of the associated dynamical system (8.20) of the Type-I Gamma distributed process in model (8.31), that is, if we take

$$\ddot{y}(t) + a\dot{y}(t) - \left(\alpha - \frac{1}{2}\right)\frac{1}{y} + \frac{y}{2} = w(t) \qquad (8.33)$$

we have the same density distribution $p(y)$ as the associated variable y of the Type-I Gamma process. By combining, with (8.32), the variable transformation

$$x(t) = \frac{\beta y(t)^2}{2}, \qquad (8.34)$$

we can have another Gamma-distributed process (Type-III Gamma-distributed process) $x(t)$ that has a vibration mechanism. Some of the Type-III Gamma distributed processes simulated by the locally linearized time series models with $\alpha = 4$, $\beta = 1$, and $\Delta t = 0.05$ are shown in Figure 8.5a.

From (8.33) and (8.34), we have, for the Type-III Gamma-distributed process $x(t)$, the following nonlinear oscillation model:

$$\ddot{x} + a(x, \dot{x})\dot{x} + b(x)x = \sqrt{2\beta x}w(t),$$

where

$$a(x, \dot{x}) = \left(a - \frac{\dot{x}}{2x}\right)$$

and

$$b(x) = 1 - \frac{\alpha\beta}{x}.$$

The model is rewritten as

$$\ddot{x} + a(x, \dot{x})\dot{x} + b(x, w)x = 0,$$

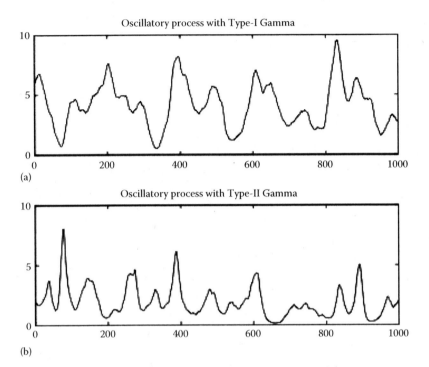

FIGURE 8.5
(a) Type-3 Gamma simulation data. (b) Type-4 Gamma simulation data.

where

$$b(x,w) = \left\{ 1 - \frac{\alpha\beta}{x} - \frac{\sqrt{2\beta}}{\sqrt{x}} w(t) \right\}.$$

The model shows that for large positive value of x, it behaves like a linear oscillation model, but when $x(t)$ approaches to zero a special nonlinear mechanism of $a(x, \dot{x})$ and $b(x, w)$ starts functioning to stop $x(t)$ going negative. The interplay of these mechanisms yields to positive-valued oscillation behavior of $x(t)$.

8.4.3 Type-IV Gamma-Distributed Process

If we replace $b(x)x$ in model (8.31) by $f(.)$ of the dynamical system,

$$f(y) = \frac{(\alpha-1)\beta}{\sqrt{2\beta}} - \frac{\exp(\sqrt{2\beta}y)}{\sqrt{2\beta}},$$

of the process $y(t)$ in the Type-II Gamma process model (8.28), that is, if we use

$$\ddot{y}(t) + a\dot{y}(t) - \frac{\alpha\beta}{\sqrt{2\beta}} + \exp\left\{-\sqrt{2\beta}y(t)\right\} = w(t), \tag{8.35}$$

we have the same marginal distribution $p(y)$ as the associated variable y of the Type-II Gamma process (8.28). By combining (8.35) and the variable transformation,

$$x(t) = \exp\left(\sqrt{2\beta}y(t)\right), \tag{8.36}$$

we can have another Gamma-distributed process (Type-IV) $x(t)$, which has a nonlinear vibration mechanism. Some of the Type-IV Gamma distributed processes simulated by the locally linearized time series model with $\alpha = 4$, $\beta = 1$, and $\Delta t = 0.05$ are shown in Figure 8.5b.

From (8.35) and (8.36), we have the following nonlinear oscillation model whose marginal stationary distribution is Gamma type,

$$\ddot{x}(t) + a(x, \dot{x})\dot{x}(t) + b(x)x(t) = \sqrt{2\beta}x(t)w(t), \tag{8.37}$$

where $a(x, \dot{x}) = \left(a - \dfrac{\dot{x}}{x}\right)$, and $b(x) = \left(\sqrt{2\beta} - \dfrac{\alpha\beta}{x}\right)$. The model (8.37) shows that for a large $x(t) > 0$, it behaves like a linear oscillation model since $a(x, \dot{x})$ and $b(x)$ are almost constant. However, when $x(t)$ approaches zero the nonlinear mechanism of $a(x, \dot{x})$ and $b(x)$ starts functioning to stop $x(t)$ going negative. The interplay of theses mechanisms creates a positive-valued oscillatory process of $x(t)$.

The model is rewritten as an oscillation model whose second coefficient $b(x)$ is disturbed by a Gaussian white noise $w(t)$ as

$$\ddot{x}(t) + a(x, \dot{x})\dot{x}(t) + b(x, w)x(t) = 0,$$

where

$$b(x, w) = \sqrt{2\beta} - \frac{\alpha\beta}{x} - \sqrt{2\beta}w(t).$$

It implies that even though the basic oscillation dynamics are deterministic, if the coefficients of the deterministic model are regulated by an unknown external force, it can lead to a nondeterministic stochastic system.

8.4.4 Variable Transformation and the Innovation Approach

So far, we have seen several possible mechanical structures behind a homogeneous stochastic process with a certain marginal distribution. For the Gamma distribution, for instance, we have seen four different Gamma-distributed processes:

$$
1. \quad
\left(
\begin{aligned}
x &= \frac{\beta y^2}{2}, \\
\dot{y} &= \left(\left(\alpha - \frac{1}{2} \right) \frac{1}{y} - \frac{y}{2} \right) dt + w(t),
\end{aligned}
\right.
\tag{8.38}
$$

$$
2. \quad
\left(
\begin{aligned}
x(t) &= \exp\left\{ \sqrt{2\beta} y \right\}, \\
\dot{y} &= \frac{(\alpha - 1)\beta}{\sqrt{2\beta}} - \frac{\exp(\sqrt{2\beta} y)}{\sqrt{2\beta}} + w(t),
\end{aligned}
\right.
\tag{8.39}
$$

$$
3. \quad
\left(
\begin{aligned}
x &= \frac{\beta y^2}{2}, \\
\ddot{y} + a\dot{y} &- \left(\alpha - \frac{1}{2} \right) \frac{1}{y} + \frac{y}{2} = w(t),
\end{aligned}
\right.
\tag{8.40}
$$

$$
4. \quad
\left(
\begin{aligned}
x &= \exp\left\{ \sqrt{2\beta} y \right\}, \\
\ddot{y} + a\dot{y} &- \frac{(\alpha - 1)\beta}{\sqrt{2\beta}} + \exp\left\{ \sqrt{2\beta} y \right\} + w(t)
\end{aligned}
\right.
\tag{8.41}
$$

where each of them has its own dynamics specified by the own associated dynamical system driven by the Gaussian white noise $w(t)$ together with its own variable transformation.

Note that each of the models shows its own way of transforming the Gaussian white noise $w(t)$ into Gamma-distributed temporally dependent process $x(t)$. In time-series data analysis, the model (8.38) suggests taking the square root transformation of the positive-valued time-series data before fitting a nonlinear time series model, while the model (8.39) suggests taking the log-transformation before fitting a nonlinear time series model.

It is interesting to see that the use of these variable transformations, commonly seen in empirical non-Gaussian time-series data analysis (see Ozaki and Iino, 2001), is suggested from our theoretical analysis. In some cases (Type-I Gamma process with $\alpha = 1/2$), the process is transformed to a Gaussian process by a memoryless transformation, where a linear modeling

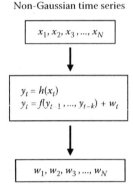

$$x_1, x_2, x_3, \ldots, x_N$$

$$y_t = h(x_t)$$
$$y_t = f(y_{t-1}, \ldots, y_{t-k}) + w_t$$

$$w_1, w_2, w_3, \ldots, w_N$$

FIGURE 8.6
Whitening with a variable-transformation and a dynamic model.   Gaussian white noise

is sufficient. This means that memoryless variable transformations are as important as nonlinear time series models in non-Gaussian time-series data analysis. From the late 1980s, some general approaches to non-Gaussian time series modeling have attracted time-series analysts' attention, where researchers are interested in using computationally expensive numerical methods for integrating non-Gaussian conditional distributions of the data without looking into the innovations. Certainly it is not recommended, if the time-series data are of Type-I Gamma process of $\alpha = 1/2$, to use such a general non-Gaussian time series modeling method for analyzing the data without paying attention to the possibility of transforming the data to white innovations by a simple variable transformation plus linear modeling.

If we look at $x(t)$ and $w(t)$ in the opposite way, we can see that these four models show four different ways of whitening the temporally dependent Gamma-distributed process $x(t)$ into Gaussian white noise $w(t)$. In experimental science, we have a single observation data, and what we are interested in is to find a proper model explaining the data. We call the problem "inverse problem" in contrast to the "forward problem," where the researcher's interest is in finding properties of $x(t)$ from an assumed theoretical model. Obviously, the time-series analysis is classified as an inverse problem, but its characteristic point is that the number of data is finite. Using the time-series data of finite length, we need to make an inference on the model behind the data. Here we would like to remark again that the innovation approach is valid and useful for the present situation. What is required is to find a variable transformation $y(t) = h(x(t))$ and the associated dynamical system in terms of the transformed variable, that is, $\dot{y} = f(y)$ from the time-series data, $x_1, x_2, \ldots, x_N$ (Figure 8.6).

9

Bridge between Continuous-Time Models and Discrete-Time Models

9.1 Four Types of Dynamic Models

9.1.1 Why We Need Them

We have seen that there are two types of dynamic models for prediction, one is a class of deterministic dynamic models and another is a class of nondeterministic (or stochastic) dynamic models. Note that a deterministic model can be regarded as a limit of nondeterministic model in the sense that the prediction error of the nondeterministic model is zero. Although the two groups of models, deterministic models and stochastic models, are closely related, the tools and theories developed for each of them are quite different.

A well-known powerful tool for continuous-time deterministic models is the classic calculus developed since the era of Newton and Laplace. For continuous-time stochastic models, these classic tools for deterministic models have some technical problems and need to be extended to stochastic calculus.

Although there exist two ways of extending ordinary calculus into stochastic calculus, that is, Ito calculus (Ito 1942, 1944, 1951) and Stratonovich calculus (Stratonovich 1966), this does not yield any difficult problem, because there is a one-to-one correspondence between the two calculi. Each of the theories of stochastic calculus is consistent and self-contained (Mortensen 1979).

We will see that the difference between the two calculi becomes visible in the definition of a stochastic differential equation model, with nonconstant diffusion coefficient, of the type $dx = a(x)dt + b(x)dW(t)$. However, both calculi have their own consistent way of defining a variable transformation, $y = h(x)$, so that the model can be transformed, using its own rule of variable transformation, to the so-called Langevin type model with transformed variable y, that is, $dy = f(y)dt + dW(t)$, where $y = h(x)$. We have seen some examples of this type of variable transformations in the previous chapter (see Section 8.3). Thus, we can have a one-to-one correspondence between any model defined in Ito calculus or in Stratonovich calculus through the Langevin type model form, as long as the functions used for defining the stochastic differential equation model are smooth analytic functions.

Another way of classifying prediction models is to find whether the dynamics are specified in a continuous-time model or in a discrete-time model. All the dynamic models considered in Chapters 3 through 6 are classified into two groups, continuous-time models and discrete-time models. Theoretical tools such as power spectral theories, auto covariance functions, Doob decompositions, etc., have been developed in parallel, for both continuous-time models and for discrete-time models.

All prediction models may be classified in terms of these two criteria: (1) whether the model is deterministic or stochastic, and (2) whether the model is described in continuous time or in discrete time. Then all prediction models may be classified into one of four groups of dynamic model: (1) continuous-time deterministic models, (2) continuous-time stochastic models, (3) discrete-time deterministic models, and (4) discrete-time stochastic models.

Models in all four groups may be used for predicting the same phenomena. Naturally, models from all groups shares similarities being dynamic models for the same dynamic phenomena. For example, the frequency properties of dynamic phenomena are characterized by the power spectrum both for continuous-time models and discrete-time models (see Sections 3.2 and 5.2).

On the other hand, models from the different groups have their own advantages and disadvantages over the other. One of the significant advantages of discrete-time nonlinear stochastic dynamic models (i.e., time series models) over continuous-time stochastic dynamic models is that, since time series models are statistical models, they can be easily estimated and identified from observed time series data of finite length.

Continuous-time stochastic dynamical system models, on the other hand, were developed originally as mathematical models without consideration of statistical identification of the models using discrete-time data. Their statistical identification is, unlike discrete-time models, not so simple.

However, being continuous-time mathematical models gives them an advantage over discrete-time dynamic models in characterizing stochastic dynamic phenomena, especially for non-Gaussian Markov processes, with powerful analytical tools developed in the theory of probability and Markov diffusion processes, as we saw in Chapter 8.

Unfortunately, however, continuous-time Markov diffusion process models and discrete-time nonlinear time series models have rarely been discussed together. This is because most time series analysts are not interested in continuous-time models, and most mathematicians and probabilists are not interested in discrete-time stochastic time series models. The observation data for any dynamic phenomena are usually given as time series data with an equidistant time interval. The discrete-time models have been playing a main role in time series analysis. Continuous-time models have been studied mostly by applied mathematicians who are mainly interested in deriving mathematical properties of the models either by theoretical or by numerical means. The estimation

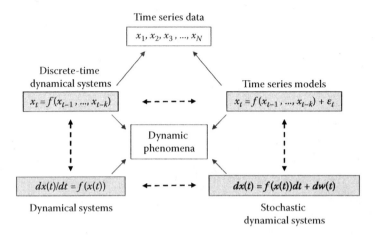

FIGURE 9.1
Four models and the four bridges.

of continuous-time dynamic models has been discussed by some statisticians without paying attention to similar discrete-time models in the subject of time series analysis. Whether we like it or not, it has been a tendency that both types of model were studied separately without paying much attention to the similarities or differences between the two.

One of our objectives, in the present book, is to promote a defiant approach towards this tendency of "sectionalism." Here we would like to start from time series data and look into useful models and theories, whether continuous-time dynamic models or discrete-time models. For this purpose we would like to find bridges connecting the 4 groups of dynamic models (see Figure 9.1). If we have some reasonable bridges, they provide us with more powerful methods, than those constructed only for a single group of models, for the data analysis of time series in general (Ozaki 1990, Ruelle 1987, Sugihara 1995, Takens 1981).

9.1.2 Discretization Bridge and Stability

Before going to the discussion of building bridges between the continuous-time general nonlinear dynamic models and discrete-time general nonlinear dynamic models, let us start from looking at four simple typical linear dynamic models, where we can easily see clear relations between the models:

1. Continuous-time linear deterministic dynamical system model,

$$dx/dt = -ax \tag{9.1}$$

2. Continuous-time linear stochastic dynamical system model,

$$dx = -axdt + dW(t) \tag{9.2}$$

3. Discrete-time linear deterministic dynamical system model,

$$x_{t+1} = \phi x_t \tag{9.3}$$

4. Discrete-time linear stochastic dynamical system model,

$$x_{t+1} = \phi x_t + w_{t+1} \tag{9.4}$$

The relation between (9.1) and (9.2) is obvious and has been commonly used in sciences, where the stochastic model,

$$dx = -axdt + dW(t),$$

is also written informally as $\dot{x} = ax + W(t)$, and is called Langevin equation derived from the dynamical system $\dot{x} = -ax$.

The model (9.4) is a first-order AR model, and the relation between (9.3) and (9.4) is also obvious, although (9.3) is hardly ever used by itself as a prediction model. Deterministic discrete-time models have been studied more often for the nonlinear case related to the study of chaos. However, the model (9.3) is often discussed together with (9.4) for the purpose of characterizing the stability and the stationarity of the stochastic process defined by (9.4).

From these relations, our attention will naturally be directed to the relation between a and ϕ of the four models. For the deterministic case the relation is naturally given as $\phi = \exp(-a\Delta t)$, where Δt is the time interval between the nominal time points, t and $t + 1$, determined when the time series are sampled from the continuous-time data. This is because the analytic solution for $\dot{x}(t) = -ax$ is given, with the initial value x_0 at $t = 0$, by $x(t) = \exp(-at)x_0$. Then the time series, $x_0, x_{\Delta t}, \ldots, x_{k\Delta t}$, defined by

$$x_{t+\Delta t} = \exp(-a\Delta t)x_t,$$

will coincide with the trajectory $x(t) = \exp(-at)x_0$ of $dx(t)/dt = -ax(t)$ at the equidistant time points, $t = 0, t = \Delta t, t = \Delta t, \ldots$.

From the earlier heuristic discussion, we see that the four models are closely connected by the two types of the bridges (see Figure 9.2), that is, a bridge connecting the continuous-time model and discrete-time model (which we call Bridge C-D) and a bridge connecting the deterministic model and the stochastic model (which we call Bridge D-S). Incidentally, we note that we have already seen an example of Bridge C-D when we introduced the continuous-time spectral density function for the multivariate stochastic linear dynamical system models in Section 5.2.

An important point here is that the stability condition of the discrete-time model $x_{t+\Delta t} = \exp(-a\Delta t)x_t$ is specified automatically by the condition for the stability of the original continuous-time model, $dx/dt = -ax$. As long as this

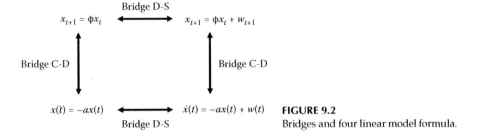

FIGURE 9.2
Bridges and four linear model formula.

condition (specified by "the real part of the eigenvalue(s) of the transition coefficient (matrix) $-a$" is negative) is satisfied, the discretized model, $x_{t+\Delta t} = \exp(-a\Delta t)x_t$ is always stable, since its trajectory is exactly on the trajectory given by the analytic solution of $dx/dt = -ax$.

An argument of the stability conditions for the stochastic case goes in parallel to the deterministic case. The stationarity condition of the stochastic version of the linear model, that is, $dx/dt = -ax + w(t)$, is equivalent to the stability condition of $dx/dt = -ax$, that is, $a > 0$, and the stationarity condition for the stochastic version of the discrete-time linear model, $x_{t+1} = \phi x_t + w_{t+1}$, is equivalent to the stability condition, that is, $|\phi| < 1$.

9.1.3 Bridge for the Stochastic Case

The integration and differentiation of stochastic processes can be realized by the two definitions of stochastic calculus, i.e., Ito calculus and Stratonovich calculus. By both definitions of stochastic calculus, the integrated solution of $dx = -axdt + dW(t)$ at time point t is given by

$$x(t) = x(0) + \int_0^t -ax(u)du + \int_0^t dW(u),$$

where the third term of the right-hand side is the stochastic integral defined as a limit of

$$\lim_{n\to\infty} \sum_{i=0}^{n-1} \{W(t_{i+1}) - W(t_i)\},$$

where the interval $[0, t]$ is divided into n subintervals with

$$0 \le t_0 < t_1 < \cdots < t_n = t.$$

Suppose the time series, $x_1, x_2, ..., x_N$, are sampled with a reasonably small interval Δt to trace a smooth change of the trajectory of $x(t)$. Then, it will be reasonable to define a discrete-time trajectory of the process as

$$x(t + \Delta t) = x(t) + \int_{t}^{t+\Delta t} -ax(u)du + \int_{t}^{t+\Delta t} dW(u)$$

$$= \exp(-a\Delta t)x(t) + w_{t+\Delta t},$$

where

$$w_{t+\Delta t} = \int_{t}^{t+\Delta t} dW(u),$$

$$E[w_{t+\Delta t}] = 0,$$

$$E[(w_{t+\Delta t})(w_{t+\Delta t})'] = \sigma_w^2 \left[\frac{\exp(-2a\Delta t) - 1}{-2a} \right].$$

Thus, we have a stochastic dynamic model of an AR model type,

$$x_{t+1} = \phi x_t + w_{t+1},$$

as a discrete-time counterpart of the stochastic continuous-time dynamic model,

$$dx/dt = -ax + w(t),$$

where

$$\phi = \exp(-a\Delta t),$$

$$E[w_{t+1}^2] = \sigma_w^2 \{\exp(-2a\Delta t) - 1\}/(-2a),$$

and σ_w^2 is a variance of the continuous-time white noise defined by,

$$E[W(t + \Delta t) - W(t)]^2 = \sigma_w^2 \Delta t.$$

This discretization scheme can be used in an opposite direction, that is, from discrete-time model to continuous-time model, if the coefficient (matrix) satisfies certain stability conditions. For example, if the coefficient (matrix) ϕ of a discrete-time model, $x_{t+1} = \phi x_t$, satisfies the inequalities, $0 < \phi < 1$, a natural counterpart of the continuous-time model for this discrete-time model will be $dx/dt = -ax$, with $a = -\log(1 - \phi)$. This means that we can have a natural bridge connecting the continuous-time linear model, $dx/dt = -ax$ and discrete-time linear model, $x_{t+1} = \phi x_t$, under the aforementioned stability conditions. Similarly, from the stochastic discrete-time dynamic model, $x_{t+1} = \phi x_t + w_{t+1}$, with $0 < \phi < 1$, we have a continuous-time version as $dx/dt = -ax + w(t)$, with $a = -\log(1 - \phi)$.

9.1.4 How to Choose Δt

When we discuss the relation of the continuous-time stochastic model, $dx = f(x)dt + dW(t)$, and the time series $x_1, x_2, ..., x_N$, the choice of sampling interval Δt is always important. Here we have to notice that there are two different types of problems for choosing Δt: the difference of the two types is typically seen in the simple example of a linear stochastic dynamical system model,

$$dx = -axdt + dW(t). \tag{9.5}$$

1. How to choose Δt in order to generate sampled data, $x_t, x_{t+\Delta t}, ..., x_{t+k\Delta t}$, from the continuous-time model (9.5)
2. How to choose Δt of the model

$$x_{t+\Delta t} = x_t - \Delta t a x_t + w_{t+\Delta t} \tag{9.6}$$

for a given time series $x_1, x_2, ..., x_N$.

In (1) a is fixed but the sample data, $x_t, x_{t+\Delta t}, ..., x_{t+k\Delta t}$, are not fixed. In (2) sample data, $x_1, x_2, ..., x_N$, are fixed, but we do not know a and Δt and are trying to estimate them from the data. In (1) we can change the data by changing Δt, where the data $x_t, x_{t+\Delta t}, ...,$ change more smoothly if we make Δt smaller. However in (2), the time series data are fixed and are not affected by the choice of Δt. Here we have

$$x_{t+1} = (1 - \Delta t a)x_t + w_{t+1}.$$

What we can estimate from the given data is the autoregressive coefficient, $(1 - \Delta t a)$, of the model and it is more or less uniquely estimated from the data by minimizing the sum of squares of prediction errors. Here we have the estimated coefficient $\hat{\phi} = 1 - a\Delta t$. Then if we change Δt from 0.01 to 0.001, it affects only the scaling unit of a since $a = (1 - \hat{\phi})/\Delta t$.

9.2 Local Linearization Bridge

9.2.1 Bridge for Nonlinear Dynamic Models

Our next interest will be naturally on the nonlinear generalization of these relations and bridges. Before we go to the discussion of general nonlinear models, $dx/dt = f(x)$, with general nonlinear function $f(x)$, we first consider

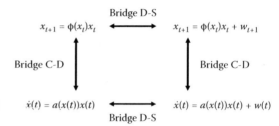

FIGURE 9.3
Bridges and four nonlinear model formula.

bridges for a special class of nonlinear functions that behaves similarly to a linear function when x is close to zero. We consider first bridges for those $f(x)$'s satisfying $f(0) = 0$ (see Figure 9.3). This type of nonlinear functions is written, using some nonlinear function $a(x)$, as $f(x) = a(x)x$, where we have $|a(0)| < \infty$.

The set of nonlinear dynamical systems satisfying the conditions mentioned earlier includes the following four examples of typical nonlinear dynamical systems:

Example 1:

$$\frac{dx}{dt} = -ax^3$$

Example 2:

$$\frac{dx}{dt} = 2x - x^3$$

Example 3:

$$\frac{dx}{dt} = -6x + 5.5x^3 - x^5$$

Example 4:

$$\frac{dx}{dt} = -\tanh\left(\sqrt{2}x\right)$$

For the nonlinear dynamical system model, $\dot{x} = a(x)x$, we have its stochastic counterpart as $dx = a(x)xdt + dW(t)$. Then an interesting question is whether we have its discretized version $x_{t+1} = \phi(x_t)x_t$, and $x_{t+1} = \phi(x_t)x_t + w_{t+1}$ as for the linear case. If we have such a discretized version, then we want to know how the two functions, $a(x)$ and $\phi(x_t)$, are related to each other. In other words, our question is whether we can have such $\phi(x_t)$ that allows having the bridge (see Figure 9.3) connecting the two dynamic models (9.7) and (9.9) and also the bridge connecting (9.8) and (9.10) of the four different types of models, which are specified as

1. Continuous-time nonlinear deterministic dynamical system model,

$$\dot{x} = a(x)x \tag{9.7}$$

2. Continuous-time linear stochastic dynamical system model,

$$dx = a(x)xdt + dW(t) \tag{9.8}$$

3. Discrete-time nonlinear deterministic dynamical system model,

$$x_{t+1} = \phi(x_t)x_t \tag{9.9}$$

4. Discrete-time nonlinear stochastic dynamical system model,

$$x_{t+1} = \phi(x_t)x_t + w_{t+1} \tag{9.10}$$

9.2.2 Discretization and the Instability Problem: Deterministic Cases

Let us start from a simple example of a nonlinear deterministic dynamical system.

Example 1

$$\dot{x} = -ax^3. \tag{9.11}$$

If we can have, for (9.11), an analytic solution,

$$x(t + t_0) = \exp\{L(x_t)\Delta t\}x(t_0),$$

with some $L(x_t)$ on the interval $[t_0, t_0 + t)$ as for the linear case, we may have a natural discrete-time model,

$$x_{t+\Delta t} = \phi(x_t, \Delta t)x_t,$$

where

$$\phi(x_t, \Delta t) = \exp\{L(x_t)\Delta t\}.$$

Unfortunately, we do not have any such analytic solution for the nonlinear dynamical systems. Such an obvious exact relation between the linear continuous-time model and the linear discrete-time model is not available. Then to obtain a discrete-time model, in the general case of nonlinear dynamic models, we may look for an approximate relation. We may first obtain an approximate model by discretizing the model, that is, by replacing dx/dt by $(x_{t+\Delta t} - x_t)/\Delta t$, which yields

$$x_{t+\Delta t} = x_t - \Delta t a x_t^3. \tag{9.12}$$

This discretizing scheme is called the Euler scheme and has been commonly used in the computer simulation of nonlinear dynamical systems in most textbooks of numerical mathematics (see, e.g., Henrici 1962). However, the derived discretized model has a serious problem for a statistical time series model, that is, the instability problem.

For example, if the initial value x_0 of (9.12) is $|x_0| > \sqrt{(2/\Delta t)}$, then $(1 - \Delta t |x_0|^2) < -1$, so that the trajectory $x_{k\Delta t}$ of the discrete-time dynamical system model,

$$x_{(k+1)\Delta t} = x_{k\Delta t} - \Delta t a x_{k\Delta t}^3 = \left(1 - \Delta t a x_{k\Delta t}^2\right) x_{k\Delta t}, \tag{9.13}$$

shows oscillatory explosive behavior for $k \to \infty$. In the numerical analysis, researchers are interested in solving the continuous-time equation, so that they can choose any small Δt, satisfying $-1 < (1 - \Delta t |x_0|^2) < 1$, which guarantees the stability of the discretized solution. However, in time series analysis, Δt is fixed. We cannot adjust it for stabilizing the discretized solution. Therefore, this discretization scheme transforms the model with a finite non explosive trajectory into a model (9.13) of unstable explosive trajectory.

What we really need is a discretization scheme guaranteeing the stability of the trajectory for any initial value x_0 for a given "fixed" Δt. In the field of numerical analysis, there are some more refined (with higher order approximation) discretization schemes of nonlinear differential equations, apart from the Euler scheme. The Runge–Kutta method is one of them and has been used in the discretization of complex nonlinear differential equations. With the Runge–Kutta method, the continuous-time model (9.11) is discretized into

$$y_{t+\Delta t} = y_t + \frac{\Delta t}{6}(k_1 + 2k_2 + 2k_3 + k_4) = p_{80}(y_t)y_t, \tag{9.14}$$

where

$$k_1 = -y_t^3, \quad k_2 = -\left(y_t + \frac{\Delta t}{2}k_1\right)^3, \quad k_3 = -(y_t + \Delta t k_3)^3$$

and $p_{80}(x_t)$ is an 80th order polynomial of x_t. We can see however, that this does not solve the instability problem, because the 80th order polynomial, $p_{80}(x_t)$, goes to infinity when $|x_t| \to \infty$ for any fixed value of Δt.

9.2.3 Discretization and Instability Problem: Stochastic Cases

The instability problem is more serious for the stochastic case where we need to build a bridge C-D between the continuous-time stochastic dynamical system,

$$dx = a(x)xdt + dW(t),$$

and the discrete-time stochastic dynamical system, that is, nonlinear time series model,

$$x_{t+\Delta t} = \phi(x_t)x_t + w_{t+\Delta t}.$$

In the stochastic case, the difficulty comes not only from the initial value of the process x_t, but it is also caused by the driving noise $w_{t+\Delta t}$. If $\phi(x_t)$ is a polynomial function of x_t, then since the driving noise $w_{t+\Delta t}$ is Gaussian distributed the probability of pushing the point x_t out of a feasible region is always greater than zero. Because of this, the Markov chain process defined by such a polynomial type nonlinear AR model is always explosive and non-ergodic (Jones 1978), and we cannot have any finite-variance discrete-time process with a stationary steady state distribution for this type of Markov chain processes. The bridge C-D built by such a discretization scheme is not useful for our purpose, because the stochastic process defined by

$$dx = -ax^3 dt + dW(t)$$

is known to be a finite-variance process with a stationary distribution of the leptokurtic type,

$$Q(x) = Q_0 \exp\left(-\frac{x^4}{2\sigma^2}\right).$$

What we are interested in is a bridge connecting a non explosive stochastic dynamical system model, $dx = f(x)dt + dW(t)$, into a non explosive discrete-time model, so that the processes defined on both sides of the bridge are of "finite-variance" property. Such a bridge could be useful for the statistical analysis of the time series observed from nonlinear but non explosive dynamic phenomena. In this situation, what we need is a discretization scheme transforming the "finite-variance" continuous-time process into "finite-variance" discrete-time process (i.e., stationary ergodic Markov chain defined on the Euclidian state space).

9.2.4 LL Scheme for Deterministic Systems

A useful discretization scheme comes from the following idea of local linearization. The original dynamical system satisfies $\dot{x}(s) = f(x(s))$.

By taking the derivative of $\dot{x}$, we have

$$\ddot{x}(s) = J(x(s))\dot{x}(s)$$

on any interval, $-\infty < s < \infty$ where $J(x(s)) = \dfrac{\partial f(x(s))}{\partial x(s)}$. This is because we have

$\ddot{x}(s) = \dfrac{\partial}{\partial s} \dot{x}(s) = \dfrac{\partial}{\partial s} f(x(s)) = \dfrac{\partial f(x)}{\partial x} \dfrac{\partial x(s)}{\partial s} = J(x(s))\dot{x}(s)$. However, we consider,

instead of

$$\ddot{x}(s) = J(x(s))\dot{x}(s),$$

an approximate system, which is valid only on the short interval $[t_0, t_0 + \Delta t]$, satisfying

$$\ddot{x}(s) = J_{t_0}\dot{x}(s). \tag{9.15}$$

The solution of (9.15) for $[0, \tau]$ is

$$\dot{x}(t_0 + \tau) = e^{J_{t_0}\tau}\dot{x}(t_0), \tag{9.16}$$

which is valid for $\tau \in [0, \Delta t]$. If we integrate this on $\tau \in [0, \Delta t]$, we have

$$x(t_0 + \Delta t) = x(t_0) + J_{t_0}^{-1}(e^{J_{t_0}\Delta t} - 1)f(x(t_0)). \tag{9.17}$$

Note that, when $f(x) = a(x)x$, we can have a discrete-time model for $\dot{x} = f(x)$ as

$$x_{t+\Delta t} = \phi(x_t)x_t,$$

where $\phi(x_t)$ is

$$\phi(x_t) = 1 + J_t^{-1}(e^{J_t\Delta t} - 1)f(x_t)/x_t$$

$$= 1 + J_t^{-1}(e^{J_t\Delta t} - 1)a(x_t).$$

The scheme is called the Local Linearization (LL) scheme. It has been shown (Ozaki 1985a,b) that for most nonlinear functions $f(x)$ yielding to a stable nonlinear dynamical system, $\dot{x} = f(x)$, we have

$$|\phi(x_t)| < 1 \quad \text{for } |x_t| \to \infty,$$

and so it does not have the instability problems displayed by the Euler scheme, Runge–Kutta scheme, and other conventional schemes. We can see this using some of the examples in the previous sections.

The example 1 in Section 9.2.1, that is,

$$\dot{x} = -ax^3,$$

is transformed by the LL scheme to the following discrete-time model:

$$x_{t+\Delta t} = \left[1 + \frac{1}{3}\left\{e^{-3x_t^2} - 1\right\}\right]x_t$$

$$= \left(\frac{2}{3} + \frac{1}{3}e^{-3x_t^2}\right)x_t.$$

Here,

$$\left|\frac{2}{3} + \frac{1}{3}e^{-3x_t^2}\right| < 1$$

for any large $|x_t|$, so that the trajectory does not show oscillatory explosions no matter what initial value it starts from.

The example 3 in Section 9.2.1, that is,

$$\dot{x} = -6x + 5.5x^3 - x^5,$$

is transformed by the LL scheme to the following discrete-time model:

$$x_{t+\Delta t} = \phi(x_t)x_t,$$

where

$$\phi(x_t) = \left\{1 + J_t^{-1}\left(e^{J_t \Delta t} - 1\right)\left(-6 + 5.5x_t^2 - x_t^4\right)\right\}$$

and

$$J_t = -6 + 16.5x_t^2 - 5x_t^4.$$

Since we have, from this $\phi(x_t)$,

$$|\phi(x_t)| \to e^{-6\Delta t} < 1, \quad \text{for } |x_t| \to \infty,$$

we have again obtained a stationary finite-variance time series process by the LL discretization.

9.2.5 LL Scheme for Stochastic Systems

The Local Linearization scheme can be extended to the stochastic case where the continuous-time dynamic model is specified by a stochastic differential equation of the type

$$\dot{x} = f(x) + \sigma w(t). \tag{9.18}$$

Here we assume that $f(x)$ is a smooth nonlinear function so that we have its time differential as

$$\dot{f} = J(x)\dot{x}.$$

With the local linearity assumption, that is, $\partial f / \partial x$ is constant, $J_{x_{t_0}}$, for the interval $[t_0, t_0 + \Delta t]$, we have $\dot{f} = J_{x_{t_0}}\dot{x}$. Then, $f(x(t))$ can be integrated for the interval, $t_0 \leq \tau < t_0 + \Delta t$, as

$$f(x(\tau)) \approx f(x_{t_0}) + J_{t_0}(x(\tau) - x_{t_0}) = J_{t_0}x(\tau) + f(x_{t_0}) - J_{t_0}x_{t_0}.$$

Using this approximate $f(x(\tau))$, we have, from

$$\dot{x}(\tau) = f(x(\tau)) + \sigma w(\tau),$$

the following approximate linear stochastic differential equation on each short time interval, $[t_0, t_0 + \Delta t)$,

$$\dot{x}(\tau) = J_{t_0}x(\tau) + N_{t_0} + \sigma w(\tau),$$

where

$$N_{t_0} = f(x_{t_0}) - J_{t_0}x_{t_0}.$$

By integrating this for the interval, $[t_0, t_0 + \Delta t)$, we have

$$x_{t_0 + \Delta t} = \exp(J_{t_0}\Delta t)x_{t_0}$$

$$+ N_{t_0} \int_{t_0}^{t_0 + \Delta t} \exp\{J_{t_0}(t_0 + \Delta t - u)\}du$$

$$+ \int_{t_0}^{t_0 + \Delta t} \exp\{J_{t_0}(t_0 + \Delta t - u)\}\sigma w(u)du$$

$$= x_{t_0} + \left(\frac{1}{J_{t_0}}\right)\{\exp(J_{t_0}\Delta t) - 1\}f(x_{t_0})$$

$$+ \sigma \int_{t_0}^{t_0 + \Delta t} \exp(J_{t_0}(t_0 + \Delta t - u))w(u)du. \tag{9.19}$$

The variance of the third term on the right-hand side of (9.19) is

$$\sigma^2\left\{\exp(2J_{t_0}\Delta t)-1\right\}/(2J_{t_0}).$$

This suggests that a discrete-time stochastic counterpart model for (9.18) is

$$x_{t+\Delta t}=\phi(x_t)x_t+\sigma B_t w_{t+\Delta t},$$

where

$$\phi(x_t)=1+(1/J_t)\{\exp(J_t\Delta t)-1\}f(x_t)/x_t,$$

$$B_t=\sqrt{\{\exp(2J_{t_0}\Delta t)-1\}/(2J_{t_0})},$$

and the variance of the discrete-time Gaussian white noise $w_{t+\Delta t}$ is 1. Note that $B_t \approx \sqrt{\Delta t}$ for sufficiently small Δt. If $f(x)$ is of the type $f(x) = a(x)x$, then we have

$$\phi(x_t)=1+(1/J_t)\left\{\exp(J_t\Delta t)-1\right\}a(x_t).$$

With this discretization scheme, we have, from the model of Example 1,

$$\dot{x}=-ax^3+\sigma w(t), \tag{9.20}$$

the following discrete-time model,

$$x_{t+\Delta t}=\left(\frac{2}{3}+\frac{1}{3}e^{-3x_t^2}\right)x_t+\sigma\sqrt{\Delta t}w_{t+\Delta t}. \tag{9.21}$$

Note that the model is a special case of the first-order ExpAR model, which we saw in Section 6.1, given by

$$x_{t+1}=(\phi+\theta e^{-\gamma x_t^2})x_t+n_{t+\Delta t}. \tag{9.22}$$

The time series process defined by the ExpAR model (9.22) is known to be ergodic and stationary when $|\phi| < 1$. That means the LL scheme transforms the finite-variance continuous-time process defined by (9.20) into a finite-variance discrete-time process of (9.21).

We can easily see that a more general finite-variance process $x(t)$ defined by

$$\dot{x} = a_1 x(t) + a_2 x(t)^2 + \cdots + a_r x(t)^r + \sigma w(t)$$

is also transformed by the LL scheme into a stationary finite-variance process. Sufficient conditions for the nonlinear function, $f(x)$, of $\dot{x} = f(x) + \sigma w(t)$ to be transformed into LL-discretized finite-variance process are discussed in Ozaki (1985a).

9.2.6 LL Scheme for Stochastic Systems with an Exogenous Input

The Local Linearization scheme can be extended to the case where the continuous-time dynamic model is driven by an exogenous input variable $y(t)$ as well as Gaussian white noise,

$$\dot{x} = f(x) + \sigma w(t) + cy(t). \tag{9.23}$$

Here, in the same way as for the stochastic systems in Section 9.2.5, we can assume that $f(x)$ is a smooth nonlinear function so that we have its time differential as

$$\dot{f} = J(x)\dot{x}.$$

With the local linearity assumption, that is, $\partial f/\partial x$ is constant, $J_{x_{t_0}}$, for the interval $[t_0, t_0 + \Delta t]$, we have $\dot{f} = J_{x_{t_0}}\dot{x}$. Then, $\dot{f}(x(t))$ can be integrated for the interval, $t_0 \leq \tau < t_0 + \Delta t$, as

$$f(x(\tau)) \approx f(x_{t_0}) + J_{t_0}(x(\tau) - x_{t_0}) = J_{t_0}x(\tau) + f(x_{t_0}) - J_{t_0}x_{t_0}.$$

Using this approximate $f(x(\tau))$, we have, from

$$\dot{x}(\tau) = f(x(\tau)) + \sigma w(\tau) + cy(\tau),$$

the following approximate linear stochastic differential equation on each short time interval, $[t_0, t_0 + \Delta t)$,

$$\dot{x}(\tau) = J_{t_0}x(\tau) + N_{t_0} + \sigma w(\tau) + cy(\tau),$$

where

$$N_{t_0} = f(x_{t_0}) - J_{t_0}x_{t_0}.$$

By integrating this for the interval, $[t_0, t_0 + \Delta t)$, we have

$$x_{t_0+\Delta t} = \exp(J_{t_0}\Delta t)x_{t_0}$$

$$+ N_{t_0} \int_{t_0}^{t_0+\Delta t} \exp\{J_{t_0}(t_0 + \Delta t - u)\}du$$

$$+ \int_{t_0}^{t_0+\Delta t} \exp\{J_{t_0}(t_0 + \Delta t - u)\}\sigma w(u)du$$

$$+ \int_{t_0}^{t_0+\Delta t} \exp\{J_{t_0}(t_0 + \Delta t - u)\}cy(u)du.$$

When $y(t)$ is a "non-stochastic" smooth function of t, a natural approximation for the fourth term is given by

$$\int_{t_0}^{t_0+\Delta t} \exp\{J_{t_0}(t_0 + \Delta t - u)\}cy(u)du \approx (1/J_{t_0})\{\exp(J_{t_0}\Delta t) - 1\}cy(t_0).$$

Then we have a discrete-time stochastic counterpart for (9.23), which is a discrete-time nonlinear autoregressive model with an exogenous variable,

$$x_{t+\Delta t} = \phi(x_t)x_t + \sigma B_t w_{t+\Delta t} + C_t y_t,$$

where

$$\phi(x_t) = 1 + (1/J_t)\{\exp(J_t\Delta t) - 1\}f(x_t)/x_t,$$

$$B_t = \sqrt{\{\exp(2J_{t_0}\Delta t) - 1\}/(2J_{t_0})},$$

$$C_t = (1/J_t)\{\exp(J_t\Delta t) - 1\}c.$$

Application of the present model to the estimation and simulation of nonlinear riverflow time series models in stochastic hydrology is found in Ozaki (1985b).

9.3 LL Bridges for Higher Order Linear/Nonlinear Processes

9.3.1 Linear Processes

So far we have considered time discretization of first-order dynamical systems and their stochastic versions. The oscillatory behavior of some time series cannot be explained by these. Here we need second or higher order differential equation models in continuous-time modeling and second or higher order AR models in discrete-time modeling.

A typical and simple continuous-time model for an oscillatory process is the following second-order linear stochastic differential equation model:

Example-L1

$$\ddot{x}(t) + c\dot{x}(t) + \omega^2 x(t) = n(t), \tag{9.24}$$

where $n(t)$ is a continuous-time Gaussian white noise of variance σ_n^2. The model is equivalent to the bivariate stochastic dynamical system model,

$$\begin{pmatrix} \ddot{x}(t) \\ \dot{x}(t) \end{pmatrix} = \begin{pmatrix} -a & -b \\ 1 & 0 \end{pmatrix} \begin{pmatrix} \dot{x}(t) \\ x(t) \end{pmatrix} + \begin{pmatrix} n(t) \\ 0 \end{pmatrix},$$

which we write as

$$\dot{z}(t) = Jz(t) + w(t),$$

where

$$z(t) = \begin{pmatrix} \dot{x}(t) \\ x(t) \end{pmatrix}, \quad w(t) = \begin{pmatrix} n(t) \\ 0 \end{pmatrix}, \quad J = \begin{pmatrix} -a & -b \\ 1 & 0 \end{pmatrix}.$$

Since the dynamic model $\dot{z}(t) = Jz(t)$ is linear, we have its analytic solution $z(t) = e^{Jt}z(0)$ for the interval, $[t, t + \Delta t]$, and we can write down the discrete-time version of the model as

$$z_{t+\Delta t} = e^{J\Delta t}z_t + w_{t+\Delta t}, \tag{9.25}$$

where

$$w_{t+\Delta t} = \int_t^{t+\Delta t} e^{J(t+\Delta t - s)}w(s)ds.$$

Here the prediction error of $z_{t+\Delta t}$, given z_t, $z_{t-\Delta t}$, ..., is

$$w_{t+\Delta t} = z_{t+\Delta t} - E[z_{t+\Delta t} | z_t, ...] = z_{t+\Delta t} - e^{J\Delta t} z_t. \tag{9.26}$$

The variance of $w_{t+\Delta t}$ is

$$\Sigma_{w_{t+\Delta t}} = E\left[\left(\int_t^{t+\Delta t} e^{J(t+\Delta t-s)} w(s) ds \right) \left(\int_t^{t+\Delta t} e^{J(t+\Delta t-s)} w(s) ds \right)^t \right]$$

$$= \sigma_n^2 \begin{pmatrix} \sigma_{1,1} & \sigma_{1,2} \\ \sigma_{2,1} & \sigma_{2,2} \end{pmatrix}, \tag{9.27}$$

where

$$\sigma_{1,1} = \frac{1}{(\mu_1 - \mu_2)^2} \left[\frac{\mu_1}{2} \left(e^{2\mu_1 \Delta t} - 1 \right) - \frac{2\mu_1 \mu_2}{(\mu_1 + \mu_2)} \left(e^{(\mu_1 + \mu_2)\Delta t} - 1 \right) + \frac{\mu_2}{2} \left(e^{2\mu_2 \Delta t} - 1 \right) \right]$$

$$\sigma_{2,2} = \frac{1}{(\mu_1 - \mu_2)^2} \left[\frac{1}{2\mu_1} \left(e^{2\mu_1 \Delta t} - 1 \right) - \frac{2}{(\mu_1 + \mu_2)} \left(e^{(\mu_1 + \mu_2)\Delta t} - 1 \right) + \frac{1}{2\mu_2} \left(e^{2\mu_2 \Delta t} - 1 \right) \right]$$

$$\sigma_{1,2} = \sigma_{2,1} = \frac{1}{(\mu_1 - \mu_2)^2} \left[\frac{1}{2} \left(e^{2\mu_1 \Delta t} - 1 \right) - \left(e^{(\mu_1 + \mu_2)\Delta t} - 1 \right) + \frac{1}{2} \left(e^{2\mu_2 \Delta t} - 1 \right) \right].$$

Here μ_1, μ_2 are eigenvalues of J (see Ozaki 1986 for details).

We note that, although the variance matrix of continuous-time noise $w(t)$ is rank 1, the variance matrix of the discrete time noise $w_{t+\Delta t}$ is rank 2 as a result of the integration over $[t, t + \Delta t]$. $\Sigma_{w_{t+\Delta t}}$ is decomposed as

$$\Sigma_{w_{t+\Delta t}} = \sigma_n^2 \begin{pmatrix} \sigma_{1,1} & \sigma_{1,2} \\ \sigma_{2,1} & \sigma_{2,2} \end{pmatrix} = \sigma_n^2 U \begin{pmatrix} \lambda_1 & 0 \\ 0 & \lambda_2 \end{pmatrix} U^t$$

using the unitary matrix U, whose elements are given by

$$u_{1,1} = \sigma_{1,2} \Big/ \sqrt{\sigma_{1,2}^2 + (\lambda_1 - \sigma_{1,1})^2}$$

$$u_{1,1} = \sigma_{1,2} \Big/ \sqrt{\sigma_{1,2}^2 + (\lambda_2 - \sigma_{2,2})^2}$$

$$u_{1,2} = (\lambda_1 - \sigma_{1,1}) \Big/ \sqrt{\sigma_{1,2}^2 + (\lambda_1 - \sigma_{1,1})^2}$$

$$u_{2,1} = u_{1,2}.$$

The eigenvalues λ_1 and λ_2 are given as the roots of

$$\lambda^2 - (\sigma_{1,1} + \sigma_{2,2})\lambda + \sigma_{1,1}\sigma_{2,2} - \sigma_{1,2}^2 = 0.$$

This means that, from (9.25), we finally have a bivariate discrete-time model

$$z_{t+\Delta t} = Az_t + Bv_{t+\Delta t}, \tag{9.28}$$

where $v_{t+\Delta t}$ is a bivariate discrete-time Gaussian white noise of variance $\sigma_n^2 I$,

$$A = e^{J\Delta t}$$

and

$$B = U \begin{pmatrix} \sqrt{\lambda_1} & 0 \\ 0 & \sqrt{\lambda_2} \end{pmatrix}.$$

When the observation data of the second-order differential equation model (9.24) is a scalar time series, $x_t, x_{t+\Delta t}, x_{t+2\Delta t} \ldots$, without velocity observations, $\dot{x}_t, \dot{x}_{t+\Delta t}, \dot{x}_{t+2\Delta t} \ldots$, the model (9.28) needs to be converted to a state space model, with the state $z_t = (\dot{x}_t, x_t)'$,

$$z_{t+\Delta t} = Az_t + Bv_{t+\Delta t}$$
$$x_t = Cz_t, \tag{9.29}$$

where

$$A = e^{J\Delta t},$$

$$B = U \begin{pmatrix} \sqrt{\lambda_1} & 0 \\ 0 & \sqrt{\lambda_2} \end{pmatrix},$$

and

$$C = (0, 1).$$

The 2 dimensional linear state space model (9.29) can be converted to

$$x_t - \phi_1 x_{t-\Delta t} - \phi_2 x_{t-2\Delta t} = CBv_t + C(A - \phi_1 I)Bv_{t-\Delta t},$$

using the Cayley–Hamilton theorem, where

$$A^2 - \phi_1 A - \phi_2 I = 0.$$

Let $w_t = CBv_t$; then with the one-dimensional Gaussian white noise w_t, we have the following ARMA(2,1) representation:

$$x_t = \phi_1 x_{t-\Delta t} + \phi_2 x_{t-2\Delta t} + w_t + \theta w_{t-\Delta t}. \tag{9.30}$$

Here ϕ_1 and ϕ_2 are functions of a and b since

$$A = e^{J\Delta t}, \quad J = \begin{pmatrix} -a & -b \\ 1 & 0 \end{pmatrix}.$$

Also, θ and $\sigma_w^2 = E[w_t^2]$ are written (see Ozaki 1986 for the derivation) as functions of μ_1, μ_2 and σ_n^2 as the following:

$$\theta = \frac{(\mu_1 - \mu_2)\left\{\mu_2 e^{\mu_2 \Delta t} - \mu_2 e^{\mu_1 \Delta t} - e^{(\mu_1 + \mu_2)\Delta t}\left(\mu_2 e^{\mu_1 \Delta t} - \mu_1 e^{\mu_2 \Delta t}\right)\right\}}{\left(\mu_2 e^{\mu_1 \Delta t} - \mu_1 e^{\mu_2 \Delta t}\right)^2 + \mu_1 \mu_2 \left(e^{\mu_1 \Delta t} - e^{\mu_2 \Delta t}\right)^2 - (\mu_1 - \mu_2)^2}.$$

$$\sigma_w^2 = \frac{\left\{\left(\mu_2 e^{\mu_1 \Delta t} - \mu_1 e^{\mu_2 \Delta t}\right)^2 + \mu_1 \mu_2 \left(e^{\mu_1 \Delta t} - e^{\mu_2 \Delta t}\right)^2 - (\mu_1 - \mu_2)^2\right\}}{2\mu_1 \mu_2 (\mu_1^2 - \mu_2^2)(\mu_1 - \mu_2)} \sigma_n^2.$$

Here μ_1 and μ_2 are eigenvalues of J, so that they are also functions of a and b.

Example-L2: General pth order differential equation model

The derivation of the discrete-time model for the process $x(t)$ generated from the pth order stochastic differential equation model,

$$x^{(p)}(t) + b_1 x^{(p-1)}(t) + \cdots + b_{p-1} x^{(1)}(t) + b_p x(t) = n(t), \tag{9.31}$$

depends on how the observation data are presented. It will not be likely that all the higher order derivatives $x^{(k)}(t)$ ($k = 1, 2, \ldots, p - 1$) of $x(t)$ are available as observed time series data. The most likely situation is that only the scalar discrete observations, $x_{\Delta t}, x_{2\Delta t}, \ldots, x_{N\Delta t}$, are available. Then the dynamics of the data is characterized by the state space model, where the observation matrix is given by a special $1 \times p$ matrix C as

$$\dot{z}(t) = Jz(t) + w(t)$$

$$x_t = Cz(t), \tag{9.32}$$

where

$$C = \begin{pmatrix} 0 & \cdots & 0 & 1 \end{pmatrix},$$

$$z(t) = \begin{pmatrix} x^{(p-1)}(t) \\ x^{(p-2)}(t) \\ \cdots \\ \dot{x}(t) \\ x(t) \end{pmatrix}, \quad w(t) = \begin{pmatrix} n(t) \\ 0 \\ \cdots \\ 0 \\ 0 \end{pmatrix},$$

and

$$J = \begin{pmatrix} -b_1 & -b_2 & \cdots & -b_{p-1} & -b_p \\ 1 & 0 & \cdots & 0 & 0 \\ 0 & 1 & \cdots & 0 & 0 \\ \cdots & \cdots & \cdots & \cdots & \cdots \\ 0 & 0 & \cdots & 1 & 0 \end{pmatrix}.$$

Since the p-dimensional dynamic model $\dot{z}(t) = Jz(t)$ is linear, we have its analytic solution $z(t + \Delta t) = e^{J\Delta t}z(t)$, for the time interval $[t, t + \Delta t)$, and we can write down the discrete-time version of the model as

$$z_{t+\Delta t} = e^{J\Delta t}z_t + w_{t+\Delta t}$$

$$w_{t+\Delta t} = \int_t^{t+\Delta t} e^{J(t+\Delta t - s)}w(s)ds,$$

which we write, together with the observation equation, as the following discrete-time state space representation,

$$z_{t+\Delta t} = e^{J\Delta t}z_t + Bv_{t+\Delta t}$$

$$x_t = Cz_t, \tag{9.33}$$

where elements of $p \times p$ matrix B can be written as functions of $b_1, \ldots, b_p$, in the same way as for the second-order case of the previous example 1, although we need more laborious work of linear algebra. The model (9.33) can be converted, using the Cayley–Hamilton theorem, to the equivalent ARMA(p,p–1) model,

$$x_t = \phi_1 x_{t-\Delta t} + \cdots + \phi_p x_{t-p\Delta t} + CBv_t + C(A - \phi_1 I)Bv_{t-\Delta t} + \cdots$$

$$+ C(A^{p-1} - \phi_1 A^{p-2} - \cdots - \phi_{p-1}I)Bv_{t-(p-1)\Delta t},$$

where $\phi_1, \ldots, \phi_p$ are characteristic coefficients of the matrix $A = e^{JAt}$, that is,

$$A^p - \phi_1 A^{p-1} - \cdots - \phi_p A - \phi_p I = 0.$$

By introducing $w_t = CBv_t$, we have a standard ARMA($p,p-1$) representation,

$$x_t = \phi_1 x_{t-\Delta t} + \cdots + \phi_p x_{t-p\Delta t} + w_t + \theta_1 w_{t-\Delta t} + \cdots + \theta_{p-1} w_{t-(p-1)\Delta t},$$

where $\phi_1, \ldots, \phi_p, \theta_1, \ldots, \theta_{p-1}$ and $\sigma_w^2 = E[w_t^2] = E[(CBv_t)^2]$ are finally written down as functions of $b_1, \ldots, b_p$ and σ_n^2.

Example L3. Spectral decomposition model

The spectral decomposition model in Section 5.3.5 falls in the same category as the previous example of the pth order stochastic differential equation model, except that the matrices specifying the linear state space representation are different.

A typical example of the use of the present model is for EEG spectral decomposition, where the EEG signal x_t is considered, under the assumption of the superposition principle, as a sum of rhythms of different oscillations, such as delta $S_1(t)$, theta $S_2(t)$, alpha $S_3(t)$, beta $S_4(t)$, and gamma $S_5(t)$:

$$x_t = S_1(t) + S_2(t) + S_3(t) + S_4(t) + S_5(t). \tag{9.34}$$

Each cyclic component can be represented by the second-order differential equation model, where

$$(D^2 + \omega_1^2)S_1(t) = u_1(t): \text{ delta } (\omega_1 : 1 < \omega_1 < 3 \text{ Hz})$$

$$(D^2 + \omega_2^2)S_2(t) = u_2(t): \text{ theta } (\omega_2 : 3 < \omega_2 < 7 \text{ Hz})$$

$$(D^2 + \omega_3^2)S_3(t) = u_3(t): \text{ alpha } (\omega_3 : 7 < \omega_3 < 12 \text{ Hz})$$

$$(D^2 + \omega_4^2)S_4(t) = u_4(t): \text{ beta } (\omega_4 : 12 < \omega_4 < 30 \text{ Hz})$$

$$(D^2 + \omega_5^2)S_5(t) = u_5(t): \text{ gamma } (\omega_5 : 30 < \omega_5 < 80 \text{ Hz}).$$

$u_i(t)$ ($i = 1, 2, \ldots, 5$) is a white noise with variance σ_i^2. A state space representation for the time series x_t is

$$\dot{z}(t) = A^{(para)}z(t) + u(t)$$
$$x_t = Cz(t), \tag{9.35}$$

where

$$z(t) = (\dot{S}_1(t),\ S_1(t),\ \dot{S}_2(t),\ S_2(t),\ldots,\dot{S}_5(t),\ S_5(t))',$$

$$C = (0,1,0,1,0,1,0,1,0,1),$$

$$A^{(para)} = \begin{pmatrix} A_1 & O & \cdots & O \\ O & A_2 & \cdots & O \\ \cdots & \cdots & \cdots & \cdots \\ O & O & \cdots & A_5 \end{pmatrix}$$

$$A_i = \begin{pmatrix} 0 & \omega_i^2 \\ -1 & 0 \end{pmatrix}, \quad (i = 1, 2,\ldots, 5),$$

$$u(t) = (u_1(t),\ 0,\ u_2(t),\ 0,\ldots,\ u_5(t),\ 0)'$$

with the variance–covariance matrix $\Sigma_u = \text{diag}\ (\sigma_1^2,\ 0,\ \sigma_2^2,\ 0,\ldots,\ \sigma_5^2,\ 0)$. Since the 10-dimensional dynamic model $\dot{z}(t) = A^{(para)}z(t)$ is linear, we have its analytic solution $z(t + \Delta t) = e^{A(para)\Delta t}z(t)$, for the time interval $[t, t + \Delta t)$, and we can write the discrete-time version of the model as

$$z_{t+\Delta t} = \exp(A^{(para)}\Delta t)z_t + w_{t+\Delta t}$$

$$w_{t+\Delta t} = \int_t^{t+\Delta t} \exp(A^{(para)}(t + \Delta t - s))w(s)ds,$$

which we write, together with the observation equation, as the following discrete-time state space representation,

$$z_{t+\Delta t} = \exp(A^{(para)}\Delta t)z_t + Bv_{t+\Delta t}$$

$$x_t = Cz_t, \tag{9.36}$$

where elements of the 10×10 matrix B can be written as functions of parameters of the original continuous-time model, in the same way as for the previous example L2. Since the 10-dimensional state space model (9.36) is linear, it can also be converted to an ARMA(10,9) model for the time series x_t using the Cayley–Hamilton theorem.

Example L4. Tank model

The tank model in Chapter 5 also falls in the same category as the pth order differential equation model (9.31). With the assumption of the scalar time series observation, the continuous-time tank model is written as a state space model,

$$\dot{z}(t) = Jz(t) + Bu(t)$$
$$x_t = Cz(t),$$

(9.37)

where $u(t)$ is a scalar input, which could be a stimulus $s(t)$ plus white noise $n(t)$, whose variance is σ_n^2, as

$$u(t) = s(t) + n(t),$$

$$J = \begin{pmatrix} -\lambda_1 & 0 & 0 & \cdots & 0 & 0 & \cdots & 0 & 0 \\ 1 & -\lambda_2 & 0 & \cdots & 0 & 0 & \cdots & 0 & 0 \\ 0 & 1 & -\lambda_2 & \cdots & 0 & 0 & \cdots & 0 & 0 \\ \cdots & \cdots & \cdots & \cdots & \cdots & \cdots & \cdots & \cdots & \cdots \\ 0 & 0 & 0 & \cdots & -\lambda_k & 0 & \cdots & 0 & 0 \\ 0 & 0 & 0 & \cdots & 1 & -\lambda_k & \cdots & 0 & 0 \\ \cdots & \cdots & \cdots & \cdots & \cdots & \cdots & \cdots & \cdots & \cdots \\ 0 & 0 & 0 & \cdots & 0 & 0 & \cdots & -\lambda_k & 0 \\ 0 & 0 & 0 & \cdots & 0 & 0 & \cdots & 1 & -\lambda_k \end{pmatrix},$$

$$z(t) = \{z_1(t), dz_1(t)/dt, z_2(t), \ldots, d^{k-1}z_k(t)/dt^{k-1},$$

$$d^{k-2}z_k(t)/dt^{k-2}, \ldots, dz_k(t)/dt, z_k(t)\}',$$

$C = (1, 0, 1, \ldots, 0, \ldots, 0, 1)$, $B = (b_1, b_2, 0, \ldots, b_k, 0, \ldots, 0, 0)'$. Here $n(t)$ could be zero and $u(t) = s(t)$ if the Tank model is assumed to be deterministic, like in the SPM model used in Friston et al. (1994) and Worsley et al. (2002).

The discretization of the state space model (9.37) is derived in the same way as for the pth order differential equation model of example L2. The only a difference is that the model (9.37) has an exogenous variable $s(t)$ driving the state. Then a discretization scheme converts the continuous-time state space model with exogenous input

$$\dot{z}(t) = Jz(t) + Bu(t)$$
$$x_t = Cz(t)$$

(9.38)

into a discrete state space model with exogenous variable input $s_t, s_{t-\Delta t}, \ldots$ as the following,

$$z_{t+\Delta t} = Az_t + Bs_t + Bn_t$$
$$x_t = Cz_t,$$

(9.39)

which is written, using the Cayley–Hamilton theorem, as the following ARMAX model form,

$$x_t = \phi_1 x_{t-\Delta t} + \cdots + \phi_p x_{t-p\Delta t} + w_t + \theta_1 w_{t-\Delta t} + \cdots + \theta_{p-1} w_{t-(p-1)\Delta t}$$
$$+ \psi_1 s_{t-\Delta t} + \cdots + \psi_{p-1} s_{t-(p-1)\Delta t},$$

where the parameters $\phi_1, \ldots, \phi_p, \theta_1, \ldots, \theta_{p-1}, \psi_1, \ldots, \psi_{p-1}$ and $\sigma_w^2 = E[w_t^2]$ are written as functions of $\lambda_1, \ldots, \lambda_k, b_1, \ldots, b_r$ and σ_n^2.

9.3.2 Instability Problem of Nonlinear Oscillatory Processes

The discretization of a multivariate continuous-time nonlinear dynamic model can be realized, essentially, in the same way as for the linear case. However, we need to be careful, for nonlinear cases, about the possible instability problem caused by the discretization. We consider how the instability is solved in our local linearization method with some examples below.

Example 1: Stochastic Duffing type oscillation model

A stochastic Duffing model is defined, with a continuous-time Gaussian white noise $n(t)$ of variance σ_n^2, by

$$\ddot{x}(t) + c\dot{x}(t) + \alpha x(t) + \beta x^3(t) = n(t).$$

If we discretize the deterministic version of the model, that is,

$$\ddot{x}(t) + c\dot{x}(t) + \alpha x(t) + \beta x^3(t) = 0,$$

by replacing $\dot{x}(t)$ by

$$\frac{x(t) - x(t - \Delta t)}{\Delta t}$$

and replacing $\ddot{x}(t)$ by

$$\left\{\left(\frac{x_{t+\Delta t}-x_t}{\Delta t}\right)-\left(\frac{x_t-x_{t-\Delta t}}{\Delta t}\right)\right\}\Big/\Delta t,$$

respectively, we have

$$\left\{\left(\frac{x_{t+\Delta t}-x_t}{\Delta t}\right)-\left(\frac{x_t-x_{t-\Delta t}}{\Delta t}\right)\right\}\Big/\Delta t+a\frac{x_t-x_{t-\Delta t}}{\Delta t}+bx_t+\beta x_t^3=0,$$

which leads to the following deterministic nonlinear polynomial AR model,

$$x_{t+\Delta t}=\left\{2-a\Delta t-b(\Delta t)^2\right\}x_t-\beta(\Delta t)^2 x_t^3-(1-a\Delta t)x_{t-\Delta t}.$$

This implies the following second-order nonlinear AR model,

$$x_{t+\Delta t}=\phi_1(x_t)x_t+\phi_2 x_{t-\Delta t}+w_{t+\Delta t}, \tag{9.40}$$

where

$$\phi_1(x_t)=\left\{2-a\Delta t-b(\Delta t)^2\right\}-\beta(\Delta t)^2 x_t^2,$$

$$\phi_2=-(1-a\Delta t),$$

as a stochastic version of the Duffing type nonlinear oscillation process x_t driven by a Gaussian white noise w_t. Note that the model (9.40) can be written in the form of a bivariate Markov chain model,

$$\begin{pmatrix} x_{t+\Delta t} \\ x_t \end{pmatrix}=\begin{pmatrix} \phi_1(x_t) & \phi_2 \\ 1 & 0 \end{pmatrix}\begin{pmatrix} x_t \\ x_{t-\Delta t} \end{pmatrix}+\begin{pmatrix} w_{t+\Delta t} \\ 0 \end{pmatrix}. \tag{9.41}$$

Here, $\phi_1(x_t)$, the element of the transition matrix of the model, is a second-order polynomial function of x_t, and so the eigenvalues of the matrix can be outside the unit circle for $|x_t| \to \infty$ (see Figure 9.4)

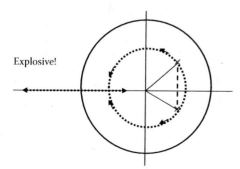

Explosive!

FIGURE 9.4
Polynomial AR: An eigenvalue shifts out of the unit circle.

9.3.3 ExpAR(2) Model and Nonlinear Oscillatory Processes

A model family of ExpAR models was introduced to stop its trajectory becoming explosive as the polynomial type AR models. Here the nonlinear AR coefficient $\phi_1(x_t)$ is parameterized with a bounded nonlinear function such as

$$\phi_1(x_t) = \phi_{1,0} + \phi_{1,1} \exp(-\gamma x_t^2),$$

where $|\phi_1(x_t)| < \infty$ for $|x_t| \to \infty$ (see Figure 9.5a).

Let λ_∞ and $\bar{\lambda}_\infty$ be the roots of

$$\Lambda^2 - \phi_{1,0}\Lambda - \phi_2 = 0,$$

and λ_0 and $\bar{\lambda}_0$ be the roots of

$$\Lambda^2 - (\phi_{1,0} + \phi_{1,1})\Lambda - \phi_2 = 0$$

and $0 < -\phi_2 < 1$, and then the roots of the eigenvalues of the transition matrix have constant absolute values and its argument shifts between λ_0 and λ_∞ depending on x_t (see Figure 9.5b). This mechanism generates amplitude-dependent frequency shift phenomena typically observed by Duffing type nonlinear oscillations.

ExpAR model is also known to characterize the van der Pol oscillation type stochastic limit cycle behavior of time series by parameterizing $\phi_2(x_t)$ with a bounded nonlinear function such as

$$\phi_2(x_t) = \phi_{2,0} + \phi_{2,1} \exp\left(-\gamma x_t^2\right).$$

Here if $0 < -\phi_{2,0} < 1$ (see Figure 9.6a), the eigenvalues of the transition matrix of the model

$$\begin{pmatrix} x_{t+1} \\ x_t \end{pmatrix} = \begin{pmatrix} \phi_1 & \phi_2(x_t) \\ 1 & 0 \end{pmatrix}\begin{pmatrix} x_t \\ x_{t-1} \end{pmatrix} + \begin{pmatrix} w_{t+1} \\ 0 \end{pmatrix} \qquad (9.42)$$

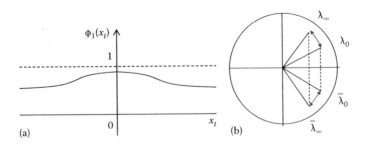

(a)   (b)

FIGURE 9.5
(a) ϕ_1-function of Duffing-type ExpAR(2). (b) Shift of eigenvalues of Duffing type ExpAR(2).

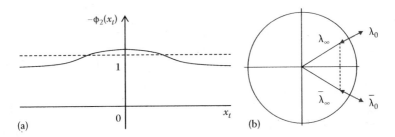

FIGURE 9.6
(a) ϕ_2-function of van der Pol type ExpAR(2). (b) Shift of eigenvalues of van der Pol type ExpAR(2).

stay inside the unit circle for $|x_t| \to \infty$, and the ExpAR model defines a non explosive and stationary finite-variance discrete-time process.

Here, even though the roots λ_0 and $\bar{\lambda}_0$ of $\Lambda^2 - \phi_1\Lambda - (\phi_{2,0} + \phi_{2,1}) = 0$ go outside the unit circle, the process does not explode and the model defines a finite-variance process. This mechanism (see Figure 9.6a and b) yields stochastic limit cycle phenomena typically observed in stochastic van der Pol type non-linear oscillations (Haggan and Ozaki 1981, Ozaki 1985a).

9.3.4 LL Discretization of Nonlinear Oscillatory Models

We will see next how the instability problem of nonlinear oscillatory models is solved systematically by the LL discretization method using some typical examples.

Example 1: Stochastic nonlinear oscillation model

A stochastic Duffing model represented by

$$\ddot{x}(t) + c\dot{x}(t) + \alpha x(t) + \beta x^3(t) = n(t)$$

and a stochastic van der Pol model represented by

$$\ddot{x} - \alpha(1 - x^2)\dot{x} + x = n(t)$$

are special cases of a general nonlinear oscillation model,

$$\ddot{x} + a(x)\dot{x} + b(x)x = n(t). \tag{9.43}$$

Here, $n(t)$ is a continuous-time Gaussian white noise. Both models can be formulated as a bivariate continuous-time dynamical system model,

$$\dot{z} = f(z) + \xi(t), \tag{9.44}$$

where

$$z = \begin{pmatrix} \dot{x} \\ x \end{pmatrix}, \quad f(z) = \begin{pmatrix} -a(x)\dot{x} \\ -b(x)x \end{pmatrix}, \quad \xi(t) = \begin{pmatrix} n(t) \\ 0 \end{pmatrix}.$$

For the stochastic Duffing equation model, $a(x)$ is constant, $a(x) = c$, while $b(x) = \alpha + \beta x^2$. For the stochastic van der Pol model, $b(x) = 1$ and the restoring force is linear, while the damping force is not a constant but a nonlinear function of x, that is, $a(x) = -\alpha(1 - x^2)$. The discretization scheme, which we introduced for the one-dimensional nonlinear dynamical system in the previous section, can be applied to the multidimensional nonlinear dynamic model (9.44) without an essential change.

The nonlinear bivariate dynamical system,

$$\dot{z}(t) = f(z(t)) + w(t),$$

could be converted, by the LL scheme, into a discrete-time model

$$z_{t+\Delta t} = z_t + J_t^{-1}(e^{J_t \Delta t} - 1)f(z_t) + w_{t+\Delta t}, \tag{9.45}$$

where $e^{J_t \Delta t}$ is a matrix defined by

$$e^{J_t \Delta t} = \sum_{k=0}^{\infty} \frac{1}{k!}(J_t \Delta t)^k,$$

and

$$w_{t+\Delta t} = \int_{t}^{t+\Delta t} e^{J_t(t+\Delta t - u)}w(u)du.$$

The variance of $w_{t+\Delta t}$ is

$$\Sigma_{w_{t+\Delta t}} = E\left[\left(\int_{t}^{t+\Delta t} e^{J_t(t+\Delta t - s)}w(s)ds\right)\left(\int_{t}^{t+\Delta t} e^{J_t(t+\Delta t - s)}w(s)ds\right)'\right]$$

$$= \sigma_n^2 \begin{pmatrix} \sigma_{t,1,1} & \sigma_{t,1,2} \\ \sigma_{t,2,1} & \sigma_{t,2,2} \end{pmatrix}, \tag{9.46}$$

where

$$\sigma_{t,1,1} = \frac{1}{\left(\mu_{t,1}-\mu_{t,2}\right)^2}\left[\frac{\mu_{t,1}}{2}\left(e^{2\mu_{t,1}\Delta t}-1\right)-\frac{2\mu_{t,1}\mu_{t,2}}{\left(\mu_{t,1}+\mu_{t,2}\right)}\left(e^{(\mu_{t,1}+\mu_{t,2})\Delta t}-1\right)+\frac{\mu_{t,2}}{2}\left(e^{2\mu_{t,2}\Delta t}-1\right)\right]$$

$$\sigma_{t,2,2} = \frac{1}{\left(\mu_{t,1}-\mu_{t,2}\right)^2}\left[\frac{1}{2\mu_{t,1}}\left(e^{2\mu_{t,1}\Delta t}-1\right)-\frac{2}{\left(\mu_{t,1}+\mu_{t,2}\right)}\left(e^{(\mu_{t,1}+\mu_{t,2})\Delta t}-1\right)+\frac{1}{2\mu_{t,2}}\left(e^{2\mu_{t,2}\Delta t}-1\right)\right]$$

$$\sigma_{t,1,2} = \sigma_{t,2,1} = \frac{1}{\left(\mu_{t,1}-\mu_{t,2}\right)^2}\left[\frac{1}{2}\left(e^{2\mu_{t,1}\Delta t}-1\right)-\left(e^{(\mu_{t,1}+\mu_{t,2})\Delta t}-1\right)+\frac{1}{2}\left(e^{2\mu_{t,2}\Delta t}-1\right)\right].$$

Here $\mu_{t,1}$, $\mu_{t,2}$ are eigenvalues of J_t (see Ozaki 1986 for the detail). The rank of the variance matrix of the discrete-time white noise $w_{t+\Delta t}$ is full rank 2, even though the rank of the variance matrix of original continuous-time Gaussian white noise $w(t)$ is 1. However, our main interest is not on the precise derivation of the stochastic part but on the deterministic part,

$$z_{t+\Delta t} = A_t z_t, \tag{9.47}$$

of the discretized models obtained from the stochastic Duffing type oscillations and stochastic van der Pol type oscillations.

When we can write as $f(z_t) = F(z_t)z_t$, the model (9.45) is rewritten as a discrete-time state space model form,

$$z_{t+\Delta t} = A_t z_t + B n_{t+\Delta t}$$
$$x_t = (0,1)z_t, \tag{9.48}$$

where $B = (1,0)'$ and

$$A_t = I + J_t^{-1}(e^{J_t\Delta t}-1)F(z_t)$$

Here the eigenvalues of the transition matrix A_t of (9.48) are x_t-dependent. Since the LL-discretized models are supposed to be preserving the finite-variance property of the original continuous-time process, we may expect that they may stay inside the unit circle for $|x_t| \to \infty$. In fact the behavior of the x_t-dependent eigenvalues of the LL-discretized model (9.48) for the Duffing type oscillations and the x_t-dependent eigenvalues of the ExpAR model (9.41) is very similar. Numerical examples of a typical pattern of

x_t-dependent absolute values of the eigenvalues of the transition matrix A_t of an LL-discretized Duffing type model (9.48) are present in Ozaki (1986).

The similarity of the LL-discretized model and the ExpAR model is also seen for the van der Pol oscillations. The stochastic van der Pol oscillations can be modeled by

$$\ddot{x} + a(x)\dot{x} + b(x)x = n(t),$$

where $a(x) = -a_1(1 - x^2)$ and $b(x) = b_1$. Then the LL-discretized model is given by (9.48) with

$$J_t = \begin{pmatrix} a_1(1-x^2) & -2a_1 x\dot{x} - b_1 \\ 1 & 0 \end{pmatrix}$$

and

$$F_t = \begin{pmatrix} a_1(1-x^2) & -b_1 \\ 1 & 0 \end{pmatrix}.$$

A typical pattern of the absolute value of the x_t-dependent complex conjugate eigenvalues of the transition matrix A_t is shown in Figure 9.7. Here we need to point out the clear similarity to the Figure 9.6a for the ExpAR model, heuristically introduced for the modeling of van der Pol type oscillating time series.

Apart from these simple and typical nonlinear oscillation models, there are many other more complex nonlinear dynamic models known in science and engineering. For example, we have seen, in Chapter 5, the Hodgkin-Huxley model, the Zetterberg Model, the Balloon Model, and the DCM model in neurosciences. All of these models are discretized into multidimensional discrete-time dynamic models by applying the LL scheme. All the information about the complex nonlinear dynamics of these models is contained in the state-dependent eigenvalues of the high-dimensional transition matrices of the model. How to use the discretized models for identifying the nonlinear model structures and estimating the model parameters will be discussed in Chapters 10 and 12.

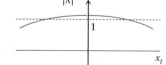

FIGURE 9.7
Absolute values of the eigenvalues of the LL-van der Pol model.

9.4 LL Bridges for Processes from the Pearson System

So far we have seen that we can have a reasonable bridge C-D connecting the continuous-time model classes of $\dot{x} = a(x)x$ and of $\dot{x} = a(x)x + \sigma w(t)$ type and their discrete-time counterparts, $x_{t+\Delta t} = \phi(x_t)x_t$ and $x_{t+\Delta t} = \phi(x_t)x_t + \sigma\sqrt{\Delta t}w_{t+\Delta t}$. Here we assumed that $a(x)$ is a polynomial type nonlinear function as

$$a(x) = a_1 + a_2 x + \cdots + a_r x^{r-1}.$$

However, we know that many other types of nonlinear dynamical systems $dx/dt = f(x)$, which are not necessarily polynomials of x and $f(0) \neq 0$, have been used in various areas of sciences.

Our next important question is whether the current LL scheme can be extended and is valid to a more general family of nonlinear functions. In order to see this, let us consider next how we get a finite-variance discrete-time process from the continuous-time process introduced from the Pearson system, which includes more general nonlinear function families as seen in Chapter 8.

We know, from the discussion in Chapter 8, that 2 types of continuous-time stochastic dynamic models are involved behind the stochastic process $x(t)$ (see Figure 8.3) introduced from the Pearson system. These models are a stochastic differential equation model with nonconstant diffusion coefficient,

$$\dot{x} = a(x) + b(x)w(t),$$

and the associated stochastic differential equation model with a constant diffusion coefficient (which is sometimes called the Langevin equation model),

$$\dot{y} = f(y) + \sigma w(t),$$

where $y(t) = h(x(t))$.

Here we have two possible ways of making a bridge between continuous-time models and discrete-time models.

(1) One is building a bridge directly between the stochastic process $x(t)$ and the discrete-time process x_t, using a discretization scheme for

$$\dot{x} = a(x) + b(x)w(t).$$

(2) Another is to transform $x(t)$ to $y(t) = h(x(t))$ and build a bridge between the continuous-time process $y(t)$ and the discrete-time process y_t using a discretization scheme, and then we can obtain a discrete-time version of $x(t)$ by transforming it back to x_t by $x_t = h^{-1}(y_t)$.

Note that the diffusion term $b(x)$ of the first model is dependent on x, so that the prediction error variance of the discretized process x_t is dependent on the value of x_t at each time point t, while the prediction error variance for $y(t)$ in (ii) is constant, once a proper variable transformation $y = h(x)$ is found. Therefore, (ii) will be more convenient for us to make a bridge between the continuous-time model families and the discrete-time model families. Then we can separate the problem into two: (1) the problem of finding a proper transformation, $y = h(x)$, and (2) the problem of finding a proper discretization scheme transforming the finite-variance process $y(t)$ of

$$\dot{y} = f(y) + \sigma w(t)$$

into a discrete-time finite-variance process $y_t = h(x_t)$ defined by a discrete-time dynamic model obtained by a proper discretization scheme.

9.4.1 Discretized Cauchy-Distributed Process

Example P1: Discretized Cauchy-distributed process
Let us see how we can introduce a discrete-time Cauchy distributed process from the continuous-time Cauchy-distributed process of Section 8.3.4. The Cauchy-distributed process $x(t)$ defined by

$$\dot{x} = -(2\alpha - 1)x + \sqrt{2(1 + x^2)}\, w(t)$$

is transformed by

$$y = \frac{1}{\sqrt{2}} \sinh^{-1} x$$

into another fat-tailed distributed process $y(t)$ defined by a stochastic differential equation with a constant diffusion coefficient as

$$\dot{y} = -\sqrt{2}\alpha \tanh(\sqrt{2}y) + w(t).$$

The model is transformed, by the LL-discretization scheme, into a discrete-time model,

$$y_{t+\Delta t} = \Phi(y_t) + \sqrt{\Delta t}\, w_{t+\Delta t}, \tag{9.49}$$

where

$$\Phi(y_t) = y_t + \left[\exp\left\{ \frac{-2\alpha\Delta t}{\cosh^2\left(\sqrt{2}y_t\right)} \right\} - 1 \right] \frac{\cosh\left(\sqrt{2}y_t\right)\sinh\left(\sqrt{2}y_t\right)}{\sqrt{2}}.$$

Since we have

$$\Phi(y_t)/y_t \to e^{-2\alpha\Delta t} \quad \text{for } y_t \to 0,$$

the model (9.49) is written as

$$y_{t+\Delta t} = \Phi(y_t)y_t + \sqrt{\Delta t}\,w_{t+\Delta t}0, \tag{9.50}$$

with

$$\Phi(y_t) = \begin{cases} \left[1 + \left[\exp\left\{ \dfrac{-2\alpha\Delta t}{\cos h^2\left(\sqrt{2}y_t\right)} \right\} - 1 \right] \dfrac{\cos h\left(\sqrt{2}y_t\right)\sin h\left(\sqrt{2}y_t\right)}{\sqrt{2}y_t} \right] & \text{for } y_t \neq 0 \\[2em] e^{-2\alpha\Delta t} & \text{for } y_t = 0 \end{cases}$$

The function $\Phi(y_t)$ of (9.50) is a bounded smooth function and satisfies

$$\text{(i)} \quad |\Phi(y_t)| < 1$$

and

$$\text{(ii)} \quad \Phi(y_t) \to 1 \quad \text{for } |y_t| \to \infty.$$

9.4.2 Discretized Beta-Distributed Process

Remember that some of the nonlinear function $f(y)$ of the processes derived from the Pearson system do not satisfy $f(0) = 0$. For example, $f(y)$ of the Beta-distributed processes does not satisfy $f(0) = 0$, if $\alpha \neq \gamma$.

Example 5: Discretized Beta-distributed process
We have seen in Section 8.3.5 that the Beta-distributed process $x(t)$ defined by

$$\dot{x} = \{(\alpha - \gamma) - (\alpha + \gamma + 2)x\} + \sqrt{2(1 - x^2)}\,w(t)$$

is transformed by

$$y = \frac{1}{\sqrt{2}}\sin^{-1}x$$

into another fat-tailed distributed process $y(t)$ defined by a stochastic differential equation with a constant diffusion coefficient as

$$\dot{y} = \left\{ \frac{\alpha - \gamma}{\sqrt{2}} \frac{1}{\cos\left(\sqrt{2}y\right)} - \frac{\alpha + \gamma + 1}{\sqrt{2}} \tan\left(\sqrt{2}y\right) \right\} + w(t).$$

The model is transformed by the LL-discretization scheme into a discrete-time model,

$$y_{t+\Delta t} = \Phi(y_t) + \sqrt{\Delta t}\, w_{t+\Delta t}, \tag{9.51}$$

where

$$\Phi(y_t) = y_t + J_t^{-1}(\exp^{J_t \Delta t} - 1) f(y_t),$$

$$f(y_t) = \frac{\alpha - \gamma}{\sqrt{2}\cos\left(\sqrt{2}y_t\right)} - \frac{\alpha + \gamma + 1}{\sqrt{2}} \tan\left(\sqrt{2}y_t\right),$$

and

$$J_t = \frac{(\alpha - \gamma)\sin\left(\sqrt{2}y_t\right) - (\alpha + \gamma + 1)}{\cos^2\left(\sqrt{2}y_t\right)}.$$

When $\alpha \neq \gamma$, we have $f(0) \neq 0$ and we cannot write (9.51) in the form of

$$y_{t+\Delta t} = \phi(y_t) y_t + \sqrt{\Delta t}\, w_{t+\Delta t} \tag{9.52}$$

since $\Phi(0) = \dfrac{\alpha - \gamma}{\sqrt{2}} \neq 0$. However, we can rewrite it as

$$y_{t+\Delta t} = \Phi(y_t) + \sqrt{\Delta t}\, w_{t+\Delta t}.$$

$$= \phi(y_t) y_t + \Phi(0) + \sqrt{\Delta t}\, w_{t+\Delta t}. \tag{9.53}$$

Here the bounded smooth function $\phi(y_t)$, defined by

$$\phi(y_t) = \frac{\{\Phi(y_t) - \Phi(0)\}}{y_t},$$

satisfies

$$\text{(i)} \ |\phi(y_t)| < 1$$

and

$$\text{(ii)} \ \phi(y_t) \to 1 \quad \text{for } |y_t| \to \infty.$$

9.4.3 Discretized Type I Gamma-Distributed Process

Example P3: Type I Gamma-distributed process
For the type I Gamma-distributed process $x(t)$, we have

$$\left(\begin{array}{l} y = \dfrac{\sqrt{2\beta x}}{\beta} \\[3mm] dy = \left\{ \left(\alpha - \dfrac{1}{2} \right) \dfrac{1}{y} - \dfrac{y}{2} \right\} dt + dw. \end{array} \right.$$

Here we have a nonlinear function

$$f(y_t) = \frac{\left(\alpha - \dfrac{1}{2} \right)}{y_t} - \frac{y_t}{2},$$

which behaves discontinuously near $y_t = 0$. For a small $\varepsilon > 0$, we have $f(\varepsilon) \to +\infty$ and $f(-\varepsilon) \to -\infty$ as $\varepsilon \to 0$. The dynamic model for $y(t)$ is transformed, by the LL-discretization scheme, into a discrete-time model,

$$y_{t+\Delta t} = \Phi(y_t) + \sqrt{\Delta t}\, w_{t+\Delta t},$$

where

$$\Phi(y_t) = y_t + J_t^{-1}(e^{J_t \Delta t} - 1) f(y_t),$$

$$f(y_t) = \frac{\left(\alpha - \dfrac{1}{2} \right)}{y_t} - \frac{y_t}{2},$$

and

$$J_t = \frac{-\left(\alpha - \dfrac{1}{2} \right)}{y_t^2} - \frac{1}{2}.$$

Note that $f(y_t)$ and J_t are undefined at $y_t = 0$. Also, for $|y_t| \to 0$, we have

$$f(y_t) = \frac{\left(\alpha - \dfrac{1}{2}\right)}{y_t} - \frac{y_t}{2} = \frac{\left(\alpha - \dfrac{1}{2}\right) - \dfrac{1}{2}y_t^2}{y_t} \to \pm\infty,$$

$$J_t = \frac{-\left(\alpha - \dfrac{1}{2}\right)}{y_t^2} - \frac{1}{2} \to -\infty,$$

$$e^{J_t \Delta t} \to 0$$

and

$$J_t^{-1} f(y_t) = \frac{\left(\alpha - \dfrac{1}{2}\right) - \dfrac{1}{2}y_t^2}{-\left(\alpha - \dfrac{1}{2}\right) - \dfrac{1}{2}y_t^2}\, y_t \to 0.$$

However, we have

$$J_t^{-1}\left(e^{J_t \Delta t} - 1\right) f(y_t) \approx y_t + o(y_t),$$

where

$$o(y_t)/y_t \to 0 \quad \text{as } |y_t| \to 0.$$

Then, we have

$$\Phi(y_t) \to 2y_t \quad \text{for } |y_t| \to 0.$$

That means when $\alpha \neq 1/2$, we can rewrite (9.51) as

$$y_{t+\Delta t} = \phi(y_t)y_t + \sqrt{\Delta t}\, w_{t+\Delta t},$$

where the bounded smooth function $\phi(y_t)$ is defined by

$$\phi(y_t) = \begin{cases} \dfrac{\Phi(y_t)}{y_t} & \text{for } y_t \neq 0 \\ 2 & \text{for } y_t = 0 \end{cases},$$

which satisfies

$$\text{(i)} \quad \phi(y_t) \to \phi_0 < 1 \quad \text{for } |y_t| \to \infty.$$

9.4.4 Discretized Type II Gamma-Distributed Process

Example P4: Type II Gamma-distributed process
For the Type II Gamma-distributed process $x(t)$, we have

$$
\left(
\begin{aligned}
y &= \frac{1}{\sqrt{2\beta}} \log x \\
dy &= \left\{ \left(\frac{\alpha\beta}{\sqrt{2\beta}} - \frac{\exp\left(\sqrt{2\beta}y\right)}{\sqrt{2\beta}} \right) \right\} dt + dW
\end{aligned}
\right)'
$$

where $f(0) = \dfrac{\alpha\beta - 1}{\sqrt{2\beta}} \neq 0$ unless $\alpha\beta = 1$. The model is transformed, by the LL-discretization scheme, into a discrete-time model,

$$
y_{t+\Delta t} = \Phi(y_t) + \sqrt{\Delta t}\, w_{t+\Delta t}, \tag{9.54}
$$

where

$$
\Phi(y_t) = y_t + J_t^{-1}(\exp^{J_t\Delta t} - 1) f(y_t),
$$

$$
f(y_t) = \left\{ \frac{\alpha\beta - \exp\left(\sqrt{2\beta}y_t\right)}{\sqrt{2\beta}} \right\},
$$

and

$$
J_t = -\exp\left(\sqrt{2\beta}y_t\right).
$$

Note that when $\alpha\beta \neq 1$ we have $\Phi(y_t) \to (1 - e^{-\Delta t})\dfrac{\alpha\beta - 1}{\sqrt{2\beta}}$ for $|y_t| \to 0$. That means, when $\alpha\beta \neq 1$, we can rewrite (9.54) as

$$
\begin{aligned}
y_{t+\Delta t} &= \Phi(y_t) + \sqrt{\Delta t}\, w_{t+\Delta t} \\
&= \phi(y_t)y_t + \Phi(0) + \sqrt{\Delta t}\, w_{t+\Delta t}. \tag{9.55}
\end{aligned}
$$

Here, we have

$$
\Phi(0) = (1 - e^{-\Delta t})\frac{\alpha\beta - 1}{\sqrt{2\beta}},
$$

and the bounded smooth function $\phi(y_t)$ defined by

$$\phi(y_t) = \{\Phi(y_t) - \Phi(0)\}/y_t$$

satisfies

$$\text{(i) } |\phi(y_t)| \leq 1$$

and

$$\text{(ii) } \phi(y_t) \to \phi_0 < 1 \quad \text{for } y_t \to \infty.$$

9.4.5 $f(y)$ and $\phi(y_t)$

So far we have seen several examples of the finite-variance discrete-time processes introduced from continuous-time finite-variance processes of various non-Gaussian marginal distributions. All of them can be written in one of the following forms:

$$y_{t+\Delta t} = \phi(y_t)y_t + \sqrt{\Delta t}\, w_{t+\Delta t}$$

or

$$y_{t+\Delta t} = \phi(y_t)y_t + \Phi(0) + \sqrt{\Delta t}\, w_{t+\Delta t},$$

where $\Phi(0)$ is a constant and $\phi(y_t)$ is a bounded smooth function satisfying

$$\phi(y_t) \to \phi_0 \leq 1 \quad \text{for } y_t \to \pm\infty.$$

An interesting correspondence between the nonlinear functions $-f(y)$ of $dy/dt = f(y)$ and $\phi(y_t)$ of $y_{t+1} = \phi(y_t)y_t$ is noticed when we compare the shape of the nonlinear functions for the 4 typical examples of the nonlinear functions of Examples 1, 2, 3, and 4 in Section 9.2.1.

The typical nonlinear patterns of $-f(y)$ of $dy/dt = f(y)$ and $\phi(y_t)$ of $y_{t+1} = \phi(y_t)y_t$ are shown in Figure 9.8a and b respectively. By comparing the shapes of the nonlinear function $-f(y)$ in Figure 9.8a and the shapes of the smooth bounded function $\phi(y_t)$ derived from $-f(y)$ in Figure 9.8b, we understand why the distributional behavior of the time series of the three ExpAR models of Examples (8.2) through (8.4) of Section 8.1.1 are similar to those of Examples 1, 2, and 4 of Section 9.2.1. By looking at these figures, we notice that $\phi(y_t)$ characterizes

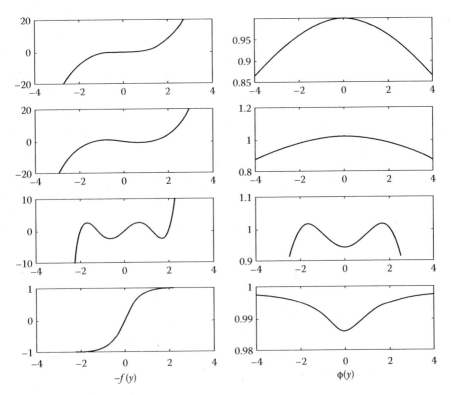

FIGURE 9.8
Examples of nonlinear functions and their φ-functions.

the gradient information of the $-f(y)$, although the relation between them is not as straightforward as the relation between $-f(y)$ of the dynamical system, $\dot{y} = f(y)$, and its potential function,

$$V(y) = -\int^{y} f(\eta)d\eta.$$

9.4.6 Discussions

So far we have seen that there is a natural bridge C-D connecting the continuous-time (deterministic and stochastic) model space, and the discrete-time (deterministic and stochastic) model space. (see Figure 9.3). These bridges provide us with a much wider scope and tools when we analyze time series. By using the bridges, we can analyze the data with any tools developed in those areas related to dynamical systems and Markov diffusion processes (see Figure 8.3) apart from those developed in time series analysis and statistics.

The LL-discretized model,

$$x_t = \phi(x_{t-1}|\theta)x_{t-1} + \Phi_0(\theta) + \sigma w_t, \tag{9.56}$$

which is derived as an approximation of a continuous-time model,

$$\dot{x} = f(x|\theta) + \sigma w(t), \tag{9.57}$$

can be regarded as a discrete-time model parameterized in the continuous-time model (9.57) with the parameter vector θ. An interesting question is what is the most suitable parametric nonlinear time series model for approximating the LL-discretized model (9.56). The answer to this question may be obvious. Appropriate nonlinear time series models are those models with a "smooth" and "bounded" nonlinear function $\phi(x_{t-1}|\theta)$ of x_{t-1}. Both ExpAR models and RBF-AR models have this property, where $\phi(x_{t-1}|\theta)$ is parameterized using Gaussian Radial Basis functions, satisfying

$$|\phi(x_{t-1}|\theta)| \rightarrow \phi_0 \leq 1 \quad \text{for } |x_{t-1}| \rightarrow \infty.$$

9.4.7 Crossing the Bridge from Discrete to Continuous

An interesting question is whether we can obtain a continuous-time dynamic model $dx/dt = f(x|\phi) + w(t)$ from a discrete-time model $x_{t+1} = A(x_t|\phi)x_t + w_{t+1}$, where ϕ is the parameter to characterize the discrete-time model.

To see this we try to obtain, from $A\Delta t(x|\phi)$ of the model,

$$x_{t+\Delta t} = A_{\Delta t}(x|\phi)x_t + w_{t+\Delta t}, \tag{9.58}$$

a function $f_{\Delta t}(x|\phi)$ whose local linearization is equal to $A_{\Delta t}(x|\phi)$, that is, the function that satisfies the local linearization relationship

$$f_{\Delta t}(x|\phi) = \left\{ A_{\Delta t}(x|\phi) - 1 \right\} \frac{\{\partial f_{\Delta t}(x|\phi)/\partial x\}x}{\exp[\{\partial f_{\Delta t}(x|\phi)/\partial x\}\Delta t] - 1}. \tag{9.59}$$

In general, we cannot give an explicit form of $f_{\Delta t}(x|\phi)$. Instead, we can obtain it by the following numerical iterative procedure at each point, x_i, x_{i+1} $(=x_i + \Delta x)$, x_{i+2} $(=x_i + 2\Delta x)$, ... on any finite interval $[X_1, X_2]$, .

$$f_{\Delta t}^{(k+1)}(x_i|\phi) = (1 - \Delta s)f_{\Delta t}^{(k)}(x_i|\phi) + \Delta s\{A_{\Delta t}(x|\phi) - 1\}$$

$$\times \frac{[\{f_{\Delta t}^{(k)}(x_{i+1}|\phi) - f_{\Delta t}^{(k)}(x_i|\phi)\}/\Delta x]x_i}{\exp[\{f_{\Delta t}^{(k)}(x_{i+1}|\phi) - f_{\Delta t}^{(k)}(x_i|\phi)\}\Delta t/\Delta x] - 1}. \tag{9.60}$$

This works because $f_{\Delta t}(x|\phi)$ can be regarded as a limiting function $f_{\Delta t}(x, \infty|\phi)$, characterized by

$$\frac{\partial f_{\Delta t}(x,s|\phi)}{\partial s} = 0,$$

of a deterministic spatial process $f_{\Delta t}(x, s|\phi)$ defined by the partial differential equation,

$$\frac{\partial f_{\Delta t}(x,s|\phi)}{\partial s} = -f_{\Delta t}(x,s|\phi) + \{A_{\Delta t}(x|\phi) - 1\} \frac{\{\partial f_{\Delta t}(x,s|\phi)/\partial x\}x}{\exp[\{\partial f_{\Delta t}(x,s|\phi)/\partial x\}\Delta t] - 1}.$$

We have $f_{\Delta t}^{(k+1)}(x_i|\phi) \to f_{\Delta t}(x|\phi)$ for $k \to \infty$ if Δs of (9.60) is sufficiently small compared with Δx. For the initial function $f_{\Delta t}^{(0)}(x_i|\phi)$ for the iteration (9.60), the following function,

$$f_{\Delta t}^{(0)}(x_i|\phi) = \frac{1}{\Delta t}\{A_{\Delta t}(x|\phi) - 1\}x,$$

which is obtained from the local linearization relation (9.59) by employing the approximation

$$\exp[\{\partial f_{\Delta t}(x|\phi)/\partial x\}\Delta t] \approx 1 + \{\partial f_{\Delta t}(x|\phi)/\partial x\}\Delta t,$$

will be useful. Of course, the function $f_{\Delta t}(x|a)$ obtained from $f(x|a)$ through the locally linearized $A_{\Delta t}(x|a)$,

$$A_{\Delta t}(x|a) = 1 + \left\{\frac{\partial f(x|a)}{\partial x}\right\}^{-1}\left[\exp\left\{\frac{\partial f(x|a)}{\partial x}\Delta t\right\} - 1\right]f(x|a),$$

using the previous procedure, is not equal to the original $f(x|a)$, since the local linearity approximation is used in deriving $A_{\Delta t}(x|a)$. We can easily see how fast $f_{\Delta t}(x|a)$ approaches the true $f(x|a)$ for $\Delta t \to 0$ by employing the previous numerical procedure. Figure 9.9 shows $f(x|a)$ and $f_{\Delta t}(x|a)$ of the model in Example 3 for several different Δt's, that is, $\Delta t = 0.04$, $\Delta t = 0.03$, $\Delta t = 0.02$, $\Delta t = 0.01$, where the smaller Δt, the better $f_{\Delta t}(x/a)$ approximates $f(x/a)$.

9.5 LL Bridge as a Numerical Integration Scheme

9.5.1 Numerical Integration Scheme and Required Properties

In the previous sections we have tried to build a bridge between the continuous-time dynamic model and discrete-time dynamic model, where we have found that the LL scheme for the deterministic and stochastic dynamical system is useful for transforming a finite-variance continuous-time process into a finite-variance discrete-time process. The scheme can be regarded as a numerical method for the integration of the stochastic differential equations, which has been one of the important topics in the field of computational

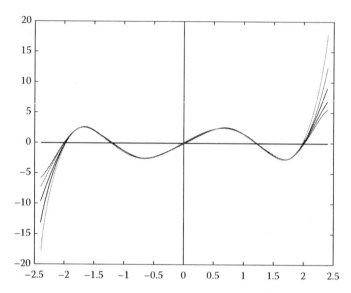

FIGURE 9.9
Different approximate function for different Δt's.

mathematics and numerical analysis, where many other conventional methods such as the Euler scheme and Runge–Kutta schemes are known (Kloeden and Platen 1995).

For nonlinear differential equations, we do not have any analytic solution, whether they are deterministic or stochastic, and all the solutions are approximate solutions obtained by numerical integration schemes. The goodness of the scheme is assessed by checking the solutions $x_{t_0}, x_{t_1}, \dots, x_{t_N}$ on a finite time interval $[t_0, T]$ from several criteria, where important criteria for an integration scheme are whether it

(1) is computationally fast.
(2) is consistent, that is, $E\left[\left|x_{t_N} - x(T)\right|\right] \to 0$ for $\Delta t \to 0$.
(3) has a high order of the convergence for $\Delta t \to 0$, where the order m is defined by

$$\left\{ E_0 \left[x(T) - x_{t_N} \right]^2 \mid x_{t_0} = x_0 \right\}^{1/2} = o(\Delta t^m).$$

(4) is stable, that is, the scheme transforms a finite-variance continuous-time process into a finite-variance discrete-time process.

Naturally, a scheme that has high order m is important. However, a more important criterion is stability.

9.5.2 A-Stability

Researchers in sciences and engineering are not much interested in nonlinear differential equations that have explosive trajectories. Researchers are interested in solving not only those stochastic differential equations whose trajectories are converging to limit points but also those stochastic differential equations producing bounded trajectories such as limit cyclic oscillations or chaotic dynamics. Numerical integration schemes are essential for studying the trajectories of such differential equations in many scientific fields.

Unfortunately, the most explicit schemes for nonlinear differential equations have a numerical instability problem. They are consistent and their convergence rate may be high, but their asymptotic behavior is unstable and often explodes to infinity for $t \to \infty$ for a fixed Δt, even though their theoretical solution is known to be bounded. Researchers want to have a reliable stable scheme that inherits the qualitative properties of the original dynamical system. Researchers are interested in having a similar non explosive asymptotic trajectory of the original dynamical system (whether their limits are equilibrium points, limit cycles, or chaos attractor surfaces), independent of the initial value x_0. A scheme that guarantees these properties is called an A-stable scheme.

The "mathematical" definition of the A-stability is rather informal and indirect, in most textbooks of numerical analysis (see Kloeden and Platen 1995), where the A-stability is discussed in terms of the stability of an associated linear dynamical system with eigenvalues with negative real parts. However, the limiting behavior of nonlinear systems may be very different from that of linear equations. Then the most reliable definition is the "descriptive" one given in the previous paragraph.

9.5.3 Implicit Schemes

Most schemes known for the integration of the deterministic equation, such as the Euler scheme and the Runge–Kutta scheme, are also used for the integration of stochastic differential equations. However, as we saw in Section 9.2.6, these schemes have serious problems of numerical instability. To guarantee the stability of the discretization, implicit schemes are often used for the integration of stiff equations.

Mathematically, an explicit method is written with some function $F(.)$ as

$$x(t + \Delta t) = F(x(t)),$$

whereas the explicit scheme is defined, at each step, as a solution of an equation satisfying some conditions, which we write as

$$G(x(t), x(t + \Delta t)) = 0.$$

Here, in order to obtain $x(t + \Delta t)$ from $x(t)$, one needs to solve this equation using some iterative numerical root-finding method.

For example, we can see the difference between the explicit scheme and the implicit scheme, in the following:

$$\dot{x} = -x^r, x \in [0, a],$$

where we have to calculate numerical solutions using an explicit scheme and an implicit scheme, with the initial condition $x(0) = 1$. A solution by an explicit Euler scheme is obtained as

$$x_{k+1} = \left(1 - \Delta t x_k^{r-1}\right) x_k,$$

which comes from considering the approximation

$$\frac{x_{k+1} - x_k}{\Delta t} = -x_k^r.$$

A solution by an implicit Euler scheme could be obtained by considering the root x_{k+1} as the solution for the algebraic equation

$$x_{k+1} + \Delta t x_{k+1}^r = x_k,$$

whose roots can be obtained by an iterative root-finding algorithm, such as Newton's method.

The implicit method has been used for the solution of stiff equations because the scheme is A-stable. However, the cost we pay for the stability is that it requires an extra computation (solving an algebraic equation by an iterative method) at each time point, and it can be much harder to implement. If we use the explicit method, we do not need to have such a computationally expensive iterative numerical root-finding method. However, in order to avoid computational instability and keep the errors bounded, the explicit method requires impractically small time steps Δt. The advantage of the LL scheme is that it is A-stable and explicit at the same time, and it does not need to make Δt very small to prevent the trajectory exploding to infinity.

9.5.4 LL Scheme with Ito Calculus

If the LL scheme has so much advantage in integrating the stochastic differential equations, it may be worth investigating its characteristics further. Shoji (1998) suggested to use Ito calculus, instead of Stratonovich calculus, for the Local Linear approximation of the drift function. The idea of the local linearization method comes from approximating the drift function $f(x)$ in

$$dx = f(x)dt + \sigma dW(t) \tag{9.61}$$

by a linear function of x for a short time interval $[t_0, t_0 + \Delta t)$. Here the local behavior of the drift function $f(x)$ is specified by

$$df = (\partial f / \partial x)dx. \tag{9.62}$$

This is interpreted as a differential of $f(x)$ with the Stratonovich stochastic calculus. If we employ the Ito stochastic calculus, we obtain another representation for df as follows:

$$df = \{\partial f / \partial x\}dx + \left\{(\sigma^2/2)\partial^2 f / \partial x^2\right\}dt. \tag{9.63}$$

Since df of (9.63) includes second-order derivatives of $f(.)$, we may need to employ an extra assumption (A2) together with (A1) as follows:

(A1) $\partial f / \partial x$ is constant for the interval,

  i.e., $\partial f / \partial x = J_{t_0}$ for the interval $[t_0, t_0 + \Delta t)$.

(A2) $\partial^2 f / \partial x^2$ is constant for the interval,

  i.e., $\partial^2 f / \partial x^2 = H_{t_0}$ for the interval $[t_0, t_0 + \Delta t)$.

With these assumptions, (9.63) can be solved for $t_0 \le \tau < t_0 + \Delta t$ as

$$f(x_\tau) - f\left(x_{t_0}\right) = \left\{(\sigma^2/2)H_{t_0}\right\}(\tau - t_0) + J_{t_0}\left(x_\tau - x_{t_0}\right).$$

Then we have to focus our attention, instead of (9.61), only on the following approximate linear stochastic differential equation (9.64) on each short time interval $[t_0, t_0 + \Delta t)$,

$$dx = J_{t_0}xdt + M_{t_0}dt + (\sigma^2/2)H_{t_0}tdt + \sigma dW(t), \tag{9.64}$$

where

$$M_{t_0} = f\left(x_{t_0}\right) - J_{t_0}x_{t_0} - \left\{(\sigma^2/2)H_{t_0}\right\}t_0.$$

The solution of this linear stochastic differential equation on the interval $[t_0, t_0 + \Delta t]$ leads to

$$x(t_0 + \Delta t) = \exp(J_{t_0}\Delta t)x(t_0)$$

$$+ M_{t_0} \int_{t_0}^{t_0 + \Delta t} \exp(J_{t_0}(t_0 + \Delta t - u))du$$

$$+ \frac{\sigma^2 H_{t_0}}{2} \int_{t_0}^{t_0 + \Delta t} u\exp(J_{t_0}(t_0 + \Delta t - u))du$$

$$+ \sigma \int_{t_0}^{t_0 + \Delta t} \exp(J_{t_0}(t_0 + \Delta t - u)dW(u)). \qquad (9.65)$$

The second and the third term of the right hand side of (9.65) are given, respectively, by

$$\frac{M_{t_0}}{J_{t_0}}\left\{\exp(J_{t_0}\Delta t) - 1\right\}$$

and

$$\left(\frac{\sigma^2 H_{t_0}}{2}\right)\left(\frac{1}{J_{t_0}^2}\right)\left\{\exp\left((J_{t_0}\Delta t) - 1 - J_{t_0}\Delta t\right)\right\}.$$

Then we have, from (9.65), the following scheme:

$$x(t_0 + \Delta t) = x(t_0) + \frac{1}{J_{t_0}}\left\{\exp(J_{t_0}\Delta t) - 1\right\}f(x_{t_0})$$

$$+ \frac{\sigma^2 H_{t_0}}{2J_{t_0}^2}\left\{\exp\left((J_{t_0}\Delta t) - 1 - J_{t_0}\Delta t\right)\right\}$$

$$+ \int_{t_0}^{t_0 + \Delta t} \exp(J_{t_0}(t_0 + \Delta t - u)\sigma dW(u)). \qquad (9.66)$$

The variance of the third term of the right-hand side of (9.66) is given, as in the previous section, by

$$\frac{\left\{\exp(2J_{t_0}\Delta t) - 1\right\}}{(2J_{t_0})}.$$

Thus we obtain the following discrete-time model:

$$x_{t_0+\Delta t} = A(x_t)x_t + B(x_t)n_{t_0+\Delta t} \tag{9.67}$$

$$A(x_t) = 1 + \frac{f(x_t)}{J_t x_t}\left\{\exp(J_t\Delta t)-1\right\}$$

$$+ \frac{\sigma^2 H_{t_0}}{2J_{t_0}^2 x_t}\left\{\exp(J_{t_0}\Delta t)-1-J_{t_0}\Delta t\right\}$$

$$B(x_t) = \left\{\frac{\exp(2J_t\Delta t)-1}{(2J_t)}\right\}^{1/2}.$$

Thus, we have obtained a slightly different scheme when we employ the Ito stochastic calculus at the stage of local approximation for linearization.

It has been shown that the present method (which is called the New Local Linearization (NLL) scheme) inherits the A-stability of the original LL scheme, and at the same time it improves the parameter estimation performance in the quasi-maximum likelihood method applied to the numerical simulation studies (Shoji and Ozaki 1997, 1998, Shoji 1998, Jimenez et al. 1999).

9.5.5 Order of Convergence

All the schemes are naturally convergent, that is,

$$E_0\left[|x(T)-x_{t_N}|\,\big|\,x_{t_0}=x_0\right] \xrightarrow[\Delta t\to 0]{} 0.$$

Then what interests researchers is the speed of the convergence, which is characterized by the order m of the convergence defined by

$$\left\{E_0\left[x(T)-x_{t_N}\right]^2\big|\,x_{t_0}=x_0\right\}^{1/2} = o(\Delta t^m).$$

Naturally, a scheme that has high order m is important. It has been shown that the order of the Euler method is 0.5. Several explicit schemes of higher order convergence have been introduced. Milstein (1994), by using the Ito-Taylor expansion of the process, introduced a scheme:

$$x_{n+1} = x_n + a(x_n)\Delta t + b(x_n)\Delta W + \frac{1}{2}b(x_n)b(x_n)'\{(\Delta W)^2-\Delta t\}.$$

The scheme includes $(\Delta W)^2$ as well as ΔW. The order of the convergence of the scheme is known to be 1.0, which is higher than the Euler scheme.

However, we note that the LL scheme attains the same order as the Milstein scheme without including the quadratic term w_{t+1}^2.

In general, a higher order convergence can be expected if we include a higher order of the multiple integrals of the Ito-Taylor expansion of the process (Kloeden and Platen 1995, Milstein 1994). The Platen-Wagner-1.5 Scheme includes w_{n+1} and w_{n+1}^2 as

$$x_{n+1} = x_n + a(x_n)\Delta t + b(x_n)w_{n+1} + \frac{1}{2}b(x_n)b(x_n)'\left\{w_{n+1}^2 - \Delta t\right\}$$

$$+ a(x_n)'b(x_n)\Delta Z + \frac{1}{2}\left(a(x_n)a(x_n)' + \frac{1}{2}b(x_n)^2 a(x_n)''\right)(\Delta t)^2$$

$$+ \left(a(x_n)b(x_n)' + \frac{1}{2}b(x_n)^2 b(x_n)''\right)\{w_{n+1}\Delta t - \Delta Z\}$$

$$+ \frac{1}{2}b(x_n)\left(b(x_n)b(x_n)'' + (b(x_n)')^2\right)\left\{\frac{1}{3}w_{n+1}^2 - \Delta t\right\}w_{n+1}.$$

Here, ΔZ is given by

$$\Delta Z = \int_{\tau_n}^{\tau_{n+1}}\int_{\tau_n}^{s_2} dW(s_1)ds_2,$$

which is known to have zero mean and variance,

$$E[(\Delta Z)^2] = \frac{1}{3}(\Delta t)^2.$$

The scheme is known to have order 1.5 convergence. We note that the order of convergence of the NLL scheme is also 1.5, even though the NLL scheme is much simpler without the quadratic term w_{n+1}^2. Another scheme of convergence of order 2.0, where x_{n+1} is given as an explicit function of $w_{n+1}, w_{n+1}^2, w_{n+1}^3$ and w_{n+1}^4, is found in Kloeden and Platen (1995).

However, the order of convergence and the numerical stability are different problems. Improving the order of convergence does not mean an improvement in the stability problem. For the user of numerical integration schemes for the stochastic differential equations, an A-stable scheme with a high order of convergence is the most desirable. De la Cruz et al. (2006, 2007) introduced a sophisticated method (called HOLL) to improve the order of convergence of the LL scheme to any higher order of convergence for highly nonlinear ordinary differential equations, where the A-stability of the original LL schemes is inherited in the generalized scheme.

9.5.6 Discussions

9.5.6.1 Integration Schemes and Innovation Approach

Let us see the integration scheme from the viewpoint of the innovation approach. All the explicit schemes are specifying the one-step ahead value $x_{t+\Delta t}$ of the process as an explicit function of the past values, x_t, $x_{t-\Delta t}$ ..., and the present and past values of Gaussian white noise, $w_{t+\Delta t}$, w_t, $w_{t-\Delta t}$ ..., as follows:

(i) Euler scheme: $x_{t+\Delta t} = \phi^{(E)}(x_t, w_{t+\Delta t})$

(ii) LL scheme: $x_{t+\Delta t} = \phi^{(LL)}(x_t, w_{t+\Delta t})$

(iii) NLL scheme: $x_{t+\Delta t} = \phi^{(NLL)}(x_t, w_{t+\Delta t})$

(iv) Milstein scheme-1.0: $x_{t+\Delta t} = \phi^{(M)}\left(x_t, w_{t+\Delta t}, w_{t+\Delta t}^2\right)$

(v) Wagner-Platen-1.5: $x_{t+\Delta t} = \phi^{(PW)}\left(x_t, w_{t+\Delta t}, w_{t+\Delta t}^2\right)$

(vi) Kloeden-Platen-2.0: $x_{t+\Delta t} = \phi^{(KP)}\left(x_t, w_{t+\Delta t}, w_{t+\Delta t}^2, w_{t+\Delta t}^3, w_{t+\Delta t}^4\right)$.

We note that predictors, $E[x_{t+\Delta t}|x_t]$, of only 3 schemes, the Euler scheme, LL scheme, and NLL scheme, yield Gaussian prediction errors,

$$w_{t+\Delta t} = x_{t+\Delta t} - E[x_{t+\Delta t}|x_t],$$

which matches our basic idea of the innovation approach to the time series analysis. For the schemes, (iv), (v), and (vi), calculation of $w_{t+\Delta t}$ from the past observation x_t is not as easy as in the other schemes. The properties of the schemes are compared in Table 9.1, where the order k of the power of the

TABLE 9.1

Comparison of the Six Schemes

	k of $w_{t+\Delta t}^k$	m of (Δt^m)	A-Stability
Euler	1	0.5	No
LL	1	1.0	Yes
NLL	1	1.5	Yes
Milstein-1.0	2	1.0	No
Wagner-Platen-1.5	2	1.5	No
Kloeden-Platen-2.0	4	2.0	No

white noise term $w_{t+\Delta t}^k$ used, order m of the convergence, and whether the scheme is A-stable or not are shown.

9.5.6.2 Multidimensional LL Schemes

We have discussed the Local Linearization scheme mostly for the scalar case. The extension of the scheme to the multivariate case is straightforward. For a multivariate dynamical system, $dx/dt = f(x)$, we have a local linearized discrete-time system,

$$x_{t+\Delta t} = x_t + J_t^{-1}\left(e^{J_t\Delta t} - 1\right)f(x_t),$$

where J_t is a Jacobian matrix defined by

$$J_t = \left(\frac{\partial f(x)}{\partial x}\right)_{x=x_t},$$

and e^J is a matrix exponential function of J. Then the multidimensional LL scheme and NLL scheme transform $dx = f(x) + \sigma n(t)$, or equivalently, $\dot{x} = f(xdt + \sigma dW(t))$, into a discrete-time stochastic dynamic model,

$$x_{t+\Delta t} = a(x_t) + \eta_{t+\Delta t}.$$

Here,

$$\eta_{t+\Delta t} = \int_t^{t+\Delta t} \exp\{J_t(t-u)\}\sigma dW(u).$$

$a(x_t)$ for the LL scheme is given by

$$a(x_t) = x_t + J_t^{-1}\left(e^{J_t\Delta t} - 1\right)f(x_t) + \eta_{t+\Delta t},$$

and $a(x_t)$ for the NLL scheme is given by

$$a(x_t) = x_t + J_t^{-1}\left(e^{J_t\Delta t} - 1\right)f(x_t) + \frac{1}{2}J_t^{-2}\left(e^{J_t\Delta t} - 1 - J_t\Delta t\right)M(x_t).$$

Here, $M(x_t)$ is

$$M(x_t) = \{tr(\sigma\sigma'H_1), \ldots, tr(\sigma\sigma'H_k)\}', H_i = \left(\frac{\partial^2 f_i(x)}{\partial x_r \partial x_s}\right)_{1\le r,s\le k}.$$

9.5.6.3 Related Topics

The most commonly studied topics about the stochastic differential equation models are classified into two: (1) numerical integration of the models for simulations, and (2) estimation of the models from time series observations. Since the models are useful for describing many dynamic phenomena in sciences and engineering, they have been studied in many different fields in science and engineering. For example, a group (Milstein 1974, Kloeden and Platen 1995, Talay 1995, Schurz 2002 etc.) in the field of numerical analysis have laid emphasis on the development of the discretization method for simulating stochastic differential equation models. A group in applied probability and Markov Chain theory (Tweedie 1975, 1983, Roberts and Tweedie 1996, Stramer and Tweedie 1999 etc.) have emphasized the topic related to the ergodicity and stationarity of the discretized Markov chains. The study by a group in statistics, econometrics, and probability theory (Prakasa Rao 1999, Singer 2002, Gallant and Long 1997, Bergstrom 1966, Robinson 1976, Hansen 1982, Kutoyantz 1984, Yoshida 1992, Ait-Sahalia 2002, Sorensen 2004 etc.) is naturally focused on the estimation of the stochastic differential equation models from time series data. However, the simulation of complex stiff nonlinear differential equation models is not in their scope. Both estimations and simulations of stochastic differential equation models have been studied by a time series and dynamical systems group (Ozaki, Shoji, Jimenez, Biscay, De La Cruz, Carbonell, Singer, etc.), where the emphasis is on the practical applicability of the dynamic models in time series data analysis.

TABLE 9.2

Comparison of Different Approaches from Different Fields

Who?	Which Field?	Simulation or Estimation?	Kalman Filter Approach?
Tweedie (1983), Roberts and Stramer (2001).	Probability and Markov chain theory	Simulation and estimation	No
G.N. Milstein (1974), Kloeden and Platen (1995), Talay (1995), Schurz (2002).	Numerical methods for stochastic differential equations	Simulation	No
Y. Kutoyantz (1984), Yoshida (1992), Sorensen (2004).	Probability and stochastic processes	Estimation	No
A.R. Bergstrom (1966), Robinson (1976), Hansen (1982).	Econometrics and time series analysis	Estimation	No
B.L.S. Prakasa Rao (1999), Gallant and Tauchen (1997).	Statistics and time series analysis	Estimation	No
Ozaki (1985a, 1993b), Shoji and Ozaki (1997), Jimenez et al. (1999), Biscay et al. (1995), Carbonell et al. (2002), De La Cruz et al. (2007), Singer (2002).	Time series analysis and dynamical systems	Simulation and estimation	Yes

In applications, most time series data are more or less non-Gaussian, possibly with nonlinear dynamics. When the stochastic differential equation is non-Gaussian with nonlinear dynamics, analytical treatment of the models is impossible, and we need to employ some kind of approximations whether it is for simulations or estimations. In such situations, it is hard to say theoretically which approximation is the best without much numerical experience in applications. Another important point in applications of stochastic differential equation models is that the dimension of the model describing the dynamic phenomena may be multidimensional, but the observation data are often one-dimensional time series data contaminated with observation errors. Most estimation methods developed so far are not able to handle such situations except the innovation approach with the Kalman filter techniques or the EM algorithm (we discuss these in Chapters 11–13). It is unfortunate that much interaction has not been seen between the researchers in different research fields of stochastic differential equation modeling. To help readers understand what is going on in these different research fields, we give a summary in Table 9.2. It is hoped that more mutual understanding and communication between the scientists in different disciplines are seen in the future.

10

Likelihood of Dynamic Models

So far we have seen, in Chapters 3 through 6, Part I, that there are many types of dynamic models for prediction. These models have been developed and used not only in neuroscience but also in many scientific fields such as physics, electrical engineering, chemistry, biology, genetics, and macroeconomics. In this chapter, we see how the dynamic models treated in Chapters 3 through 6, Part I, are estimated and identified from observed time series data.

10.1 Innovation Approach

Identification and estimation of stochastic or deterministic dynamical system models from observed time series data has been a much-studied topic since the era of Wiener and Kolmogorov. Many methods have been introduced and discussed since the 1930s. Among them, one approach called the innovation approach (or equivalently prediction error approach) may be specially interesting and useful for applied scientists, since the guideline principle is intuitively simple, the computational algorithm and statistical diagnostic checking are straightforward and easy to perform, and its theoretical properties (such as consistency and efficiency) have been confirmed under reasonable assumptions.

The guiding principle of the innovation approach is useful for modelers in various stages of modeling dynamics, either at the microscopic level or macroscopic level, whenever the dynamic phenomena are measured in time series data. Intuitively small prediction error is preferred, and this is one of the reasons why the least squares method is widely used in the statistical analysis of time series. However, this choice is also mathematically supported by a Markov process theory, summarized in the following theorem.

Theorem (Doob-Feller)
For any sample-continuous Markov process x_t, the prediction error,

$$w_t = x_t - E[x_t | x_{t-\Delta t}, x_{t-2\Delta t}, \ldots]$$

converges to a Gaussian white noise for $\Delta t \to 0$.

The theorem shows that the maximum likelihood estimation method of models for Markov processes and the least squares method, based on the Gaussian prediction errors, are essentially equivalent when the

prediction errors are temporally homogeneous. Remember that all the stochastic dynamic models that we saw in Chapters 3 through 6, Part I, are
local Gauss Markov models, in which the prediction errors are somehow
Gaussian and homoscedastic (i.e., temporally homogeneous), and the deterministic dynamic models are regarded as limits of some stochastic local
Gauss Markov models with the driving noise variance $\sigma^2 \to 0$.

This means that we can identify most of these models by minimizing
the sum of squares of the prediction errors or maximizing the likelihood
of these models using the Gaussian innovations. Even though the original
process is not Gaussian we can except the prediction error to be almost,
if not exactly, Gaussian so that it is still appropriate to use the Gaussian
likelihood. In this case, the estimates are called Gaussian estimates and
are known to share the same attractive asymptotic properties as maximum
likelihood estimates (Whittle 1962). However, this is on the assumption that
the (possibly vector) variable x_t of the dynamic model is identical to the
observed time series variable. If the observed time series is a scalar variable and the model explaining the time series is a multivariate dynamic
model, the previous estimation method cannot be used since the prediction errors cannot possibly be given explicitly as a function of the observed
time series data. In such situations, Frost–Kailath's theorem (see Section
7.3.4) is very useful. The theorem enables us to represent the time series x_t
in terms of innovations that are guaranteed to have the same probabilistic
properties as the errors derived from the unobserved system driving noise
and the observation noise ε_t. A concrete algorithm for deriving the innovations is realized by using the Kalman filter method, which is discussed in
Chapter 11. This innovation representation will prove central to the development of the likelihood for the state space models with various multivariate continuous-time nonlinear dynamic models discussed in Chapter 5.

If the dynamic model is presented from the beginning as a discrete-time
local Gauss model, such as those nonlinear AR models in Chapter 3, and
the process is observed without observation noise, the innovations of the
series is also given explicitly as functions of observations using the dynamic
model, without using Kalman filter. Thus, the likelihood function is even
easier to calculate than the case for the continuous-time models.

Since the computation and maximization of the likelihood function of
dynamic models is the most important for the actual modeling of time series
data, we see, in this chapter, how the log-likelihood functions of the dynamic
models treated in Chapters 3 through 6, Part I, both in continuous time and
discrete time, are calculated for the situations where all the variables of the
models are observed without observation errors. We discuss the calculation
of the likelihood for more complicated situations in Chapters 11 through 13,
Part III, where the dimension of observed data could be less than the dimension of the process vector, that is., $\dim(x_t) \leq \dim(z_t)$, and x_t could include an
additional observation noise ε_t, that is, $x_t = Cz_t + \varepsilon_t$ with a possibly rectangular observation matrix C.

10.2 Likelihood for Continuous-Time Models

10.2.1 Innovation-Based Likelihood

The key idea of the computation of the log-likelihood of temporally dependent time series is to whiten the series into independent series. If we could assume that the time series $(x_1, \ldots, x_N)'$ is observed from a temporally dependent continuous-time Markov process model $M(\theta, \sigma)$ where the sampling interval is sufficiently small, then the series can be whitened into prediction errors, $(w_1, \ldots, w_N)'$, using the model $M(\theta, \sigma)$, as follows

$$w_t = x_t - E[x_t | x_{t-1}, x_{t-2}, \ldots, x_1 | \theta, \sigma] \quad (t = 1, 2, \ldots, N).$$

Here the log-likelihood of the model can be written as

$$p(x_1, \ldots, x_N | \theta, \sigma) = p(w_1, \ldots, w_N | \theta, \sigma) \det\left(\frac{\partial w}{\partial x}\right)$$

$$= \prod_{t=1}^{N} p(w_t | \theta, \sigma),$$

where $(\partial w / \partial x)$ is the Jacobian matrix of the transformation from $(x_1, \ldots, x_N)'$ to $(w_1, \ldots, w_N)'$. Since w_t is uncorrelated with the past $x_{t-1}, x_{t-2}, \ldots$, we have

$$\left[\begin{array}{ll} \dfrac{\partial w_i}{\partial x_j} = 1 & \text{for } i = j \\[2mm] \dfrac{\partial w_i}{\partial x_j} x = 0 & \text{for } i < j \end{array}\right.$$

so that the determinant of the Jacobian matrix is 1, that is, $\det(J) = 1$. We know from Doob's basic theorem that the prediction error w_t of the continuous Markov process is approximately Gaussian white, and we can write the (-2)log-likelihood as

$$(-2)\log p(x_1, \ldots, x_N | \theta, \sigma) = (-2)\log p(w_1, \ldots, w_N | \theta, \sigma)$$

$$= \sum_{t=1}^{N} \log |\Sigma_{w_t}| + \sum_{t=1}^{N} w_t^T \Sigma_{w_t}^{-1} w_t + N \log 2\pi.$$

We know intuitively that it is better for the prediction error to be small, and this is one of the reasons why the Least Squares method is widely used in the statistical analysis of time series. This is also mathematically supported by a Markov process theory, summarized in the Theorem (Doob-Feller) (see Doob 1953 or Feller 1966 for details).

The maximum likelihood method for the continuous-time data, $x(t)$, ($t_0 \leq t \leq T$), rather than discrete-time data, x_1, x_2, ..., x_N, is also discussed in statistical inference theory for Markov diffusion processes (see, e.g., Prakasa Rao 1999). In most scientific and engineering applications, however, observations are recorded and stored in computers as discrete-time data with a sufficiently small sampling interval. Therefore, in the present book, we consider and discuss, the estimation problem of a continuous-time model, using only discrete-time observation data.

From the computational viewpoint, the most important and critical point is how to construct the conditional expectation, $E[x_t|x_{t-1}, ...]$, that is, the predictor of x_t for the continuous-time dynamic models. We will see next how $E[x_t|x_{t-1}, ...]$ and the likelihood functions are constructed for the continuous-dynamic models using the examples that we discussed in Chapter 5.

10.2.2 Examples

10.2.2.1 Linear Stochastic Dynamical System

Example 1:

$$\dot{x} = ax + \sigma w(t).$$

Here we assume that the variance of the Gaussian white noise $w(t)$ is 1. The unknown parameters to estimate from time series observations, x_1, x_2, ..., x_N, are thus the coefficient a and the noise variance σ^2. Using the discretized model,

$$x_t = \exp(a\Delta t)x_{t-1} + B_t w_t,$$

with $B_t = \sqrt{\{\exp(2a\Delta t) - 1\}/2a}$, and Δt chosen as indicated in Section 9.1.4, we have the prediction error,

$$w_t = x_t - E[x_t|x_{t-1}, x_{t-2}, ...]$$

$$= x_t - \exp(a\Delta t)x_{t-1},$$

and the error variance,

$$\sigma_t^2 = E[w_t^2]$$

$$= \sigma^2\{\exp(2a\Delta t) - 1\}/(2a)$$

$$\approx \sigma^2 \Delta t.$$

Then, we can calculate (−2)log-likelihood of the model as

$$(-2)\log p(x_1, \ldots, x_N | a, \sigma^2) = (-2)\log p(x_2, \ldots, x_N | x_1, a, \sigma^2) + (-2)\log p(x_1 | a, \sigma^2)$$

$$= (-2)\log p(w_2, \ldots, w_N | x_1, \theta) + (-2)\log p(x_1 | a, \sigma^2)$$

$$\approx \sum_{t=2}^{N} \log \sigma_t^2 + \sum_{t=2}^{N} \frac{w_t^2}{\sigma_t^2} + (N-1)\log 2\pi.$$

Here we assume that the number of data points, N, is sufficiently large so that the initial value effect by (−2)log $p(x_1 | a, \sigma^2)$ can be ignored.

10.2.2.2 Linear Oscillation Models

Example 2:

$$\ddot{x}(t) + c\dot{x}(t) + \omega^2 x(t) = n(t) \tag{10.1}$$

$$n(t) : N(0, \sigma_n^2).$$

The second-order stochastic differential equation model (10.1) can be rewritten using the two-dimensional stochastic dynamical system model,

$$\begin{pmatrix} \ddot{x}(t) \\ \dot{x}(t) \end{pmatrix} = \begin{pmatrix} -a & -b \\ 1 & 0 \end{pmatrix} \begin{pmatrix} \dot{x}(t) \\ x(t) \end{pmatrix} + \begin{pmatrix} n(t) \\ 0 \end{pmatrix},$$

which we write as

$$\dot{z}(t) = Jz(t) + w(t), \tag{10.2}$$

where

$$z(t) = \begin{pmatrix} \dot{x}(t) \\ x(t) \end{pmatrix}, \quad w(t) = \begin{pmatrix} n(t) \\ 0 \end{pmatrix}, \quad J = \begin{pmatrix} -a & -b \\ 1 & 0 \end{pmatrix}.$$

The likelihood function of the model is dependent on how the observed time series data are given, that is, it depends on whether (i) they are bivariate time series observations, $z_{\Delta t} = (x_{\Delta t}, \dot{x}_{\Delta t})'$, $z_{2\Delta t} = (x_{2\Delta t}, \dot{x}_{2\Delta t})'$, ..., $z_{N\Delta t} = (x_{N\Delta t}, \dot{x}_{N\Delta t})'$, or (ii) they are given as scalar time series observations, $x_{\Delta t}, x_{2\Delta t}, \ldots, x_{N\Delta t}$.

1. The case when $z_{\Delta t} = (x_{\Delta t}, \dot{x}_{\Delta t})'$, $z_{2\Delta t} = (x_{2\Delta t}, \dot{x}_{2\Delta t})'$, ..., $z_{N\Delta t} = (x_{N\Delta t}, \dot{x}_{N\Delta t})'$ are observed:

   To write the likelihood of the model with the observations, $z_{\Delta t}, z_{2\Delta t}, ...,$ $z_{N\Delta t}$, we need to calculate the prediction errors for $z_{t+\Delta t}$ with z_t and its prediction error variance. Since the dynamic model is linear, we have, for the time interval $[t, t + \Delta t]$,

$$z_{t+\Delta t} = e^{J\Delta t} z_t + w_{t+\Delta t} \tag{10.3}$$

$$w_{t+\Delta t} = \int_t^{t+\Delta t} e^{J(t+\Delta t - s)} w(s) ds$$

   and the prediction error of $z_{t+\Delta t}$, given $z_t, z_{t-\Delta t}, ...,$ is

$$w_{t+\Delta t} = z_{t+\Delta t} - E[z_{t+\Delta t} | z_t, ...] = z_{t+\Delta t} - e^{J\Delta t} z_t. \tag{10.4}$$

   The variance of $w_{t+\Delta t}$ is given by, the local linearization method (see equation (9.27)). We note that, although the variance matrix of continuous-time noise $w(t)$ is rank one, the variance matrix of the discrete-time noise $w_{t+\Delta t}$ is rank two as a result of the integration over $[t, t + \Delta t]$. All the elements of the matrix $\Sigma_{w_{t+\Delta t}}$ are given as functions of the model parameters, as we saw in equations (9.24) through (9.29). This means that, from (10.3), we can have a discrete-time model

$$z_{t+\Delta t} = A z_t + B v_{t+\Delta t}, \tag{10.5}$$

   where $v_{t+\Delta t}$ is a Gaussian white noise of variance σ_n^2,

$$A = e^{J\Delta t},$$

$$B = U \begin{pmatrix} \sqrt{\lambda_1} & 0 \\ 0 & \sqrt{\lambda_2} \end{pmatrix},$$

   where the elements of matrix U are given as functions of parameter values, a, b, and σ_n^2. Then for the bivariate discrete-time observation data, $z_{\Delta t}, z_{2\Delta t}, ..., x_{N\Delta t}$, the (-2)log-likelihood of the 2-dimensional model,

$$\dot{z}(t) = J z(t) + w(t), \tag{10.6}$$

   where $E[w(t)w(t)'] = \begin{pmatrix} \sigma_n^2 & 0 \\ 0 & 0 \end{pmatrix}$, is given approximately by

$$(-2)\log p(z_{\Delta t}, ..., z_{N\Delta t} | \theta) = (-2)\log p(w_{\Delta t}, ..., w_{N\Delta t} | \theta)$$

$$\approx \sum_{s=1}^N \log|\Sigma_{w_{s\Delta t}}| + \sum_{s=1}^N w_{s\Delta t}' \Sigma_{w_{s\Delta t}}^{-1} w_{s\Delta t} + N \log 2\pi. \tag{10.7}$$

2. The case when only scalar time series $x_{\Delta t}, x_{2\Delta t}, \ldots, x_{N\Delta t}$ are observed (without velocity observations, $\dot{x}_{\Delta t}, \dot{x}_{2\Delta t}, \dot{x}_{3\Delta t}, \ldots$):
Here we need to specify the predictor $E[x_t | x_{t-\Delta t}, \ldots]$ in order to derive the prediction error,

$$w_t = x_t - E[x_t | x_{t-\Delta t}, \ldots],$$

and its variance, $E[(x_t - E[x_t | x_{t-\Delta t}, \ldots])^2] = \sigma_{w_t}^2$. With the specified w_t and σ_w^2, we can write the (-2)log-likelihood as

$$\sum_{t=1}^{N} \log \sigma_w^2 + \sum_{t=1}^{N} w_t^2 / \sigma_{w_t}^2 + N \log 2\pi. \tag{10.8}$$

The prediction error $w_t = x_t - E[x_t | x_{t-\Delta t}, \ldots]$ and its variance can be calculated by applying the Kalman filter to the state space representation

$$\begin{aligned} z_{t+\Delta t} &= A z_t + B v_{t+\Delta t} \\ x_t &= C z_t \end{aligned} \tag{10.9}$$

for the scalar time series, $x_{\Delta t}, x_{2\Delta t}, \ldots, x_{N\Delta t}$, where $v_{t+\Delta t}$ is a Gaussian white noise of variance σ_n^2, as given in (10.5). The details of the Kalman filter method will be explained in Chapter 11.

Note that there is an alternative method for calculating the prediction error $w_t = x_t - E[x_t | x_{t-\Delta t}, \ldots]$ of the model without using the Kalman filter. Here we can calculate the prediction error by using the fact that the model (10.9) is in fact converted to the equivalent ARMA(2,1) model,

$$x_t = \phi_1 x_{t-\Delta t} + \phi_2 x_{t-2\Delta t} + w_t + \theta w_{t-\Delta t} \tag{10.10}$$

where ϕ_1, ϕ_2, and θ are written as functions of a and b. We have seen the derivation of ARMA(2,1) representation (10.10) in the Example L1 in Section 9.3.1.

10.2.2.3 General pth Order Differential Equation Models

Example 3:

$$x^{(p)}(t) + b_1 x^{(p-1)}(t) + \cdots + b_{p-1} x^{(1)}(t) + b_p x(t) = n(t) \tag{10.11}$$

$$n(t) : N(0, \sigma_n^2)$$

The likelihood of the model (10.11) depends on how the observation data is given. The most realistic situation is that only the scalar discrete observations, $x_{\Delta t}$, $x_{2\Delta t}$, ..., $x_{N\Delta t}$, are available as observations, and higher order derivatives $x^{(k)}(t)$ are not available as observed data. Then the problem of the specification of the likelihood of the model (10.11) reduces to the likelihood specification of the state space model where the observation matrix is given by a special $1 \times p$ matrix C as

$$\dot{z}(t) = Jz(t) + w(t)$$
$$x_t = Cz(t),$$
(10.12)

where $C = (0, ..., 0, 1)$,

$$z(t) = \begin{pmatrix} x^{(p-1)}(t) \\ x^{(p-2)}(t) \\ ... \\ \dot{x}(t) \\ x(t) \end{pmatrix}, \quad w(t) = \begin{pmatrix} n(t) \\ 0 \\ ... \\ 0 \\ 0 \end{pmatrix}, \quad \text{and} \quad J = \begin{pmatrix} -b_1 & -b_2 & ... & -b_{p-1} & -b_p \\ 1 & 0 & ... & 0 & 0 \\ 0 & 1 & ... & 0 & 0 \\ ... & ... & ... & ... & ... \\ 0 & 0 & ... & 1 & 0 \end{pmatrix}.$$

The log-likelihood function of the state space model (10.12) is specified as

$$\sum_{t=1}^{N} \log \sigma_w^2 + \sum_{t=1}^{N} w_t^2 / \sigma_{w_t}^2 + N \log 2\pi$$
(10.13)

if we specify the predictor $x_{t|t-\Delta t} = E[x_t|x_{t-\Delta t}, ...]$ to derive the prediction error, $w_t = x_t - x_{t|t-\Delta t}$ and its variance, $E[(x_t - x_{t|t-\Delta t})^2] = \sigma_{w_t}^2$. A general Kalman filtering approach to calculating the prediction error w_t and its variance $\sigma_{w_t}^2$ for obtaining the likelihood of a possibly nonlinear state space model is discussed in Chapters 12 and 13, Part III.

When the model is linear, however, we have an alternative approach, that is, the ARMA modeling approach, as we saw in Example–L2 in Section 9.3.1, where the model (10.12) was discretized into

$$z_{t+\Delta t} = e^{J\Delta t} z_t + Bv_{t+\Delta t}$$
$$x_t = Cz_t.$$
(10.14)

This state space model is converted further to an ARMA($p,p-1$) representation,

$$x_t = \phi_1 x_{t-\Delta t} + \cdots + \phi_p x_{t-p\Delta t} + CBv_t + C(A - \phi_1 I)Bv_{t-\Delta t} + \cdots$$

$$+ C(A^{p-1} - \phi_1 A^{p-2} - \cdots - \phi_{p-1} I)Bv_{t-(p-1)\Delta t}.$$

which is rewritten as

$$x_t = \phi_1 x_{t-\Delta t} + \cdots + \phi_p x_{t-p\Delta t} + w_t + \theta_1 w_{t-\Delta t} + \cdots + \theta_{p-1} w_{t-(p-1)\Delta t}$$

where $w_t = CBv_t$, and $\phi_1, \ldots, \phi_p, \theta_1, \ldots, \theta_{p-1}, \sigma_w^2 = E[w_t^2] = E[(CBv_t)^2]$ are functions of $b_1, \ldots, b_p$ and σ_w^2. Then the (-2)log-likelihood is calculated in the same way as the case for the general ARMA(p,q) model, which is explained later in Section 10.3.2.

10.2.2.4 Tank Models

Example 4: With the assumption of the scalar time series observation, the continuous-time Tank model is written as a state space model,

$$\dot{z}(t) = Jz(t) + Bu(t)$$
$$x_t = Cz(t) + \xi_t \tag{10.15}$$

$$J = \begin{pmatrix} -\lambda_1 & 0 & 0 & \cdots & 0 & 0 & \cdots & 0 & 0 \\ 1 & -\lambda_2 & 0 & \cdots & 0 & 0 & \cdots & 0 & 0 \\ 0 & 1 & -\lambda_2 & \cdots & 0 & 0 & \cdots & 0 & 0 \\ \cdots & \cdots & \cdots & \cdots & \cdots & \cdots & \cdots & \cdots \\ 0 & 0 & 0 & \cdots & -\lambda_k & 0 & \cdots & 0 & 0 \\ 0 & 0 & 0 & \cdots & 1 & -\lambda_k & \cdots & 0 & 0 \\ \cdots & \cdots & \cdots & \cdots & \cdots & \cdots & \cdots & \cdots \\ 0 & 0 & 0 & \cdots & 0 & 0 & \cdots & -\lambda_k & 0 \\ 0 & 0 & 0 & \cdots & 0 & 0 & \cdots & 1 & -\lambda_k \end{pmatrix}, \; z(t) = \begin{pmatrix} z_1^{(1)}(t) \\ z_1^{(2)}(t) \\ z_2^{(2)}(t) \\ \cdots \\ z_1^{(k)}(t) \\ z_2^{(k)}(t) \\ \cdots \\ z_{k-1}^{(k)}(t) \\ z_k^{(k)}(t) \end{pmatrix}, \; B = \begin{pmatrix} b_1 \\ b_2 \\ 0 \\ \cdots \\ b_k \\ 0 \\ \cdots \\ 0 \\ 0 \end{pmatrix},$$

where $u(t)$ is a scalar input, which could be a sum of a stimulus $s(t)$ and a white noise $n(t)$ whose variance is σ_n^2, that is, $u(t) = s(t) + n(t)$. Here if the Tank model is assumed to be deterministic, the white noise $n(t)$ is zero. $C = (1, 0, 1, \ldots, 0, \ldots, 0, 1)$, and ξ_t is a Gaussian white observation noise.

The likelihood of the linear continuous-time state space model (10.15) is derived in the same way as for the (p)th order differential equation model of Example 3. The only difference is that the driving input of the model (10.15) includes an exogenous variable $s(t)$. Suppose the dimension of the state $z(t)$ is p. Then a discretization scheme converts (in the same way as for the one-dimensional case in Section 9.2.6) the continuous-time state space model with exogenous input $s(t)$,

$$\dot{z}(t) = Jz(t) + Bs(t) + Bn(t)$$
$$x_t = Cz(t) + \xi_t \tag{10.16}$$

into an equivalent ARMAX(p, p, $p-1$) model (note that when $\xi_t = 0$, the MA order is $p-1$.), that is,

$$x_t = \phi_1 x_{t-\Delta t} + \cdots + \phi_p x_{t-p\Delta t} + w_t + \theta_1 w_{t-\Delta t} + \cdots + \theta_p w_{t-p\Delta t}$$
$$+ \psi_1 s_{t-\Delta t} + \cdots + \psi_{p-1} s_{t-(p-1)\Delta t}.$$

(10.17)

Here the parameters $\phi_1, \ldots, \phi_p, \theta_1, \ldots, \theta_p, \psi_1, \ldots, \psi_{p-1}$, and $\sigma_w^2 = E[w_t^2]$ are written as functions of $\lambda_1, \ldots, \lambda_k, b_1, \ldots, b_k, \sigma_\xi^2$, and σ_n^2. Then (-2)log-likelihood of the model (on the condition that the exogenous variable observations, $s_1, \ldots, s_N$, are given) can be calculated in the same way as for the discrete-time ARMAX model, which is explained later in Section 10.3.3.

10.2.2.5 Spectral Decomposition Model

Example 5: The EEG spectral decomposition model is another example of a continuous-time structural model, where the EEG signal x_t is considered as the sum of rhythms of different oscillations, such as delta $S_1(t)$, theta $S_2(t)$, alpha $S_3(t)$, beta $S_4(t)$, and gamma $S_5(t)$, that is,

$$x_t = S_1(t) + S_2(t) + S_3(t) + S_4(t) + S_5(t).$$

(10.18)

Each cyclic component $S_i(t)$ ($i = 1, \ldots, 5$) can be represented by the second-order differential equation model,

$$\ddot{S}_i(t) = -\omega_i^2 S_i(t) + u_i(t),$$

where
  ω_i characterizes the frequency of the oscillations
  $u_i(t)$ is a Gaussian white noise ($i = 1, 2, \ldots, 5$) with variance σ_i^2

Then the time series observation x_t is modeled by a parallel type state space representation,

$$\dot{z}(t) = A^{(para)} z(t) + u(t)$$
$$x_t = Cz(t),$$

(10.19)

where $z(t) = (\dot{S}_1(t), S_1(t), \dot{S}_2(t), S_2(t), \ldots, \dot{S}_5(t), S_5(t))'$, $C = (0, 1, 0, 1, 0, 1, 0, 1, 0, 1)$,

$$A^{(para)} = \begin{pmatrix} A_1 & O & \cdots & O \\ O & A_2 & \cdots & O \\ \cdots & \cdots & \cdots & \cdots \\ O & O & \cdots & A_5 \end{pmatrix}, \quad A_i = \begin{pmatrix} 0 & \omega_i^2 \\ -1 & 0 \end{pmatrix}, \quad (i = 1, 2, \ldots, 5),$$

$$u(t) = (u_1(t), 0, u_2(t), 0, \ldots, u_5(t), 0)',$$

with variance matrix, $\Sigma_u = \text{diag}(\sigma_1^2, 0, \sigma_2^2, 0, \ldots, \sigma_5^2, 0)$. Since the model is linear, the likelihood of the model is calculated in the same way (i.e., through an ARMA representation) as for the Tank model. An alternative method for writing down the likelihood using the innovations directly obtained from the state space, without going through the ARMA model, is given in Chapters 11 through 13.

Note that all the previous examples, Examples 1 through 5, are linear dynamic models. This is the reason why we can write down the likelihood of the models through the ARMA representations, even though we observe only a scalar time series variable and some of the variables are not observed. In Chapter 5, Part I, we have seen many other high-dimensional continuous-time "nonlinear" dynamical systems such as the Hodgkin-Huxley model and the Zetterberg model, etc. Although the dynamics of these models are nonlinear, the observation data are often scalar time series or a lower (than the state dimension) dimensional time series x_t. In these cases, we cannot use ARMA type representations for writing down the likelihood. However, we can calculate the innovations and the innovation-based likelihood by using Kalman filter techniques, which are discussed later in Chapter 12.

10.3 Likelihood of Discrete-Time Models

If we could assume, from the start, that the time series $(x_1, \ldots, x_N)'$ is observed from a discrete-time Markov model, $M(\theta)$, with parameter vector θ, driven by a Gaussian white noise, the derivation of the likelihood function would be more direct and straightforward. The time series can be whitened into prediction errors, $(w_1, \ldots, w_N)'$, using the dynamic model $M(\theta)$ with a parameter vector θ, as

$$w_t = x_t - E[x_t | x_{t-1}, x_{t-2}, \ldots, x_1 | \theta]. \quad (t = 1, 2, \ldots, N).$$

Since the Jacobian of the variable transformation, from $(x_1, \ldots, x_N)'$ to $(w_1, \ldots, w_N)'$, is always a triangular matrix such as

$$J = \left(\frac{\partial w_i}{\partial x_j} \right) = \begin{pmatrix} 1 & * & \cdots & * \\ 0 & 1 & \cdots & * \\ \cdots & \cdots & \cdots & \cdots \\ 0 & 0 & \cdots & 1 \end{pmatrix},$$

and $\det(J) = 1$, we can write down the (-2)likelihood of the discrete-time model $M(\theta)$ as

$$(-2)\log p(x_1, \ldots, x_N | \theta) = (-2)\log p(w_1, \ldots, w_N | \theta)$$

$$= \sum_{t=1}^{N} \log \sigma_{w_t}^2 + \sum_{t=1}^{N} w_t^2 / \sigma_{w_t}^2 + N \log 2\pi.$$

Note that when w_t is temporally homogeneous, that is, with the constant variance σ_w^2, we have

$$\text{Min}\left\{(-2)\log p(x_1, \ldots, x_N | \theta)\right\} = \text{Min}\left\{\sum_{t=1}^{N} \log \sigma_{w_t}^2 + \sum_{t=1}^{N} w_t^2 / \sigma_{w_t}^2 + N \log 2\pi\right\}$$

$$= N \log \sigma_{w_t}^2 + N + N \log 2\pi. \tag{10.20}$$

When the time series x_t is d-dimensional, this becomes

$$\text{Min}\left\{(-2)\log p(x_1, \ldots, x_N | \theta)\right\} = \text{Min}\left\{\sum_{t=1}^{N} \log |\Sigma_{w_t}| + \sum_{t=1}^{N} w_t' \Sigma_{w_t}^{-1} w_t + Nd \log 2\pi\right\}$$

$$= N \log |\Sigma_w| + Nd + Nd \log 2\pi. \tag{10.21}$$

These equations imply that minimization of σ_w^2 or $|\Sigma_w|$ (i.e., determinant of Σ_w) is essentially equivalent to maximization of the log-likelihood.

10.3.1 Likelihood of Discrete-Time Nonlinear Dynamic Models

10.3.1.1 Likelihood of Univariate Nonlinear Time Series Models

The likelihood of most nonlinear dynamic models is characterized, in the same way as for linear AR models, through the innovations. For example, the discrete-time dynamic models given in Table 6.1 are all nonlinear AR type models of lag order two, and can be written using the general nonlinear function, $f(x_{t-1}, x_{t-2} | \theta)$, as

$$x_t = f(x_{t-1}, x_{t-2} | \theta) + w_t.$$

Here the innovations are given by

$$w_t = x_t - f(x_{t-1}, x_{t-2} | \theta).$$

By separating the distribution of the initial values, x_1, x_2, and the distribution of the rest of the data by using the Gaussian likelihood of the innovations, we can write down the (−2)log-likelihood function as

$$(-2)\log p(x_1, x_2, \dots, x_N | \theta, \sigma_w^2) = (-2)\log p(x_3, x_4, \dots, x_N | x_1, x_2, \theta, \sigma_w^2)$$

$$+ (-2)\log p(x_1, x_2 | \theta, \sigma_w^2)$$

$$= (-2)\log p(w_3, w_4, \dots, w_N | x_1, x_2, \theta, \sigma_w^2)$$

$$+ (-2)\log p(x_1, x_2 | \theta, \sigma_w^2)$$

$$= (N-2)\log \sigma_w^2 + \sum_{t=3}^{N} w_t^2 / \sigma_w^2 + (N-2)\log 2\pi$$

$$+ \log p(x_1, x_2 | \theta, \sigma_w^2),$$

where σ_w^2 is the innovation variance.

For linear AR models, the initial values are also Gaussian, and we could explicitly specify $p(x_1, x_2 | \theta, \sigma_w^2)$ as an analytic function of the parameter vector θ. However if the model is nonlinear, the distribution of the process is non-Gaussian. In order to estimate the model from the given time series data, we have to put up with the approximate likelihood approach, whether we like it or not. Here we assume that the contribution from $(-2)\log p(x_1, x_2 | \theta, \sigma_w^2)$ to (−2)log-likelihood is negligibly small for a sufficiently large N, and we have

$$(-2)\log p(x_1, x_2, \dots, x_N | \theta, \sigma_w^2) = (-2)\log p(x_3, x_4, \dots, x_N | x_1, x_2, \theta, \sigma_w^2)$$

$$+ (-2)\log p(x_1, x_2 | \theta, \sigma_w^2)$$

$$\approx (-2)\log p(w_3, w_4, \dots, w_N | x_1, x_2, \theta, \sigma_w^2)$$

$$= (N-2)\log \sigma_w^2 + \sum_{t=3}^{N} w_t^2 / \sigma_w^2 + (N-2)\log 2\pi.$$

$$(10.22)$$

The minimization of (10.22) is attained by $\hat{\sigma}_w^2 = \underset{\theta}{\text{Min}} \sum_{t=3}^{N} w_t^2$, that is, the least squares estimate of the model. The minimized (−2)log-likelihood is given, with $\hat{\sigma}_w^2$, as follows:

$$\text{Min}\{(-2)\log-\text{likelihood}\} \approx (N-2)\log \hat{\sigma}_w^2 + (N-2)(\log 2\pi + 1).$$

10.3.1.2 Nonlinear Parameters and Linear Parameters

It is a well-known fact that the most discrete-time nonlinear dynamic models, although very flexible, incur a huge computational costs for the numerical minimization of $\sum_{t=3}^{N} w_t^2$, which is essential for estimating their model parameters.

Example: Single layer RBF neural network model

An RBF is a multidimensional function, which depends on the distance $r = \|X - \varsigma\|$ (where $\|.\|$ denotes a vector norm) between a d-dimensional input vector X and a "center" ς. One of the simplest approaches to the approximation of a nonlinear function is to represent it by a linear combination of fixed nonlinear basis functions $B_i(.)$ as

$$f(X) = \sum_{i=1}^{m} c_i B_i \left(\|X - \varsigma\|_{h_i}^2 \right),$$

where an RBF network is specified by three sets of parameters: the centers ς_i, the width or distance scaling parameters h_i, and the synaptic weights c_i ($i = 1, 2, \ldots, m$). The performance of an RBF network depends critically on the chosen centers ς_i ($i = 1, 2, \ldots, m$). Several heuristic methods for determining the centers have been proposed (Moody and Darken 1989, Leonard and Kramer 1991). Once the cluster centers ς_i have been determined, the distance scaling parameters h_i are determined from the r nearest neighbors heuristically by

$$h_i = \frac{1}{r} \sum_{j=1}^{r} \left\{ \|\varsigma_i - x_j\| \right\}^{21/2}$$

where x_j are the r nearest neighbors of ς_i.

Having obtained centers ς_i and the distance scaling parameters h_i, we are able to adjust the synaptic weights $c_{i,k}$ for the RBF-AR model, by minimizing the sum of squares of prediction errors, $\sum_{t=p+1}^{N} \varepsilon_t^2$, where

$$\varepsilon_t = x_t - \left\{ c_{0,0} + \sum_{k=1}^{m} c_{0,k} B_k \left(\|x_{t-1} - \varsigma_k\| \right) \right\} - \sum_{i=1}^{p} \left\{ c_{i,0} + \sum_{k=1}^{m} c_{i,k} B_k \left(\|x_{t-1} - \varsigma_k\| \right) \right\} x_{t-i}.$$

This method of obtaining $c_{i,k}$'s is equivalent to the least squares estimation method for $c_{i,k}$'s of the statistical linear regression model,

$$x_t = \left\{ c_{0,0} + \sum_{k=1}^{m} c_{0,k} B_k \left(\|x_{t-1} - \varsigma_k\| \right) \right\} + \sum_{i=1}^{p} \left\{ c_{i,0} + \sum_{k=1}^{m} c_{i,k} B_k \left(\|x_{t-1} - \varsigma_k\| \right) \right\} x_{t-i} + \varepsilon_t,$$

where the functional form of the basis function $B_k(\|x_{t-1} - \varsigma_k\|)$'s is given in advance. Then the calculation of the coefficients $c_{i,k}$ ($i = 1, \ldots, p$, $k = 1, \ldots, m$) may be performed very efficiently, in contrast to the slow back-propagation algorithm used for multilayer neural network modeling.

In the previous neural network example, parameters are classified into two classes, linear parameters, that is, the coefficients, $c_{i,k}$ ($i = 1, \ldots, p$, $k = 1, \ldots, m$), and the nonlinear, that is, the parameters of the basis function, that is, centers, ς_i ($i = 1, \ldots, m$) and scaling parameters, h_i ($i = 1, \ldots, m$). This shows that if the prediction model is formulated with general but efficient basis function regressors with known nonlinear parameters, the reconstruction of the predictor can be performed much more efficiently.

The same thing can be said for most parametric nonlinear AR models. The parameters are classified into two sets, the linear parameter set and the nonlinear parameters set. If the nonlinear parameters are fixed in advance, the minimization of the prediction errors is calculated very efficiently without an iterative numerical optimization procedure. If the nonlinear parameters are unknown and need to be estimated, we need to do numerical iterative optimization only in terms of the nonlinear parameters. Here the linear parameters are uniquely given by solving the normal equation for the linear regression model with the nonlinear base functions.

For example, for those nonlinear AR models in Table 6.1 the parameters are classified into two classes, nonlinear parameters and linear parameters as follows.

Example 1 Single layer neural network model:

$$y_t = c_0 + \pi_1 e^{-\|Y_{t-1} - \varsigma_1\|_{h_1}^2} + \pi_2 e^{-\|Y_{t-1} - \varsigma_2\|_{h_2}^2} + \varepsilon_t$$

$$= c_0 + \pi_1 e^{-\gamma_1^{(1)}(y_{t-1} - \varsigma_1^{(1)})^2 - \gamma_1^{(2)}(y_{t-2} - \varsigma_1^{(2)})^2} + \pi_2 e^{-\gamma_2^{(1)}(y_{t-1} - \varsigma_2^{(1)})^2 - \gamma_2^{(2)}(y_{t-2} - \varsigma_2^{(2)})^2} + \varepsilon_t.$$

Here $\gamma_1^{(1)}, \gamma_1^{(2)}, \gamma_2^{(1)}, \gamma_2^{(2)}, \varsigma_1^{(1)}, \varsigma_1^{(2)}, \varsigma_2^{(1)}, \varsigma_2^{(2)}$ are nonlinear parameters, and c_0, π_1, π_2 are linear parameters.

Example 2 ExpAR(2) model:

$$y_t = \left(\phi_1 + \pi_1 e^{-\gamma y_{t-1}^2}\right) y_{t-1} + \left(\phi_2 + \pi_2 e^{-\gamma y_{t-1}^2}\right) y_{t-2} + \varepsilon_t.$$

Here γ is a nonlinear parameter, and $\phi_1, \pi_1, \phi_2, \pi_2$ are linear parameters.

Example 3 STAR(2) model:

Since the STAR(2) model is equivalent to the ExpAR(2) model

$$y_t = \left\{\phi_1 y_{t-1} + \phi_2 y_{t-2}\right\} e^{-\gamma y_{t-1}^2} + \left\{\pi_1 y_{t-1} + \pi_2 y_{t-2}\right\}\left(1 - e^{-\gamma y_{t-1}^2}\right) + \varepsilon_t$$

$$= \left(\pi_1 + (\phi_1 - \pi_1)e^{-\gamma y_{t-1}^2}\right) y_{t-1} + \pi_2 + \left((\phi_2 - \pi_2)e^{-\gamma y_{t-1}^2}\right) y_{t-2},$$

the parameters of the STAR model may also be classified into nonlinear parameters and linear parameters.

Example 4 RBF-AR(2) model of Vesin-type (see Vesin 1993):

$$
\begin{aligned}
y_t &= \left\{\phi_1 + \pi_1^{(1)}e^{-\|Y_{t-1}-\varsigma_1\|_{in}^2}\right\}y_{t-1} + \left\{\phi_2 + \pi_2^{(1)}e^{-\|Y_{t-1}-\varsigma_2\|_{in2}^2}\right\}y_{t-2} + \varepsilon_t \\
&= \left\{\phi_1 + \pi_1^{(1)}e^{-\gamma_1^{(1)}(y_{t-1}-\varsigma_1^{(1)})^2 - \gamma_1^{(2)}(y_{t-2}-\varsigma_1^{(2)})^2}y_{t-1}\right\} \\
&\quad + \left\{\phi_2 + \pi_2^{(1)}e^{-\gamma_2^{(1)}(y_{t-1}-\varsigma_2^{(1)})^2 - \gamma_2^{(2)}(y_{t-2}-\varsigma_2^{(2)})^2}\right\}y_{t-2} + \varepsilon_t.
\end{aligned}
$$

Here $\gamma_1^{(1)}$, $\gamma_1^{(2)}$, $\gamma_2^{(1)}$, $\gamma_2^{(2)}$, $\varsigma_1^{(1)}$, $\varsigma_1^{(2)}$, $\varsigma_2^{(1)}$, $\varsigma_2^{(2)}$ are nonlinear parameters, and ϕ_1, $\pi_1^{(1)}$, ϕ_2, $\pi_2^{(1)}$ are linear parameters.

Example 5 RBF-AR(2) model of Ozaki-type (see Ozaki et al. 1997):

$$
\begin{aligned}
y_t &= \left\{\phi_1 + \pi_1^{(1)}e^{-\|Y_{t-1}-\varsigma_1\|_{in}^2}\right\}y_{t-1} + \left\{\phi_2 + \pi_2^{(1)}e^{-\|Y_{t-1}-\varsigma_2\|_{in2}^2}\right\}y_{t-2} + \varepsilon_t \\
&= \left\{\phi_1 + \pi_1^{(1)}e^{-\gamma_1^{(1)}(y_{t-1}-\varsigma_1^{(1)})^2 - \gamma_1^{(2)}(\Delta y_{t-1}-\varsigma_1^{(2)})^2}\right\}y_{t-1} \\
&\quad + \left\{\phi_2 + \pi_2^{(1)}e^{-\gamma_2^{(1)}(y_{t-1}-\varsigma_2^{(1)})^2 - \gamma_2^{(2)}(\Delta y_{t-1}-\varsigma_2^{(2)})^2}\right\}y_{t-2} + \varepsilon_t.
\end{aligned}
$$

Here, $\gamma_1^{(1)}$, $\gamma_1^{(2)}$, $\gamma_2^{(1)}$, $\gamma_2^{(2)}$, $\varsigma_1^{(1)}$, $\varsigma_1^{(2)}$, $\varsigma_2^{(1)}$, $\varsigma_2^{(2)}$, are nonlinear parameters, and ϕ_1, $\pi_1^{(1)}$, ϕ_2, $\pi_2^{(1)}$ are linear parameters.

Example 6 Linear threshold AR(2) model:

$$
y_t = \begin{pmatrix} \phi_1^{(1)}y_{t-1} + \phi_2^{(1)}y_{t-2} + \varepsilon_t & \text{for } y_{t-1} \geq \gamma \\ \phi_1^{(2)}y_{t-1} + \phi_2^{(2)}y_{t-2} + \varepsilon_t & \text{for } y_{t-1} < \gamma \end{pmatrix}.
$$

Here, γ is a nonlinear parameter, and $\phi_1^{(1)}$, $\phi_2^{(1)}$, $\phi_1^{(2)}$, $\phi_2^{(2)}$ are linear parameters.

Example 7 Nonlinear threshold AR model (Ozaki 1981a):

$$
y_t = \begin{pmatrix} \left\{\phi_1^{(1)} + \pi_1^{(1)}y_{t-1}^2\right\}y_{t-1} + \left\{\phi_2^{(1)} + \pi_2^{(1)}y_{t-1}^2\right\}y_{t-2} + \varepsilon_t & \text{for } |y_{t-1}|^2 < \gamma^2 \\ \left\{\phi_1^{(2)} + \pi_1^{(2)}\gamma^2\right\}y_{t-1} + \left\{\phi_2^{(2)} + \pi_2^{(2)}\gamma^2\right\}y_{t-2} + \varepsilon_t & \text{for } |y_{t-1}|^2 \geq \gamma^2 \end{pmatrix}.
$$

Here, γ is a nonlinear parameter, and $\phi_1^{(1)}$, $\phi_2^{(1)}$, $\phi_1^{(2)}$, $\phi_2^{(2)}$, $\pi_1^{(1)}$, $\pi_2^{(1)}$, $\pi_1^{(2)}$, $\pi_2^{(2)}$ are linear parameters.

Example 8 RBF-ARX(2,1) model of Ozaki type (Ozaki et al. 2004):

$$y_t = \left\{ \phi_1 + \pi_1^{(1)} e^{-\gamma_1^{(1)}(y_{t-1}-\varsigma_1^{(1)})^2 - \gamma_1^{(2)}(\Delta y_{t-1}-\varsigma_1^{(2)})^2} y_{t-1} \right\}$$

$$+ \left\{ \phi_2 + \pi_2^{(1)} e^{-\gamma_2^{(1)}(y_{t-1}-\varsigma_2^{(1)})^2 - \gamma_2^{(2)}(\Delta y_{t-1}-\varsigma_2^{(2)})^2} \right\} y_{t-2} + \psi_1 u_{t-1} + \varepsilon_t.$$

Here, $\gamma_1^{(1)}$, $\gamma_1^{(2)}$, $\gamma_2^{(1)}$, $\gamma_2^{(2)}$, $\varsigma_1^{(1)}$, $\varsigma_1^{(2)}$, $\varsigma_2^{(1)}$, $\varsigma_2^{(2)}$, are nonlinear parameters, and ϕ_1, $\pi_1^{(1)}$, ϕ_2, $\pi_2^{(1)}$, and ψ_1 are linear parameters.

10.3.1.3 Nonlinear Multivariate RBF-AR Models for Nonlinear Feedback Systems

In Section 6.5.3, we saw how the RBF-AR models are generalized into the multivariate cases and used for the modeling of nonlinear feedback systems. The computation of the log-likelihood of these multivariate nonlinear AR models is done in the way similar to that of the previous examples of scalar nonlinear AR models. For example, we can generalize the linear ARX model (4.62) into the following RBF-ARX model:

$$x_t = \phi_{1,1}^{(1)}(RBF)x_{t-1} + \phi_{1,1}^{(2)}(RBF)x_{t-2} + \phi_{1,1}^{(3)}(RBF)x_{t-3}$$

$$+ \phi_{1,2}^{(1)}y_{t-1} + \phi_{1,2}^{(2)}y_{t-2} + \phi_{1,2}^{(3)}y_{t-3} + \phi_{1,3}^{(1)}z_{t-1} + \phi_{1,3}^{(2)}z_{t-2} + \phi_{1,3}^{(3)}z_{t-3}$$

$$+ \psi_{1,1}^{(1)}u_{t-1} + \psi_{1,1}^{(2)}u_{t-2} + \psi_{1,2}^{(1)}v_{t-1} + \psi_{1,2}^{(2)}v_{t-2} + \varepsilon_t^{(x)} \tag{10.23}$$

$$y_t = \phi_{2,2}^{(1)}(RBF)y_{t-1} + \phi_{2,2}^{(2)}(RBF)y_{t-2} + \phi_{2,2}^{(3)(RBF)}y_{t-3}$$

$$+ \phi_{2,3}^{(1)}z_{t-1} + \phi_{2,3}^{(2)}z_{t-2} + \phi_{2,3}^{(3)}z_{t-3} + \phi_{2,1}^{(1)}x_{t-1} + \phi_{2,1}^{(2)}x_{t-2} + \phi_{2,1}^{(3)}x_{t-3}$$

$$+ \psi_{2,1}^{(1)}u_{t-1} + \psi_{2,1}^{(2)}u_{t-2} + \psi_{2,2}^{(1)}v_{t-1} + \psi_{2,2}^{(2)}v_{t-2}$$

$$+ \theta_{2,1}x_t + \varepsilon_t^{(y)} \tag{10.24}$$

$$z_t = \phi_{3,3}^{(1)}(RBF)z_{t-1} + \phi_{3,3}^{(2)}(RBF)z_{t-2} + \phi_{3,3}^{(3)}(RBF)z_{t-3}$$

$$+ \phi_{3,1}^{(1)}x_{t-1} + \phi_{3,1}^{(2)}x_{t-2} + \phi_{3,1}^{(3)}x_{t-3} + \phi_{3,2}^{(1)}y_{t-1} + \phi_{3,2}^{(2)}y_{t-2} + \phi_{3,2}^{(3)}y_{t-3}$$

$$+ \psi_{3,1}^{(1)}u_{t-1} + \psi_{3,1}^{(2)}u_{t-2} + \psi_{3,2}^{(1)}v_{t-1} + \psi_{3,2}^{(2)}v_{t-2}$$

$$+ \theta_{3,1}x_t + \theta_{3,2}y_t + \varepsilon_t^{(z)}. \tag{10.25}$$

Here, the $\phi_{i,j}^{(k)}(RBF)$'s are specified, for example, as

$$\phi_{i,j}^{(k)}(RBF) = \phi_{i,j}^{(k)} + \sum_{m=1}^{M} \pi_{i,j,m}^{(k)} Exp\left\{ -\left\| R_{t-1} - \varsigma_m \right\|^2 / h_m \right\}$$

or

$$\phi_{i,j}^{(k)}(RBF) = \phi_{i,j}^{(k)} + \sum_{m=1}^{M} \pi_{i,j,m}^{(k)} Exp\left\{ -\left\| \frac{R_{t-1} - \varsigma_m}{\Delta R_{t-1}} \right\|_Q^2 / h_m \right\}.$$

R_{t-1} is a variable that is known to cause the change in the characteristics of the system dynamics, that could be one of the state variables, x_{t-1}, y_{t-1}, z_{t-1}, or one of the exogenous variables, u_{t-1} and v_{t-1}.

By separating the initial values, x_1, x_2, x_3, and the distribution of the rest of the data by using the Gaussian likelihood of the innovations, we can write down the (-2)log-likelihood function, for x_1, x_2, ..., x_N, as

$$(-2)\log p(x_1, x_2, \ldots, x_N | \theta, \sigma_{w^{(x)}}^2)$$

$$= (-2)\log p(x_4, x_5, \ldots, x_N | x_1, x_2, x_3, \theta, \sigma_w^2)$$

$$+ (-2)\log p(x_1, x_2, x_3 | \theta, \sigma_w^2)$$

$$= (-2)\log p(w_4^{(x)}, w_5^{(x)}, \ldots, w_N^{(x)} | x_1, x_2, x_3, \theta, \sigma_w^2)$$

$$+ (-2)\log p(x_1, x_2, x_3 | \theta, \sigma_w^2)$$

$$\approx (N-3)\log \sigma_{w^{(x)}}^2 + \sum_{t=4}^{N} \left(w_t^{(x)} \right)^2 / \sigma_{w^{(x)}}^2 + (N-3)\log 2\pi.$$

Similarly, we have for y_1, y_2, ..., y_N,

$$(-2)\log p(y_1, y_2, \ldots, y_N | \theta, \sigma_{w^{(y)}}^2) \approx (N-3)\log \sigma_{w^{(y)}}^2 + \sum_{t=4}^{N} \left(w_t^{(y)} \right)^2 / \sigma_w^2 + (N-3)\log 2\pi$$

and for z_1, z_2, ..., z_N,

$$(-2)\log p(z_1, z_2, \ldots, z_N | \theta, \sigma_{w^{(z)}}^2) \approx (N-3)\log \sigma_{w^{(z)}}^2 + \sum_{t=4}^{N} \left(w_t^{(z)} \right)^2 / \sigma_w^2 + (N-3)\log 2\pi.$$

10.3.1.4 Multivariate RBF-AR Model

The three models of (10.23), (10.24), and (10.25) are converted into a single 3-dimensional RBF-AR model,

$$\begin{pmatrix} x_t \\ y_t \\ z_t \end{pmatrix} = A_1(RBF)\begin{pmatrix} x_{t-1} \\ y_{t-1} \\ z_{t-1} \end{pmatrix} + A_2(RBF)\begin{pmatrix} x_{t-2} \\ y_{t-2} \\ z_{t-2} \end{pmatrix} + A_3(RBF)\begin{pmatrix} x_{t-3} \\ y_{t-3} \\ z_{t-3} \end{pmatrix}$$

$$+ B_1\begin{pmatrix} u_{t-1} \\ v_{t-1} \end{pmatrix} + B_2\begin{pmatrix} u_{t-2} \\ v_{t-2} \end{pmatrix} + \begin{pmatrix} \varepsilon_t^{(x)} \\ \varepsilon_t^{(y)} \\ \varepsilon_t^{(z)} \end{pmatrix}, \tag{10.26}$$

with RBF-AR coefficient matrices, $A_1(RBF)$, $A_2(RBF)$, $A_3(RBF)$, which are specified by the RBF coefficient matrices of (10.23), (10.24), (10.25) by

$$A_1(RBF) = (I - \theta)^{-1}\begin{pmatrix} \phi_{1,1}^{(1)}(RBF) & \phi_{1,2}^{(1)} & \phi_{1,3}^{(1)} \\ \phi_{2,1}^{(1)} & \phi_{2,2}^{(1)}(RBF) & \phi_{2,3}^{(1)} \\ \phi_{3,1}^{(1)} & \phi_{3,2}^{(1)} & \phi_{3,3}^{(1)}(RBF) \end{pmatrix},$$

$$A_2(RBF) = (I - \theta)^{-1}\begin{pmatrix} \phi_{1,1}^{(2)}(RBF) & \phi_{1,2}^{(2)} & \phi_{1,3}^{(2)} \\ \phi_{2,1}^{(2)} & \phi_{2,2}^{(2)}(RBF) & \phi_{2,3}^{(2)} \\ \phi_{3,1}^{(2)} & \phi_{3,2}^{(2)} & \phi_{3,3}^{(2)}(RBF) \end{pmatrix},$$

$$A_3(RBF) = (I - \theta)^{-1}\begin{pmatrix} \phi_{1,1}^{(3)}(RBF) & \phi_{1,2}^{(3)} & \phi_{1,3}^{(3)} \\ \phi_{2,1}^{(3)} & \phi_{2,2}^{(3)}(RBF) & \phi_{2,3}^{(3)} \\ \phi_{3,1}^{(3)} & \phi_{3,2}^{(3)} & \phi_{3,3}^{(3)}(RBF) \end{pmatrix},$$

$$\theta = \begin{pmatrix} 0 & 0 & 0 \\ \theta_{21} & 0 & 0 \\ \theta_{31} & \theta_{32} & 0 \end{pmatrix},$$

and matrices B_1 and B_2 are given by

$$B_1 = (I - \theta)^{-1}\begin{pmatrix} \psi_{1,1}^{(1)} & \psi_{1,2}^{(1)} \\ \psi_{2,1}^{(1)} & \psi_{2,2}^{(1)} \\ \psi_{3,1}^{(1)} & \psi_{3,2}^{(1)} \end{pmatrix}, \quad B_2 = (I - \theta)^{-1}\begin{pmatrix} \psi_{1,1}^{(2)} & \psi_{1,2}^{(2)} \\ \psi_{2,1}^{(2)} & \psi_{2,2}^{(2)} \\ \psi_{3,1}^{(2)} & \psi_{3,2}^{(2)} \end{pmatrix}.$$

$(\xi_t, \eta_t, \varsigma_t)'$ is a 3-dimensional Gaussian white noise. By introducing the state vector variable $X_t = (x_t, y_t, z_t)'$, $U_t = (u_t, v_t)'$, and $W_t = (1 - \theta)^{-1}\left(\varepsilon_t^{(x)}, \varepsilon_t^{(y)}, \varepsilon_t^{(z)}\right)'$, the model (10.26) is written as

$$X_t = A_1(RBF)X_{t-1} + A_2(RBF)X_{t-2} + A_3(RBF)X_{t-3} + B_1U_{t-1} + B_2U_{t-2} + W_t.$$

which can be converted into a (lag order 1) Markov representation as

$$
\begin{pmatrix} X_t \\ X_{t-1} \\ X_{t-2} \end{pmatrix} = \begin{pmatrix} A_1(RBF) & A_2(RBF) & A_3(RBF) \\ I & O & O \\ O & I & O \end{pmatrix} \begin{pmatrix} X_{t-1} \\ X_{t-2} \\ X_{t-3} \end{pmatrix} + \begin{pmatrix} B_1 & B_2 \\ O & O \\ O & O \end{pmatrix} \begin{pmatrix} U_{t-1} \\ U_{t-2} \end{pmatrix} + \begin{pmatrix} W_t \\ O \\ O \end{pmatrix}.
$$

$$(10.27)$$

We note that the eigenvalues of the transition matrix,

$$
A = \begin{pmatrix} A_1(RBF) & A_2(RBF) & A_3(RBF) \\ I & O & O \\ O & I & O \end{pmatrix},
$$

are a function of the regulating variable R_{t-1}.

The success of the multivariate RBF-AR(p) modeling depends on how we parameterize the AR coefficient matrices, $A_1(RBF)$, ..., $A_p(RBF)$. There are so many ways of parameterization even if we restrict our choice of radial basis functions to the Gaussian radial basis functions. For example, we have, for the parameterization of $\phi_{i,j}^{(k)}(RBF)$, the following two candidates:

$$
\phi_{i,j}^{(k)}(RBF) = \phi_{i,j}^{(k)} + \sum_{m=1}^{M} \pi_{i,j,m}^{(k)} Exp\left\{-\left\|R_{t-1} - \varsigma_m\right\|^2 / h_m\right\},
$$

and

$$
\phi_{i,j}^{(k)}(RBF) = \phi_{i,j}^{(k)} + \sum_{m=1}^{M} \pi_{i,j,m}^{(k)} Exp\left\{-\left[\left\|\frac{R_{t-1} - \varsigma_m}{\Delta R_{t-1}}\right\|_Q^2 \middle/ h_m\right]\right\},
$$

while the Gaussian function families are only one of the many possible Radial Basis Function families (see Section 6.2.2). Certainly, of experts ideas and experience-based suggestions will be useful in choosing an appropriate functional form of the RBF.

A typical example of a nonlinear feedback system is the thermal power plant system where keeping the steam temperature of the boiler at around a fixed level is one of the important issues for control engineers. The system is nonlinear in the sense that the system characteristics change depending on the electric power that the system generates. Figure 10.1a through d show how the time series patterns change when the plant is generating power according to demand, 110, 175, 250, and 330 MW. Figure 10.2 shows the steam temperature data in Figure 10.1, for each of the four different megawatt demands, MWD = 110, MWD = 175, MWD = 250, and MWD = 330, in one graph. Figure 10.3

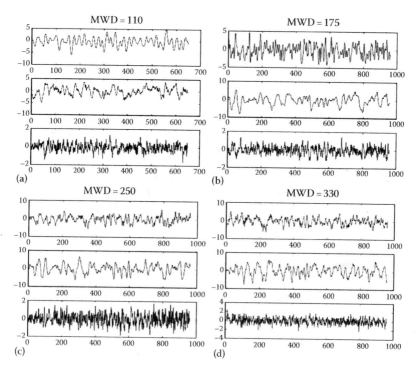

FIGURE 10.1
(a through d) Four segments of 3-dimensional power plant data.

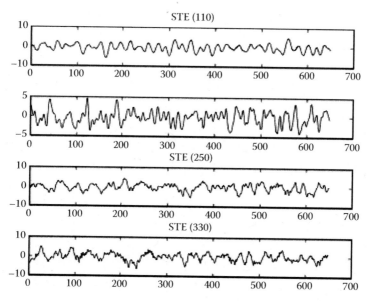

FIGURE 10.2
Four segments of 3-dimensional power plant steam temperature data.

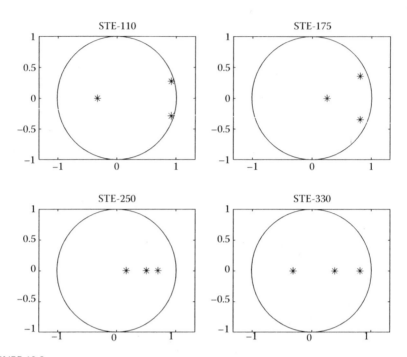

FIGURE 10.3
Eigenvalues of four linear multivariate AR models.

shows the three eigenvalues of the four linear AR models for the segments of steam temperature data. Here the characteristic roots are obtained from $\phi_{1,1}^{(1)}$, $\phi_{1,1}^{(2)}$, and $\phi_{1,1}^{(3)}$ of

$$x_t = \phi_{1,1}^{(1)}x_{t-1} + \phi_{1,1}^{(2)}x_{t-2} + \phi_{1,1}^{(3)}x_{t-3}$$

$$+ \phi_{1,2}^{(1)}y_{t-1} + \phi_{1,2}^{(2)}y_{t-2} + \phi_{1,2}^{(3)}y_{t-3} + \phi_{1,3}^{(1)}z_{t-1} + \phi_{1,3}^{(2)}z_{t-2} + \phi_{1,3}^{(3)}z_{t-3}$$

$$+ \psi_{1,1}^{(1)}u_{t-1} + \psi_{1,1}^{(2)}u_{t-2} + \psi_{1,2}^{(1)}v_{t-1} + \psi_{1,2}^{(2)}v_{t-2} + \varepsilon_t^{(x)}.$$

This type of model is fitted to each of the four segments of the series in Figure 10.1 separately.

Figure 10.4 shows the four sets of three eigenvalues of the single RBF-AR model,

$$x_t = \phi_{1,1}^{(1)}(MWD)x_{t-1} + \phi_{1,1}^{(2)}(MWD)x_{t-2} + \phi_{1,1}^{(3)}(MWD)x_{t-3}$$

$$+ \phi_{1,2}^{(1)}y_{t-1} + \phi_{1,2}^{(2)}y_{t-2} + \phi_{1,2}^{(3)}y_{t-3} + \phi_{1,3}^{(1)}z_{t-1} + \phi_{1,3}^{(2)}z_{t-2} + \phi_{1,3}^{(3)}z_{t-3}$$

$$+ \psi_{1,1}^{(1)}u_{t-1} + \psi_{1,1}^{(2)}u_{t-2} + \psi_{1,2}^{(1)}v_{t-1} + \psi_{1,2}^{(2)}v_{t-2} + \varepsilon_t^{(x)}$$

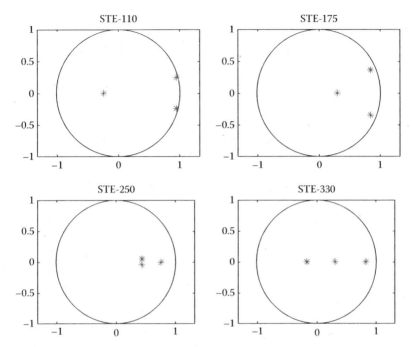

FIGURE 10.4
Eigenvalues of multivariate RBF-AR.

$$\phi_{i,j}^{(k)}(MWD) = \phi_{i,j}^{(k)} + \sum_{m=1}^{M} \pi_{i,j,m}^{(k)} Exp\left\{-\left\|\frac{MWD_{t-1} - c_m}{\Delta MWD_{t-1}}\right\|_Q^2 / h_m\right\},$$

where the three eigenvalues are obtained from $\phi_{1,1}^{(1)}(MWD)$, $\phi_{1,1}^{(2)}(MWD)$, and $\phi_{1,1}^{(3)}(MWD)$, by making MWD = 110, 175, 250, and 330. Here the single multi-variate RBF-AR model was estimated from the joined time series (each segment of the 3-dimensional time series is given in Figure 10.1). Even though there are some small deviations, the eigenvalues of both the models are quite similar, and the single RBF-AR model characterizes the four different system characteristics specified by the four different linear models reasonably well.

10.3.2 Likelihood of Discrete-Time Linear Dynamic Models

10.3.2.1 ARMA Models

An advantage of ARMA modeling,

$$x_t = \varphi_1 x_{t-1} + \cdots + \varphi_p x_{t-p} + \theta_1 w_{t-1} + \cdots + \theta_q x_{t-q} + w_t,$$

over AR modeling is that the transfer function and the power spectrum have more parsimonious representations. As we saw in Chapter 3, the functional form of the power spectrum density function of an AR-MA(p,q) model is a rational function type,

$$p(f) = \frac{\left|1 + \theta_1 e^{-i2\pi f} + \cdots + \theta_q e^{-i2\pi fq}\right|^2}{\left|1 - \phi_1 e^{-i2\pi f} - \cdots - \phi_p e^{-i2\pi fp}\right|^2} \sigma_w^2$$

so that it does not need as many parameters as an AR model, especially when the power spectrum has a steep trough in some frequency bands.

The disadvantage is that the computational cost for the estimation of the ARMA model parameters is much more expensive than for AR models. It requires a numerical iterative nonlinear optimization procedure, where a risk of computational explosion, possibly caused by breaking the invertibility condition of MA parameters, always exists at each stage of iterations during the parameter value search. Subroutine tools for ARMA modeling are found in toolboxes of modern numerical computing packages, but many of them do not always provide satisfactory solutions to this problem. Many of them do not converge sufficiently well, a problem, which seems to be caused by too much protection against numerical explosions when the order of models, p and q, is high. Because of this, general order ARMA(p,q) modeling, ($p \geq 2$, $q \geq 2$), has not been so frequently used in practical applications, as general order AR(p) modeling.

However in some applications in science and engineering, the advantage of the ARMA(p,q) model is significant, so people may not want to miss it. In this section, we briefly review the possible computational problems associated with ARMA modeling. It should be helpful for the readers who are trying to solve these problems during the numerical computation of model identification.

To write down the likelihood of an ARMA model, we consider the whitening of the time series following the general guideline principle of the innovation approach. Here, we transform the dependent time series $x_1, \ldots, x_N$ to Gaussian white noise $w_1, \ldots, w_N$, by building the predictor $E[x_t | x_{t-1}, \ldots]$ with the model, so that the prediction error,

$$w_t = x_t - E[x_t | x_{t-1}, \ldots],$$

is a Gaussian white noise and (-2)log-likelihood of the model, with parameter α, is written down using the Gaussian likelihood as

$$(-2)\log p(x_1, \ldots, x_N | \alpha) = (-2)\log p(w_1, \ldots, w_N | \alpha)$$

$$= \sum_{t=1}^{N} \log \sigma_w^2 + \sum_{t=1}^{N} \frac{w_t^2}{\sigma_w^2} + N \log 2\pi. \qquad (10.28)$$

The predictor $E[x_t | x_{t-1}, \ldots]$ can be formed and calculated at each time step using the model as

$$E[x_t | x_{t-1}, \ldots] = \phi_1 x_{t-1} + \cdots + \phi_p x_{t-p} + \theta_1 w_{t-1} + \cdots + \theta_q w_{t-q},$$

assuming $w_{t-1}, \ldots, w_{t-q}$ are already calculated at the time point t from $x_{t-1}, \ldots, x_{t-p}$ in the previous time points, $t-1, t-2, \ldots$. Obviously, the scheme needs some modification near the starting point $t = 1$, where we cannot form a proper predictor since we don't have $x_0, \ldots, x_{1-p}$ and $w_0, \ldots, w_{1-q}$.

Several methods for the calculation of approximate likelihood have been presented since the late 1960s.

1. One of the most famous is the Box and Jenkins (1970) approach, in which they use a technique for generating initial values $x_0, x_{-1}, \ldots$, and $w_0, w_{-1}, \ldots$, using the backward prediction model for the ARMA model.

2. A more simple approximation method is to use the approximation (10.28) assuming the initial conditions, $x_0 = 0$, $x_{-1} = 0$, ..., and $w_0 = 0$, $w_{-1} = 0$, ...

3. Another approximate method is to discard the first r initial values and to start from time point $t = r + 1$, with assumptions, $w_r = 0, w_{r-1} = 0, \ldots$, $w_{r-q} = 0$, where $r = \text{Max}(p,q)$.
   Then we have, for $N \gg r$,

$$(-2) \log p(x_1, \ldots, x_N | \theta)$$

$$\approx (-2) \log p(w_{r+1}, \ldots, w_N | \theta)$$

$$= (N - r) \log \sigma_w^2 + \sum_{t=r+1}^{N} w_t^2 / \sigma_w^2 + (N - r) \log 2\pi. \tag{10.29}$$

4. Instead of the approximate likelihoods, we can use the exact likelihood of the ARMA model. Here all the relevant initial effects can be included inside the likelihood by introducing intermediate variables, $w_0, w_{-1}, \ldots, w_{1-p}, w_{1-p-1}, \ldots, w_{1-p-q}$, which can be expressed in terms of $\phi_1, \ldots, \phi_p, \theta_1, \ldots, \theta_q$. (see Newbold 1974). Another way of exact likelihood computation is by Jones (1980) where the Akaike's canonical state space representation (see Section 3.4.4) for the ARMA model is employed.

After writing down the likelihood, we need to search for the parameter values attaining the maximum of the likelihood, using some numerical optimization method such as the Newton-Raphson method. This numerical part of the maximum likelihood computation of ARMA models requires some expertise, which can be acquired only by empirical studies. Some useful guidance for the numerical maximization of the log-likelihood is given in the following:

1. The characteristic roots of the MA parameters of the ARMA model are not supposed to go outside the unit circle. For example, for the first-order MA model

$$x_t = (1 - \theta B)w_t,$$

θ is supposed to be $-1 < \theta < 1$. If $\theta > 1$, we have

$$w_t = x_t / (1 - \theta B)$$

$$= (1 + \theta B + \theta^2 B^2 + \cdots)x_t.$$

Since the coefficients of x_{t-k} ($k = 1, 2, \ldots$) are θ^k and $\theta^k \to \infty$, for $k \to \infty$, we have

$$\sum_{t=1}^{N} w_t^2 = \sum_{t=1}^{N} (x_t + \theta x_{t-1} + \theta^2 x_{t-2} + \cdots)^2 \to \infty \quad \text{for } N \to \infty.$$

For finite N, the value may stay finite but will surely take a very large value. If we start the optimization procedure of the maximization of the likelihood function, say $lk(\theta)$, with an initial value θ in the admissible region, that is, $-1 < \theta < 1$, we will never normally attain such a $\theta > 1$ by the numerical optimization procedure. However, if θ is close to the unit circle, the risk of $(\theta + \Delta\theta)$ going out of the unit circle becomes high when the optimization program is using the numerical gradient, $\{lk(\theta + \Delta\theta) - lk(\theta)\}/\Delta\theta$ with a fixed $\Delta\theta$, in the optimization algorithm, in order to find a direction to further improve the objective function. For $k \to \infty$, the speed of increase in $(\theta_1)^k$ is very fast for some $\theta_1 > 1$, so that the computation of the likelihood may break down in the middle of the numerical optimization procedure. To avoid this numerical explosion problem, it is recommended to insert a checking routine for the eligibility of the MA parameters in the iterative search procedure of the likelihood optimization.

2. The characteristic roots of the AR parameters of the ARMA model are not supposed to go outside the unit circle either. If we use, rather than the least squares method, the exact likelihood function, we never obtain such AR coefficients, $\phi_1, \ldots, \phi_p$, yielding the characteristic roots $\lambda_1, \ldots, \lambda_p$, lying on the unit circle. This is because if one of the roots defined by the coefficients approaches the unit circle, the determinant of the covariance matrix of the process goes to zero so that the log-likelihood goes to minus infinity. The numerical maximization procedure will stop, before one of the characteristic roots

reaches the unit circle. Thus, the maximum likelihood method with the exact likelihood always attains eligible stationary parameters, no matter how close the true characteristic roots are to the unit circle. However, this is, of course, at the cost of the computational load required when we use the exact likelihood of the ARMA model.

3. Sometimes, we may obtain a pair of AR characteristic roots and MA characteristic roots that are very similar. When we encounter this kind of a situation, we need to pay special attention to the overfitting problem. For example, the ARMA(2,2) model,

$$(1 - \phi_1 B - \phi_2 B^2)x_t = (1 - \theta_1 B - \theta_2 B^2)w_t,$$

is almost equivalent to the white noise process,

$$x_t = w_t,$$

when $\phi_1 \approx \theta_1$ and $\phi_2 \approx \theta_2$. This kind of situation typically occurs when both the AR and MA roots are close to the unit circle, implying that there is a peak and trough in the power spectrum at very close (may or may not be equivalent) frequencies. However, the unit of the metric in the parameter space (closeness of ϕ and θ) and the unit of the metric of the log-likelihood value space are not the same. Even though the parameters look similar, the values of the log-likelihood could be significantly different (see, e.g., the results of ARMA modeling for Series B of Box-Jenkins data in Ozaki 1977). Therefore, we need to check the significance of the extra terms by comparing the (−2)log-likelihood and AIC of the models before we make a decision on the overfitting. If AIC of ARMA(2,2) is smaller than the AIC of the white noise model, it means that there are some dynamic activities near the frequency specified by the AR parameters or MA parameters.

4. One way to obviate the numerical difficulty associated with the maximum likelihood estimation of the ARMA model is to parameterize in the continuous-time models. Remember we saw, in Section 9.3.1, that the ARMA(2,1) model obtained from the second-order differential equation model,

$$\ddot{x}(t) + c\dot{x}(t) + \omega^2 x(t) = n(t),$$

by the local linearization scheme is equivalent to that of an ARMA(2,1) model, where the MA parameter is dependent on the AR parameters, thus giving only two independent parameters. From the pth order differential equation model, we have an ARMA($p,p-1$) model, but again the MA parameters are specified as functions of the AR parameters. Here, there is no chance of having an ARMA model with very sharp troughs at the numerical search stage for MA parameters in the likelihood maximization.

10.3.2.2 Univariate AR Models

Next, we look into the computational method for the maximum likelihood estimation of the AR models. The AR model is one of the most commonly used discrete-time prediction models. Many estimation methods of AR models are known to applied scientists in various scientific fields. In order to explain and clarify the meaning of our innovation approach for the estimation of prediction models, we first see the relations between the maximum likelihood method and the least squares method, using the scalar AR model as a simple and typical example of the prediction model.

The ordinary maximum likelihood estimates of the AR(p) model

$$x_t = \phi_1 x_{t-1} + \cdots + \phi_p x_{t-p} + w_t$$

are obtained by minimizing the (−2)log-likelihood function,

$$(-2)\log p(x_1, x_2, \ldots, x_N | \phi_1, \ldots, \phi_p, \sigma_w^2)$$

$$= (-2)\log p(x_{p+1}, x_{p+2}, \ldots, x_N | x_1, \ldots, x_p, \phi_1, \ldots, \phi_p, \sigma_w^2)$$

$$+ (-2)\log p(x_1, \ldots, x_p | \phi_1, \ldots, \phi_p, \sigma_w^2)$$

$$= (-2)\log p(w_{p+1}, w_{p+2}, \ldots, w_N | x_1, \ldots, x_p, \phi_1, \ldots, \phi_p, \sigma_w^2)$$

$$+ (-2)\log p(x_1, \ldots, x_p | \phi_1, \ldots, \phi_p, \sigma_w^2)$$

$$= (N-p)\log 2\pi + (N-p)\log \sigma_w^2 + \sigma_w^{-2} \sum_{t=p+1}^{N} w_t^2$$

$$+ (-2)\log p(x_1, \ldots, x_p | \phi_1, \ldots, \phi_p, \sigma_w^2). \tag{10.30}$$

When p is much smaller than N, that is, the data length N is very large compared with the model order p, the contribution of $(-2)\log p(x_1, \ldots, x_p | \phi_1, \ldots, \phi_p, \sigma_w^2)$ which comes from the initial values, $x_1, \ldots, x_p$, to the (−2)log-likelihood function is negligibly small, and people often use the following approximate (−2)log-likelihood function,

$$(-2)\log p(x_1, x_2, \ldots, x_N | \phi_1, \ldots, \phi_p, \sigma_w^2)$$

$$\approx (N-p)\log 2\pi + (N-p)\log \sigma_w^2 + \frac{1}{\sigma_w^2} \sum_{t=p+1}^{N} w_t^2, \tag{10.31}$$

in the computation of the maximum likelihood estimates. The parameter estimates obtained by maximizing the approximate likelihood (10.31) are called approximate maximum likelihood estimates or quasi-maximum likelihood estimates.

Exact maximum likelihood estimates of the AR(p) model

$$x_t = \phi_1 x_{t-1} + \cdots + \phi_p x_{t-p} + w_t$$

are obtained by minimizing the exact (-2)log-likelihood function, which is obtained from (10.30) as follows:

$$(-2)\log p(x_1, x_2, \ldots, x_N | \phi_1, \ldots, \phi_p, \sigma_w^2)$$

$$= (N-p)\log \sigma_w^2 + \sigma_w^{-2} \sum_{t=p+1}^{N} w_t^2$$

$$+ \log |R(p-1)| + \sum_{i,j}^{p} s_{i,j} x_i x_j + N \log 2\pi.$$

Here, $s_{i,j}$ is the (i,j)th element of $R(p-1)^{-1}$, the inverse of the $p \times p$ Toeplitz matrix of autocovariances (see 10.40). Here, contribution

$$(-2)\log p(x_1, \ldots, x_p | \phi_1, \ldots, \phi_p, \sigma_w^2)$$

$$= (-2)\log \left[\frac{1}{(2\pi)^{p/2} \{\det R(p-1)\}^{1/2}} \exp\left\{ -\frac{1}{2} X_p' R(p-1)^{-1} X_p \right\} \right]$$

$$= p \log 2\pi + \log |R(p-1)| + X_p' R(p-1)^{-1} X_p$$

from the initial values, $x_1, \ldots, x_p$, is taken into account. For $p = 1$, the exact (-2) log-likelihood is written as

$$(-2)\log p(x_1, \ldots, x_N | \phi, \sigma_w^2) = N \log 2\pi + (N-1)\log \sigma_w^2 + \frac{1}{\sigma_w^2} \sum_{t=2}^{N} (x_t - \phi x_{t-1})^2$$

$$+ \log \frac{\sigma_w^2}{1-\phi^2} + \frac{1-\phi^2}{\sigma_w^2} x_1^2.$$

The computational cost of the exact maximum likelihood estimations is of course higher compared with the approximate method. However, as we saw for the ARMA model case, the exact method has several advantages. One is that the stationarity of the estimated model is guaranteed. Another is that the asymptotic efficiency of the parameter estimation is guaranteed, that is, the convergence of the estimated parameters to the true parameters is the fastest in the asymptotic sense.

10.3.2.3 Linear Multivariate AR Models

The r-dimensional AR(p) model,

$$x_t = A_1 x_{t-1} + \cdots + A_p x_{t-p} + w_t,$$

where w_t is an r-dimensional Gaussian white noise with variance matrix Σ_w, can also be estimated by the maximum likelihood method using the following (−2)log–likelihood:

$$(-2)\log p(x_1, x_2, \ldots, x_N | A_1, \ldots, A_p, \Sigma_w)$$

$$= (-2)\log p(x_{p+1}, x_{p+2}, \ldots, x_N | x_1, \ldots, x_p, A_1, \ldots, A_p, \Sigma_w)$$

$$+ (-2)\log p(x_1, \ldots, x_p | A_1, \ldots, A_p, \Sigma_w)$$

$$= (-2)\log p(w_{p+1}, w_{p+2}, \ldots, w_N | x_1, \ldots, x_p, A_1, \ldots, A_p, \Sigma_w)$$

$$+ (-2)\log p(x_1, \ldots, x_p | A_1, \ldots, A_p, \Sigma_w)$$

$$= \left[(N-p)r \log 2\pi + (N-p)\log|\Sigma_w| + \sum_{t=p+1}^{N} w_t^T \Sigma_w^{-1} w_t \right]$$

$$+ (-2)\log p(x_1, \ldots, x_p | A_1, \ldots, A_p, \Sigma_w).$$

The maximum likelihood method with the exact likelihood needs a numerical iterative optimization method. When the dimension r increases, the number of parameters to estimate increases. The computation for the parameter estimation is time consuming, especially for high lag order multivariate AR models. Although this method is time consuming compared with the method we see later in Section 10.4.2, unlike multivariate ARMA cases, it works without any numerical problem. In practice, however, the least squares solution with Yule–Walker equations (4.3) and (4.4) or multivariate Levinson schemes are often used, instead of the maximum likelihood method. These methods are useful when the number of data points is sufficiently large. We see these methods other than the maximum likelihood method in Section 10.4.

10.3.2.4 Linear Multivariate ARMA Models

We can formulate, in principle, the maximum likelihood computation method for the multivariate ARMA models, in the framework of the innovation approach. However, unlike the multivariate AR cases, we will have serious numerical difficulties with this approach. The difficulties can be obviated only by the use of a proper parametric representation of the process in the framework of state space representation. Leaving this discussion to Chapters 11 through 13, we will see what the problem is and how this difficulty could arise in multivariate time series data analysis.

Let x_t be generated from the d-dimensional ARMA(p,q) model,

$$x_t = A_1 x_{t-1} + \cdots + A_p x_{t-p} + B_1 w_{t-1} + \cdots + B_q w_{t-q} + w_t.$$

The prediction error w_t is given by

$$w_t = x_t - (A_1 x_{t-1} + \cdots + A_p x_{t-p} + B_1 w_{t-1} + \cdots + B_q w_{t-q}).$$

Assuming $x_t = 0$ for $t = 0, -1, \ldots, 1 - p$ and $w_s = 0$ for $s = 0, -1, \ldots, 1 - q$, we have

$$\log p(x_1, x_2 \ldots, x_N) \approx \log p(w_1, w_2 \ldots, w_N).$$

Here we assume that N is large enough that the effect of the initial values for the log-likelihood is negligibly small. Then we can approximately write the (-2)log-likelihood function as

$$(-2)\log p(x_1, x_2, \ldots, x_N)$$

$$\approx Nd\log(2\pi) + N\log|G| + Ntr(C_0 G^{-1}),$$

where $G = E[w_t w_t']$ and $|G|$ denotes the determinant of G, and $C_0 = \dfrac{1}{N}\sum_{i=1}^{N} \tilde{x}_t \tilde{x}_t'$. The maximized (-2)log-likelihood is given by

$$N\log|C_0| + Nd(\log 2\pi + 1).$$

Here, C_0 is in fact given by $C_0 = \dfrac{1}{N}\sum_{t=1}^{N} w_t w_t'$, where w_t is the prediction error obtained from the coefficient parameters, that is, $A_1, \ldots, A_p, B_1, \ldots, B_q$, attaining the minimum of $N\log|C_0|$.

The innovation-based maximum likelihood method works without any numerical problems for scalar AR Models, for scalar MA models, for scalar ARMA models, and for multivariate AR models, but not for multivariate ARMA models. Even though the likelihood is written in terms of the coefficient parameters, for the multivariate ARMA models the maximum point may not be numerically well defined. The Hessian matrix near the true parameters may be singular. This numerical problem of the singularity of the Hessian matrix has a close relation to the way we parameterize the multivariate ARMA model. This point makes a clear difference between the AR type (whether linear or nonlinear) models where the coefficient parameters are guaranteed to be estimated by the least squares method or the MLE method without any numerical problems.

For example, we have no numerical problem in maximizing the innovation-based log-likelihood of a 3-dimensional AR(2) model

$$
\begin{pmatrix} x_t^{(1)} \\ x_t^{(2)} \\ x_t^{(3)} \end{pmatrix} = \begin{pmatrix} a_{1,1}^{(1)} & a_{1,2}^{(1)} & a_{1,3}^{(1)} \\ a_{2,1}^{(1)} & a_{2,2}^{(1)} & a_{2,3}^{(1)} \\ a_{3,1}^{(1)} & a_{3,2}^{(1)} & a_{3,3}^{(1)} \end{pmatrix} \begin{pmatrix} x_{t-1}^{(1)} \\ x_{t-1}^{(2)} \\ x_{t-1}^{(3)} \end{pmatrix} + \begin{pmatrix} a_{1,1}^{(2)} & a_{1,2}^{(2)} & a_{1,3}^{(2)} \\ a_{2,1}^{(2)} & a_{2,2}^{(2)} & a_{2,3}^{(2)} \\ a_{3,1}^{(2)} & a_{3,2}^{(2)} & a_{3,3}^{(2)} \end{pmatrix} \begin{pmatrix} x_{t-2}^{(1)} \\ x_{t-2}^{(2)} \\ x_{t-2}^{(3)} \end{pmatrix} + \begin{pmatrix} w_t^{(1)} \\ w_t^{(2)} \\ w_t^{(3)} \end{pmatrix}
$$

in terms of all the coefficients of the two 3×3 matrices. On the other hand for a 3-dimensional ARMA(2,1) model,

$$
\begin{pmatrix} x_t^{(1)} \\ x_t^{(2)} \\ x_t^{(3)} \end{pmatrix} = \begin{pmatrix} a_{1,1}^{(1)} & a_{1,2}^{(1)} & a_{1,3}^{(1)} \\ a_{2,1}^{(1)} & a_{2,2}^{(1)} & a_{2,3}^{(1)} \\ a_{3,1}^{(1)} & a_{3,2}^{(1)} & a_{3,3}^{(1)} \end{pmatrix} \begin{pmatrix} x_{t-1}^{(1)} \\ x_{t-1}^{(2)} \\ x_{t-1}^{(3)} \end{pmatrix} + \begin{pmatrix} a_{1,1}^{(2)} & a_{1,2}^{(2)} & a_{1,3}^{(2)} \\ a_{2,1}^{(2)} & a_{2,2}^{(2)} & a_{2,3}^{(2)} \\ a_{3,1}^{(2)} & a_{3,2}^{(2)} & a_{3,3}^{(2)} \end{pmatrix} \begin{pmatrix} x_{t-2}^{(1)} \\ x_{t-2}^{(2)} \\ x_{t-2}^{(3)} \end{pmatrix}
$$

$$
+ \begin{pmatrix} b_{1,1}^{(1)} & b_{1,2}^{(1)} & b_{1,3}^{(1)} \\ b_{2,1}^{(1)} & b_{2,2}^{(1)} & b_{2,3}^{(1)} \\ b_{3,1}^{(1)} & b_{3,2}^{(1)} & b_{3,3}^{(1)} \end{pmatrix} \begin{pmatrix} w_{t-1}^{(1)} \\ w_{t-1}^{(2)} \\ w_{t-1}^{(3)} \end{pmatrix} + \begin{pmatrix} w_t^{(1)} \\ w_t^{(2)} \\ w_t^{(3)} \end{pmatrix},
$$

we can write the innovation-based log-likelihood in terms of all the coefficients of the three 3×3 matrices, but certainly we will have numerical difficulty in maximizing the likelihood with respect to the parameters because of the ill-conditioning caused by the redundant parameterization.

The possibility of redundant parameterization and nonunique parameterization of the multivariate ARMA model are the main obstacles for the models to be used and applied in multivariate time series analysis in neurosciences and many other scientific research areas. As we saw in Chapter 4, one way of obviating the nonuniqueness of the parameterization of the multivariate ARMA model is to use the canonical realization method of state space representation for the multivariate ARMA process. The inferential problems of general state space models, including multivariate ARMA models, are discussed later in Chapters 11 through 13.

10.3.2.5 Linear Spatial Time Series Models

We saw in Section 4.5.2, that the innovation approach is useful for introducing parametric models for spatial time series. The approach is also useful for the estimation of the spatial time series models. When a set of fMRI data is given for $64 \times 64 \times 36$ grid points, the data are regarded as a $64 \times 64 \times 36$-dimensional spatial time series. For an NN-ARX model,

$$x_t^{(v)} = a_1^{(v)} x_{t-1}^{(v)} + \cdots + a_p^{(v)} x_{t-p}^{(v)} + \frac{1}{6}\left(\sum_{v' \in N(v)} b_{v'}^{(v)} x_{t-1}^{(v')} \right) + \theta_1^{(v)} s_{t-1} + \cdots + \theta_r^{(v)} s_{t-r} + \varepsilon_t^{(v)}, \quad (10.32)$$

the (-2)log-likelihood is specified as

$$(-2)\log p(x_1^{(1,1,1)}, \ldots, x_1^{(64,64,36)}, \ldots, x_N^{(1,1,1)}, \ldots, x_N^{(64,64,36)} | \varphi)$$

$$= (-2)\log \det(L^{-1}) + (-2)\log p(y_1^{(1,1,1)}, \ldots, y_1^{(64,64,36)}, \ldots, y_N^{(1,1,1)}, \ldots, y_N^{(64,64,36)} | \varphi)$$

$$= (-2)\log \det(L^{-1}) + (-2)\{\log p(y_1^{(1,1,1)}, \ldots, y_1^{(64,64,36)} | \varphi)$$

$$+ \log p(y_2^{(1,1,1)}, \ldots, y_2^{(64,64,36)}, \ldots, y_N^{(1,1,1)}, \ldots, y_N^{(64,64,36)} | \varphi, y_1^{(1,1,1)}, \ldots, y_1^{(64,64,36)})\}$$

$$= (-2)\log \det(L^{-1}) + (-2)\{\log p(y_1^{(1,1,1)}, \ldots, y_1^{(64,64,36)} | \varphi)$$

$$+ \log p(\varepsilon_2^{(1,1,1)}, \ldots, \varepsilon_2^{(64,64,36)}, \ldots, \varepsilon_N^{(1,1,1)}, \ldots, \varepsilon_N^{(64,64,36)} | \varphi, y_1^{(1,1,1)}, \ldots, y_1^{(64,64,36)})\},$$

where L is the spatial whitening operator (see Section 4.5.2.2).

When the number of time points N is sufficiently large, the first term is negligibly small compared with the rest and we have an approximate (-2) log-likelihood as

$$(-2)\log p(x_1^{(1,1,1)}, \ldots, x_1^{(64,64,36)}, \ldots, x_N^{(1,1,1)}, \ldots, x_N^{(64,64,36)} | \varphi)$$

$$\approx \log p(\varepsilon_2^{(1,1,1)}, \ldots, \varepsilon_2^{(64,64,36)}, \ldots, \varepsilon_N^{(1,1,1)}, \ldots, \varepsilon_N^{(64,64,36)} | \varphi, y_1^{(1,1,1)}, \ldots, y_1^{(64,64,36)})$$

$$+ (-2)\log \det(L^{-1})$$

$$= \sum_{v \in V} \left[\sum_{t=2}^{T} \left\{ \log \sigma_{\varepsilon_t^{(v)}}^2 + \frac{(\varepsilon_t^{(v)})^2}{\sigma_{\varepsilon_t^{(v)}}^2} \right\} \right] + (-2)\log \det(L^{-1}) + C_{d,N}$$

$$\varepsilon_t^{(v)} = y_t^{(v)} - \left\{ a_1^{(v)} y_{t-1}^{(v)} + \ldots + a_p^{(v)} y_{t-p}^{(v)} + \frac{1}{6}\left(\sum_{v' \in N(v)} b_{v'}^{(v)} y_{t-1}^{(v')} \right) + \theta_1^{(v)} s_{t-1} + \ldots + \theta_r^{(v)} s_{t-r} \right\}.$$

Here, $V = \{(1, 1, 1), \ldots, (64, 64, 36)\}$, the number of elements of V is $d = 64 \times 64 \times 36$, and $C_{d,N}$ is a constant, $C_{d,N} = d(N - 1)\log 2\pi$.

In principle, the maximum likelihood estimates of the parameters, that is, the linear coefficients, $a_1^{(v)}, \ldots, a_p^{(v)}$, $b_{v'}^{(v)}$, ($v' \in Neighbor(v)$), $\theta_1^{(v)}, \ldots, \theta_r^{(v)}$, and the noise variances, $\sigma_{\varepsilon_t^{(v)}}^2$ ($v \in V$), are obtained by minimizing the (-2)log-likelihood function in terms of the parameters. The likelihood of the model is assessed by using the attained minimum value of the (−2)log-likelihood function. Here, we note that the number of parameters is huge. Even for the simplest NN-ARX model of $p = 1$, and $r = 1$, the number of the parameters is $d \times (p + r + 1) = 64 \times 64 \times 36 \times 3$. The numerical minimization of this function in terms of 440,000 parameters is not realistic even with modern super computers. On the other hand, if we take advantage of the temporal and spatial independence of the innovations it is quite easy to derive an approximate computational method for obtaining approximate maximum likelihood estimates of these huge numbers of parameters.

If $\varepsilon_t(v)$ and $\varepsilon_t(v')$ are independent for $v \neq v'$, then the minimization of

$$\sum_{v \in V} \left[\sum_{t=2}^{T} \left\{ \log \sigma^2_{\varepsilon_t^{(v)}} + \frac{(\varepsilon_t^{(v)})^2}{\sigma^2_{\varepsilon_t^{(v)}}} \right\} \right]$$

is achieved, for each $v \in V$, by minimizing

$$\sum_{t=2}^{T} \left\{ \log \sigma^2_{\varepsilon_t^{(v)}} + \frac{(\varepsilon_t^{(v)})^2}{\sigma^2_{\varepsilon_t^{(v)}}} \right\}$$

independently from the rest of v's in V. The minimization is actually done by solving, at each $v \in V$, a linear equation derived from the linear regression model,

$$y_t^{(v)} = \left\{ a_1^{(v)} y_{t-1}^{(v)} + \cdots + a_p^{(v)} y_{t-p}^{(v)} + \frac{1}{6} \left(\sum_{v' \in N(v)} b_{v'}^{(v)} y_{t-1}^{(v')} \right) + \theta_1^{(v)} s_{t-1} + \cdots + \theta_r^{(v)} s_{t-r} \right\} + \varepsilon_t^{(v)}.$$

$$(10.33)$$

The minimized value of

$$\sum_{t=2}^{T} \left\{ \log \sigma^2_{\varepsilon_t^{(v)}} + \frac{(\varepsilon_t^{(v)})^2}{\sigma^2_{\varepsilon_t^{(v)}}} \right\}$$

is given, at each voxel (v), by

$$\text{Min} \sum_{t=2}^{T} \left\{ \log \sigma^2_{\varepsilon_t^{(v)}} + \frac{(\varepsilon_t^{(v)})^2}{\sigma^2_{\varepsilon_t^{(v)}}} \right\} = (N-1) \log \hat{\sigma}^2_{\varepsilon_t^{(v)}} + (N-1),$$

where $\hat{\sigma}^2_{\varepsilon_t^{(v)}}$ is the least squares estimate of the variance of $\varepsilon_t^{(v)}$ of (10.33). Then we obtain the minimum of the (−2)log-likelihood function by

$$\text{Min} \sum_{v \in V} \left[\sum_{t=2}^{T} \left\{ \log \sigma^2_{\varepsilon_t^{(v)}} + \frac{(\varepsilon_t^{(v)})^2}{\sigma^2_{\varepsilon_t^{(v)}}} \right\} \right] = \sum_{v \in V} (N-1) \log \hat{\sigma}^2_{\varepsilon_t^{(v)}} + d(N-1).$$

Here all the estimated values are given as explicit functions of the observation data, and the time consuming numerical nonlinear optimization method is not needed for the minimization of (−2)log-likelihood.

Our computational method is, so far, based on the assumption of the independence (temporal and spatial) of the innovations. To validate the results obtained by the present computational method, we need to make sure that the innovations really are independent. This is always the case for any estimation method, including the simplest scalar AR model. After fitting the model, we need to check whether the residuals are really independent, homogeneous and Gaussian, using some statistical method, such as a portmanteau test, Kolmogorov–Smirnov test, etc. (see Box and Jenkins 1970).

10.3.3 Likelihood of Linear/Nonlinear Time Series Models with Exogenous Inputs

Likelihood of linear/nonlinear time series models with exogenous input variables can be derived in the same way as for the cases of ordinary time series models. For the time series where the exogenous time series, u_{t-1}, u_{t-2}, ..., has useful information for the prediction of x_t, we simply whiten the series x_t into the prediction error, w_t, by subtracting the predictor, $x_{t|t-1}$, that is, the conditional expectation of x_t calculated using all the information available at the time point $t - 1$; thus,

$$w_t = x_t - x_{t|t-1}$$

$$= x_t - E[x_t | x_{t-1}, x_{t-2}, \ldots, u_{t-1}, u_{t-2}, \ldots].$$

Then the (−2)log-likelihood of the model with exogenous variables, whether linear dynamic models (such as ARX and ARMAX models) or nonlinear dynamic models (such as RBF-ARX models) with parameter vector θ, is approximately written by taking advantage of the Gaussianity of w_t as

$$(-2)\log p(x_1, \ldots, x_N | \theta, u_1, \ldots, u_N)$$

$$= (-2)\log p(x_{p+1}, \ldots, x_N | x_1, \ldots, x_p, \theta, u_1, \ldots, u_N)$$

$$+ (-2)\log p(x_1, \ldots, x_p | \theta, u_1, \ldots, u_N)$$

$$= (-2)\log p(w_{p+1}, \ldots, w_N | x_1, \ldots, x_p, \theta, u_1, \ldots, u_N)$$

$$+ (-2)\log p(x_1, \ldots, x_p, \theta, u_1, \ldots, u_N)$$

$$\approx (-2)\log p(w_{p+1}, \ldots, w_N | x_1, \ldots, x_p, \theta, u_1, \ldots, u_N)$$

$$= (N - p)\log \sigma_w^2 + \sum_{t=p+1}^{N} w_t^2 / \sigma_w^2 + (N - p)\log 2\pi.$$

Here, the prediction error w_t is specified by

$$w_t = x_t - E[x_t | x_{t-1}, x_{t-2}, \ldots, u_{t-1}, u_{t-2}, \ldots]$$

$$= x_t - \left(\phi_0 + \sum_{i=1}^{p} \phi_i x_{t-i} + \sum_{i=1}^{q} \varphi_i u_{t-i} \right)$$

when the model is an ARX model,

$$x_t = \phi_0 + \sum_{i=1}^{p} \phi_i x_{t-i} + \sum_{i=1}^{q} \varphi_i u_{t-i} + w_t,$$

and the prediction error w_t is specified by

$$w_t = x_t - E[x_t | x_{t-1}, x_{t-2}, \ldots, u_{t-1}, u_{t-2}, \ldots]$$

$$= x_t - \left(\phi_0 + \sum_{i=1}^{p} \phi_i(X_{t-1}) x_{t-i} + \sum_{i=1}^{q} \varphi_i u_{t-i} \right)$$

when the model is an RBF-ARX model,

$$x_t = \phi_0 + \sum_{i=1}^{p} \phi_i(X_{t-1}) x_{t-i} + \sum_{i=1}^{q} \varphi_i u_{t-i} + w_t,$$

where

$$X_{t-1} = (x_{t-1}, \ldots, x_{t-r})',$$

$$\varsigma_l = (\varsigma_l^{(0)}, \varsigma_l^{(1)}, \ldots, \varsigma_l^{(r)})',$$

$$\phi_i(X_{t-1}) = c_{j,0} + \sum_{l}^{m} c_{j,l} \exp\left\{ -\|X_{t-1} - \varsigma_l\|_{H_l}^2 \right\},$$

and

$$\|X_{t-1} - \varsigma_l\|_{H_l}^2 = \sum_{k=1}^{r} \frac{\left(\Delta^{k-1} x_{t-j} - \varsigma_l^{(k)} \right)^2}{h_l^{(k)}}.$$

10.4 Computationally Efficient Methods and Algorithms

Although the innovation-based maximum likelihood method is valid and powerful in most time series modeling problems, in some multivariate and high lag order model situations, the computational cost for the numerical maximization of the innovation-based likelihood is too expensive and also time-consuming. Several computationally efficient alternative methods and algorithms have been developed for linear AR models both in the univariate and multivariate cases. We see these in the next section.

10.4.1 Least Squares Methods for Univariate AR Models

10.4.1.1 Yule–Walker Equation (Moment Method)

Let us assume that a univariate time series is generated from the following AR(p) model:

$$x_t = \phi_1 x_{t-1} + \phi_2 x_{t-2} + \cdots + \phi_p x_{t-p} + w_t. \tag{10.34}$$

If we multiply the both sides of (10.34) by x_{t-k} and take expectation, we have

$$\gamma_k = \phi_1 \gamma_{k-1} + \phi_2 \gamma_{k-2} + \cdots + \phi_p \gamma_{p-k}, \tag{10.35}$$

where $\gamma_k = E[x_t x_{t-k}]$. If we do this for $k = 1, \ldots, p$, we have the following Yule–Walker equation:

$$
\begin{pmatrix}
\gamma_0 & \gamma_1 & \cdots & \gamma_{p-1} \\
\gamma_1 & \gamma_0 & \cdots & \gamma_{p-2} \\
\cdots & \cdots & \cdots & \cdots \\
\gamma_{p-1} & \gamma_{p-2} & \cdots & \gamma_0
\end{pmatrix}
\begin{pmatrix}
\phi_1 \\
\phi_2 \\
\cdots \\
\phi_p
\end{pmatrix}
=
\begin{pmatrix}
\gamma_1 \\
\gamma_2 \\
\cdots \\
\gamma_p
\end{pmatrix}. \tag{10.36}
$$

This equation shows that we can obtain an estimate of the AR coefficients, $(\hat{\phi}_1, \hat{\phi}_2, \ldots, \hat{\phi}_p)'$, uniquely by solving the Yule–Walker equation (10.36) replacing the auto covariances, $\gamma_0, \gamma_1, \gamma_2, \ldots, \gamma_p$, by their estimates, $\hat{\gamma}_0, \hat{\gamma}_1, \hat{\gamma}_2, \ldots, \hat{\gamma}_p$, as

$$
\begin{pmatrix}
\hat{\phi}_1 \\
\hat{\phi}_2 \\
\cdots \\
\hat{\phi}_p
\end{pmatrix}
=
\begin{pmatrix}
\hat{\gamma}_0 & \hat{\gamma}_1 & \cdots & \hat{\gamma}_{p-1} \\
\hat{\gamma}_1 & \hat{\gamma}_0 & \cdots & \hat{\gamma}_{p-2} \\
\cdots & \cdots & \cdots & \cdots \\
\hat{\gamma}_{p-1} & \hat{\gamma}_{p-2} & \cdots & \hat{\gamma}_0
\end{pmatrix}^{-1}
\begin{pmatrix}
\hat{\gamma}_1 \\
\hat{\gamma}_2 \\
\cdots \\
\hat{\gamma}_p
\end{pmatrix}.
$$

The residual variance σ_w^2 is obtained by using another relation

$$\gamma_0 = \phi_1 \gamma_1 + \phi_2 \gamma_2 + \cdots + \phi_p \gamma_p + \sigma_w^2, \tag{10.37}$$

which comes from (10.34). This method is a kind of moment method, where the model parameters, $(\phi_1, \phi_2, \ldots, \phi_p)'$ and σ_w^2, are uniquely derived from the estimated second-order moments, $\hat{\gamma}_0, \hat{\gamma}_1, \hat{\gamma}_2, \ldots, \hat{\gamma}_p$. Estimation of the auto covariance function could be given either by

$$\hat{\gamma}_k = \frac{1}{N-k} \sum_{t=k+1}^{N} x_t x_{t-k} \tag{10.38}$$

or by

$$\hat{\gamma}_k = \frac{1}{N} \sum_{t=k+1}^{N} x_t x_{t-k}. \tag{10.39}$$

Since the computation of the auto covariance function is straightforward, the present parameter estimation method is computationally quite efficient compared with the maximum likelihood method.

The matrix $R(p-1)$ defined by

$$R(p-1) = \begin{pmatrix} \gamma_0 & \gamma_1 & \cdots & \gamma_{p-1} \\ \gamma_1 & \gamma_0 & \cdots & \gamma_{p-2} \\ \cdots & \cdots & \cdots & \cdots \\ \gamma_{p-1} & \gamma_{p-2} & \cdots & \gamma_0 \end{pmatrix} \tag{10.40}$$

has Toeplitz matrix from and is known to play an important role in the characterization of AR models. We note that the auto covariances, $\gamma_0, \gamma_1, \gamma_2, \ldots,$ of an AR model are specified by the parameters $\phi_1, \ldots, \phi_p$ and σ_w^2 (see Section 3.2.4). For example $\gamma_0, \gamma_1, \ldots, \gamma_p$ are obtained by solving the following linear equation (which comes from 10.36 and 10.37):

$$\begin{pmatrix} \gamma_0 \\ \gamma_1 \\ \gamma_2 \\ \cdots \\ \gamma_{p-1} \\ \gamma_p \end{pmatrix} = \begin{pmatrix} 1 & -\phi_1 & -\phi_2 & \cdots & -\phi_{p-1} & -\phi_p \\ -\phi_1 & 1-\phi_2 & -\phi_3 & \cdots & -\phi_p & 0 \\ -\phi_2 & -\phi_1-\phi_3 & 1-\phi_4 & \cdots & 0 & 0 \\ \cdots & \cdots & \cdots & \cdots & \cdots & \cdots \\ -\phi_{p-1} & -\phi_{p-2}-\phi_p & -\phi_{p-3} & \cdots & 1 & 0 \\ -\phi_p & -\phi_{p-1} & -\phi_{p-2} & \cdots & -\phi_1 & 1 \end{pmatrix}^{-1} \begin{pmatrix} \sigma_w^2 \\ 0 \\ 0 \\ \cdots \\ 0 \\ 0 \end{pmatrix}, \tag{10.41}$$

and $\gamma_{p+1}, \gamma_{p+2}, \ldots$ are obtained recursively by the linear relation, $\gamma_k = \phi_1\gamma_{k-1} + \cdots + \phi_p\gamma_{k-p}$ for $k \geq p + 1$. Thus, for any estimated auto covariance function, $\hat{\gamma}_0, \hat{\gamma}_1, \hat{\gamma}_2, \ldots, \hat{\gamma}_p, \hat{\gamma}_{p+1}, \ldots, \hat{\gamma}_{N-1}$, which comes from time series data, $x_1, x_2, \ldots, x_N$, we can have estimates, $\hat{\phi}_1, \ldots, \hat{\phi}_p$ and $\hat{\sigma}_w^2$, using the relations, (10.41) and (10.35), for the second moments of the process. However, this is true only when the estimated auto covariance function is positive definite. Therefore, we need to be careful when we choose a computational scheme for the estimation of the auto covariance function. For example, the estimated auto covariance function $\hat{\gamma}_0, \hat{\gamma}_1, \hat{\gamma}_2, \ldots, \hat{\gamma}_p$ with (10.39) is known to be positive definite, and the use of (10.41) is valid. However, the estimate $\hat{\gamma}_0, \hat{\gamma}_1, \hat{\gamma}_2, \ldots, \hat{\gamma}_p$ by (10.38) is not necessarily positive definite, and we may end up with an incomprehensive solution if we mechanically use (10.41) with the illegitimate nonpositive definite $\hat{\gamma}_0, \hat{\gamma}_1, \hat{\gamma}_2, \ldots, \hat{\gamma}_p$.

10.4.1.2 Least Squares Estimates

Remember that an approximate (-2)log-likelihood function of AR(p) model was given by

$$(-2)\log p(x_1, x_2, \ldots, x_N | \phi_1, \ldots, \phi_p, \sigma_w^2) \approx (N-p)\log 2\pi + (N-p)\log \sigma_w^2 + \frac{1}{\sigma_w^2}\sum_{t=p+1}^{N} w_t^2.$$

(10.42)

If we minimize this approximate (-2)log-likelihood function with the given data using some iterative numerical optimization method, parameters start converging to the true values. However, as the parameters approach to the true value, the convergence becomes slower and slower and the Hessian matrix, the second derivative of the objective function, approaches a singular matrix. This is because there exists a relation (dependency) between the parameters at the minimized point (if not at its neighbor) of the approximate (-2)log-likelihood function $(= l(\phi_1, \ldots, \phi_p, \sigma_w^2))$. Here we have the relation

$$\left.\frac{\partial l(\phi_1, \ldots, \phi_p, \sigma_w^2)}{\partial \sigma_w^2}\right|_{\theta=\hat{\theta}} = \frac{N-p}{\hat{\sigma}_w^2} - \frac{\sum_{t=p+1}^{N} w_t^2}{\hat{\sigma}_w^4} = 0$$

at the point $\hat{\theta} = (\hat{\phi}_1, \ldots, \hat{\phi}_p, \hat{\sigma}_w^2)'$. Thus, the (quasi) maximum likelihood estimate of $\theta = (\phi_1, \phi_2, \ldots, \phi_p)'$ is obtained as $\hat{\theta} = (\hat{\phi}_1, \hat{\phi}_2, \ldots, \hat{\phi}_p)'$, which attains the minimum $\hat{\sigma}_w^2$. Then this is equivalent to the least squares estimate of the AR model, which is obtained by solving the following linear equation:

$$Y = X\theta, \tag{10.43}$$

where

$$Y = \begin{pmatrix} x_{p+1} \\ x_{p+2} \\ \cdots \\ x_N \end{pmatrix}, \quad X = \begin{pmatrix} x_p & \cdots & x_1 \\ x_{p+1} & \cdots & x_2 \\ \cdots & \cdots & \cdots \\ x_{N-1} & \cdots & x_{N-p} \end{pmatrix} \quad \text{and} \quad \theta = \begin{pmatrix} \phi_1 \\ \cdots \\ \phi_p \end{pmatrix}.$$

Then the least squares estimate $\hat{\theta}$ is obtained by

$$\hat{\theta} = (X'X)^{-1}X'Y \qquad (10.44)$$

The least squares estimate can also be obtained by dropping σ_w^2 from the set of independent parameters of the objective function by using the relation satisfied at the minimized point and minimizing the sum of squares,

$$\sum_{t=1}^{N} \{w_t(\phi_1, \ldots, \phi_p)\}^2,$$

$$w_t(\phi_1, \ldots, \phi_p) = x_t - (\phi_1 x_{t-1} + \cdots + \phi_p x_{t-p}),$$

with respect to $\phi_1, \ldots, \phi_p$ by a numerical optimization procedure. Naturally, this is much more time consuming and inefficient than the usual least squares method.

We note that $X'X$ in (10.43) is written as

$$X'X = \begin{pmatrix} C_F(1,1) & C_F(2,1) & \cdots & C_F(p,1) \\ C_F(1,2) & C_F(2,2) & \cdots & C_F(p,2) \\ \cdots & \cdots & \cdots & \cdots \\ C_F(1,p) & C_F(2,p) & \cdots & C_F(p,p) \end{pmatrix}$$

and $X'Y$ in (10.43) is written as

$$X'Y = \begin{pmatrix} C_F(0,1) \\ C_F(0,2) \\ \cdots \\ C_F(0,p) \end{pmatrix}.$$

Here, $C_F(i,j)$'s are given by

$$C_F(i,j) = \frac{1}{(N-p)} \sum_{t=p+1}^{N} x_{t-i} x_{t-j} \quad \text{for } i = 0, \ldots, p \text{ and } j = 1, \ldots, p.$$

This means that the least squares estimate is equivalent to the solution obtained by solving the Yule–Walker equation,

$$\begin{pmatrix} \gamma_0 & \gamma_1 & \cdots & \gamma_{p-1} \\ \gamma_1 & \gamma_0 & \cdots & \gamma_{p-2} \\ \cdots & \cdots & \cdots & \cdots \\ \gamma_{p-1} & \gamma_{p-2} & \cdots & \gamma_0 \end{pmatrix} \begin{pmatrix} \phi_1 \\ \phi_2 \\ \cdots \\ \phi_p \end{pmatrix} = \begin{pmatrix} \gamma_1 \\ \gamma_2 \\ \cdots \\ \gamma_p \end{pmatrix},$$

by replacing $\gamma_{|i-j|}$ by $C_F(i,j)$, that is,

$$
\begin{pmatrix}
C_F(1,1) & C_F(2,1) & \cdots & C_F(p,1) \\
C_F(1,2) & C_F(2,2) & \cdots & C_F(p,2) \\
\cdots & \cdots & \cdots & \cdots \\
C_F(1,p) & C_F(2,p) & \cdots & C_F(p,p)
\end{pmatrix}
\begin{pmatrix}
\phi_1 \\
\phi_2 \\
\cdots \\
\phi_p
\end{pmatrix}
=
\begin{pmatrix}
C_F(0,1) \\
C_F(0,2) \\
\cdots \\
C_F(0,p)
\end{pmatrix}.
$$

Note that $C_F(i,i) \neq C_F(j,j)$ for $i \neq j$. Note also that $C_F(i,j)$ is consistent, that is,

$$
C_F(i,j) \underset{N\to\infty}{\to} \gamma_{|i-j|}.
$$

However, when N is finite, the use of $C_F(i,j)$ often causes a problem. The matrix $C_F(i,j)$ is not necessarily positive definite unlike the theoretical Toeplitz matrix (10.40). This means that the AR model determined by the least squares estimated parameters is not necessarily a stationary AR process. Then the use of the parametric spectral representation of the AR process defined by the least squares estimated parameters cannot be always justified.

We note that the estimates obtained by solving the Yule–Walker equation using the biased auto covariance estimates given by

$$
\hat{\gamma}_k = \frac{1}{N} \sum_{t=k+1}^{N} x_t x_{t-k} \quad (k = 0,1,\ldots,p)
$$

instead of the unbiased estimates, $C_F(0,k) = \dfrac{1}{(N-k)} \sum_{t=k+1}^{N} x_t x_{t-k}$, as

$$
\begin{pmatrix}
\hat{\phi}_1 \\
\hat{\phi}_2 \\
\cdots \\
\hat{\phi}_p
\end{pmatrix}
=
\begin{pmatrix}
\hat{\gamma}_0 & \hat{\gamma}_1 & \cdots & \hat{\gamma}_{p-1} \\
\hat{\gamma}_1 & \hat{\gamma}_0 & \cdots & \hat{\gamma}_{p-2} \\
\cdots & \cdots & \cdots & \cdots \\
\hat{\gamma}_{p-1} & \hat{\gamma}_{p-2} & \cdots & \hat{\gamma}_0
\end{pmatrix}^{-1}
\begin{pmatrix}
\hat{\gamma}_1 \\
\hat{\gamma}_2 \\
\cdots \\
\hat{\gamma}_p
\end{pmatrix},
$$

yield a stationary AR process and have been used by many authors (see Hannan 1969b, Akaike 1973b). Here the Yule–Walker solution with $\hat{\gamma}_k$ is equivalent to the least squares solution with an extra initial condition, that is, $x_t = 0$; $t = 0, -1, -2,$ and an extra end condition, $x_t = 0$ for $t = N + 1, N + 2, \ldots$.

We can easily see that for $\xi = (\xi_0, \ldots, \xi_L)' \neq (0, \ldots, 0)'$,

$$
(\xi_0 \quad \xi_1 \quad \cdots \quad \xi_L)
\begin{pmatrix}
\hat{\gamma}_0 & \hat{\gamma}_1 & \cdots & \hat{\gamma}_L \\
\hat{\gamma}_1 & \hat{\gamma}_0 & \cdots & \hat{\gamma}_{L-1} \\
\cdots & \cdots & \cdots & \cdots \\
\hat{\gamma}_L & \hat{\gamma}_{L-1} & \cdots & \hat{\gamma}_0
\end{pmatrix}
\begin{pmatrix}
\xi_0 \\
\xi_1 \\
\cdots \\
\xi_L
\end{pmatrix}
$$

$$
= \frac{1}{N} \sum_{t=1}^{N+L} (\xi_0 x_t + \xi_1 x_{t-1} + \cdots + \xi_L x_{t-L})^2 > 0,
$$

which guarantees that the matrix $R(L) = (\hat{\gamma}_{i,j})$ is positive definite.

When the AR process is close to being nonstationary, that is, when one of the characteristic roots is close to one, the initial condition and the end condition are rather too restrictive because the process is likely to drift away from the origin more often than ordinary cases. This is one of the explanations why the Yule–Walker method tends to produce biased parameter estimates compared with other methods for short sample data.

10.4.1.3 Levinson Estimate

Because of its structural simplicity, the AR model can enjoy some more computational advantage, that is, the recursive computation of the parameter estimation. This is especially useful when we do not know the true order of the model and we need to calculate the model parameters of many different orders before we finally choose the best model order. Remember (see Section 3.2.4) that we have an equivalence relation,

$$
\{\phi_{k,1}, \phi_{k,2}, \ldots, \phi_{k,k}, \sigma_w^2(k)\} \leftrightarrow \{\gamma_0, \gamma_1, \gamma_2, \ldots, \gamma_k\},
$$

between the model parameters and the auto covariance function of the AR(k) model, so that $\phi_{k,k}$ and $\sigma^2(k)$, $(k = 0, 1, 2, \ldots, p)$ are defined by the solutions of

$$
E[(x_t - \phi_{k,1} x_{t-1} - \cdots - \phi_{k,k} x_{t-k}) x_{t-l}] = 0 \quad \text{for } l = 1, 2, \ldots, k,
$$

and

$$
\sigma_w^2(k) = E[(x_t - \phi_{k,1} x_{t-1} - \cdots - \phi_{k,k} x_{t-k}) x_t] \quad \text{for } k = 0, 1, 2, \ldots, p
$$

This means that they are obtained by solving the Yule–Walker equation (10.36) and

$$
\sigma_w^2(k) = \gamma_0 - \phi_{k,1} \gamma_1 - \cdots - \phi_{k,k} \gamma_k
$$

for $k = 0, 1, 2, \ldots, p$.

In time series analysis, we are given a data set, $x_1, x_2, ..., x_N$, but we do not know the true order of the model behind the data. We need to calculate models of several orders and compare the likelihood and AIC of these models. Then we could calculate an estimate of the auto covariance function, up to lag $N-1$, that is, $\{\hat{\gamma}_0, \hat{\gamma}_1, \hat{\gamma}_2, ..., \hat{\gamma}_{N-1}\}$. Note that if we have a time series data, $x_1, x_2, ..., x_N$, we can have an estimate of auto covariance function, of lag order possibly up to $N-1$, that is, $\{\hat{\gamma}_0, \hat{\gamma}_1, \hat{\gamma}_2, ..., \hat{\gamma}_{N-1}\}$. Then with this estimated auto covariance function, we have the relations

$$\{\hat{\phi}_{1,1}, \hat{\sigma}_w^2(1)\} \leftrightarrow \{\hat{\gamma}_0, \hat{\gamma}_1\},$$

$$\{\hat{\phi}_{2,1}, \hat{\phi}_{2,2}, \hat{\sigma}_w^2(2)\} \leftrightarrow \{\hat{\gamma}_0, \hat{\gamma}_1, \hat{\gamma}_2\},$$

$$...$$

$$\{\hat{\phi}_{N-2,1}, \hat{\phi}_{N-2,2}, ..., \hat{\phi}_{N-2,N-2}, \hat{\sigma}_w^2(N-2)\} \leftrightarrow \{\hat{\gamma}_0, \hat{\gamma}_1, \hat{\gamma}_2, ..., \hat{\gamma}_{N-2}\},$$

$$\{\hat{\phi}_{N-1,1}, \hat{\phi}_{N-1,2}, ..., \hat{\phi}_{N-1,N-1}, \hat{\sigma}_w^2(N-1)\} \leftrightarrow \{\hat{\gamma}_0, \hat{\gamma}_1, \hat{\gamma}_2, ..., \hat{\gamma}_{N-1}\}.$$

In these computations, we can save computational load by taking advantage of the theoretical Levinson's relations (see 3.8 through 3.12, Section 3.4.5), between $\{\hat{\phi}_{1,1}, \hat{\sigma}_w^2(1)\}$ and $\{\hat{\phi}_{2,1}, \hat{\phi}_{2,2}, \hat{\sigma}_w^2(2)\}$, between $\{\hat{\phi}_{2,1}, \hat{\phi}_{2,2}, \hat{\sigma}_w^2(2)\}$ and $\{\hat{\phi}_{3,1}, \hat{\phi}_{3,2}, \hat{\phi}_{3,3}, \hat{\sigma}_w^2(3)\}$, and so on. Here we do not need to solve the Yule–Walker equations of each order independently. We can use the first order coefficient to calculate the second-order coefficients using these explicit relations.

The estimated AR coefficients, $\hat{\phi}_{k,1}, \hat{\phi}_{k,2} ... \hat{\phi}_{k,k}$, and the estimated residual variance, $\hat{\sigma}_w^2(k)$, from $k = 0$ to $k = p$ are obtained in this way using the recursive Levinson scheme (3.8) through (3.12), where γ_k ($k = 0, 1, 2, ..., p$) are to be replaced by a positive definite sample auto covariance function, $\hat{\gamma}_k$ ($k = 0, 1, 2, ..., p$). The recursively obtained AR coefficients, $\hat{\phi}_{p,1}, \hat{\phi}_{p,2} ... \hat{\phi}_{p,p}$, and residual variance, $\hat{\sigma}_w^2(p)$, are the estimates of the AR(p) model, and they are equivalent to the Yule–Walker estimates.

Levinson scheme shares the same deficit with the Yule–Walker estimates since they are equivalent. Remember that the Yule–Walker solution with the sample auto covariance function, $\hat{\gamma}_k$ ($k = 0, 1, ...$), is equivalent to the least squares solution with an extra initial condition, that is, $x_t = 0$ for $t = 0, -1, -2, ...$, and an extra end condition, $x_t = 0$ for $t = N+1, N+2, ...$ Because of these artificial conditions, the estimated parameters are sometimes very much biased, especially when some of the characteristic roots of the AR process are close to one.

10.4.1.4 Burg Algorithm and MEM Estimate

As we saw in Section 10.4.1.2, the least squares method sometimes yields a nonstationary AR model, where the characteristic roots may not necessarily be inside the unit circle. However, the estimated parameters are known to be less biased than the Yule–Walker estimate. On the other hand, the Yule–Walker estimate is known to lead to a stationary model since the method employs a positive definite auto covariance function. However, it is known that the estimated parameter by the Yule–Walker method is rather biased compared with the least squares estimate, especially when the number of data points is small or when a very low-frequency component is included in the time series data. A method that compensates for the deficits of the two methods and implements the merits of both the methods is the so-called Burg algorithm (Burg 1967). The Burg algorithm always provides us with a stationary model with a less biased estimate.

In order to describe the Burg algorithm here, let us use the same notations used, in Section 3.2.4, for explaining partial autocorrelations, that is, $\phi_{1,1}, \ldots,$ $\phi_{k,k}$ and $\sigma_w^2(k)$, the residual variance. They are specified as a solution of Yule–Walker equations for the kth order AR model. Burg (1967) suggests estimating these parameters by minimizing the sum of the residual variances of the forward model and the backward model, that is,

$$\frac{1}{2}\left[\frac{1}{N-k}\sum_{t=k+1}^{N}\left\{\left(x_t - \phi_{k,1}x_{t-1} - \cdots - \phi_{k,k}x_{t-k}\right)^2 + \left(x_{t-k} - \phi_{k,1}x_{t-k+1} - \cdots - \phi_{k,k}x_t\right)^2\right\}\right]$$

$$=\frac{1}{2}\left(\sigma_F^2(k)+\sigma_B^2(k)\right). \tag{10.45}$$

Obviously, the key idea of the Burg algorithm is in this objective function (10.45), where the reduction of the residual variances, $\sigma_F^2(k)$ and $\sigma_F^2(k)$, is achieved at the same time, so that it can avoid the problem of biased estimates of parameters. This method is free from the extra assumptions of initial values and end values of the sample data. Note that the exact likelihood for the forward model and the exact likelihood for the backward model are equivalent, so that there is no need to consider such an ad-hoc objective function as (10.45), when we use the maximum likelihood method with the exact likelihood computation. Computationally, however, the Burg algorithm is quite efficient, and at the same time the Burg estimates have been recognized to be statistically as efficient as the exact maximum likelihood estimates. The algorithm has been used as a part of MEM (Maximum Entropy Method). which is known to be an efficient general method for estimating the power spectrum of a stationary time series.

The Burg algorithm and MEM are often found in time series textbooks for geophysicists and engineers. A point often missed in their presentations is that the Burg algorithm is valid only for a scalar AR model. As we saw in Section 4.1, the forward model and the backward model are not the same for the multivariate AR models. Therefore, Burg's objective function (10.45) cannot be used for multivariate cases.

MEM is a method of power spectrum estimation which it takes advantage of the one-to-one relationship between the power spectrum and the auto covariance of the Gaussian stationary processes. Remember that we saw in Section 3.2.4, that there are five different ways for characterizing Gaussian AR(p) process:

1. By the AR coefficients, ϕ_1, ϕ_2, ..., ϕ_p, and the noise variance σ_w^2
2. By the auto covariance function, γ_k ($k = 0, 1, 2, ...$)
3. By the power spectrum function, $p(f)$, ($-0.5 \le f \le 0.5$)
4. By the characteristic roots, λ_1, λ_2, ..., λ_p and the residual variance σ_w^2
5. By the partial auto correlations, ϕ_k, ($k = 1, 2, ..., p$), and the residual variances, $\sigma_w^2(k)$ ($k = 0, 1, 2, ..., p$)

Finding the power spectrum $p(f)$ of the process is one way of characterizing the AR(p) process, and finding the auto covariances, γ_0, γ_1, ..., γ_N, γ_{N+1}, ..., is another way. MEM suggests estimating the power spectrum of the process through the "properly" estimated auto covariance function. Here, MEM provides a logical base and a concrete method of properly fixing the rest of the auto covariances, γ_{k+1}, γ_{k+2}, ..., with the first $k + 1$ positive definite auto covariances, γ_0, γ_1, ..., γ_k.

However, what MEM suggests is nothing but the use of the parametric AR model spectrum, where the proper (positive definite) higher order lag auto covariances, γ_{k+1}, γ_{k+2}, ..., are obtained using the Burg algorithm.

MEM is computationally efficient and quite useful for estimating the power spectrum in applied time series analysis. The only problem with MEM is that people are often confused by its name, especially Maximum Entropy. The role played by Maximum Entropy is the deduction of the AR(k) model structure of the process from the given γ_0, γ_1, ..., γ_k.

Answers to more important statistical questions such as the selection of the model order and whether the parameterization in the AR model family is more efficient than MA and ARMA model families do not come from the way MEM implements the entropy. "Entropy" and "Maximum" are used differently in Akaike's approach to the statistical model identification, which we will see in Section 10.5.

10.4.2 Least Squares Method for Multivariate AR Models

As we saw in Section 10.3.2.3, the multivariate r-dimensional AR(p) model,

$$x_t = A_1 x_{t-1} + \cdots + A_p x_{t-p} + w_t,$$

can be estimated, in principle, by the maximum likelihood. The maximum likelihood method with the exact likelihood needs a numerical iterative optimization method. This is time-consuming, especially for high-order and high-dimensional AR models.

A method that is computationally much more efficient than the maximum likelihood method is the multivariate version of the Yule–Walker method, which is given in the following section.

10.4.2.1 Yule–Walker Estimates

The multivariate version of the Yule–Walker method is computationally much more efficient than the maximum likelihood method. It is defined by the following two stages:

1. First, estimate the sample auto covariances, $\hat{\Gamma}_1 \dots \hat{\Gamma}_p$, from the sample data, $x_1, x_2, \dots, x_N$, by

$$\hat{\Gamma}_k = \frac{1}{N} \sum_{t=1}^{N-k} x_{t+k} x_t'. \quad (k = 0, 1, \dots, p). \tag{10.46}$$

2. Then the Yule–Walker estimates of the parameters, $A_1, \dots, A_p$ and Σ_w, are obtained by solving the multivariate version of the Yule–Walker equation (see Section 4.1.2) together with the estimated auto covariances, as

$$\left(\hat{A}_1 \quad \dots \quad \hat{A}_p \right) = \left(\hat{\Gamma}_1 \quad \dots \quad \hat{\Gamma}_p \right) \begin{pmatrix} \hat{\Gamma}_0 & \dots & \hat{\Gamma}_{p-1} \\ \dots & \dots & \dots \\ \hat{\Gamma}_{-p+1} & \dots & \hat{\Gamma}_0 \end{pmatrix}^{-1} \tag{10.47}$$

and

$$\hat{\Sigma}_W = \hat{\Gamma}_0 - \sum_{j=1}^{p} \hat{A}_j \hat{\Gamma}_j'.$$

Note that the estimated auto covariance function by (10.46) is positive definite, so that the estimated model specified by the estimated parameters is always stationary, and the model has a parametric spectral representation that plays an important role in the causality analysis between the variables of the multivariate time series (see Chapter 14).

10.4.2.2 Levinson-Whittle Estimate

The Levinson-Whittle method is another computationally efficient method for the estimation of parameters of the multivariate AR model. The method is based on the recursive properties of the parameters defined by the auto covariances of the AR process, which we saw in Section 4.1.5. The method consists of the following two stages:

1. The sample auto covariances, $\hat{\Gamma}_0$, $\hat{\Gamma}_1$, ..., $\hat{\Gamma}_{n-1}$, are estimated from the data by

$$\hat{\Gamma}_k = \frac{1}{N} \sum_{t=1}^{N-k} x_{t+k}\, x_t', \quad (k = 0, 1 \ldots, p_{n-1}).$$

2. Then the Levinson-Whittle estimates of the parameters are given by using the recursive scheme (4.15) of Section 4.1.5, as

$$\hat{F}_k^{(n+1)} = \hat{F}_k^{(n)} - \hat{\Sigma}_{V_n}^{-1} \hat{G}_n \hat{B}_{n+1-k}^{(n)} \quad k = 1, \ldots, n$$

$$\hat{F}_{n+1}^{(n+1)} = -\hat{\Sigma}_{V_n}^{-1} \hat{G}_n$$

$$\hat{B}_k^{(n+1)} = \hat{B}_k^{(n)} - \hat{\Sigma}_{W_n}^{-1} \hat{H}_n \hat{F}_{n+1-k}^{(n)} \quad k = 1, \ldots, n$$

$$\hat{B}_{n+1}^{(n+1)} = -\hat{\Sigma}_{W_n}^{-1} \hat{H}_n,$$

where

$$\hat{G}_n = \hat{\Gamma}_{-n-1} + \hat{\Gamma}_{-n} \hat{F}_1^{(n)} + \cdots + \hat{\Gamma}_{-1} \hat{F}_n^{(n)}$$

$$\hat{H}_n = \hat{\Gamma}_{n+1} + \hat{\Gamma}_n \hat{B}_1^{(n)} + \cdots + \hat{\Gamma}_1 \hat{B}_n^{(n)}$$

$$\hat{\Sigma}_{W_n} = \hat{\Gamma}_0 + \hat{\Gamma}_1 \hat{F}_1^{(n)} + \cdots + \hat{\Gamma}_n \hat{F}_n^{(n)}$$

$$\hat{\Sigma}_{V_n} = \hat{\Gamma}_0 + \hat{\Gamma}_1 \hat{B}_1^{(n)} + \cdots + \hat{\Gamma}_n \hat{B}_n^{(n)}$$

with initial values

$$\hat{F}^{(0)} = \hat{B}^{(0)} = I, \quad \hat{\Sigma}_{W_0} = \hat{\Sigma}_{V_0} = \hat{\Gamma}_0, \hat{G}_0 = \hat{\Gamma}_{-1}, \quad \text{and} \quad \hat{H}_0 = \hat{\Gamma}_1.$$

Here we remember (see Section 4.1.5) that $F_k^{(n)}$ ($k = 1, \ldots, n$) are forward filters such that

$$X(t) + \sum_{i=1}^{n} (F_i^{(n)})' X(t - i) = W_n(t),$$

so that $F_k^{(n)} = -A_k'$, $(k=1, \ldots, n)$, by definition, for the (n)th order multivariate AR model,

$$X(t) = A_1 X_{t-1} + \cdots A_n X_{t-n} + W_n(t).$$

As long as the estimated auto covariance function is eligible, that is, positive definite, the derived model defines a stationary AR process.

10.4.2.3 Least Squares Estimates

Multivariate (r-dimensional) AR modeling has been used first in engineering applications since the late 1960s. In the 1960s and 1970s, the computer power was not as strong as today's, so the recursive Levinson scheme was used for the estimation of models instead of the least squares method or quasi maximum likelihood method. They were successfully used mostly in areas where the number of data points are large. Although the scheme yields eligible stationary solutions, a disadvantage is that when the data length is short, it leads to biased estimates because of the initial value problems.

The likelihood of a multivariate AR model is given by (see Section 10.3.2.3)

$$(-2)\log p(x_1, x_2, \ldots, x_N | A_1, \ldots, A_p, \Sigma_w)$$

$$= (-2)\log p(x_{p+1}, x_{p+2}, \ldots, x_N | x_1, \ldots, x_p, A_1, \ldots, A_p, \Sigma_w)$$

$$+ (-2)\log p(x_1, \ldots, x_p | A_1, \ldots, A_p, \Sigma_w)$$

$$\approx (-2)\log p(x_{p+1}, x_{p+2}, \ldots, x_N | x_1, \ldots, x_p, A_1, \ldots, A_p, \Sigma_w)$$

$$= (-2)\log p(w_{p+1}, w_{p+2}, \ldots, w_N | x_1, \ldots, x_p, A_1, \ldots, A_p, \Sigma_w).$$

When the number of time points of the data is large, the contribution of $(-2)\log p(x_1, \ldots, x_p | A_1, \ldots, A_p, \Sigma_w)$ to the (-2)log-likelihood is small, and we can use the approximate (-2)log-likelihood,

$$(-2)\log p(x_{p+1}, x_{p+2}, \ldots, x_N | x_1, \ldots, x_p, A_1, \ldots, A_p, \Sigma_w)$$

$$= (-2)\log p(w_{p+1}, w_{p+2}, \ldots, w_N | x_1, \ldots, x_p, A_1, \ldots, A_p, \Sigma_w)$$

$$= (N-p)\log|\Sigma_w| + \sum_{t=p+1}^{N} w_t' \Sigma_w^{-1} w_t + (N-p)r \log 2\pi.$$

Here, the initial values are fixed and their distributions are neglected. Then, maximization of the approximate likelihood of the r-dimensional AR model is equivalent to minimization of

$$(N-p)\log|\Sigma_w| + \sum_{t=p+1}^{N} w_t' \Sigma_w^{-1} w_t,$$

which is equivalent to minimization of $|C_w|$, where

$$C_w = \frac{1}{(N-p)} \sum_{t=p+1}^{N} (x_t - A_1 x_{t-1} - \cdots - A_p x_{t-p})(x_t - A_1 x_{t-1} - \cdots - A_p x_{t-p})'.$$

This method is a least squares method and is known to have a much smaller initial effect and to be not so biased compared with the estimate obtained by the multivariate version of the Yule–Walker method.

The minimization of $|C_w|$ in terms of the parameters, that is, all the elements $a_{i,j}^{(k)}$ $(i, j = 1, \ldots, r)$ of A_k $(k = 1, \ldots, p)$, is not as simple a numerical task as for the scalar case. Several approximate but computationally efficient methods have been introduced for calculating the least squares estimates of the multivariate AR model.

One simple idea is to treat each variable of the r-variate time series as a scalar time series with $(r-1)$ exogenous variables and minimize the sum of squares of the residuals of the scalar ARX model channel by channel, which is computationally much less demanding than the aforementioned orthodox least squares method, where the determinant of the multivariate residuals is minimized.

The limitation of this method is that the method implicitly assumes that the residual $w_t = (w_t^{(1)}, w_t^{(2)}, \ldots, w_t^{(r)})'$ have a diagonal variance–covariance matrix, $E[w_t w_t'] = \Sigma_w$, which is not very likely in the real-world of time series data. A heuristic method for taking care of this problem of instantaneous correlation between $w_t^{(i)}$ and $w_t^{(j)}$ has been presented (see Geweke 1982) by introducing extra terms such as $c x_t^{(i)}$ for the prediction of $x_t^{(j)}$.

1. *Example 1* 2-dimensional case

   For example, with a simple 2-dimensional first-order AR model, for a 2-dimensional time series, $(x_t \; y_t)'$ $(t = 1, 2, \ldots, N)$, we can think of estimating the coefficients, a_{11} and a_{12}, by minimizing the sum of squares of $\delta_t(x) = x_t - (a_{1,1} x_{t-1} + a_{1,2} y_{t-1})$ and estimating a_{21} and a_{22} by minimizing the sum of squares of $\delta_t(y) = y_t - (c x_t + a_{2,1} x_{t-1} + a_{2,2} y_{t-1})$, separately. Here we consider the following equations, with the assumption of the mutual independence of the noises $\delta_t(x)$ and $\delta_t(y)$.

   $$\begin{aligned} x_t &= a_{11} x_{t-1} + a_{12} y_{t-1} + \delta_t^{(x)} \\ y_t &= c x_t + a_{21} x_{t-1} + a_{22} y_{t-1} + \delta_t^{(y)}. \end{aligned} \tag{10.48}$$

   This is equivalent to assuming the following parametric bivariate AR model,

   $$\begin{pmatrix} x_t \\ y_t \end{pmatrix} = \begin{pmatrix} 1 & 0 \\ -c & 1 \end{pmatrix}^{-1} \begin{pmatrix} a_{11} & a_{12} \\ a_{21} & a_{22} \end{pmatrix} \begin{pmatrix} x_{t-1} \\ y_{t-1} \end{pmatrix} + \begin{pmatrix} 1 & 0 \\ -c & 1 \end{pmatrix}^{-1} \begin{pmatrix} \delta_t^{(x)} \\ \delta_t^{(y)} \end{pmatrix}, \tag{10.49}$$

instead of the usual bivariate AR model,

$$x_t = b_{11}x_{t-1} + b_{12}y_{t-1} + w_t^{(x)}$$
$$y_t = b_{21}x_{t-1} + b_{22}y_{t-1} + w_t^{(y)}, \qquad (10.50)$$

for the bivariate time series, $(x_t\, y_t)'$ $(t = 1, 2, ..., N)$. The model (10.49) is rewritten as

$$
\begin{pmatrix} x_t \\ y_t \end{pmatrix} = \begin{pmatrix} 1 & 0 \\ -c & 1 \end{pmatrix}^{-1} \begin{pmatrix} a_{11} & a_{12} \\ a_{21} & a_{22} \end{pmatrix} \begin{pmatrix} x_{t-1} \\ y_{t-1} \end{pmatrix} + \begin{pmatrix} 1 & 0 \\ -c & 1 \end{pmatrix}^{-1} \begin{pmatrix} \delta_t^{(x)} \\ \delta_t^{(y)} \end{pmatrix}
$$

$$
= \begin{pmatrix} 1 & 0 \\ c & 1 \end{pmatrix} \begin{pmatrix} a_{11} & a_{12} \\ a_{21} & a_{22} \end{pmatrix} \begin{pmatrix} x_{t-1} \\ y_{t-1} \end{pmatrix} + \begin{pmatrix} 1 & 0 \\ c & 1 \end{pmatrix} \begin{pmatrix} \delta_t^{(x)} \\ \delta_t^{(y)} \end{pmatrix}
$$

$$
= \begin{pmatrix} a_{11} & a_{12} \\ a_{21} + ca_{11} & a_{22} + ca_{12} \end{pmatrix} \begin{pmatrix} x_{t-1} \\ y_{t-1} \end{pmatrix} + \begin{pmatrix} 1 & 0 \\ c & 1 \end{pmatrix} \begin{pmatrix} \delta_t^{(x)} \\ \delta_t^{(y)} \end{pmatrix}. \qquad (10.51)
$$

Although we assume that the noises $\delta_t(x)$ and $\delta_t(y)$ are not correlated, the driving noise for x_t and the driving noise for y_t can be correlated because of the presence of the parameter c in (10.51). On the other hand, the ordinary AR model is parameterized as

$$
\begin{pmatrix} x_t \\ y_t \end{pmatrix} = \begin{pmatrix} b_{11} & b_{12} \\ b_{21} & b_{22} \end{pmatrix} \begin{pmatrix} x_{t-1} \\ y_{t-1} \end{pmatrix} + \begin{pmatrix} w_t^{(x)} \\ w_t^{(y)} \end{pmatrix},
$$

where the variance matrix of the bivariate residual, $(w_t^{(x)}, w_t^{(y)})'$, is not necessarily diagonal. Here we do not expect $E[w_t^{(x)}w_t^{(y)}] = 0$, and we have an extra parameter characterizing the correlation between $w_t^{(x)}$ and $w_t^{(y)}$.

2. *Example 2* 3-dimensional AR model
   For the 3-dimensional AR(p) model for the 3-dimensional time series, $X_t = (x_t, y_t, z_t)'$, we can think of the coefficients of the following linear ARX models for each series, $x_1, x_2, ..., x_N, y_1, y_2, ..., y_N$ and $z_1, z_2, ..., z_N$, by minimizing the sum of squares of the prediction errors of each variable separately:

$$
\begin{aligned}
x_t = {} & a_{1,1}x_{t-1} + \cdots + a_{1,p}x_{t-p} \\
& + b_{1,1}y_{t-1} + \cdots + b_{1,p}y_{t-p} \\
& + c_{1,1}z_{t-1} + \cdots + c_{1,p}z_{t-p} \\
& + \varepsilon_t^{(x)}
\end{aligned}
$$

$$y_t = b_{2,1}y_{t-1} + \cdots + b_{2,p}y_{t-p}$$
$$+ c_{2,1}z_{t-1} + \cdots + c_{2,p}z_{t-p}$$
$$+ a_{2,1}x_{t-1} + \cdots + a_{2,p}x_{t-p}$$
$$+ \theta_{2.1}x_t + \varepsilon_t^{(y)}$$

$$z_t = c_{3,1}z_{t-1} + \cdots + c_{3,p}z_{t-p}$$
$$+ a_{3,1}x_{t-1} + \cdots + a_{3,p}x_{t-p}$$
$$+ b_{3,1}y_{t-1} + \cdots + b_{3,p}y_{t-p}$$
$$+ \theta_{3.1}x_t + \theta_{3.2}y_t + \varepsilon_t^{(z)}.$$

With $X_t = (x_t, y_t, z_t)'$, $\varepsilon_t = (\varepsilon_t^{(x)}, \varepsilon_t^{(y)}, \varepsilon_t^{(z)})'$

and

$$\theta = \begin{pmatrix} 0 & 0 & 0 \\ \theta_{2,1} & 0 & 0 \\ \theta_{3,1} & \theta_{3,2} & 0 \end{pmatrix}, \quad \alpha_1 = \begin{pmatrix} a_{1,1} & b_{1,1} & c_{1,1} \\ a_{2,1} & b_{2,1} & c_{2,1} \\ a_{3,1} & b_{3,1} & c_{3,1} \end{pmatrix}, \quad \ldots, \quad \alpha_p = \begin{pmatrix} a_{1,p} & b_{1,p} & c_{1,p} \\ a_{2,p} & b_{2,p} & c_{2,p} \\ a_{3,p} & b_{3,p} & c_{3,p} \end{pmatrix},$$

we rewrite the 3 scalar ARX models by a single 3-dimensional AR(p) model as

$$X_t = \theta X_t + \alpha_1 X_{t-1} + \cdots + \alpha_p X_{t-p} + \varepsilon_t,$$

which can be transformed to

$$X_t = A_1 X_{t-1} + \cdots + A_p X_{t-p} + w_t,$$

where

$$A_i = (I - \theta)^{-1}\alpha_i \quad (i = 1, 2, \ldots, p)$$
$$w_t = (I - \theta)^{-1}\varepsilon_t.$$

Here the variance of the noise w_t is not diagonal and is given by

$$E[w_t w_t'] = (I - \theta')^{-1}\Sigma_\delta(I - \theta')^{-1}.$$

10.4.3 The Least Squares Method for Multivariate ARX Models

In feedback systems modeling, some exogenous input variables are included only for the objective of improving the prediction and control of some important variables. For example if the two exogenous variables u_t and v_t are known to be useful for the prediction of the three main variables, x_t, y_t, and

z_t, we can think of a 3-dimensional ARX model with two exogenous variables, u_t and v_t, for the modeling of the 3-dimensional time series as follows:

$$
\begin{pmatrix} x_t \\ y_t \\ z_t \end{pmatrix} = \begin{pmatrix} \phi_{1,1}^{(1)} & \phi_{1,2}^{(1)} & \phi_{1,3}^{(1)} \\ \phi_{2,1}^{(1)} & \phi_{2,2}^{(1)} & \phi_{2,3}^{(1)} \\ \phi_{3,1}^{(1)} & \phi_{3,2}^{(1)} & \phi_{3,3}^{(1)} \end{pmatrix} \begin{pmatrix} x_{t-1} \\ y_{t-1} \\ z_{t-1} \end{pmatrix} + \begin{pmatrix} \phi_{1,1}^{(2)} & \phi_{1,2}^{(2)} & \phi_{1,3}^{(2)} \\ \phi_{2,1}^{(2)} & \phi_{2,2}^{(2)} & \phi_{2,3}^{(2)} \\ \phi_{3,1}^{(2)} & \phi_{3,2}^{(2)} & \phi_{3,3}^{(2)} \end{pmatrix} \begin{pmatrix} x_{t-2} \\ y_{t-2} \\ z_{t-2} \end{pmatrix} + \begin{pmatrix} \phi_{1,1}^{(3)} & \phi_{1,2}^{(3)} & \phi_{1,3}^{(3)} \\ \phi_{2,1}^{(3)} & \phi_{2,2}^{(3)} & \phi_{2,3}^{(3)} \\ \phi_{3,1}^{(3)} & \phi_{3,2}^{(3)} & \phi_{3,3}^{(3)} \end{pmatrix} \begin{pmatrix} x_{t-3} \\ y_{t-3} \\ z_{t-3} \end{pmatrix}
$$

$$
+ \begin{pmatrix} \psi_{1,1}^{(1)} & \psi_{1,2}^{(1)} \\ \psi_{2,1}^{(1)} & \psi_{2,2}^{(1)} \\ \psi_{3,1}^{(1)} & \psi_{3,2}^{(1)} \end{pmatrix} \begin{pmatrix} u_{t-1} \\ v_{t-1} \end{pmatrix} + \begin{pmatrix} \psi_{1,1}^{(2)} & \psi_{1,2}^{(2)} \\ \psi_{2,1}^{(2)} & \psi_{2,2}^{(2)} \\ \psi_{3,1}^{(2)} & \psi_{3,2}^{(2)} \end{pmatrix} \begin{pmatrix} u_{t-2} \\ v_{t-2} \end{pmatrix} + \begin{pmatrix} w_t^{(x)} \\ w_t^{(y)} \\ w_t^{(z)} \end{pmatrix}, \tag{10.52}
$$

where $w_t = (w_t^{(x)}, w_t^{(y)}, w_t^{(z)},)'$ is a Gaussian white noise whose variance matrix could be nondiagonal. Here the dynamics of the exogenous variable are not of our interest.

This type of multivariate ARX model for linear feedback systems can be estimated efficiently by the channel-wise least squares method discussed here for the multivariate AR models, and the least squares estimate is regarded as an approximation to the maximum likelihood estimate of the multivariate ARX model. The difference is that the ARX model has extra terms of exogenous variables added to the case of the AR model. The 3-dimensional ARX model (10.52), for example, yields the 3 independent least squares problems, with the following regression models:

$$
x_t = \phi_{1,1}^{(1)} x_{t-1} + \phi_{1,1}^{(2)} x_{t-2} + \phi_{1,1}^{(3)} x_{t-3}
$$
$$
+ \phi_{1,2}^{(1)} y_{t-1} + \phi_{1,2}^{(2)} y_{t-2} + \phi_{1,2}^{(3)} y_{t-3} + \phi_{1,3}^{(1)} z_{t-1} + \phi_{1,3}^{(2)} z_{t-2} + \phi_{1,3}^{(3)} z_{t-3}
$$
$$
+ \psi_{1,1}^{(1)} u_{t-1} + \psi_{1,1}^{(2)} u_{t-2} + \psi_{1,2}^{(1)} v_{t-1} + \psi_{1,2}^{(2)} v_{t-2} + \varepsilon_t^{(x)} \tag{10.53}
$$

$$
y_t = \phi_{2,2}^{(1)} y_{t-1} + \phi_{2,2}^{(2)} y_{t-2} + \phi_{2,2}^{(3)} y_{t-3}
$$
$$
+ \phi_{2,3}^{(1)} z_{t-1} + \phi_{2,3}^{(2)} z_{t-2} + \phi_{2,3}^{(3)} z_{t-3} + \phi_{2,1}^{(1)} x_{t-1} + \phi_{2,1}^{(2)} x_{t-2} + \phi_{2,1}^{(3)} x_{t-3}
$$
$$
+ \psi_{2,1}^{(1)} u_{t-1} + \psi_{2,1}^{(2)} u_{t-2} + \psi_{2,2}^{(1)} v_{t-1} + \psi_{2,2}^{(2)} v_{t-2}
$$
$$
+ \theta_{2,1} x_t + \varepsilon_t^{(y)} \tag{10.54}
$$

$$
z_t = \phi_{3,3}^{(1)} z_{t-1} + \phi_{3,3}^{(2)} z_{t-2} + \phi_{3,3}^{(3)} z_{t-3}
$$
$$
+ \phi_{3,1}^{(1)} x_{t-1} + \phi_{3,1}^{(2)} x_{t-2} + \phi_{3,1}^{(3)} x_{t-3} + \phi_{3,2}^{(1)} y_{t-1} + \phi_{3,2}^{(2)} y_{t-2} + \phi_{3,2}^{(3)} y_{t-3}
$$
$$
+ \psi_{3,1}^{(1)} u_{t-1} + \psi_{3,1}^{(2)} u_{t-2} + \psi_{3,2}^{(1)} v_{t-1} + \psi_{3,2}^{(2)} v_{t-2}
$$
$$
+ \theta_{3,1} x_t + \theta_{3,2} y_t + \varepsilon_t^{(z)}. \tag{10.55}
$$

Here, $\sum_{t=4}^{N}(\varepsilon_t^{(x)})^2$, $\sum_{t=4}^{N}(\varepsilon_t^{(y)})^2$ and $\sum_{t=4}^{N}(\varepsilon_t^{(z)})^2$ are minimized independently.

Then with the estimated coefficients, we can form the 3-variate ARX model as

$$
\begin{pmatrix} x_t \\ y_t \\ z_t \end{pmatrix} = A_1 \begin{pmatrix} x_{t-1} \\ y_{t-1} \\ z_{t-1} \end{pmatrix} + A_2 \begin{pmatrix} x_{t-2} \\ y_{t-2} \\ z_{t-2} \end{pmatrix} + A_3 \begin{pmatrix} x_{t-3} \\ y_{t-3} \\ z_{t-3} \end{pmatrix} + B_1 \begin{pmatrix} u_{t-1} \\ v_{t-1} \end{pmatrix} + B_2 \begin{pmatrix} u_{t-2} \\ v_{t-2} \end{pmatrix} + \begin{pmatrix} \xi_t \\ \eta_t \\ \varsigma_t \end{pmatrix}, \qquad (10.56)
$$

with the nondiagonal variance matrix Σ_n of the noise $w_t = (\xi_t, \eta_t, \varsigma_t)' = (I - \theta)^{-1}(\varepsilon_t^{(x)}, \varepsilon_t^{(y)}, \varepsilon_t^{(z)})'$ given by

$$
\Sigma_w = (I - \theta)^{-1} \begin{pmatrix} \sigma_{\varepsilon^{(x)}}^2 & 0 & 0 \\ 0 & \sigma_{\varepsilon^{(y)}}^2 & 0 \\ 0 & 0 & \sigma_{\varepsilon^{(z)}}^2 \end{pmatrix} (I - \theta')^{-1},
$$

$$
\theta = \begin{pmatrix} 0 & 0 & 0 \\ \theta_{21} & 0 & 0 \\ \theta_{31} & \theta_{32} & 0 \end{pmatrix}.
$$

The coefficient matrices, A_1, A_2, A_3, B_1 and B_2, are related to the coefficient matrices of (10.56) by

$$
\theta = \begin{pmatrix} 0 & 0 & 0 \\ \theta_{2,1} & 0 & 0 \\ \theta_{3,1} & \theta_{3,2} & 0 \end{pmatrix}, \quad
A_1 = (I - \theta)^{-1} \begin{pmatrix} \phi_{1,1}^{(1)} & \phi_{1,2}^{(1)} & \phi_{1,3}^{(1)} \\ \phi_{2,1}^{(1)} & \phi_{2,2}^{(1)} & \phi_{2,3}^{(1)} \\ \phi_{3,1}^{(1)} & \phi_{3,2}^{(1)} & \phi_{3,3}^{(1)} \end{pmatrix}, \quad
A_2 = (I - \theta)^{-1} \begin{pmatrix} \phi_{1,1}^{(2)} & \phi_{1,2}^{(2)} & \phi_{1,3}^{(2)} \\ \phi_{2,1}^{(2)} & \phi_{2,2}^{(2)} & \phi_{2,3}^{(2)} \\ \phi_{3,1}^{(2)} & \phi_{3,2}^{(2)} & \phi_{3,3}^{(2)} \end{pmatrix},
$$

$$
A_3 = (I - \theta)^{-1} \begin{pmatrix} \phi_{1,1}^{(3)} & \phi_{1,2}^{(3)} & \phi_{1,3}^{(3)} \\ \phi_{2,1}^{(3)} & \phi_{2,2}^{(3)} & \phi_{2,3}^{(3)} \\ \phi_{3,1}^{(3)} & \phi_{3,2}^{(3)} & \phi_{3,3}^{(3)} \end{pmatrix}, \quad
B_1 = (I - \theta)^{-1} \begin{pmatrix} \psi_{1,1}^{(1)} & \psi_{1,2}^{(1)} \\ \psi_{2,1}^{(1)} & \psi_{2,2}^{(1)} \\ \psi_{3,1}^{(1)} & \psi_{3,2}^{(1)} \end{pmatrix}, \quad
B_2 = (I - \theta)^{-1} \begin{pmatrix} \psi_{1,1}^{(2)} & \psi_{1,2}^{(2)} \\ \psi_{2,1}^{(2)} & \psi_{2,2}^{(2)} \\ \psi_{3,1}^{(2)} & \psi_{3,2}^{(2)} \end{pmatrix}.
$$

This method is easily extended to the nonlinear multivariate RBF-ARX models, which have been applied in modeling and controlling complex systems such as power plant control systems (Toyoda et al. 1997, Peng et al. 2003, Haggen–Ozaki et al. 2009).

10.5 Log-Likelihood and the Boltzmann Entropy

So far we have seen that the innovation approach provides us with a useful guideline for developing estimation methods, such as the (exact or approximate) maximum likelihood method, least squares method, and moment method, for the estimation of the model parameters.

Among these estimation methods, the advantage of maximum likelihood estimates, in terms of asymptotic efficiency, is widely known both to theoretical statisticians (Bayesians or frequentists) and applied statisticians. A problem here is that none of these estimation methods provides us with an answer to the question of which model structure to choose from so many possible prediction models available for one and the same time series data. An answer to such a practical question as "which type of nonlinear model, chaos model or neural network model, should we choose for a given time series data?" has not been seriously considered in the traditional framework of statistics.

From time to time, the model identification problem has been treated in some time series textbooks after the 1980s, but it is discussed in a much more limited sense such as the order determination problem of linear AR models under the very unrealistic assumption that the true model is a finite-order AR model. Obviously, if an assumption behind the theorem is unrealistic, none of these theoretical results are useful in applications, no matter how rigorously theorems are proved.

Unfortunately, comparing the goodness of different types of prediction models other than linear AR models, under an assumption that the true model is not in our candidate model structures, has been out of question for most statisticians. Comparing the maximized log-likelihood values of different types of model structures was an odd practice in traditional statistics until Akaike (1973a, 1974c, 1985). However, if we are concerned about the prediction problem, it is quite natural to pay attention and compare the prediction performance (i.e., the estimated prediction error variance) of different models even though their model structure is different. Certainly, a smaller prediction model looks preferable for most users of prediction models.

In 1973, after his intensive study and attempt (Akaike 1969, 1971) to find an optimal way of order determination of multivariate AR models for designing a predictive control system for engineering processes, Akaike introduced a new paradigm for the statistical model identification problem in general, where the true model structure is unknown and it is not necessarily included in the set of candidate models we are trying to estimate. A principle that plays an essential role in the new paradigm is called the "Entropy Maximization Principle."

10.5.1 Akaike's Entropy Maximization Principle

Akaike's *Entropy Maximization Principle (E.M.P.)* (Akaike 1977, p. 30, 1985) suggests solving the model identification problem as follows:

"Formulate the object of statistical inference as the estimation of a probability density distribution $q(x)$ from the data z and try to find $p(x|\theta(x))$ which will maximize the expected entropy,

$$E_z B(q; p) = \int B(q; p) q(z) dz."$$

Here $B(q; p)$ is given by

$$B(q; p) = -\int q(x) \log \frac{q(x)}{p(x|\theta(x))} dx$$

$$= \int q(x) \log p(x|\theta(z)) dx - \int q(x) \log q(x) dx$$

and is called the Boltzmann entropy.

When an observation z of x is obtained, $\log p(x|\theta(z))$ provides a natural estimate of

$$\int q(x) \log p(x|\theta(z)) dx.$$

When $p(x|\theta(z))$ is our only candidate model, the maximum likelihood model, $p(x|\hat{\theta}(z))$, is chosen as the best estimated model from the E.M.P. (Entropy Maximization Principle). $\hat{\theta}(z)$ is such that it gives

$$\underset{\theta}{\text{Max}}\, p(z|\theta(z)).$$

The true density $q(x)$ is always unknown except for artificial simulation data, but we can always obtain a solution, $p(x|\hat{\theta}(z))$, without specifying $q(x)$. This shows that the Entropy Maximization Principle provides us with an objective justification of the use of likelihood as a "measure of rational belief," which was advocated by Fisher (1956).

When we have several candidate models, we can choose the model that gives the largest maximum likelihood among the candidate models. If only we have models presented to specify the data distribution, it is possible to compare the likelihood of non-nested models or of Gaussian models and non-Gaussian models. The entropy principle applies to any statistical model specifying the distribution of the observed data. Hence, the goodness of Bayes models with several different priors can be assessed on the same basis as other non-Bayesian statistical models.

Although the maximum likelihood model gives a simple and useful estimate of

$$\int q(x) \log p(x|\theta(z)) dx,$$

and hence of $E_x B(q; p)$, it sometimes presents a very important practical problem to data analysis. In statistical modeling, when a model is modified into a more sophisticated one with more parameters it tends to yield a larger likelihood. A typical example is the AR (autoregressive) modeling of time series where all the models, from zero-order model to $(n - 1)$-order model, are nested. The highest order AR model always gives the largest likelihood value. We cannot avoid this problem when we try to identify a parametric model for real data. This problem, however, can be solved when we introduce a more sophisticated estimate of $E_x B(q;p)$. The log-likelihood is known to be a rather crude estimate of $E_x B(q;p)$ and has a bias. Taking account of this bias, Akaike (1973a) first presented the criterion AIC given by

$$AIC = (-2)\log(\text{Maximized likelihood}) + 2k$$

as an unbiased estimate of $(-2)E_x B(q;p)$. Here $k = \dim(\theta)$ is the number of independent parameters within the model $p(x|\theta)$ to be fitted to the data and adjusted to attain the maximum of the likelihood.

AIC is a function including k, the number of parameters, as well as N, the number of observations. By choosing a model with smaller AIC, an EMP model may be found even when some of our candidate models are non-nested. Hence, the EMP could be regarded as a quantitative implementation of Occam's razor or the Principle of parsimony in statistical model identification. Other refinements of the estimate of $(-2)E_x B(q;p)$ were introduced by Ishiguro et al. (1994) and Konishi and Kitagawa (1996). However, we have to be careful in applications that the "refinement of the criterion" does not compensate for the "shortage" of the models we were thinking to apply to the data. Inventing and developing new efficient model structures is the only way to solve difficult problems in time series analysis in practice.

10.5.2 Prevailing Confusions about the Entropy and AIC

Because of its great strength in applications, the EMP has been widely accepted and used by many scientists in many inferential problems, inspiring experimental scientists to further study in many areas. However, several sources of confusions are seen among some statisticians and applied scientists. Confusion seems to come from the very meaning of entropy, which goes back to Boltzmann and Gibbs in statistical mechanics. Boltzmann (1877) introduced the well-known H-function,

$$H(p) = -\sum_{i=1}^{k} p_i \log p_i,$$

as the "logarithm of the probability of obtaining the distribution $p = \{p_1, p_2, ..., p_k\}$ from a uniform distribution." Gibbs (1902) introduced a function, called the Gibbs' measure, that is,

$$\Sigma = \int p(x)\log p(x)dx,$$

as a measure for the deviation of a given distribution from a canonical distribution. Gibb's measure appears similar to Bottzmann's *H*-function but the difference between the two forms are not so small as they appear. Shannon later introduced

$$H(p) = -\sum_{i=1}^{k} p_i \log p_i$$

as a measure of uncertainty in communication theory (Shannon (1948)). Both Shannon's and Gibbs' measures lost the original probabilistic meaning of Boltzmann's entropy that is, "the logarithm of probability of obtaining the distribution $p = \{p_0, p_1, p_2, ..., p_k\}$ from a uniform distribution."

Let us see how Boltzmann obtained the H-function from the probabilistic consideration of the energy state of molecules of gas. He specified the energy state of a gas, with a total number of molecules n and the total energy of gas λ, by $\{n_0, n_1, ..., n_k\}$. Since the total number of molecules in the gas is n, it holds that $n_0 + n_1 +, ..., + n_k = n$ and since the number of molecules having kinetic energy $i\varepsilon$ is n_i, it holds that $n_1 + 2n_2 +, ..., + kn_k = \lambda$.

Boltzmann said that, when the total energy $\lambda\varepsilon$ is allocated equally to 0ε, ε, 2ε, ..., $k\varepsilon$, the probability of obtaining the distribution $\{n_0, n_1, ..., n_k\}$ is

$$W = Q/J,$$

where J is the number of all the possible ways of allocating the energy, and Q is the number of all possible ways of allocation yielding $\{n_0, n_1, ..., n_k\}$, that is,

$$Q = \frac{n!}{n_0! n_1! ... n_k!}.$$

Boltzmann pointed out that the state that maximizes W is the most likely distribution of the states in thermodynamic entropy.

Using Stirling's formula

$$\log n_i! = n_i \log n_i - n_i$$

with the assumption that n_i is sufficiently large and $p_i = n_i/n$, we obtain

$$\log Q = n\left(-\sum_{k=0}^{k} p_i \log p_i\right),$$

hence Boltzmann's *H*-function.

10.5.3 Einstein's Inductive Use of Boltzmann Entropy

T. and P. Ehrenfest (1912, p. 77) refered to the work of Boltzmann's Entropy used by Einstein, as follows:

> ...the trick which Einstein uses systematically deserves special atten-
> tion. He retains the relationship:
> Entropy=Logarithm of the "probability"

and then, reversing Boltzmann's procedure, he calculates the relative "probability" of two states from the experimentally determined values of the entropy. In this way he calculates —always appropriately adapting the meaning of "probability" — the average values, in time or otherwise, of the parameters of the state of a system. In those applications where we know from experience that there is a violation of the theorem of the equipartition of kinetic energy, this procedure goes essentially beyond the range of validity of the methods of Boltzmann."

This remark by T. and P. Ehrenfest (1912) shows that Einstein was using essentially the same idea as the maximum likelihood method, where he uses the Boltzmann entropy in an inductive way, that is, in an opposite direction to the deduction. In the succeeding part of the article of T. and P. Ehrenfest (1912), we can see how they and some of the physicists at the time were excited about the potential and possible impact of the present inductive method on the science. The modern paradigm of statistical modeling by Akaike, which was built on the fusion of the ideas of Boltzmann and Fisher, is a realization of what these physicists dreamed at the beginning of the twentieth century.

Akaike (1977) revised the original meaning of Boltzmann's entropy in a more general situation where the uniform distribution of energy is replaced by a general distribution $q = \{q_1, q_2, ..., q_k\}$.

The probability of getting $n = \{n_0, n_1, n_2, ..., n_k\}$ sampled from the distribution $q = \{q_0, q_1, q_2, ..., q_k\}$, instead of the uniform distribution, is

$$R = \frac{n!}{n_0!n_1!...n_k!} q_0^{n_0} q_1^{n_1} ... q_k^{n_k}.$$

If we use Stirling's formula with $p_i = n_i/n$, we have

$$\log R = N\left[-\sum_{i=0}^{k} p_i \log\left(\frac{p_i}{q_i}\right) \right],$$

which leads to

$$B(q;p) = -\sum_i p_i \log\left(\frac{p_i}{q_i}\right).$$

Akaike (1977) called this quantity Boltzmann entropy to acknowledge his original contribution. The continuous space version of the Boltzmann entropy is given by

$$B(q;p) = -\int p(x)\log\left(\frac{p(x)}{q(x)}\right)dx.$$

The importance of the probabilistic meaning of Boltzmann's entropy has been pointed out by many distinguished physicists such as Max Planck (1904), Albert Einstein (1905), and P. and T. Ehrenfest (1912). Actually, it was Max Planck who gave us the formula

$$S = k \log W,$$

which symbolizes the contribution of Boltzmann. Here S is thermodynamic entropy, W is the probability of the state, and k is the so-called Boltzmann constant. Einstein called the relation

Entropy = logarithm of "the probability of the state"

Boltzmann's principle and used it systematically in his work on Brownian motion.

T. and P. Ehrenfest (1912) repeatedly stressed the importance of the probabilistic implication and superiority of Boltzmann's entropy over Gibbs' measure. Interestingly, already at the beginning of the twentieth century, soon after Einstein's work, Ehrenfest (see T. and P. Ehrenfest (1912), p. 77) foresaw the possible close relationship of the Boltzmann entropy to the biological statistics then being developed by Pearson (1900) and others. Here what he pointed out was the great potential and wide range of validity of the Boltzmann's method in many branches of statistics, and this is exactly what Akaike rediscovered and promoted since 1970s through his Entropy Maximization Principle.

After so many decades and after all these acknowledgements, however, misconception and confusion do not seem to be over. The following story (Tribus 1988, p. 32) of Tribus' conversation with Shannon shows how strongly the confusion has prevailed among academics: "The same function (Shannon's entropy) appears in statistical thermodynamics and, on the advice of John Von Neumann, Claude Shannon called it entropy. I talked with Dr. Shannon about this, asking him why he had called his function by a name that was already in use in another field. I said it was bound to cause some confusion between the theory of information and thermodynamics. He said that Von Neumann had told him: No one really understands entropy. Therefore, if you know what you mean by it and you use it when you are in an argument, you will win every time." Incidentally, this was not the only time when Von Neumann caused misunderstandings, for he made another more serious and harmful mistake, in his work on quantum mechanics, which misguided physicists for decades (see Gribbin 1995). Misconception is also found among statistical physicists. For example, Jaynes (1988, p. 16) says, "Until the discovery of Shannon's theorem two, it was not possible to understand just what we were doing in statistical mechanics, nor to have any confidence in it for the prediction of irreversible processes. However, we can now see that statistical mechanics is a much more powerful tool than physicists had realized."

Confusion is also seen among statisticians. In statistics, a measure of the deviation of distribution $q(x)$ from $p(x)$ was introduced by Kullback (1959) as

$$I(p;q) = \int p(x)\log(\frac{p(x)}{q(x)})dx,$$

and has been discussed by many statisticians. This quantity, which is called Kullback–Leibler information, is exactly the negative of Boltzmann entropy $B(p; q)$. However, again the probabilistic meaning of Boltzmann's entropy is completely ignored by Kullback and many (if not most) statisticians. These confusions and mistakes clearly show the weakness of the formal mathematical treatment of theory without paying much attention to its implications and to the historical background behind the theory.

Since physicists are confused with Boltzmann's entropy and Shannon's entropy, it is not surprising that some statisticians are confused with Boltzmann entropy and Kullback–Leibler information. Many physicists and engineers are excited about the implications of the MEM (Maximum Entropy Method) based on Gibbs–Shannon entropy in statistical sciences, but it should not be forgotten that statisticians are already excited about the implication of the EMP (entropy maximization principle) based on the Boltzmann entropy for the same purposes.

Part III

State Space Modeling

State space models are useful for modeling the dynamics of many variables, where some of the variables are unobservable as data and only a subset of the variables are observed as time series data. Here the inference problems are classified into two: (a) an inference problem that considers the estimation of the unobserved state variables from the partially observed time series data assuming the model is known and (b) an inference problem that considers the identification and estimation of the unknown model from time series data. We will see in Part III (Chapters 11 through 13) that the innovation approach combined with the Kalman filter provides us with useful tools for solving both inference problems (a) and (b). Chapter 13 also discusses several computational problems arising in the statistical identification of state space models. Finally, the application of state space modeling for the generalization of Granger and Akaike causalities is discussed in Chapter 14.

11

Inference Problem (a) for State Space Models

11.1 State Space Models and Innovations

In Part II, we saw that the innovation approach has been useful for the identification and estimation of stochastic or deterministic dynamical system models from observed time series data, where the use of innovation-based LSE and MLE methods is mathematically supported by a Markov process theory, summarized in the theorem, that says that for any continuous Markov process x_t, the prediction error, $w_t = x_t - E[x_t | x_{t-\Delta t}, x_{t-2\Delta t} \ldots]$, converges to a Gaussian white noise for $\Delta t \to 0$.

In Part III, we see that the innovation approach is also useful for the estimation and identification of discrete-time nonlinear state space models. In state space modeling, however, the whitening of the time series is not as simple as for other non-state space dynamic models such as AR and ARMA models. This is because, for a state space model, the dynamics of a time series x_t are characterized in terms of state variable z_t or $z(t)$, but the state variable is not directly observed. We observe z_t only through the observation equation $x_t = Cz_t + \varepsilon_t$, where the observation matrix C could be rectangular, that is, the dimension of the observed time series x_t could be smaller than the dimension of the state z_t. In addition, the observed data x_t are, in general, contaminated by the observation error ε_t. This makes the inference problem for state space models more complex, since two types of inference problem need to be considered. One is "inference on state variables," where we assume that the parameters of the state space model are specified, and another is "inference on parameters" of the state space model. In real applications, we know observed time series, $x_1, x_2, \ldots, x_N$, and we are expected to estimate, from the time series data, both the unobserved state variables and the state space model, at the same time.

11.1.1 Linear State Space Models

We have already seen several examples of linear state space models, in Chapters 3 and 5. For example, discrete-time structural models and continuous-time structural models are written in the form of a state space representation. The idea of a parallel type discrete-time structural model was introduced from the AR(p) model, (see Section 5.3.1),

$$x_t = \phi_1 x_{t-1} + \phi_2 x_{t-2} + \cdots + \phi_p x_{t-p} + w_t$$

$$= \frac{1}{(1 - \phi_1 B + \phi_2 B^2 + \cdots + \phi_p B^p)} w_t$$

$$= \frac{1}{(B^{-1} - \lambda_1)(B^{-1} - \bar{\lambda}_1) \cdots (B^{-1} - \lambda_r)(B^{-1} - \bar{\lambda}_r)(B^{-1} - \lambda_{2r+1}) \cdots (B^{-1} - \lambda_p)} w_{t-p},$$

$$(11.1)$$

where B is the backward shift operator, that is, $B^i x_t = x_{t-i}$. From (11.1), we have a parallel type linear state space model as

$$\begin{cases} z_t = A^{(para)} z_{t-1} + \eta_t^{(para)} \\ x_t = C^{(para)} z_t \end{cases}, \tag{11.2}$$

where $\lambda_1, \bar{\lambda}_1, \ldots, \lambda_r, \bar{\lambda}_r, \lambda_{2r+1}, \ldots, \lambda_p$ are characteristic roots of the model (11.1) given as the roots of the characteristic equation

$$\Lambda^p - \phi_1 \Lambda^{p-1} - \cdots - \phi_p = 0.$$

Here, we assume that characteristic roots are composed of r pairs of complex conjugate roots and $(p - 2r)$ real roots, where

$$z_t = (\xi_t^{(1)}, \quad \xi_{t-1}^{(1)}, \quad \ldots, \quad \xi_t^{(r)}, \quad \xi_{t-1}^{(r)}, \quad \xi_t^{(2r+1)}, \quad \ldots, \quad \xi_t^{(p)})',$$

$$\xi_t^{(j)} = 2\mathrm{Re}(\lambda_j)\xi_{t-1}^{(j)} - |\lambda_j|^2 \xi_{t-2}^{(j)} + \varepsilon_t^{(j)}, \quad (j = 1, \ldots, r),$$

$$\xi_t^{(k)} = \lambda_k \xi_{t-1}^{(k)} + \varepsilon_t^{(k)}, \quad (k = 2r+1, \ldots, p).$$

Then, for (11.2), we have

$$A^{(para)} = \begin{pmatrix} 2\mathrm{Re}(\lambda_1) & -|\lambda_1|^2 & \cdots & 0 & 0 & 0 & \cdots & 0 \\ 1 & 0 & \cdots & 0 & 0 & 0 & \cdots & 0 \\ \cdots & \cdots & \cdots & \cdots & \cdots & \cdots & \cdots & \cdots \\ 0 & 0 & \cdots & 2\mathrm{Re}(\lambda_r) & -|\lambda_r|^2 & 0 & \cdots & 0 \\ 0 & 0 & \cdots & 1 & 0 & 0 & \cdots & 0 \\ 0 & 0 & \cdots & 0 & 0 & \lambda_{2r+1} & \cdots & 0 \\ \cdots & \cdots & \cdots & \cdots & \cdots & \cdots & \cdots & \cdots \\ 0 & 0 & \cdots & 0 & 0 & 0 & \cdots & \lambda_p \end{pmatrix},$$

$$\eta_t^{(para)} = (\varepsilon_t^{(1)} \quad 0 \quad \cdots \quad \varepsilon_t^{(r)} \quad 0 \quad \varepsilon_t^{(2r+1)} \quad \cdots \quad \varepsilon_t^{(p)})',$$

and

$$C^{(para)} = (1, 0, \ldots, 1, 0, 1, \ldots, 1).$$

For the continuous-time case, the idea of a parallel type structural model was introduced analogously from the pth order stochastic differential equation model

$$x^{(p)}(t) + b_1 x^{(p-1)}(t) + \cdots + b_{p-1} x^{(1)}(t) + b_p x(t) = n(t),$$

and its characteristic roots, $\mu_1, \bar{\mu}_1, \ldots, \mu_r, \bar{\mu}_r, \mu_{2r+1}, \ldots, \mu_p$, of the characteristic equation

$$\rho^{(p)} + b_1 \rho^{p-1} + \cdots + b_{p-1} \rho + b_p = 0$$

as

$$\begin{cases} \dot{z}(t) = A^{(para)} z(t) + \eta^{(para)}(t) \\ x_t = C^{(para)} z_t \end{cases}, \tag{11.3}$$

where

$$z(t) = (\dot{\xi}_{(1)}(t), \xi_{(1)}(t), \ldots, \dot{\xi}_{(r)}(t), \xi_{(r)}(t), \xi_{(2r+1)}(t), \ldots, \xi_{(p)}(t))',$$

$$\ddot{\xi}_{(j)}(t) = 2\operatorname{Re}(\mu_j)\dot{\xi}_{(j)}(t) - |\mu_j|^2 \xi_{(j)}(t) + \varepsilon_{(j)}(t), \quad (j = 1, \ldots, r),$$

$$\dot{\xi}_{(k)}(t) = \mu_k \xi_{(k)}(t) + \varepsilon_{(k)}(t), \quad (k = 2r+1, \ldots, p),$$

$$\eta^{(para)}(t) = (\varepsilon_{(1)}(t), 0, \varepsilon_{(2)}(t), 0, \ldots, \varepsilon_{(r)}(t), 0, \varepsilon_{(2r+1)}(t), \ldots, \varepsilon_{(p)}(t))',$$

$$C^{(para)} = (0, 1, \ldots, 0, 1, 1, \ldots, 1),$$

and

$$A^{(para)} = \begin{pmatrix} 2\operatorname{Re}(\mu_1) & -|\mu_1|^2 & \cdots & 0 & 0 & 0 & \cdots & 0 \\ 1 & 0 & \cdots & 0 & 0 & 0 & \cdots & 0 \\ \cdots & \cdots & \cdots & \cdots & \cdots & \cdots & \cdots & \cdots \\ 0 & 0 & \cdots & 2\operatorname{Re}(\mu_r) & -|\mu_r|^2 & 0 & \cdots & 0 \\ 0 & 0 & \cdots & 1 & 0 & 0 & \cdots & 0 \\ 0 & 0 & \cdots & 0 & 0 & \mu_{2r+1} & \cdots & 0 \\ \cdots & \cdots & \cdots & \cdots & \cdots & \cdots & \cdots & \cdots \\ 0 & 0 & \cdots & 0 & 0 & 0 & \cdots & \mu_p \end{pmatrix}.$$

$\varepsilon_{(j)}(t)$ is a continuous-time Gaussian white noise with variance $\sigma_{(j)}^2$ for $j = 1, \ldots, r$ and for $j = 2r + 1, \ldots, p$.

11.1.2 Nonlinear State Space Models: Hodgkin–Huxley Model

We have also seen several examples of nonlinear state space examples related to dynamic models such as Hodgekin–Huxley model, Zetterberg model, Balloon model, and DCM, in Chapter 5. They are all well suited to be considered in the framework of the state space model since the dynamics of the models are nonlinear and some of the variables in the models are not observed as time series data. For example, the Hodgkin-Huxley model is written as

$$\dot{z} = f(z)$$
$$x_t = Cz_t + \varepsilon_t,$$

(11.4)

where ε_t is a 1-dimensional Gaussian observation noise, $C = (1, 0, 0, 0)$, $z(t) = (V(t), m(t), h(t), n(t))'$ and

$$f(z) = \begin{pmatrix} \frac{-1}{C_M}[\{\bar{g}_{Na}m(t)^3 h(t) + \bar{g}_K n(t)^4 + \bar{g}_l\}V(t) + \bar{g}_{Na}m^3 hE_{Na} \\ + \bar{g}_K n^4 E_K + \bar{g}_l E_l + I(t)] \\ -\{\alpha_m(V(t)) + \beta_m(V(t))\}m(t) + \alpha_m(V(t)) \\ -\{\alpha_h(V(t)) + \beta_h(V(t))\}h(t) + \alpha_h(V(t)) \\ -\{\alpha_n(V(t)) + \beta_n(V(t))\}n(t) + \alpha_n(V(t)) \end{pmatrix},$$

$C_M, \bar{g}_{Na}, \bar{g}_K, \bar{g}_l, E_{Na}, E_K, E_l$ are physiological constants. When we consider an r-dimensional physiological driving noise $w(t)$ and the observation noise ε_t, we obtain a stochastic version of the Hodgkin–Huxley model,

$$\dot{z} = f(z) + Bw(t)$$
$$x_t = Cz_t + \varepsilon_t,$$

where B is a $4 \times r$ matrix.

11.1.3 Whitening through ARMA(k,k)

Since the state variable is not directly observed, we need a special approach to whiten the time series with the state space model, which is explained in Section 11.2. However, if the state space model is linear, at least we can estimate the model itself by using the ARMA model estimation method, since a general k-dimensional state space model,

$$z_t = Az_{t-1} + Bw_t$$
$$x_t = Cz_t + \varepsilon_t,$$

(11.5)

with the complete knowledge of A, B, C and the variance matrices, R_1 of w_t and R_2 of ε_t.

The estimation problem for the state z_t $(1 \leq t \leq N)$ may be classified into one of the following three types:

1. Prediction problem: Find an optimal estimate of z_t from the "past" observation data, $x_1, x_2, ..., x_{t-1}$.

2. Smoothing problem: Find an optimal estimate of z_t from all the observation data, $x_1, x_2, ..., x_{t-1}, x_t, x_{t+1}, ..., x_N$, that is, including the "future" data.

3. Filtering problem: Find an optimal estimate of z_t from the observation data "up to the present" time point t, that is, $x_1, x_2, ..., x_{t-1}, x_t$.

Here the problems are characterized according to the time series data we use for the estimation, whether we use only past data, $x_1, x_2, ..., x_{t-1}$ (prediction problem), or all the data including future data, $x_1, x_2, ..., x_{t-1}, x_t, x_{t+1}, ..., x_N$, (smoothing problem), or time series data up to the present time point, $x_1, x_2, ..., x_{t-1}, x_t$ (filtering problem). We sometime denote the predicted estimate as $z_{t|t-1}$, the smoothed estimate as $z_{t|N}$, and the filtered estimates as $z_{t|t}$, respectively. Here the smoothed estimate $z_{t|N}$ may look most suitable for our purpose, that is, to obtain a solution to the inference problem (a). However, we will know, after we study the Kalman filter theory, that an optimal solution of the smoothing problem is derived from optimal solutions to the prediction problem and the filtering problem (but not the other way round).

In fact this is true only when the state space model is known, and we leave this model identification problem to the next chapter (Section 12.5) and concentrate on the inference problem (a), that is, the problem of estimating the state z_t from time series $x_1, x_2, ..., x_N$, assuming that the state space model is known.

11.2.2 Minimum Variance Principle

Suppose that we have observation data up to time point N, i.e., $x_1, x_2, ..., x_N$, from a state space model,

$$z_t = Az_{t-1} + Bw_t$$

$$x_t = Cz_t + \varepsilon_t.$$

Let us first look into the prediction and the filtering problem of the state z_t. If the filtered estimate of the state at time point $t - 1$ is given as $z_{t-1|t-1}$, the predictor of the state at time point t will naturally be $Az_{t-1|t-1}$. Thus, we have

$$z_{t|t-1} = Az_{t-1|t-1}. \tag{11.16}$$

At the next time step, we may obtain a new observation x_t. The prediction error for the new observation, that is, $v_t = x_t - Cz_{t|t-1}$, will contain new information for the estimation of the state z_t, and the new information could be used to update and improve the estimate of the state z_t from $z_{t|t-1}$ to $z_{t|t}$ as

$$z_{t|t} = z_{t|t-1} + K_t v_t, \tag{11.17}$$

where

$$v_t = x_t - Cz_{t|t-1} = C(z_t - z_{t|t-1}) + \varepsilon_t. \tag{11.18}$$

Since ε_t is independent from $(z_t - z_{t|t-1})$, in (11.18), we can write the innovation variance matrix, $\Sigma_{v_t} = E[v_t v_t']$, as

$$\Sigma_{v_t} = (CP_{t|t-1}C' + R_2), \tag{11.19}$$

where $P_{t|t-1}$ is a variance matrix of the state prediction error, that is,

$$P_{t|t-1} = E[(z_t - z_{t|t-1})(z_t - z_{t|t-1})'],$$

and R_2 is the variance matrix of the observation noise ε_t.

Now the question is how to choose K_t in (11.17) and what is the optimal K_t? Kalman showed that the "Optimal K_t minimizes the estimation error variance, $P_{t|t} = E[(z_t - z_{t|t})(z_t - z_{t|t})']$, and is given by

$$K_t = P_{t|t-1}C'(CP_{t|t-1} + R_2)^{-1}, \tag{11.20}$$

where

$$\begin{aligned}P_{t|t-1} &= E[(z_t - z_{t|t-1})(z_t - z_{t|t-1})'] \\ &= AP_{t-1|t-1}A' + BR_1B'." \end{aligned} \tag{11.21}$$

The matrix K_t is usually called the Kalman gain. If we accept that K_t of (11.20) gives the optimal filtered estimate of z_t as (11.17), then we have

$$\begin{aligned}P_{t|t} &= E[(z_t - z_{t|t})(z_t - z_{t|t})'] \\ &= E[\{z_t - z_{t|t-1} + K_t(x_t - Cz_{t|t-1})\}\{z_t - z_{t|t-1} + K_t(x_t - Cz_{t|t-1})\}'] \\ &= P_{t|t-1} - K_tCP_{t|t-1} - P_{t|t-1}C'K_t' + K_t(CP_{t|t-1}C' + R_2)K_t' \\ &= P_{t|t-1} - K_tCP_{t|t-1}. \end{aligned} \tag{11.22}$$

Therefore, if the optimal estimate $z_{0|0}$ of z_0 and its error variance $P_{0|0}$ at the starting time point are given, we can obtain $P_{1|0}$, K_1, $P_{1|1}$, $P_{2|1}$, K_2, $P_{2|2}$, ..., recursively. At the same time, with these K_1, K_2, ..., and the observed data, x_1, x_2, ..., we can obtain the optimal filtered estimates of the state, $z_{1|1}$, $z_{2|2}$, ..., with (11.17). Then the optimal predicted estimates of the state, $z_{1|0}$, $z_{2|1}$, ..., are given using the filtered estimates from (11.16).

Next, we see the K_t given by (11.18) actually minimizes

$$E[a'\tilde{z}_{t|t}]^2 = E[a'\tilde{z}_{t|t}\tilde{z}'_{t|t}a]$$

$$= a'E[\tilde{z}_{t|t}\tilde{z}'_{t|t}]a$$

$$= a'P_{t|t}a$$

for any nonzero vector a where $\tilde{z}_{t/t} = z_t - z_{t/t}$. Note that the error variance $P_{t|t}$ of the filtered estimate $z_{t|t}$ is rewritten using (11.18) and (11.17), as

$$P_{t|t} = E[(z_t - z_{t|t-1})(z_t - z_{t|t-1})']$$

$$= E[\{z_t - z_{t|t-1} - K_tC(z_t - z_{t|t-1}) - K_tv_t)\}\{z_t - z_{t|t-1} - K_tC(z_t - z_{t|t-1}) - K_tv_t)\}'].$$

$$= (I - K_tC)P_{t|t-1}(I - K_tC)' + K_tR_2K_t' \tag{11.23}$$

Using (11.23), we get

$$a'P_{t|t}a = a'[P_{t|t-1} - K_tCP_{t|t-1} - P_{t|t-1}C'K_t' + K_t\{CP_{t|t-1}C' + R_2\}K_t']a.$$

Then we rewrite $P_{t|t}$ as

$$P_{t|t} = P_{t|t-1} - P_{t|t-1}C'\{CP_{t|t-1}C' + R_2\}^{-1}CP_{t|t-1}$$

$$+ [K_t - P_{t|t-1}C'\{CP_{t|t-1}C' + R_2\}^{-1}]\{R_2 + CP_{t|t-1}C'\}.$$

$$\times [K_t - P_{t|t-1}C'\{CP_{t|t-1}C' + R_2\}^{-1}]'$$

This shows that K_t such that

$$K_t = P_{t|t-1}C'\{CP_{t|t-1}C' + R_2\}^{-1}$$

gives the minimum of $P_{t|t}$ for any nonzero vector a.

Note that Equations 11.16 through 11.22 are recurrence relations, where the optimal prediction of the next state value, that is, $z_{t+1|t}$, is obtained as a weighted sum of the present value $z_{t|t-1}$ and the present observation x_t. The weighting factor K_t is a matrix defined by Equation 11.19. Notice that this equation is easily implemented on a computer, since the past observation need not be stored; only the present value of $z_{t|t-1}$ is needed at each step. This, of course, is a consequence of the Markovian character of the state process.

Because of the optimality of $z_{t|t}$, as an estimate of z_t from $x_1, \ldots, x_t$, the filtered estimate $z_{t|t}$ is some times called the minimum variance estimate and the filter itself is called the minimum variance filter.

Note that the matrices A and B of the state equation can be functions of time, where we can write A_t and B_t. That means that the Kalman filter scheme is valid for state space models that are linear but possibly time-dependent nonstationary. Note that the minimization is local in time, and the idea of global minimization, that is,

$$\text{Min} \sum_{t=1}^{N} (z_{t+1} - z_{t+1|t})(z_{t+1} - z_{t+1|t})',$$

is out of the concern of the Kalman filter, where the main interest is to solve the inference problem (*a*). The difference of the local minimization and the global minimization comes up as a delicate issue when we consider the inference problem (*b*), that is, the identification problem of the state space model. This is discussed in Chapter 12.

11.2.3 Innovations and Kalman Filter

We note that the filtered state estimate $z_{t|t}$ is expressed, from (11.17), as a linear combination of innovations, $v_1, \ldots, v_t$. It means that although the filter scheme is derived from the minimum variance principle of the state estimate, the innovation (prediction error) at each time point plays an important role.

In fact, apart from the above "minimum variance principle"–based derivation of the Kalman filter, there are many other ways to derive the Kalman filter. An innovation-based approach, which we see below, is useful in clarifying the dynamic and probabilistic nature of the Kalman filter.

Suppose we have a linear state space model,

$$z_t = Az_{t-1} + Bw_t \tag{11.24}$$

$$x_t = Cz_t + \varepsilon_t. \tag{11.25}$$

Our goal is to find an optimal filtered state estimate,

$$z_{t|t} = E[z_t | x_1, \ldots, x_t], \quad t = 1, 2, \ldots, \tag{11.26}$$

and a predictor,

$$z_{t+1|t} = E[z_{t+1} | x_1, \ldots, x_t], \quad t = 1, 2, \ldots, \tag{11.27}$$

in a form that can be implemented on a computer in real time. Let us consider the innovation process, v_t, $t = 1, 2, \ldots$, for the observation process, x_t, $t = 1, 2, \ldots$.

Since the two processes, x_t and v_t, are causally equivalent (see Section 7.2), we can replace the conditioning on $x_1, \ldots, x_t$ in (11.26) and (11.27) by conditioning on $v_1, \ldots, v_t$ so that

$$z_{t|t} = E[z_t \,|\, v_1, \ldots, v_t], \quad t = 1, 2, \ldots \tag{11.28}$$

and

$$z_{t+1|t} = E[z_{t+1} \,|\, v_1, \ldots, v_t], \quad t = 1, 2, \ldots. \tag{11.29}$$

Since all the processes involved here are zero mean Gaussian, the conditional expectation (11.26) or (11.27) must be a homogeneous linear combination, for example,

$$z_{t+1|t} = \sum_{k=1}^{t} c_{t,k} v_k \quad t = 1, 2, \ldots, \tag{11.30}$$

where $c_{t,k}$ are, so far, unknown matrices. To find their values, multiply (11.30) on the right by v_j' and take the expectation. Since the innovation process v_j is Gaussian white noise, that is,

$$E[v_t v_j'] = \begin{pmatrix} \Sigma_{v_t} & \text{if } j = t, \\ 0 & \text{if } j \neq t, \end{pmatrix}$$

we obtain for any $j = 1, 2, \ldots, t$

$$E[z_{t+1|t} v_j'] = c_{t,j} \Sigma_{v_t}. \tag{11.31}$$

However, by (11.29) for any $j = 1, 2, \ldots, t$

$$z_{t+1|t} v_j' = E[z_{t+1} \,|\, v_1, \ldots, v_t] v_j'$$

$$= E[z_{t+1} v_j' \,|\, v_1, \ldots, v_t]$$

and hence by taking expectations

$$E[z_{t+1|t} v_j'] = E[z_{t+1} v_j'].$$

Thus, (11.31) implies

$$E[z_{t+1} v_j'] = c_{t,j} \Sigma_{v_j},$$

and assuming that the matrices Σ_{v_j} are nonsingular, we find that

$$c_{t,j} = E[z_{t+1}v_j']\Sigma_{v_j}^{-1}, \quad j = 1,\ldots,t.$$

We can now substitute this into (11.30), thus obtaining

$$z_{t+1|t} = \sum_{k=1}^{t} E[z_{t+1}v_k']\Sigma_{v_k}^{-1}v_k. \quad t = 1, 2, \ldots \tag{11.32}$$

Using the state equation, $z_{t+1} = Az_t + Bw_{t+1}$, the expectation, $E[z_{t+1}v_k']$ in (11.32) becomes,

$$E[z_{t+1}v_k'] = AE[z_t v_k'] + BE[w_{t+1}v_k'], \quad k = 1,\ldots,t. \tag{11.33}$$

However, for $1 \le k \le t$, the state z_k is independent of the system noise Bw_{t+1} so that

$$E[Bw_{t+1}z_k']C' = 0 \quad \text{for } 1 \le k \le t. \tag{11.34}$$

Since we have $Cz_t = x_t - \varepsilon_t$, (11.34) leads to

$$E[Bw_{t+1}x_k'] - E[Bw_{t+1}\varepsilon_k'] = 0,$$

and since the system noise w_t's and observation noise ε_t's are assumed to be mutually independent, we have

$$E[Bw_{t+1}\varepsilon_k'] = 0.$$

Thus,

$$E[Bw_{t+1}x_k'] = 0. \quad (k = 1,\ldots,t).$$

Since $x_1, \ldots, x_t$ and $v_1, \ldots, v_t$ are causally linearly equivalent, we must also have

$$E[Bw_{t+1}v_k'] = 0 \quad \text{for } k = 1,\ldots,t.$$

Therefore, the second expectation on the right-hand side of (11.33) is zero, and we obtain from (11.32) and (11.33)

$$z_{t+1|t} = A\sum_{k=1}^{t} E[z_t v_k']\Sigma_{v_k}^{-1}v_k, \quad t = 1, 2, \ldots. \tag{11.35}$$

Now Equation 11.32 holds for all $t = 1, 2, \ldots$, in particular

$$z_{t|t-1} = \sum_{k=1}^{t-1} E[z_t v_k'] \Sigma_{v_k}^{-1} v_k,$$

so that if we separate the last term in the summation in (11.35), we can write

$$z_{t+1|t} = A[z_{t|t-1} + E[z_t v_t'] \Sigma_{v_t}^{-1} v_t]$$

or by denoting

$$K_t = E[z_t v_t'] \Sigma_{v_t}^{-1}, \quad t = 1, 2, \ldots \tag{11.36}$$

as

$$z_{t+1|t} = A(z_{t|t-1} + K_t v_t), \quad t = 1, 2, \ldots \tag{11.37}$$

Note that, by definition (11.28), we have

$$z_{t|t} = z_{t|t-1} + K_t v_t. \tag{11.38}$$

Then (11.37) is written as

$$z_{t+1|t} = A z_{t|t}. \tag{11.39}$$

Now v_t is an innovation process of the observation process x_t and is defined by

$$\begin{aligned} v_1 &= x_1 - E[x_1] \\ v_t &= x_t - E[x_t | x_1, \ldots, x_{t-1}], \quad t = 2, 3, \ldots \end{aligned} \tag{11.40}$$

Substituting for $x_t = Cz_t + \varepsilon_t$ into the conditional expectation above, we get

$$E[x_t | x_1, \ldots, x_{t-1}] = CE[z_t | x_1, \ldots, x_{t-1}] + E[\varepsilon_t | x_1, \ldots, x_{t-1}]$$

$$= C z_{t|t-1}.$$

It follows that

$$v_t = x_t - C z_{t|t-1}, \quad t = 1, 2, \ldots \tag{11.41}$$

$$z_{t+1|t} = A(z_{t|t-1} + K_t x_t - K_t C z_{t|t-1}),$$

and

$$z_{t+1|t} = A(z_{t|t-1} + K_t x_t - K_t C z_{t|t-1}).$$

To complete the filter specification, it is necessary to show how the Kalman gain sequence

$$K_t, \quad t = 1, 2, \ldots$$

is calculated. Let us denote the state prediction error

$$\tilde{z}_{t|t-1} = z_t - z_{t|t-1}, \quad t = 1, 2, \ldots,$$

and let

$$P_{t|t-1} = E[\tilde{z}_{t|t-1}\tilde{z}'_{t|t-1}], \quad t = 1, 2, \ldots, \tag{11.42}$$

be its covariance matrix. From definition (11.40) of the innovation process v_t, $t = 1, 2, \ldots$, we obtain substitution for x_t from the observation equation $x_t = C z_t + \varepsilon_t$ on the equation

$$v_t = C\tilde{z}_{t|t-1} + \varepsilon_t \quad t = 1, 2, \ldots \tag{11.43}$$

Since z_t and ε_t are independent, it follows from (11.43) that

$$E[z_t v'_t] = E[z_t \tilde{z}'_{t|t-1}]C'. \tag{11.44}$$

By the orthogonality principle, $E[z_t \tilde{z}'_{t|t-1}]$ of the right-hand side of (11.44) equals the minimum mean square prediction error $P_{t|t-1}$, so that

$$E[z_t v'_t] = P_{t|t-1}C', \quad t = 1, 2, \ldots$$

However, $\tilde{z}_{t|t-1}$ and ε_t are also independent, and so the covariance matrix of the left-hand side of (11.43) is the sum of the covariance matrices of the two random vectors on the right-hand side. Using symbols, this is written as

$$\Sigma_{v_t} = C P_{t|t-1}C' + R_2, \quad t = 1, 2, \ldots \tag{11.45}$$

Thus, we can write the Kalman gain (11.36) as

$$K_t = P_{t|t-1}C'[C P_{t|t-1}C' + R_2]^{-1}, \quad t = 1, 2, \ldots \tag{11.46}$$

The filtered estimate $z_{t|t}$ and its error variance matrix $P_{t|t}$ are recursively determined with (11.38) and (11.46) as

$$z_{t|t} = z_{t|t-1} + K_t v_t$$

$$P_{t|t} = P_{t|t-1} + K_t \Sigma_{v_t} K_t' - P_{t|t-1} C' K_t' - K_t C P_{t|t-1}$$

$$= P_{t|t-1} - K_t C P_{t|t-1}. \tag{11.47}$$

Similarly, $z_{t+1|t}$ and its error variance matrix $P_{t+1|t} = E[\tilde{z}_{t+1|t} \tilde{z}_{t+1|t}']$ are written as

$$z_{t+1|t} = A z_{t|t}$$

$$P_{t+1|n} = A P_{t|t} A' + B R_1 B'. \tag{11.48}$$

The Kalman filter scheme for the linear state space model,

$$z_t = A z_{t-1} + B w_t$$

$$x_t = C z_t + \varepsilon_t,$$

is summarized in the following seven equations:

State predictor:

$$z_{t|t-1} = A z_{t-1|t-1} \tag{11.49}$$

Filtered estimate of the state:

$$z_{t|t} = z_{t|t-1} + K_t v_t \tag{11.50}$$

Innovation:

$$v_t = x_t - C z_{t|t-1} \tag{11.51}$$

Variance matrix of innovation v_{t+1}:

$$\Sigma_{v,t+1} = C P_{t+1} C' + R_2 \tag{11.52}$$

Kalman gain:

$$K_t = P_{t|t-1} C' \{ C P_{t|t-1} C' + R_2 \}^{-1} \tag{11.53}$$

Variance matrix of $\{ z_t - z_{t|t} \}$:

$$P_{t|t} = P_{t|t-1} - K_t C P_{t|t-1} \tag{11.54}$$

Variance matrix of $\{z_{t+1} - z_{t+1|t}\}$:

$$P_{t+1|t} = AP_{t|t}A' + BR_1B' \tag{11.55}$$

Note that the parameters of the state space model, that is, A, B, C, R_1, R_2, $z_{0|0}$, $P_{0|0}$, are assumed to be known. Estimation of these parameters is out of our present concern in inference problem (*a*). The problem belongs to inference problem (*b*), which is discussed in Chapter 12.

11.2.4 Kalman Filter as a Real-Time Whitening Operator

From the above derivation, we can see that all the computations of the Kalman filter are in a recursive form, and the filter transforms the observed time series into white Gaussian innovations in real time. This is shown schematically in Figure 11.1.

This means that the time series $x_1, ..., x_N$ is transformed into white Gaussian innovations, $v_1, ..., v_N$, and their variance matrices, $\Sigma_{v_1}, \Sigma_{v_2}, ..., \Sigma_{v_N}$, by the Kalman filter scheme with the state space model,

$$z_t = Az_{t-1} + Bw_t$$

$$x_t = Cz_t + \varepsilon_t,$$

where all the parameters $\theta = (A, B, C, R_1, R_2)'$ are assumed to be given. Then the likelihood of the state space model is calculated as the following, using the innovations, $v_t = x_t - E[x_t|x_{t-1}, ..., x_1, \theta]$, recursively calculated by the Kalman filter:

$$(-2)\log p(x_1, ..., x_N|\theta) = (-2)\log p(v_1, ..., v_N|\theta)$$

$$\approx \sum_{t=1}^{N} \log \Sigma_{v_t} + \sum_{t=1}^{N} v_t'\Sigma_{v_t}^{-1}v_t + N \log 2\pi.$$

Note, however, that the white Gaussian innovations are dependent on the assumed model. From two different state space models, model-1 and model-2,

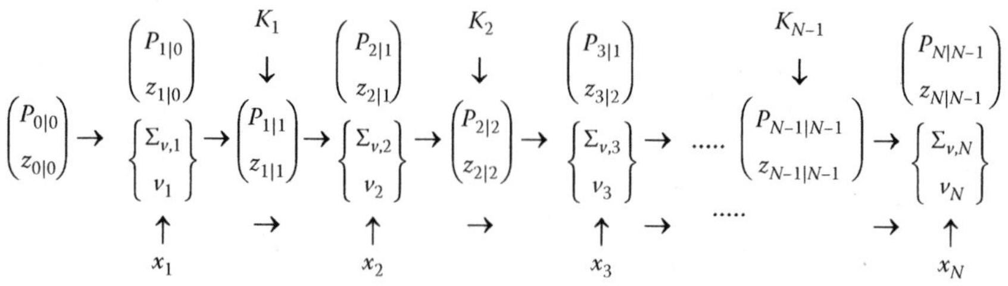

FIGURE 11.1
Kalman filter as a real-time whitening filter.

we have two different sets of innovations, $(v_1^{(1)}, v_2^{(1)}, \ldots, v_N^{(1)})$ for model-1 and $(v_1^{(2)}, v_2^{(2)}, \ldots, v_N^{(2)})$ for the model 2. Which set of innovations to choose is discussed in inference problem (b) in Chapter 12.

11.2.5 Observability and Controllability

So far we have seen that the Kalman filter provides us with an efficient computational method for solving the inference problem (a). In the derivation of the Kalman filter scheme in the previous sections, we implicitly assumed that the true model is "proper" in the sense that the solution of the problem (a) exists. However, it must be noted that if the state space model is not well presented, the Kalman filter may not work. That means even if the state process,

$$z_t = Az_{t-1} + Bw_t,$$

is simulated without any problem, and data x_t are generated from the state process z_t through the observation equation,

$$x_t = Cz_t + \varepsilon_t,$$

it may not be possible to obtain the filtered estimate $z_{t|t}$ from the data $x_1, \ldots, x_N$ with the Kalman filter scheme, if the A, B, and C of the state space model are not "properly" presented.

It is important to clarify conditions to check whether a given state space model is well presented so that the Kalman filter introduced in the previous section works and the time series are whitened into Gaussian white innovations. For the k-dimensional linear state space model,

$$z_t = Az_{t-1} + Bw_t$$
$$x_t = Cz_t + \varepsilon_t. \tag{11.56}$$

three "sufficient" conditions for the existence of the Kalman filter solution may be specified as follows (Jazwinski 1970, Aoki 1990):

(i) Rank$(O) = k$, where k is the dimension of the state, where O is

$$O = \begin{pmatrix} C \\ CA \\ \cdots \\ CA^{k-1} \end{pmatrix}. \tag{11.57}$$

Here O is called the "observability matrix," and the condition (i) is called the "observability condition."

(ii) Rank(C) = k,

$$C = \begin{pmatrix} B & AB & \cdots & A^{k-1}B \end{pmatrix}. \tag{11.58}$$

Here C is called the "controllability (reachability) matrix," and the condition (ii) is called the "controllability (reachability) condition."

(iii) The eigenvalues λ_i ($i = 1, \ldots, k$) of the transition matrix A satisfy

$$0 < |\lambda_i| < 1 \quad (i = 1, \ldots, k). \tag{11.59}$$

The matrix A satisfying this condition is called a stable matrix, and this condition (iii) is called the "stability condition."

Note that these conditions have been presented for linear state space models, but the same ideas are valid for the case of state space models with nonlinear dynamics, where the transition matrix A of (11.56) is to be substituted by some linearized matrix such as the Jacobian.

Note, however, that the stability condition (11.59) is not a "necessary" condition for nonlinear state space models to have a proper solution to the inference problem (a). For example, the Jacobian matrices of the Lorenz chaos and Hodgkin–Huxley models are functions of state variables, and they do not satisfy the condition (11.59) for some regions of the state variables.

11.2.6 Ill-Posed Inverse Problem and Observability

The EEG inverse problem is defined as the problem of estimating the primary current density J_t from a given EEG time series Y_t. The relation between J_t and Y_t is described by the equation

$$Y_t = KJ_t + \varepsilon_t. \tag{11.60}$$

The matrix K, linking the current density J_t with the measurement Y_t, is called the lead field matrix and is obtained by applying Maxwell's equations to a particular head model (Nunetz 1981). The inverse problem is often called ill-posed since the number of the dimensions of Y_t is much smaller than the number of voxels for which the primary current density needs to be estimated. This motivates people to introduce various types of constraints in order to convert the problem into a well-posed problem.

If we look at the same problem from a different angle, that is, state space modeling, we come to a very different view. The state space formulation of the inverse problem is described as an estimation of the current density J_t from the observed time series Y_t, in the following state space model,

$$J_t = AJ_{t-1} + Bw_t$$
$$Y_t = KJ_t + \varepsilon_t, \tag{11.61}$$

where K is a lead field matrix calculated in the same way as before in (11.60). Now we notice that the inverse problem can be reformulated as a well-posed problem if we set the estimation problem within the framework of the filtering problem with the proper state space model (11.61). Given a time series observation, $Y_1, Y_2, \ldots, Y_N$, we have three types of the estimate of J_t, that is, predicted estimate $J_{t|t-k}$, filtered estimate $J_{t|t}$, and the smoothed estimate $J_{t|N}$. The consistency of the filtered estimates is guaranteed by the observability condition (Kailath 1980, Aoki 1990).

With this theorem, the perspective of solving the inverse problem becomes much brighter and more promising. Since the observation matrix (i.e., lead field matrix) K connecting the unobserved variable J_t and the observed Y_t is given, we need here to identify the dynamic model, $J_t = A J_{t-1} + B w_t$ for J_t.

A difficulty that is unique to the present EEG inverse problem is centers on the size of the dimensionality, that is, the 10,000 dimensional state variable. An efficient design of a computational algorithm for solving (approximately) the 10,000 dimensional associated Riccati equation is required (see Galka et al. 2004b).

11.2.7 Kalman Filter and Input–Output System Models

Often we have an estimation problem of a dynamical system, where the dynamical system is driven by an external force u_t as well as the system noise w_t. Filtering and smoothing of the state of such input–output system models can be done in the same way as the usual state space model (Sorenson 1966).

Suppose we have a state space model with an exogenous vector variable u_t with coefficient matrix D of appropriate dimension as

$$z_t = A z_{t-1} + D u_{t-1} + B w_t.$$
$$x_t = C z_t + \varepsilon_t. \tag{11.62}$$

Then we have the following recursive scheme for the filtering of the state z_t:

$$z_{t|t-1} = A z_{t-1|t-1} + D u_{t-1} \quad \text{(State predictor)} \tag{11.63}$$

$$z_{t|t} = z_{t|t-1} + K_t v_t \quad \text{(Filtered estimate of the state)} \tag{11.64}$$

$$v_t = x_t - C z_{t|t-1} \quad \text{(Innovation)} \tag{11.65}$$

$$\Sigma v_{t+1} = C P_{t+1|t} C' + R_2 \quad \text{(Variance of innovation } v_{t+1}) \tag{11.66}$$

$$K_t = P_t C' \{C P_t C' + R_2\}^{-1} \quad \text{(Kalman gain)} \tag{11.67}$$

$$P_{t|t} = P_{t|t-1} - K_t C P_{t|t-1} \quad \text{(Variance of } (z_t - z_{t|t})) \tag{11.68}$$

$$P_{t+1|t} = AP_{t|t}A' + R_1 \quad \text{(Variance of } (z_{t+1} - z_{t+1|t})) \tag{11.69}$$

Note that all the equations except the predictor (11.63) of the state z_t are the same as the filtering equations for the usual state space model. This shows that if we have observed time series data $x_1, x_2, \ldots, x_N$ and $u_1, u_2, \ldots, u_N$ for the known state space model (11.62) with an exogenous variable, we can whiten the time series $x_1, x_2, \ldots, x_N$ into innovations $v_1, v_2, \ldots, v_N$, using the recursive schemes (11.63) through (11.69). A typical example of a state space model with an exogenous variable is a controlled system where u_t is an input for controlling the state z_t. Kalman (1960) showed a beautiful theory of the duality between the "optimal estimation of the state" and the "optimal control of the process x_t," defined by the state space model with control input u_t. Incidentally, Kalman's optimal control theory is a state space version of the dynamic programming method that was developed by Bellman (1957) for more general Markov process models.

The identification and estimation of a controlled feedback system from time series data is a very important subject for neuroscience as well as many other applied sciences. However, we are not going to discuss this subject in the present volume since the size of the volume is limited. For those who are interested in the topic of optimal control theory for state space models, we refer them to Whittle (1996) and Ozaki et al. (2012).

11.2.8 Probabilistic Meaning of Kalman Filter

The Kalman filter can be interpreted in many different ways. One important interpretation is its distributional interpretation (Ho and Lee 1964). For example, the Equations 11.65 and 11.66 show how to derive the prediction error v_{t+1} and its variance $\Sigma_{v_{t+1}}$ using the conditional mean $z_{t+1|t}$ and its error variance $P_{t+1|t}$. Here we note that $v_{t+1} = x_{t+1} - C z_{t+1|t}$, which suggests that v_{t+1} is also Gaussian and its probability density is equivalent to the conditional density of x_{t+1}, which we denote as $p_{(o)}(x_{t+1}|x_t, \ldots, x_1, \theta)$. This means, in (11.65) and (11.66), we derive the Gaussian probability density $p_{(o)}(x_{t+1}|x_t, \ldots, x_1, \theta)$ from the Gaussian conditional density $p_{(p)}(z_{t+1}|x_t, \ldots, x_1, \theta)$ of the state z_{t+1}, since the right-hand side of (11.66) is determined by the mean and the variance of the conditional density for the predicted state z_{t+1}, $p_{(p)}(z_{t+1}|x_t, \ldots, x_1, \theta)$. In the same way, we can interpret that (11.63) and (11.69) show how the Gaussian predictive conditional density is obtainable from the Gaussian conditional density distribution for the filtered state z_t, $p_{(f)}(z_t|x_t, \ldots, x_1, \theta)$. Finally, Equations 11.64 and 11.68 together with (11.67) are interpreted as showing how the Gaussian conditional density function $p_{(f)}(z_t|x_t, \ldots, x_1, \theta)$ of the filtered state is obtained from the Gaussian conditional density $p_{(p)}(z_t|x_{t-1}, \ldots, x_1, \theta)$ of the state z_t. The correspondence of the interpretations is schematically shown in Table 11.1.

TABLE 11.1

The Correspondence of Kalman Filter and Jazwinski Scheme

	Jazwinski Scheme	Kalman Filter
Specifying	$p_{(o)}(x_{t+1}\|x_t,\ldots,y_1,\theta)$	$n_{t+1} = x_{t+1} - Cz_{t+1\|t}$
$p^{(o)}(x_{t+1}\|x_t,\ldots,x_1,\theta)$	$= \int p(x_{t+1}\|z_{t+1})p_{(p)}(z_{t+1}\|x_t,\ldots,x_1,\theta)dz_{t+1}$	$\Sigma_{n_{t+1}} = CP_{t+1\|t}C' + R_2$
using $p^{(p)}(z_{t+1}\|x_t,\ldots,x_1,\theta)$		
Specifying	$p_{(p)}(z_{t+1}\|x_t,\ldots,x_1,\theta)$	$z_{t+1\|t} = A_t z_{t\|t}$
$p^{(p)}(z_{t+1}\|x_t,\ldots,x_1,\theta)$	$= \int p(z_{t+1}\|z_t)p_{(f)}(z_t\|x_t,\ldots,x_1,\theta)dz_t$	$P_{t+1\|t} = A_t P_{t\|t} A_t' + B_t R_1 B_t'$
using $p^{(f)}(z_t\|x_t,\ldots,x_1,\theta)$		
Specifying	$p_{(f)}(z_t\|x_t,\ldots,x_1,\theta)$	$z_{t\|t} = z_{t\|t-1} + K_t n_t$
$p^{(f)}(z_t\|x_t,\ldots,x_1,\theta)$	$= \dfrac{p(x_t\|z_t)p_{(p)}(z_t\|x_{t-1},\ldots,x_1,\theta)}{\int p(x_t\|\varsigma_t)p_{(p)}(\varsigma_t\|x_{t-1},\ldots,x_1,\theta)d\varsigma_t}$	$P_{t\|t} = P_{t\|t-1} - K_t C P_{t\|t-1}$
using $p^{(p)}(z_t\|x_{t-1},\ldots,x_1,\theta)$		$K_t = P_{t\|t-1}C'\left\{CP_{t\|t-1}C' + R_2\right\}^{-1}$

Each relation in Table 11.1, between the two conditional distributions, is written explicitly with integral equations as follows:

$$p_{(o)}(x_{t+1}|x_t,\ldots,x_1,\theta) = \int p(x_{t+1}|z_{t+1})p_{(p)}(z_{t+1}|x_t,\ldots,x_1,\theta)dz_{t+1}. \quad (11.70)$$

$$p_{(p)}(z_{t+1}|x_t,\ldots,x_1,\theta) = \int p(z_{t+1}|z_t)p_{(f)}(z_t|x_t,\ldots,x_1,\theta)dz_t. \quad (11.71)$$

$$p_{(f)}(z_t|x_t,\ldots,x_1,\theta) = \frac{p(x_t|z_t)p_{(p)}(z_t|x_{t-1},\ldots,x_1,\theta)}{\int p(x_t|\varsigma_t)p_{(p)}(\varsigma_t|x_{t-1},\ldots,x_1,\theta)d\varsigma_t}. \quad (11.72)$$

Here the Gaussian densities, $p(x_{t+1}|z_{t+1})$, $p(z_{t+1}|z_t)$, and $p(x_t|z_t)$, are specified by the state space model.

These recursive integral relations (given by Jazwinski 1970) provide us with the basis of the lately introduced monstrous computational Marko Chain Monte Carlo (MCMC) method for a nonlinear Gaussian filter, where the state space models are nonlinear and non-Gaussian, and, as a result, all the conditional densities are non-Gaussian. Here the validity of the computation totally depends on the accuracy and the speed of the numerical integration method and the huge dimensional numerical optimization method.

On one hand, this approach takes advantage of the Markovian nature of the process in the recursive representation of the conditional densities. On the other hand, unlike the innovation approach, it totally ignores other properties of the Markov process, that is, local Gaussianity. Here, the cost (computational load), we pay for the Jazwinski scheme, is rather large compared with the gain (i.e., local non-Gaussianity), since most dynamic phenomena are Markov in nature and the prediction errors are almost Gaussian as long as the sampling interval is reasonably small.

11.2.9 Kalman Filters in Continuous Time

So far we have seen that we can estimate the unobserved state efficiently using the Kalman filter scheme for discrete-time state space models:

$$
\begin{aligned}
z_t &= A z_{t-1} + B w_t \\
x_t &= C z_t + \varepsilon_t,
\end{aligned}
\tag{11.73}
$$

where
  w_t is a discrete-time Gaussian white noise with variance R_1
  ε_t is a discrete-time Gaussian white noise with variance R_2

A natural question may be whether the same techniques are available for state space models with continuous-time dynamic models. With the usual limiting technique, we can derive the continuous-time version of the Kalman filter equation as follows. Here we consider only linear state space models. (The nonlinear version of the continuous-time Kalman filter is discussed in Section 11.3.)

In the continuous-time case, two types of the observation model are possible: the first model has the continuous-time observation equation,

$$
\begin{aligned}
dz &= A z dt + B dW \\
dx &= C z dt + dE,
\end{aligned}
\tag{11.74}
$$

where dW and dE are increments of Brownian motions, $W(t)$ and $E(t)$, respectively; hence, both are continuous-time versions of Gaussian white noise. The second model has the discrete-time observation equation,

$$
\begin{aligned}
dz &= A z dt + B dW \\
x_t &= C z_t + \varepsilon_t,
\end{aligned}
\tag{11.75}
$$

where
  dW is a continuous-time Gaussian white noise
  ε_t is a discrete-time Gaussian white noise

The derived Kalman filter equations are slightly different for these two cases.

The continuous state space model with continuous observation is formally written using the Ito stochastic differential equation as

$$
\begin{aligned}
dz &= A z dt + B dW^{(1)} \\
dx &= C z dt + dW^{(2)},
\end{aligned}
\tag{11.76}
$$

where
$dW^{(1)}$ is an increment of a Brownian motion with variance $\Delta t R_1$
$dW^{(2)}$ is an increment of another Brownian motion with variance $\Delta t R_2$

However, it is more convenient to work on its informal equivalent state space model form:

$$\dot{z} = Az + Bw(t)$$
$$x(t) = Cz(t) + \varepsilon(t), \tag{11.77}$$

where
$w(t)$ is a continuous-time Gaussian white driving noise with variance R_1
$\varepsilon(t)$ is a continuous-time Gaussian observation noise with variance R_2

The Kalman filter scheme for the continuous-time state space model is introduced by taking the limit, $\Delta t \to 0$, of the scheme derived as a solution for the discretized version of the problem.

Let us first introduce the most natural discrete approximation of the continuous-time state equation (11.77) as

$$z_{t+\Delta t} = \exp(A\Delta t)z_t + Bw_{t+\Delta t}$$
$$x_t = Cz_t + \varepsilon_t^{(\Delta t)}. \tag{11.78}$$

Here, $w_{t+\Delta t}$ is a discrete-time Gaussian white noise defined by

$$w_{t+\Delta t} = \int_t^{t+\Delta t} \exp(A(t + \Delta t - u))dW^{(1)}(u)$$

$$= \int_t^{t+\Delta t} \exp(A(t + \Delta t - u))w(u)du.$$

The variance of $w_{t+\Delta t}$ is approximately equal to $\Delta t R_1$. We know from the discussion in Chapter 9 that the trajectory of this discretized model (11.78) is consistent with the trajectory of the continuous-time model (11.76) on the discretized time points. Here the observation x_t is defined on any points in the interval, including the discretized time points. But actual observation may be affected by the discretization effect, and the actual observation noise will be $\Delta t \varepsilon_t^{(\Delta t)}$ instead of ε_t, where ε_t is a continuous-time Gaussian white noise corresponding to the Brownian motion $W^{(2)}(t)$, that is, for small Δt, $\Delta t \varepsilon_t \approx \delta W^{(2)}(t) = W^{(2)}(t + \Delta t) - W^{(2)}(t)$. If we denote the variance of the observation noise for a fixed sampling interval Δt by $R_{2,\Delta t}$, then we have $R_{2,\Delta t} \xrightarrow[\Delta t \to 0]{} R_2 = E[\varepsilon_t \varepsilon_t']$, and the variance of $\varepsilon_t^{(\Delta t)}$ in the observation equation of the discretized state space model (11.78) is

$$E[\varepsilon_t^{(\Delta t)}(\varepsilon_t^{(\Delta t)})'] \approx R_{2,\Delta t}/\Delta t.$$

However, we have to remember that this asymptotic argument is only in theory, and it is rare that this actually happens in real-world applications.

Since the state space model (11.78) is an ordinary discrete state space model for a fixed Δt, we have the following Kalman filter scheme:

$$z_{t|t-\Delta t} = \exp(A\Delta t)z_{t-\Delta t|t-\Delta t}. \tag{11.79}$$

$$z_{t|t} = z_{t|t-\Delta t} + K_t v_t. \tag{11.80}$$

$$v_t = x_t - C z_{t|t-\Delta t}. \tag{11.81}$$

$$\Sigma_{v,t+\Delta t} = C P_{t+\Delta t} C' + R_{2,\Delta t}/\Delta t. \tag{11.82}$$

$$K_t = P_{t|t-\Delta t} C' \{C P_{t|t-\Delta t} C' + R_{2,\Delta t}/\Delta t\}^{-1}. \tag{11.83}$$

$$P_{t|t} = P_{t|t-\Delta t} - K_t C P_{t|t-\Delta t}. \tag{11.84}$$

$$P_{t+\Delta t|t} = \exp(A\Delta t)P_{t|t}\exp(A\Delta t)' + B\Delta t R_1 B'. \tag{11.85}$$

For the discrete-time filtering equation (11.79), we have

$$\exp(A\Delta t) = 1 + \Delta t A + o(\Delta t).$$

Then from (11.80) and (11.83) we have

$$z_{t+\Delta t|t+\Delta t} = (1+\Delta t A)z_{t|t} + K_{t+\Delta t}v_{t+\Delta t} + o(\Delta t)$$

$$= (1+\Delta t A)z_{t|t} + P_{t+\Delta t|t}C'(C'P_{t+\Delta t|t}C + R_{2,\Delta t}/\Delta t)^{-1}v_{t+\Delta t} + o(\Delta t)$$

$$(z_{t+\Delta t|t+\Delta t} - z_{t|t})/\Delta t = A z_{t|t} + P_{t+\Delta t|t}C'(\Delta t C'P_{t+\Delta t|t}C + R_{2,\Delta t})^{-1}v_{t+\Delta t} + o(\Delta t).$$

By taking the limit as $\Delta t \to 0$, we obtain

$$\frac{dz_{t|t}}{dt} = A z_{t|t} + P_{t|t}C'R_2^{-1}\{x(t) - C z_{t|t}\}. \tag{11.86}$$

As for the dynamics of the variance of the state estimate errors, we have, from (11.83), (11.84) and (11.85),

$$P_{t+\Delta t|t+\Delta t} = P_{t+\Delta t|t} - K_{t+\Delta t}CP_{t+\Delta t|t}$$

$$= (1+A\Delta t)P_{t|t}(1+A\Delta t)' + B\Delta t R_1 B'$$

$$- K_{t+\Delta t}C[(1+A\Delta t)P_{t|t}(1+A\Delta t)' + B\Delta t R_1 B']$$

$$= P_{t|t} + A\Delta t P_{t|t} + P_{t|t}A'\Delta t + B\Delta t R_1 B'$$

$$- \{P_{t|t} + O(\Delta t)\}C'[C\{P_{t|t}+O(\Delta t)\}C' + R_{2,\Delta t}/\Delta t]^{-1}C\{P_{t|t}+O(\Delta t)\}$$

$$= P_{t|t} + A\Delta t P_{t|t} + P_{t|t}A'\Delta t + B\Delta t R_1 B'$$

$$- \Delta t\{P_{t|t} + O(\Delta t)\}C'[\Delta t C\{P_{t|t}+O(\Delta t)\}C' + R_{2,\Delta t}]^{-1}C\{P_{t|t}+O(\Delta t)\}$$

$$(P_{t+\Delta t|t+\Delta t} - P_{t|t})/\Delta t = AP_{t|t} + P_{t|t}A' + BR_1 B'$$

$$- \{P_{t|t} + O(\Delta t)\}C'[\Delta t C\{P_{t|t}+O(\Delta t)\}C' + R_{2,\Delta t}]^{-1}C\{P_{t|t}+O(\Delta t)\}.$$

By taking the limit, as $\Delta t \to 0$, we obtain

$$\frac{dP_{t|t}}{dt} = AP_{t|t} + P_{t|t}A' + BR_1 B' - P_{t|t}C'R_2^{-1}CP_{t|t}. \tag{11.87}$$

Equations 11.86 and 11.87 specify the time evolution of the state estimation error and the error variance for the case of the continuous-time state space model. They show a nice correspondence to the discrete-time Kalman filter, but they are not really useful in applications in practice.

In practice, it is often the case that the data are given as time series with a sufficiently small sampling interval, but the dynamic model for explaining the phenomena behind the data is a continuous-time dynamic model. For the estimation of the unobserved state variables of such state space models, we can formulate, from the beginning, the state space model with a continuous-time state equation and a discrete-time observation equation. The state space model is called a "continuous–discrete" type, while the state space model with continuous-time state and the continuous-time observation is called a "continuous–continuous" type. In actual applications, the continuous–discrete type state space formulation is more likely to be useful in many scientific studies including neuroscience.

The Kalman filter scheme of the continuous–discrete state space model is slightly different from the continuous–continuous case. We first introduce a discrete approximation of the continuous-time state equation,

$$dz = Az\,dt + B\,dW$$
$$x_t = Cz_t + \varepsilon_t \tag{11.88}$$

as

$$z_{t+\Delta t} = \exp(A\Delta t)z_t + Bw_{t+\Delta t}$$
$$x_t = Cz_t + \varepsilon_t \tag{11.89}$$

where the discretized state equation is the same as that of the continuous–continuous type. The difference is in the observation equation. Here the observations are given in discrete-time points as the ordinary discrete-time Kalman filter cases.

The derivation of the Kalman filter for the "continuous–discrete" case is more or less the same as the derivation for the "continuous–continuous" case, except for the part relating to the observation equation. From the discrete-time filtering equation (11.79), we have

$$A(z_{t|t}) = \exp(A\Delta t)$$

$$= 1 + \Delta tA + o(\Delta t).$$

Then from (11.80) and (11.83) we have

$$z_{t+\Delta t|t+\Delta t} = (1 + \Delta tA)z_{t|t} + K_{t+\Delta t}v_{t+\Delta t} + o(\Delta t)$$

$$= (1 + \Delta tA)z_{t|t} + P_{t+\Delta t|t}C'(CP_{t+\Delta t|t}C' + \sigma_\varepsilon^2)^{-1}v_{t+\Delta t} + o(\Delta t). \tag{11.90}$$

Then we have

$$\frac{(z_{t+\Delta t|t+\Delta t} - z_{t|t})}{\Delta t} = Az_{t|t} + \frac{1}{\Delta t}P_{t+\Delta t|t}C'(CP_{t+\Delta t|t}C' + \sigma_\varepsilon^2)^{-1}v_{t+\Delta t} + o(\Delta t) \quad \text{for } t_{k-1} < t < t_k.$$

Since we do not have any observation between the interval $t_{k-1} < t < t_k$, we have $v_{t+\Delta t} = 0$ for $t_{k-1} < t < t_k$. Then we have

$$\left\{ Az_{t|t} + \frac{1}{\Delta t}P_{t+\Delta t|t}C'(CP_{t+\Delta t|t}C' + \sigma_\varepsilon^2)^{-1}v_{t+\Delta t} + o(\Delta t) \right\} \xrightarrow[\Delta t \to 0]{} 0.$$

For a small time interval, $t_k^- < t \le t_k^+ (=t_k + \delta t)$, we have an observation, at $t = t_k$, and we have a nonzero innovation, for $t + \Delta t = t_k^+ (=t_k + \delta t)$, as $v_{t+\Delta t} = x_{t_k^+} - Cz_{t_k^-|t_k^-}$. Then, using (11.90), the filtered state $z_{t_k^+|t_k^+}$ is updated as

$$z_{t_k^+|t_k^+} = z_{t_k^-|t_k^-} + P_{t_k^+|t_k^-}C'(CP_{t_k^+|t_k^-}C' + \sigma_\varepsilon^2)^{-1}(x_{t_k^+} - Cz_{t_k^-|t_k^-}) + \{\Delta tA + o(\Delta t)\}.$$

Therefore, by taking the limit $\Delta t \to 0$, we have from (11.90)

$$\frac{dz_{t|t}}{dt} = Az_{t|t} \quad \text{for } t_{k-1} < t < t_k \tag{11.91}$$

$$z_{t_k^+|t_k^+} = z_{t_k^-|t_k^-} + P_{t_k^+|t_k^-}C'\left(CP_{t_k^+|t_k^-}C'+\sigma_\varepsilon^2\right)^{-1}\left(x_{t_k^+} - Cz_{t_k^-|t_k^-}\right) \quad \text{for } t_k^- < t \le t_k^+(=t_k+\delta t).$$

$$(11.92)$$

As for the dynamics of the variance of the state estimation errors, we have, from (11.83 through 11.85),

$$P_{t+\Delta t|t+\Delta t} = P_{t+\Delta t|t} - K_{t+\Delta t}CP_{t+\Delta t|t}.$$

Since it holds that

$$P_{t+\Delta t|t} = (1+A\Delta t)P_{t|t}(1+A\Delta t)' + B\Delta t R_1 B',$$

we have

$$P_{t+\Delta t|t+\Delta t} = (1+A\Delta t)P_{t|t}(1+A\Delta t)' + B\Delta t R_1 B' - K_{t+\Delta t}CP_{t+\Delta t|t}$$

$$= P_{t|t} + A\Delta t P_{t|t} + P_{t|t}A'\Delta t + B\Delta t R_1 B' + o(\Delta t).$$

From this, we have

$$\frac{dP_{t|t}}{dt} = AP_{t|t} + P_{t|t}A' + BR_1B' \quad \text{(Continuous–Discrete)}. \qquad (11.93)$$

Note that the assumption of the asymptotics of the noise variance, that is, $R_2/\Delta t \underset{\Delta t \to 0}{\to} \sigma_\varepsilon^2$, does not match the reality. Therefore, we take a more realistic assumption that the noise variance σ_ε^2 is constant for the interval $(t_k + \delta t, t_{k+1}]$.

11.3 Nonlinear Kalman Filters

Although many dynamic models in sciences are approximately linear, there are some intrinsically nonlinear dynamic models, some of which we saw in Chapter 5. A typical example is the state space model for Hodgkin–Huxley equations (Hodgkin and Huxley 1952, Cronin 1987).

When we observe only voltage V_t out of the five variables, V_t, m_t, h_t, n_t and I_t, of the Hodgkin–Huxley equation, we can formulate the model and the observed time series in the form of a state space model:

$$\dot{z} = f(z)$$

$$x_t = (1,0,0,0)z_t + \varepsilon_t,$$

$$(11.94)$$

where

$$f(z) = \begin{pmatrix} \dfrac{-1}{C_M}[\{\bar{g}_{Na}m(t)^3h(t) + \bar{g}_K n(t)^4 + \bar{g}_l\}V(t) + \bar{g}_{Na}m^3hE_{Na} + \bar{g}_K n^4 E_K + \bar{g}_l E_l + I(t)] \\ -\{\alpha_m(V(t)) + \beta_m(V(t))\}m(t) + \alpha_m(V(t)) \\ -\{\alpha_h(V(t)) + \beta_h(V(t))\}h(t) + \alpha_h(V(t)) \\ -\{\alpha_n(V(t)) + \beta_n(V(t))\}n(t) + \alpha_n(V(t)) \end{pmatrix},$$

and the additive observation noise ε_t is a Gaussian white noise. For the stochastic Hodgkin–Huxley model with additive driving noise for dV/dt, we have a stochastic version of Hodgkin–Huxley state space model:

$$\dot{z} = f(z) + Bw(t)$$

$$x_t = (1,0,0,0)z_t + \varepsilon_t, \tag{11.95}$$

where $w(t)$ is a Gaussian white noise and B is a 4×1 matrix,

$$B = \begin{pmatrix} 1 \\ 0 \\ 0 \\ 0 \end{pmatrix}.$$

From the observed time series data, $x_1, x_2, \ldots, x_N$, we want to estimate state vector variables, $z_t = (V_t, m_t, h_t, n_t)$, ($t = 1, 2, \ldots, N$). Obviously, we cannot apply Kalman's linear filtering method to (11.94) since the state equation is a highly nonlinear dynamical system.

In Chapters 5 and 6, we have seen some other useful parametric nonlinear dynamical system models both in continuous time and discrete time (such as deterministic differential equation models, stochastic differential equation models, nonlinear AR models, neural network models, and chaos models). When we need to take into account the presence of observation errors and unobserved variables with these "nonlinear" dynamic models, we may naturally look for a nonlinear filtering scheme with these nonlinear dynamic models both in discrete time and in continuous time.

According to Jazwinski (1970), the nonlinear filtering problem is classified into the following three types:

1. Continuous–continuous type:

$$dz = a(z)dt + b(z)dW$$

$$dx = c(z)dt + d\eta. \tag{11.96}$$

Here the state equation and the observation equation are both in continuous time.

2. Continuous–discrete type:

$$dz = a(z)dt + b(z)dW$$
$$x_t = c(z_t) + \varepsilon_t. \tag{11.97}$$

Here the state equation is in continuous time, but the observations are given in discrete-time points.

3. Discrete–discrete type:

$$z_t = a(z_{t-1}) + b(z_{t-1})w_t$$
$$x_t = c(z_t) + \varepsilon_t. \tag{11.98}$$

Here both the state equation and the observation equation are given in discrete time.

The theoretical approach has focused mainly on the continuous–continuous type where the main issue is the characterization of the time evolution of the conditional density $p(z, t|t)$ of the state $z(t)$ given all the observations up to time t, that is, $x(t')$, $t \geq t' > t_0$. If $a(z)$ is nonlinear in (11.96) or (11.97), the process defined by the state equation is a non-Gaussian process even though the driving noise dW is a Gaussian white noise. When the state process is non-Gaussian, it involves not only the first two moments but also all the higher order moments. Thus, the characterization of the time evolution of the conditional distribution is equivalent to a characterization of an infinite dimensional dynamics. Kushner gave an elegant solution for this with a partial differential equation similar to the Fokker-Planck equation (cf. Kushner 1962, 1967, Jazwinski 1970, Sage and Melsa 1971). Unfortunately, it is, in general, not possible to solve this partial differential equation analytically. Then we need to look for an approximate solution. As long as the solution is useful and computationally efficient, whether the solution is exact or approximate is not an important matter.

In the present section, we are going to discuss the materials of nonlinear filtering from a heuristic and pragmatic approach for the benefit of applied scientists. (For those readers who are interested in more mathematical and formal treatment of the present nonlinear filtering method, we refer the readers to Jimenez 2011). Note that a common property of the state space models in (11.96 through 11.98) is that the dynamic models of the state equations are all "local Gauss," that is, dW and w_t are Gaussian distributed. Here the non-Gaussianity of the state process $z(t)$ or z_t, of (11.96) through (11.98), results from the nonlinearity of the dynamics.

The main concern in this section is, therefore, the derivation of the equations of the time evolution of the conditional distribution $p(z_t|X^t)$ of the "local Gauss" and "finite-variance" state process z_t with the condition that the observation $X^t = (x_1, x_2, \ldots, x_t)$ is given. Remember that, in Section 11.2.8, we saw that the linear Kalman filter characterizes a time evolution of the conditional distribution of $z(t)$ or z_t by the time evolution of the conditional mean and the conditional variance of the state process, both in continuous time and in discrete time. When the state process is non-Gaussian distributed, the mean and the variance are no longer sufficient statistics for the characterization of the distribution of the state. Theoretically speaking, we need higher order moments of state estimation errors in order to characterize the time evolution of the first and the second moments. Thus, we need to specify the time evolution of all the higher order moments of the state distribution as well as the mean and the variance.

11.3.1 Kushner's General Solution

Kushner (1962) gave an explicit solution for the characterization of the time evolution of the conditional distribution $p(z_t|X^t)$ of the state process of the local-Gauss nonlinear state space model:

$$dz = a(z)dt + b(z)dW(t)$$
$$dx = Cz + d\eta. \tag{11.99}$$

Here, $dW(t)$ and $d\eta(t)$ are increments of the Brownian motion $W(t)$ and $\eta(t)$, respectively. In general, the nonlinear filtering problem is considered with a nonlinear observation equation so that the state space model is

$$dz = a(z)dt + b(z)dW(t)$$
$$dx = c(z) + d\eta, \tag{11.100}$$

where $c(z)$ is a function (possibly nonlinear) of the state z. However, we first consider only the linear observation cases as Kushner did. We see the reason for this and some problems associated with nonlinear observation equations in Section 11.3.6.

The Kushner's solution to the inference problem (a) for (11.99) is given as

$$p(z, t + dt|X^{t+dt}) - p(z, t|X^t) \ (= dp(z, t|X^t))$$

$$= -a(z,t)dt\frac{\partial p(z,t|X^t)}{\partial z} + \frac{1}{2}b^2(z)\frac{\partial^2 p(z,t|X^t)}{\partial z}dt + dQ(z,t) \tag{11.101}$$

$$dQ(z,t) = p(z,t|X^t)\left\{dx - \underset{p}{E}[Cz]dt\right\}\Sigma_v^{-1}\left\{Cz - \underset{p}{E}[Cz]\right\},$$

where $\underset{p}{E}[.]$ is the conditional expectation using $p(z_t|X^t)$. Note that the time evolution of the function $p(z)$ of the space z, $-\infty < z < \infty$, of a diffusion process model, $dz = a(z)dt + b(z)dW(t)$, is described by a partial differential equation, which is an infinite dimensional dynamical system. However, since we have

$$\frac{\partial p(z|z,t)}{\partial t} = \sum_{n=1}^{\infty} \frac{(-1)^n}{n!} \frac{\partial^n}{\partial x^n} [M_n(z)p(z(t+\tau)|z,t)],$$

where

$$M_n(z) = \lim_{\tau \to 0} \frac{1}{\tau} \int_{\Omega} (z(t+\tau) - z(t))^n p(z(t+\tau)|z,\tau)dz$$

and

$$\lim_{\tau \to 0} \frac{1}{\tau} \int_{\Omega} (z(t+\tau) - z(t))p(z(t+\tau)|z(t),\tau)dz = a(z)$$

$$\lim_{\tau \to 0} \frac{1}{\tau} \int_{\Omega} (z(t+\tau) - z(t))^2 p(z(t+\tau)|z(t),\tau)dz = b(z)$$

$$\lim_{\tau \to 0} \frac{1}{\tau} \int_{\Omega} (z(t+\tau) - z(t))^n p(z(t+\tau)|z(t),\tau)dz = 0, \quad n \geq 3,$$

the time evolution of the conditional density distribution of the Markov diffusion process is described simply by

$$\frac{\partial p(z(t))}{\partial t} = -a(z(t))\frac{\partial p(z(t))}{\partial z} + \frac{1}{2}b^2(z(t))\frac{\partial^2 p(z(t))}{\partial z}, \tag{11.102}$$

which is called the Fokker–Planck equation or the Kolmogorov forward equation. Note that even though we need only $E[dz]$ and $E[(dz)^2]$, that is, the moments up to the second order of the increment dz, of the process $z(t)$, for describing the dynamics of $p(z)$, we still need an infinite dimensional dynamical system to characterize the spatial dynamics specified by (11.102).

A difference of Kushner's equation (for the conditional density distribution $p(z(t)|X^t)$ and the Fokker–Planck (Kolmogorov forward) equation is that the Kushner equation has an extra term,

$$dQ(z,t) = p(z(t)|X^t)\left\{dx - \underset{p}{E}[Cz]dt\right\}\Sigma_v^{-1}\left\{Cz - \underset{p}{E}[Cz]\right\},$$

which is a linear function of the new observation dx, driving the system of the Fokker–Planck equation (11.102). This extra term plays the role of updating the conditional density distribution $p(z(t)|X^t)$ with the new observation, that is, a similar role played by the terms with the Kalman gain in the Kalman filter scheme for the linear case.

Unfortunately, Kushner's filtering equation cannot be solved analytically but it can be solved numerically. The computational cost of (numerically) solving the Kushner equation is very large compared with the linear Kalman filter scheme, but it is much less than the computational cost for the numerical solution of the Jazwinski scheme (11.103) through (11.105):

$$p_{(o)}(x_{t+1}|x_t,\ldots,x_1,\theta) = \int p(x_{t+1}|z_{t+1})p_{(p)}(z_{t+1}|x_t,\ldots,x_1,\theta)dz_{t+1} \qquad (11.103)$$

$$p_{(p)}(z_{t+1}|x_t,\ldots,x_1,\theta) = \int p(z_{t+1}|z_t)p_{(f)}(z_t|x_t,\ldots,x_1,\theta)dz_t \qquad (11.104)$$

$$p_{(f)}(z_t|x_t,\ldots,x_1,\theta) = \frac{p(x_t|z_t)p_{(p)}(z_t|x_{t-1},\ldots,x_1,\theta)}{\int p(x_t|\varsigma_t)p_{(p)}(\varsigma_t|x_{t-1},\ldots,x_1,\theta)d\varsigma_t}. \qquad (11.105)$$

11.3.2 Extended Kalman Filters

In real-world applications, people are mostly interested in the estimation of the conditional mean and the variance of the state, that is, the first- and the second-order moments even in the non-Gaussian case. Therefore, many approximate nonlinear filtering methods developed in the 1960s–1980s concern the derivation of the time evolution of the conditional mean and the conditional variance. The most well known method is the extended Kalman filter (EKF) scheme.

There are three different definitions of EKF in the literature (see, for example, Schwartz and Stear 1968, Jazwinski 1970, Sage and Melsa 1971, Liang 1983). This is because the derivations are not free from how the nonlinear filtering problem is presented. The three different definitions of EKF are typically shown here. Here we assume that the observation equation is linear for simplicity.

1. EKF-1: For a continuous-time model with continuous-time filtering equations (Schwarz and Stear 1968, Sage and Melsa 1971, Liang 1983):

State space model:

$$\begin{cases} dz = a(z)dt + b(z)dW \\ dx = Cdz + d\eta \end{cases}$$

Filtering equations:

$$
\begin{cases}
\dfrac{dz_{t|t}}{dt} = a(z_{t|t}) + K_t\{x(t) - Cz_{t|t}\} \\[2mm]
K_t = P_{t|t}C'R_2^{-1} \\[2mm]
\dfrac{dP_{t|t}}{dt} = J_a(t)P_{t|t} + P_{t|t}J_a(t)' + b(z_{t|t})R_1 b(z_{t|t})' - P_{t|t}C'R_2^{-1}CP_{t|t}'
\end{cases}
$$

Here, $J_a(t) = \left(\dfrac{\partial a(z)}{\partial z}\right)_{z=z_{t|t}}$.

2. EKF-2: For a continuous-time model with discrete-time filtering equations (Jazwinski 1970):

State space model:

$$
\begin{cases}
dz = a(z)dt + b(z)dW \\
x_t = Cz_t + \varepsilon_t
\end{cases}
$$

Filter equations:

$$
\begin{cases}
z_{t+1|t} = \exp\{J_a(z_{t|t})\Delta t\}z_{t|t} \\
z_{t|t} = z_{t|t-1} + K_t(x_t - Cz_{t|t-1}) \\
K_t = P_{t|t-1}C'(CP_{t|t-1}C' + R_2)^{-1} \\
P_{t|t-1} = \exp\{J_a(z_{t|t})\}P_{t-1|t-1}\exp\{J_a(z_{t|t})\}' + b(z_{t-1|t-1})R_1 b(z_{t-1|t-1})' \\
P_{t|t} = P_{t|t-1} - K_tCP_{t|t-1}
\end{cases}
$$

Here, $J_a(z_{t|t}) = \left(\dfrac{\partial a(z)}{\partial z}\right)_{z=z_{t|t}}$.

3. EKF-3: For a discrete-time model with discrete-time filtering equations (Sage and Melsa 1971, Liang 1983):

State space model:

$$
\begin{cases}
z_t = A(z_{t-1}) + B(z_{t-1})w_t \\
x_t = Cz_t + \varepsilon_t
\end{cases}
$$

Filter equations:

$$
\begin{cases}
z_{t+1|t} = A(z_{t|t}) \\
z_{t|t} = z_{t|t-1} + K_t(x_t - Cz_{t|t-1}) \\
K_t = P_{t|t-1}C'(CP_{t|t-1}C' + R_2)^{-1} \\
P_{t|t-1} = J_A(z_{t|t})P_{t-1|t-1}J'_A(z_{t|t}) + b(z_{t-1|t-1})R_1 b(z_{t-1|t-1})' \\
P_{t|t} = P_{t|t-1} - K_t C P_{t|t-1}
\end{cases}
$$

Here, $J_A(z_{t|t}) = \left(\dfrac{\partial A(z)}{\partial z} \right)_{z=z_{t|t}}$.

It may look confusing that the same name is used for the three different schemes. The difference comes from the difference of the formulations of the nonlinear filtering. EKF-1 comes from the (1) continuous–continuous type formulation (11.96), which leads to continuous-time filtering equations with the continuous-time nonlinear dynamic state model. EKF-2 comes from the (2) continuous–discrete type formulation (11.97), which leads to discrete-time filtering equations with the continuous-time nonlinear dynamic state model. EKF-3 comes from the (3) discrete–discrete formulation (11.98), which leads to discrete-time filtering equations with the discrete-time nonlinear dynamic state model.

What is common between the three definitions of the EKF is that they are all concerned with "estimations of the conditional mean and the conditional variance of the state, that is, the first and the second-order moments, and their time evolution" even when the state process is non-Gaussian. This is, in a sense, reasonable, because most important nonlinear filtering problems are specified and formulated with a nonlinear dynamic state model together with the Gaussian driving noise and the Gaussian observation noise. There are some cases where the driving noise is non-Gaussian with a few outliers. The nonlinear filtering problems of such nonlinear dynamic state models with "local non-Gauss" driving noise are discussed in Section 13.5.

Note that the EKF-3 does not pay attention to the stationarity of the Markov chain defined by the discrete-time state dynamics. Many discrete-time nonlinear filtering schemes are known to have instability problems coming from this instability the of discrete-time Markov chain state process. It will not make sense to consider a solution of the filtering problem for any nonstationary explosive state space models. The filtering problem for a state space model may be only worth considering only when the state equation defines a stationary (finite variance) discrete-time Markov chain process. If the state process is always diverging to infinity, the estimation of the state and minimizing the variance of the estimation error cannot be an important problem to most scientists. A recursive filtering scheme for the "well-defined" (i.e., finite-variance) discrete-time state space model is given in Section 11.3.4.

For the above reasons, we concentrate on the discussion of the nonlinear filtering problems of the other two types, i.e., (1) continuous–continuous type:

$$dz = a(z)dt + b(z)dW$$
$$dx = Czdt + d\eta,$$

(11.106)

and, (2) continuous–discrete type

$$dz = a(z)dt + b(z)dW$$
$$x_t = Cz_t + \varepsilon_t$$

(11.107)

It must be noted that the idea of the continuous-time observed data is not a practical but a theoretical concept. In statistical data analysis, the observation data are usually available in discrete time. Even though some of the measurements may be recorded continuously with a recording machine, they need to be stored in a computer with a small discrete-time interval. Even in artificial simulation studies, the simulation data of any continuous-time differential equation model is actually generated with a very fine sampling interval. That means the process of solving the continuous–continuous type filtering problem needs to go through the discrete-time observations, at some stage, and needs to be solved using these discretized data. This means that the continuous–continuous filter problem is used in fact as a continuous–discrete type filtering problem in applications. Here the discretization approximation of the continuous-time equation to fit the discrete-time observation data affects the filtering results.

In order to get an efficient and computationally useful filtering scheme for the "local Gauss" state space model,

$$dz = a(z)dt + b(z)dW(t)$$

$$x_t = Cz_t + \varepsilon_t,$$

we need to use two types of approximations. One is the approximation (D-approximation) associated with the time discretization of continuous-time dynamical system, and another is the approximation (F-approximation) associated with the derivation of the filtering equation for the mean and the variance of the state from the theoretically infinite-dimensional, filtering equation for the conditional density distribution of the state.

For example, EK-1 employs the F-approximation first and the D-approximation after route. We could call this approach for deriving the discrete-time filter F-D type or F-D (see Figure 11.2). EKF-2 employs

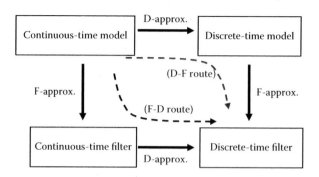

FIGURE 11.2
D-F route and F-D route.

the D-approximation first and the F-approximation after (i.e., D-F route in Figure 11.2). Most nonlinear recursive filtering schemes, for the local-Gauss state space models, are classified into these two types, D-F type and F-D type.

11.3.3 Local Gauss State Space Models and Two Types of Approximations

11.3.3.1 Scheme of F-D Route

There are so many schemes for the continuous-time nonlinear filtering for the state space model,

$$dz = a(z)dt + b(z)dW$$

$$dx = Czdt + d\eta,$$

where the dynamics of the conditional mean $z_{t|t}$ and the conditional variance $P_{t|t}$ are specified by nonlinear differential equations. One of the most well known is the EKF-1 scheme (Schwartz and Stear 1968, Sage and Melsa 1971, Liang 1983), given by

$$\frac{dz_{t|t}}{dt} = a(z_{t|t}) + K_t\left\{x(t) - Cz_{t|t}\right\}$$

$$K_t = P_{t|t}C'R_2^{-1}$$

$$\frac{dP_{t|t}}{dt} = J_a(t)P_{t|t} + P_{t|t}J_a(t)' + b(z_{t|t})R_1 b(z_{t|t})' - P_{t|t}C'R_2^{-1}CP'_{t|t}.$$

(11.108)

Here the observation is defined as continuous-time observation, but actual observations are given, in the computer, as discrete-time observations with a very small time interval. Therefore in the actual applications, the continuous-time innovation, $v(t) = x(t) - Cz_{t|t}$, needs to be replaced by its time-discretized approximation, $v_t = x_t - Cz_{t|t}$. The actual solution of the nonlinear filtering for time series data is obtained by numerically solving the differential equations (11.108) for $z_{t|t}$ and $P_{t|t}$, with some discretization method, at each discrete-time point where the time series observation is given. That means we need to numerically solve these differential equations at finely meshed time points in a certain time interval. Thus, we can have as many different schemes as the time discretization schemes for differential equations, such as the Euler method, the Heun method, or the Runge–Kutta method, etc., in the F-D root, even if we fix the F-approximation to be EKF-1 of (11.108).

11.3.3.2 Scheme of D-F Route

This approach (Jazwinski 1970, Ozaki 1993a) employs the D-approximation first to transform the continuous–discrete filtering problem,

$$dz = a(z)dt + b(z)dW$$

$$x_t = Cz_t + \varepsilon_t,$$

into a discrete–discrete filtering problem,

$$z_t = f(z_{t-1}) + G(z_{t-1})w_t$$

$$x_t = Cz_t + \varepsilon_t,$$

where $f(z_{t-1})$ and $G(z_{t-1})$ are some nonlinear functions of z_{t-1} derived from $a(.)$ and $b(.)$.

EKF-2 is regarded as one of the D-F route type schemes, where the continuous-time model is transformed to

$$z_t = \exp(J(z_{t-1}))z_{t-1} + b(z_{t-1})w_t$$

$$x_t = Cz_t + \varepsilon_t.$$

Since there are so many discretization schemes, we have as many different nonlinear filtering schemes as the time discretization schemes for any differential equation, in the D-F route as well, even if we fix the F-approximation to be one specific type.

11.3.3.3 D-F Route versus F-D Route

In EKF-1, the continuous–discrete filtering problem is solved using the continuous-time filtering equations, where the continuous-time innovation, $v(t) = x(t) - Cz_{t|t}$, needs to be calculated by an approximation of the continuous-time observation equation $v_t = x_t - Cz_{t|t}$, in actual computations.

Users of the filtering scheme must know that the F-D route schemes such as EKF-1 is not very useful for practical applications and is computationally inefficient compared with the schemes of D-F route such as EKF-2. This has been shown (Ozaki 1993a) with examples of typical polynomial type nonlinear dynamic model, $dz/dt = -6z + 5.5z^3 - z^5 + w(t)$, and the stochastic van der Pol oscillation model, $dz^2dt^2 - 3(1 - z^2)dz/dt + z = w(t)$.

This means that the D-F route associated with the continuous–continuous formulation of the nonlinear filtering is not convenient for practical users of nonlinear filtering. Incidentally, it is also known that the continuous-time observation leads to a difficult mathematical problem that cannot be obviated without introducing clumsy assumptions (see Balakrishnan 1980).

EKF-2 of the D-F route is the only EKF scheme that is free from the computational explosion problem. With the EKF-2, computational stability is guaranteed when the original state process is a nonexplosive finite-variance process. This is because the transition matrix of the discretized Markov chain has eigenvalues within the unit circle for small Δt, that is,

$$\left|eig\left\{\exp\left(\frac{\partial a(z)}{\partial z}\Delta t\right)\right\}\right| < 1,$$

if the original continuous-time process is stable and real parts of the eigenvalues of the original process are negative for large z, that is,

$$\mathrm{Re}\left\{eig\left(\frac{\partial a(z)}{\partial z}\right)\right\} < 0.$$

However, unlike the LL (Local Linearization)-based discrete-time filtering scheme (which we will see in Section 11.3.4), the scheme of EKF-2 is not consistent because

$$\frac{z_{t+\Delta t} - z_t}{\Delta t} \underset{\Delta t \to \infty}{\to} \frac{\partial a(z)}{\partial z} z \neq a(z),$$

whereas the LL-based schemes are consistent and

$$\frac{z_{t+\Delta t} - z_t}{\Delta t} \underset{\Delta t \to \infty}{\to} a(z).$$

11.3.3.4 D-F Route and Discretizations of the Local Gauss State Space Model

If we want to improve the EKF-2 scheme of the D-F route, an easy way is to replace the model discretization scheme in the EKF-2 by a more efficient scheme with desirable properties such as consistency, stability, and a higher order of convergence in terms of the discretization interval Δt.

Any discretization method of a stochastic differential equation model would transform the following nonlinear state space model

$$\dot{z} = a(z) + B(z)w(t)$$

$$x_t = Cz_t + \varepsilon_t$$

into a discrete-time state space model. The simplest example is the Euler–Maruyama scheme, which transforms this state space model to

$$z_t = z_{t-1} + \Delta t a(z_{t-1}) + B(z_{t-1})\sqrt{\Delta t}\, w_t$$

$$x_t = Cz_t + \varepsilon_t.$$

There are other schemes (see Chapter 9) such as the Milstein scheme and the Runge–Kutta type Kloeden–Platen scheme, and the LL type schemes. Each discretization scheme yields a different discrete-time state space model, defining a Markov chain process approximating the original continuous-time Markov diffusion process of the state.

If there are so many possible discrete-time state space models for a given continuous-time state space model, we need to choose the most suitable one. In the field of numerical analysis, the time discretization of a stochastic differential equation has been considered only from the viewpoint of approximating the solution, that is, the integrated trajectories, where the main concern of the approximation is the speed of the convergence of the solution defined by

$$E[|z(T) - z_{t_N}| \,|\, z(0) = z_0] = O(\Delta t)^m.$$

Here we have $T = t_N = t_0 + N\Delta t$. The speed is faster when m is larger. However as the optimality of the filtering scheme does not necessarily guarantee the computational stability of the scheme, the fast convergence of the trajectory does not necessarily lead us to the computational stability of the discretization scheme.

Note that each explicit scheme gives rise to a difference equation model,

$$z_{t+\Delta t} = \psi(z_t, w_{t+\Delta t}, w_{t+\Delta t}^2, \dots, w_{t+\Delta t}^r),$$

defining a Markov chain. As we saw in Chapter 9, the consistency and the fast speed of convergence of the trajectory do not guarantee the ergodicity and the existence of the stationary distribution of the Markov chain. A simple example is

$$z_{t+\Delta t} = z_t - \Delta t z_t^3 + w_{t+\Delta t}, \tag{11.109}$$

which is derived from $dz = -z^3 dt + dW(t)$ by the Euler–Maruyama scheme. While the process $z(t)$ defined by $dz = -z^3 dt + dW(t)$ is ergodic and stationary, the Markov chain defined by (11.109) is known to be nonergodic, and the trajectory of (11.109) is known to explode when z_t takes a large value. Obviously, the Markov chain process defined by the Euler scheme is not a finite-variance process.

It is known that A-stability (see Chapter 9) is a sufficient condition for a discretization scheme to transform a finite-variance continuous-time process into a finite-variance discrete-time Markov chain process. Remember that we saw, in Chapter 9, that only LL type schemes satisfy A-stability, guaranteeing the stability of the resulting discrete-time Markov chain process. Also, it is known that the resulting stationary distribution of the discretized Markov chain is known to be consistent and converges, geometrically quickly, to the original one as sampling interval $\Delta t \to 0$ (Tweedie 1975).

11.3.4 Local Linearization Filter

By applying the LL-based discretization schemes to the nonlinear state space model,

$$\dot{z} = a(z) + B(z)w(t)$$

$$x_t = Cz_t + \varepsilon_t,$$

we obtain a special type of discrete-time state space model as

$$z_t = \phi(z_{t-1}) + \Gamma(z_{t-1})w_t$$
$$x_t = Cz_t + \varepsilon_t \ . \tag{11.110}$$

We note that both system noise w_t and the observation noise ε_t of this discretized state space model are Gaussian so that the prediction errors, $z_t - z_{t|t-1}$, and $x_t - Cz_{t|t-1}$ are Gaussian. Taking advantage of this local Gaussianity of the state space model, we can obtain an LL-based discrete-time approximate nonlinear filtering scheme similar to the linear Kalman filter as follows:

$$z_{t+1|t} = \phi(z_{t|t})$$
$$z_{t|t} = z_{t|t-1} + K_t(x_t - Cz_{t|t-1})$$
$$K_t = P_{t|t-1}C'(CP_{t|t-1}C' + R_2)^{-1}$$
$$P_{t|t-1} = A(z_{t-1|t-1})P_{t-1|t-1}A(z_{t-1|t-1})' + \Gamma(z_{t-1|t-1})R_1\Gamma(z_{t-1|t-1})'$$
$$P_{t|t} = P_{t|t-1} - K_tCP_{t|t-1} \tag{11.111}$$
$$A(z_{t|t}) = \exp\{J_t\Delta t\}$$

$$\text{with } J_t = \left(\frac{\partial a(z)}{\partial z}\right)_{z=z_{t|t}},$$

where with the LL-based scheme, $\phi(z_{t|t})$ is given by

$$\phi(z_{t|t}) = J_t^{-1}\{\exp(J_t\Delta t) - I\}a(z_{t|t}).$$

The NLL-based nonlinear filtering scheme is given in the same way together with the NLL discretization scheme (see Section 9.5.4). Note that the evolution equation of $P_{t|t-1}$ is exact only when the state model is linear. The approximation for specifying the evolution equation of $P_{t|t-1}$ of (11.111) is a commonly used one in the derivation of an approximate nonlinear filter for the local-Gauss state space model. The exact computation of $P_{t|t-1}$ is not easy since the state prediction errors are dependent on the higher order moment, of the state estimation errors. Here we presented and discussed the LL-based filtering scheme in a heuristic way for applied scientists. For those readers who are interested in a more mathematical formal approach, we refer Jimenez and Ozaki (2002, 2003, 2005).

Note that in the ordinary discrete extended Kalman filter, that is, EKF-3 for the discrete-time state space model (11.110), $A(z_{t|t})$ is specified as

$$A(z_{t|t}) = \left(\frac{\partial\phi(z)}{\partial z}\right)_{z=z_{t|t}}.$$

An alternative heuristic way is given by Ozaki (1993a) for specifying $A(z_{t|t})$ such that $A(z_{t|t})z_{t|t} = a(z_{t|t})$, which is possible for some zero-mean processes such as the random oscillations generated from stochastic van der Pol equations or stochastic Duffing equations.

We note that the simplicity of the above LL-based nonlinear filtering schemes comes from their local linear assumption and hence the local Gaussian assumption of the state process, which is an intrinsic property of the Markov diffusion type processes defined by stochastic differential equation models. Unless the sampling interval Δt is very large, the density distribution of the increment, that is, $z_{t+\Delta t} - z_t$, of any state process defined by a stochastic differential equation is approximately Gaussian. It is unlikely that neuroscientists would conduct an experiment with an extraordinarily large sampling interval for phenomena where a continuous-time stochastic differential equation model is considered. In many cases such as modeling neural impulse data with the Hodgkin–Huxley model or modeling alpha rhythm EEG data with the Zetterberg model, the sampling interval can be assumed to be sufficiently small to satisfy the requirement of the above local step-wise Gaussianity.

11.3.5 Discussions

It is well known that the most difficult part of nonlinear filtering is computational instability. The exact solution of nonlinear filtering is given as a solution of an infinite dimensional dynamical system that is impossible to solve analytically. To obtain a practical nonlinear filtering scheme, some kind of discrete-time approximation is inevitable, and most (not all) discrete approximations might possibly have computational instability problem (see Schwartz and Stear 1968, Liang 1983). There are many reasons for the computational instabilities in nonlinear filtering.

If the state process is not finite variance and explosive, it is hard to see whether the explosion is an eligible explosion of the state process or an ineligible explosion caused by an inappropriate approximate nonlinear filter. Since we consider filtering of only finite-variance state processes, where the original state process is assumed to be stationary and have finite mean and variance, we are free from the common mistakes and confusions of interpreting the instability coming from inappropriate discrete-time model approximation as a fault of the filtering approximations.

In most textbooks of nonlinear filtering, people do not set any special condition for the state dynamic model,

$$dz = a(z)dt + b(z)dW(t),$$

except Ito's conditions for the existence of a solution of the stochastic differential equation for the state process (see Jazwinski 1970). Conventional nonlinear filtering theory considers the temporal evolution of the conditional distribution of the state, hence the evolution of moments (of the state) of all orders. However, if the process is explosive, the conditional variance of the state estimation errors also explodes. Then the state estimation error variance is also explosive. In such circumstances, nonlinear filtering does not make sense and is unnecessary for most users in applied sciences.

Confusion comes also from the fact that in the nonlinear filtering problem, two different types of approximations are involved (see Figure 11.2):

1. Filter approximation for the derivation of the dynamics of the conditional mean and the variance of the state
2. Discretization approximation of the continuous-time nonlinear dynamic equations (whether they are for $dz(t)/dt$, the original state dynamics or for $dz_{t|t}/dt$, the dynamics of the resulting filtered state)

Both approximations are employed in the process of nonlinear filtering, where the infinite dimensional dynamical system needs to be converted into a finite dimensional dynamical system. Here the computational instability problem is a very delicate issue, and it can make the nonlinear filtering problem difficult to solve. The local linearization method of Chapter 9 is valid for both routes, D-F route and F-D route (see Figure 11.2). In the D-F route case, the continuous-time state model needs to be discretized before applying the F-approximation, and in the F-D route, the continuous-time filtering equations need to be discretized after the F-approximation.

The significant superiority of the filtering performance of D-F route over F-D route is shown in Ozaki (1993a) with the numerical examples of the van der Pol equation and some other nonlinear models. Jazwinski (1970) also suggests taking essentially the same approach to Ozaki without any explicit reference to an approximate discrete-time modeling. The reason for the significant superiority of the D-F route approach over the F-D route approach is, in a sense, natural and clear. All the continuous-time equations are obtained in the limit of discrete-time equations, but the discrete-time equations are not derivable from the continuous-time equations. In the continuous-time filtering equations, all the information of order $(\Delta t)^m$ $(m \geq 2)$ are lost, even though it actually matters in the numerical studies in applications. It cannot be recovered by simply discretizing the continuous-time filtering equations. On the other hand, if we discretize the model first and find an optimal filtering equation for the discretized model after, it takes more refined care of the filtering (i.e., minimum variance estimate of the state) even though the model is approximated by a discrete-time model.

Naturally, we will expect to have a computationally efficient (i.e., minimum-variance type) recursive nonlinear filtering method for any nonlinear state models, either continuous–continuous type, continuous–discrete type, or discrete–discrete type. Unfortunately, unlike continuous-time nonlinear filtering (i.e., continuous–continuous and continuous–discrete), there have not been many useful methods developed in the discrete-time case (i.e., discrete–discrete). In engineering literature, where the early development of recursive filtering theories was found, people discuss nonlinear filtering theory mostly for continuous-time dynamic models (see Kalman and Bucy 1961, Sage and Melsa 1971, Jazwinski 1970). The main reason for this is that the probabilistic properties of the Markov chain processes defined by the discrete-time nonlinear dynamic models driven by Gaussian white noise are not well developed except in the very general case (Doeblin 1940, Kingman 1963, Tweedie 1975). For example, in the engineering literature (Liang 1983), the following general nonlinear filtering problem of a discrete-time nonlinear state space model driven by Gaussian white system noise w_{t+1} was discussed:

$$z_{t+1} = \phi(z_t) + \Gamma(z_t)w_{t+1}$$

$$x_t = Cz_t + \varepsilon_t.$$

Here no conditions are put on the stability of the state dynamic model. Then they try to introduce a nonlinear filtering scheme for the following example:

$$z_{t+1} = \phi(z_t) + \Gamma(z_t)w_{t+1}.$$

It is obvious that if $\phi(z_t)$ is a nonlinear polynomial function (such as $-z_t^3$) and if the noise w_{t+1} is Gaussian white noise, the state process defined by the model is not stationary: the state diverges as t increases, i.e., the state process z_t is an explosive Markov chain. If the state process is explosive, there is no point in making the nonlinear filtering scheme nonexplosive.

In order to avoid confusion (of the aforementioned explosive state process), we exclude those state space models with explosive nonstationary state processes. We consider only the nonlinear filtering problem of state space models with finite-variance state processes. Here the state process is stationary and has finite mean and variance. If the state process is not finite-variance and explosive, it is hard to see whether the explosion is an eligible explosion of the state process or an ineligible explosion caused by an inappropriate finite dimensional approximation of the nonlinear filter, which is, theoretically speaking, an infinite dimensional dynamical system. Most dynamic models discussed in neuroscience are known to generate nonexplosive processes even with random input, yielding finite-variance processes of special characteristic patterns. Well-known examples are the Hodgkin–Huxley model for the dynamics of potassium (Hodgkin and Huxley 1952), the Zetterberg model for the dynamics of the alpha rhythm in EEG data (Zetterberg et al. 1978), the Balloon model for hemodynamics (Friston et al. 2000), and the DCM model for neural dynamics (Friston et al. 2003).

11.3.6 Nonlinear Observation Equations

11.3.6.1 Simple Approximations

So far we have been concerned only about nonlinear filtering problems for nonlinear state space models with linear observations. Here the nonlinearities are only in the dynamics of the state equation

$$\dot{z} = a(z) + b(z)w(t)$$
$$x_t = Cz_t + \varepsilon_t. \tag{11.112}$$

In a general nonlinear filtering problem setting, the observation equation is also considered to be possibly nonlinear, so that the inference problem (a) is posed as a state estimation problem of

$$dz = a(z)dt + B(z)dw$$
$$x_t = c(z_t) + \varepsilon_t. \tag{11.113}$$

When the observation equation is nonlinear, the most commonly used non-linear filtering method may be the EKF (Liang 1983), where the nonlinear function $c(z_t)$ of the observation equation,

$$x_t = c(z_t) + \varepsilon_t,$$

is approximated by its first-order Taylor expansion; thus we use the approximation, $\{\partial c(z)/\partial z\}z$ for $\{c(z)\}$.

Obviously the idea behind this approximation is the implicit assumption that the function $c(z)$ is close to zero when the state z is close to zero and $z(t)$ remains small for the concerned time interval, $t_1 \leq t \leq t_N$.

Note that for any analytic function $c(z)$ with $c(0) = 0$, we have a Taylor expansion

$$c(z) = c'(0)z + \sum_{i=1}^{\infty} \frac{1}{(i+1)!} c^{(i+1)}(0)z^{i+1}$$

$$= C(z)z,$$

where

$$C(z) = c'(0) + \sum_{i=1}^{\infty} \frac{1}{(i+1)!} c^{(i+1)}z^{i}$$

$$= \begin{cases} c(z)/z & \text{for } z \neq 0 \\ c'(0) & \text{for } z = 0. \end{cases} \tag{11.114}$$

When the nonlinear observation is inevitable, we recommend the use of $C(z)$, which is simple and more efficient than $\partial c(z_t)/\partial z_t$ of the EKF when z is not very small.

11.3.6.2 Nonuniqueness Problem

In most cases, we recommend the user of nonlinear Kalman filter to formulate his/her model in a state space representation with a linear observation and a nonlinear state dynamic equation as far as possible. This is because there is a methodological reason why we should use an observation equation that is as simple as possible. In time series analysis, the state model and state variables are hypothetical, while the observation data is factual, and we have freedom in choosing the state variable for the given observation.

For example, the following two different canonical state space models:

1. Observer canonical form:

$$z_{t+1}^o = A_o z_t^o + B_o w_{t+1}$$
$$x_t = C_o z_t^o , \tag{11.115}$$

where

$$A_o = \begin{pmatrix} \phi_1 & 1 & 0 \\ \phi_2 & 0 & 1 \\ \phi_3 & 0 & 0 \end{pmatrix}, \quad B_o = \begin{pmatrix} 1 \\ \theta_1 \\ \theta_2 \end{pmatrix},$$

and

$$C_o = \begin{pmatrix} 1 & 0 & 0 \end{pmatrix},$$

2. Controller canonical form:

$$z_{t+1}^c = A_c z_t^c + B_c w_{t+1}$$
$$x_t = C_c z_t^c, \tag{11.116}$$

where

$$A_c = \begin{pmatrix} \phi_1 & \phi_2 & \phi_3 \\ 1 & 0 & 0 \\ 0 & 1 & 0 \end{pmatrix}, \quad B_c = \begin{pmatrix} 1 \\ 0 \\ 0 \end{pmatrix},$$

and

$$C_c = \begin{pmatrix} 1 & \theta_1 & \theta_2 \end{pmatrix}$$

define one and the same process x_t specified by the ARMA(3,2) model,

$$x_{t+3} = \phi_1 x_{t+2} + \phi_2 x_{t+1} + \phi_3 x_t + w_{t+3} + \theta_1 w_{t+2} + \theta_2 w_{t+1}. \tag{11.117}$$

Here the observation matrix of (11.115) is set to be $C_o = (1, 0, 0)$ without a parameter to estimate, whereas the observation equation of (11.116) includes MA parameters, θ_1 and θ_2 which need to be estimated from the data. This means if an ARMA model is generalized into a non-linear model where the MA parameters are state-dependent functions, we have at least two ways of implementing the nonlinearity: one is to make the observation equation nonlinear, that is, to make the C-matrix state dependent, and another is to make the driving noise of the state equation nonlinear, that is, to make the B-matrix state dependent.

Incidentally, we have two more well-known canonical state space forms (11.118) and (11.119), for the same ARMA(3,2) model (11.117), called the observability canonical form and controllability canonical form, respectively, and Akaike (1974a) and Jones (1980) employ the observability canonical form (11.118), for the modeling of an ARMA model in a state space form.

3. Observability canonical form:

$$x_{t+1}^{ob} = A_{ob}x_t^{ob} + b_{ob}u_{t+1}$$
$$y_t = C_{ob}x_t^{ob},$$

(11.118)

where

$$A_{ob} = \begin{pmatrix} 0 & 1 & 0 \\ 0 & 0 & 1 \\ \phi_3 & \phi_2 & \phi_1 \end{pmatrix}, \quad C_{ob} = \begin{pmatrix} 1 & 0 & 0 \end{pmatrix}$$

$$b_{ob} = \begin{pmatrix} 1 \\ h_1 \\ h_2 \end{pmatrix} = \begin{pmatrix} 1 & 0 & 0 \\ -\phi_1 & 1 & 0 \\ -\phi_2 & -\phi_1 & 1 \end{pmatrix}^{-1} \begin{pmatrix} 1 \\ \theta_1 \\ \theta_2 \end{pmatrix}.$$

4. Controllability canonical form:

$$x_{t+1}^{co} = A_{co}x_t^{co} + b_{co}u_{t+1}$$
$$y_t = C_{co}x_t^{co},$$

(11.119)

where

$$A_{co} = \begin{pmatrix} 0 & 0 & \phi_3 \\ 1 & 0 & \phi_2 \\ 0 & 1 & \phi_1 \end{pmatrix}, \quad b_{co} = \begin{pmatrix} 1 \\ 0 \\ 0 \end{pmatrix},$$

$$C_{co} = \begin{pmatrix} 1 & h_1 & h_2 \end{pmatrix} = \begin{pmatrix} 1 & \theta_1 & \theta_2 \end{pmatrix} \begin{pmatrix} 1 & -\phi_1 & -\phi_2 \\ 0 & 1 & -\phi_1 \\ 0 & 0 & 1 \end{pmatrix}^{-1}.$$

Sometimes the use of a specific nonlinear observation equation may be established for a state space representation for some time series data analysis. Even in such situations, we can transform the problem into a nonlinear filtering problem with a linear observation equation as follows. Suppose we are required to estimate the state variable $z(t)$ of the following nonlinear state space model

$$\begin{cases} \dot{z} = \alpha z + \beta w(t) \\ x_t = c(z_t) \end{cases} \qquad (11.120)$$

from time series observation, $x_1, x_2, \ldots, x_N$, where $c(z)$ is a nonlinear function of z. Let us introduce a new variable $y = c(z)$. Then we can write down the dynamics of y as

$$\dot{y} = \left(\frac{\partial c}{\partial z} \right) \dot{z} = \left(\frac{\partial c}{\partial z} \right) \{\alpha z + \beta w(t)\}. \qquad (11.121)$$

Here, we can consider x_t as an observation of state variable $y(t)$ whose dynamics are specified by (11.121) together with $\dot{z} = \alpha z + \beta w(t)$ of (11.120). Then we have, for x_t, another state space representation, which is a 2-dimensional nonlinear state space model with linear observation, that is,

$$\begin{cases} \dot{\varsigma} = a(\varsigma) + Bw(t) \\ x_t = C\varsigma \end{cases}$$

$$\varsigma(t) = \begin{pmatrix} z(t) \\ y(t) \end{pmatrix}, \quad a(\varsigma) = \begin{pmatrix} \alpha \\ \dfrac{\partial c(z)}{\partial z} \alpha \end{pmatrix},$$

$$B(z) = \begin{pmatrix} \beta \\ \dfrac{\partial c(z)}{\partial z} \beta \end{pmatrix}, \quad C = (0 \quad 1).$$

As we have seen in the examples considered in the previous sections, the nonlinear state space models considered in neuroscience have been formulated mostly with linear observation equations. Hereafter, in order to avoid tedious notations and superficial complication, we mostly discuss the nonlinear filtering problem with nonlinear dynamics and linear observations. Only when using a nonlinear observation is inevitable, we recommend the use of the locally linearized $C(z)$ of (11.114), instead of $\partial c(z_t)/\partial z_t$ of the EKF, together with the local linearization filter in Section 11.3.4 (see Ozaki and Thomson 2002 for details).

11.4 Other Solutions

We have considered, in the previous sections, mostly the two types of the state estimate out of the three types, that is, filtered estimate $z_{t|t}$ and the predicted estimate $z_{t|t-1}$ for the state space model

$$z_t = Az_{t-1} + Bw_t$$

$$x_t = Cz_t + \varepsilon_t.$$

Remember that there are three different types of solutions for the estimation of the state z_t from the observed data, $x_1, \ldots, x_k$. The estimates of z_t are classified according to whether $k = t$, $k < t$, and $k > t$, into the following three problems:

1. Filtering problem: Find an optimal estimate of z_t from the observation data, $x_1, x_2, \ldots, x_{t-1}, x_t$.

2. Prediction problem: Find an optimal estimate of z_t from the observation data, $x_1, x_2, \ldots, x_{t-1}$.

3. Smoothing problem: Find an optimal estimate of z_t from all the observation data, $x_1, x_2, \ldots, x_{t-1}, x_t, \ldots, x_N$.

We have not discussed so far the third type of state estimate, that is, the smoothed state estimate $z_{t|N}$. Since the smoothed estimate $z_{t|N}$ uses more information than the predicted estimate $z_{t|t-1}$ and filtered estimate $z_{t|t}$, it may be natural that people think smoothed estimate is more useful than the other two. In fact if you plot the time series of the estimated states and compare them, the plot of the smoothed estimate looks much nicer (smoother) than the plot of the time series of the filtered or predicted state estimates. However, estimating or interpolating past state variables from the present data is not the only mission of time series analysis. A more important and commonly expected mission of time series analysis in applied sciences is to predict and control the state variable concerned from the past and the present data. For these missions, smoothed state estimates do not play a very important role, whereas the predicted estimates and filtered estimates play an essential role both in model identifications and designing the optimal controller based on the identified model. In addition, we should not forget that all the results of prediction, filtering, and smoothing are uniquely given when the model is given. This means that all the information, whether it is a predictor, a filter, or a smoother, comes from the model. Leaving further discussions to Section 12.6.3, we will next see how the smoothed state estimates are derived from the predicted estimates and filtered estimates of the Kalman filter for the linear state space model.

11.4.1 Smoothers

People may think that smoothing and/or filtering of the data is essential for better prediction and better control of the dynamic system. Actually if the dynamic model is known, the optimal filter, optimal predictor, and optimal smoother are uniquely determined. In other words, model identification solves all the problems, that is, smoothing, filtering, and prediction. The following theorem shows how Kalman's optimal (minimum variance) filter leads to the optimal (minimum variance) smoother through the specified state space model (11.122).

$$z_t = Az_{t-1} + Bw_t \quad w_t \sim N(0, R_1)$$
$$x_t = Cz_t + \varepsilon_t \quad \varepsilon_t \sim N(0, R_2). \tag{11.122}$$

Theorem (Smoothing-1): (Bryson and Frazier 1963, Rauch et al. 1965, Meditch 1967). The optimal smoothed estimate of z_t for measurements, $x_1, \ldots, x_N, 1 \leq t \leq N-1$, is given, using the filtering results, $z_{t|t}, P_{t|t}, z_{t+1|t}$ and $P_{t+1|t}$, by

$$z_{t|N} = z_{t|t} + J_t \left\{ z_{t+1|N} - z_{t+1|t} \right\}, \tag{11.123}$$

where

$$J_t = P_{t|t} A' P_{t+1|t}^{-1} \tag{11.124}$$

for $t = N - 1, N-2, \ldots, 1, 0$. The corresponding smoothing error covariance matrix is given by

$$P_{t|N} = P_{t|t} + J_t \left\{ P_{t+1|N} - P_{t+1|t} \right\} J_t' \tag{11.125}$$

for $t = N-1, N-2, \ldots, 1, 0$.

Here the filter $z_{t|t}$, the predictor $z_{t+1|t}$, and the corresponding error variances $P_{t|t}$ and $P_{t+1|t}$ are those calculated from the Kalman filter equations (11.49) through (11.55) in Section 11.2.3.

As we saw in the explanation of the Kalman filter in Section 11.2, the state estimate has three faces, the minimum variance estimate, the orthogonal projection of the state to the observation space, and the conditional expectation of the state given the observations. It is always useful to remember the equivalence of these three faces, when we try to understand the implications of the recursive relations in smoothing theories (as well as the Kalman filter theories) for Gaussian linear state space cases. This equivalence is summarized in the following theorem.

Theorem (Kalman 1960): For a linear Gaussian state space model, the following three are equivalent:

1. Minimum variance estimate of a state z_k from the data $X^t = (x_1, \ldots, x_t)$
2. Orthogonal projection of the state z_k to the vector space spanned by the data $X^t = (x_1, \ldots, x_t)$, which we write as o.p. $[z_k, X^t]$
3. Conditional expectation of a state z_k given data $X^t = (x_1, \ldots, x_t)$

Let V^{k+1} be a vector space spanned by a linear combination of the innovation $v_{k+1} = x_{k+1} - E[x_{k+1} | X^k]$. Then from the properties of orthogonal projections in the Theorem (Kalman) and the nature of the vector spaces X^k, X^{k+1}, and V^{k+1}, we have the following relation between the filtered state estimate $z_{k|k}$ and the one-step smoothed estimate $z_{k|k+1}$,

$$z_{k|k+1} = o.p.[z_k, X^{k+1}]$$

$$= o.p.[z_k, X^{k+1}] + o.p.[z_k, V^{k+1}]$$

$$= z_{k|k} + S_{k+1} v_{k+1}, \tag{11.126}$$

where S_{k+1} is to be determined. With tedious calculations, we can introduce an expression of S_{k+1} in terms of $P_{k|k}$, $P_{k+1|k}$, and K_{k+1} of the Kalman filter scheme, as

$$S_{k+1} = P_{k|k} A' P_{k+1|k}^{-1} K_{k+1}$$

so that we finally obtain, from (11.126),

$$z_{k|k+1} = z_{k|k} + P_{k|k} A' P_{k+1|k}^{-1} K_{k+1} v_{k+1}. \tag{11.127}$$

Since $K_{k+1} v_{k+1}$ can be written, by the Kalman filter theory, as

$$K_{k+1} v_{k+1} = z_{k+1|k+1} - z_{k+1|k},$$

(11.127) becomes

$$z_{k|k+1} = z_{k|k} + J_k [z_{k+1|k+1} - z_{k+1|k}], \tag{11.128}$$

where

$$J_k \equiv P_{k|k} A' P_{k+1|k}^{-1}. \tag{11.129}$$

J_k is called the smoother filter gain matrix. It follows from (11.128) and (11.129) that

$$z_{k+1|k+2} = z_{k+1|k+1} + J_{k+1} [z_{k+2|k+2} - z_{k+2|k+1}] \tag{11.130}$$

and

$$J_{k+1} = P_{k+1|k+1}A'P^{-1}_{k+2|k+1}.$$ (11.131)

Again from the properties of the orthogonal projections in the Theorem (Kalman) and the nature of the vector spaces, X^{k+1}, X^{k+1}, V^{k+2}, it follows that

$$z_{k|k+2} = z_{k|k} + J_k[z_{k+1|k+1} - z_{k+1|k}] + S_{k+2}v_{k+2},$$ (11.132)

where S_{k+2} is to be determined. After similar manipulations as those for (11.127) and (11.128), this is finally written, with the smoothing filter gain matrix J_k, as

$$z_{k|k+2} = z_{k|k} + J_k[z_{k+1|k+2} - z_{k+1|k}].$$ (11.133)

We now proceed by induction. For $N \geq k + 3$, where N is an integer, we have

$$z_{k|N-1} = z_{k|k} + J_k[z_{k+1|N-1} - z_{k+1|k}]$$ (11.134)

and we seek the expression for $z_{k|N}$. From the properties of the orthogonal projection in the Theorem (Kalman) and (11.134), we have

$$z_{k|N} = o.p.[z_k, X^N]$$

$$= o.p.[z_k, X^{N-1}] + o.p.[z_k, V^N]$$

$$= z_{k|k} + J_k[z_{k+1|N-1} - z_{k+1|k}] + S_N v_N,$$

where S_N remains to be determined. The expression for S_N can be determined by a tedious calculation using the same logic as before. Then by rearranging $J_k z_{k+1|N-1}$ and $S_N v_N$, we finally obtain the recursive schemes

$$z_{k|N} = z_{k|k} + J_k[z_{k+1|N} - z_{k+1|k}]$$ (11.135)

and

$$J_k = P_{k|k}A'P^{-1}_{k+1|k}.$$ (11.136)

Then the relation between the smoothed estimate error covariance matrix

$$P_{k|N} = E[(z_k - z_{k|N})(z_k - z_{k|N})']$$

and

$$P_{k+1|N} = E[(z_{k+1} - z_{k+1|N})(z_{k+1} - z_{k+1|N})']$$

is given as

$$P_{k|N} = P_{k|k} + J_k[P_{k+1|N} - P_{k+1|k}]J_k'. \tag{11.137}$$

Thus, we have recursive smoothing equations (11.135) through (11.137) of the Theorem (Smoothing-1).

11.4.2 On-Line Smoothers

We note that the smoothing scheme (11.135) are first-order recursive equations, requiring the solution of the prediction and filtering problems as inputs. The computation is initiated at $k = N - 1$ using $z_{N|N}, z_{N-1|N-1},$ and $z_{N|N-1},$ that is,

$$z_{N-1|N} = z_{N-1|N-1} + J_{N-1}[z_{N|N} - z_{N|N-1}].$$

We also note that the smoothing filter gain matrix J depends on $P_{k|k},$ the filtered state error covariance matrix, and $P_{k+1|k},$ the predicted state error covariance matrix. Therefore in order to complete the smoothing procedure, we need to calculate and store $z_{k|k}, z_{k+1|k}, P_{k|k},$ and $P_{k+1|k}$ for $k = 0, 1, \ldots, N,$ before we start the smoothing computations.

Another smoothing procedure that has a slightly different form can be introduced from the same argument (see Meditch 1967).

Theorem (Smoothing-2) (Meditch 1967)
The optimal smoothed estimate $z_{k|N}$ of z_k $(k < N)$ for N measurements, $x_1, \ldots, x_N,$ given the optimal smoothed estimate $z_{k|N-1}$ of $z_t - z_{t-1} = w_t$ for $(N - 1)$ measurements, $x_1, \ldots, x_{N-1},$ is specified by

$$z_{k|N} = z_{k|N-1} + S_{N-1,k}[z_{N|N} - z_{N|N-1}], \tag{11.138}$$

where

$$S_{N-1,k} = \prod_{i=k}^{N-1} J_i, \tag{11.139}$$

$$J_i = P_{i|i}A'P_{i+1|i}^{-1}, \tag{11.140}$$

and

$$P_{k|N} = P_{k|N-1} - S_{N-1,k}K_N CP_{N|N-1}S_{N-1,k}', \tag{11.141}$$

for $N = k + 1, k + 2, \ldots.$

The distinction between the two Theorems is obvious. The recursion relation of *Theorem (Smoothing-1)* is between $z_{k|N}$ and $z_{k+1|N}$, whereas the recursion relation of *Theorem (Smoothing-2)* is between $z_{k|N}$ and $z_{k|N-1}$. The scheme of *Theorem (Smoothing-1)* is useful when the observed time series data is fixed, whereas the scheme of *Theorem (Smoothing-2)* is useful when the filtering and smoothing need to be calculated in an on-line environment.

In any case, we must remember that the computation of prediction and filtering is compulsory for the computation of smoothing, whereas we don't need smoothing computation for the computation of filtering and prediction.

What we are calculating in the smoothing algorithms is the conditional mean $z_{t|N} = E[z_t|x_1, \ldots, x_N]$ with the assumption that the state space model (11.122) is the true model for the observed data, $x_1, \ldots, x_N$. In real-world data analysis, however, we do not know if the model is true or not. There could be a few possible candidate models for the given observed data, and for each candidate model $M^{(i)}$ we have a corresponding smoother, that is, the conditional mean, $E^{(i)}[z_t|x_1, \ldots, x_N]$. In this sense, the smoothed estimate is not free from the model structure and the model parameters we take. If the model is hypothetical but unrealistic and poor for the data, the filter and the smoother through the model are also poor.

The smoother does not have any extra useful information than the filter for the model improvement. All the information for model improvement is in the innovations. That means the filter and predictor are sufficient tools for finding the innovations. The relation (11.125) means that the error variance of the smoothed estimate is smaller than the error variance of the filtered estimate. It means that the smoother may provide us with state estimates that are smoother and nicer looking than the filtered estimates, but it does not bring any useful extra information than the filters and the innovations for the model improvement. In early days (i.e., 1930s) of time series analysis, people (Wold, Kolmogorov, Wiener, etc.) thought that smoothed data gave useful information for identifying the dynamics behind the data. This idea drove people to search for a refinement of the smoothing method. However, what Kalman showed is that if a true state space model is known for a given Gaussian time series, the optimal smoother, optimal filter, and optimal predictor are uniquely determined by the model, i.e., the state space model. This means that model identification is the most critical problem, and all the problems of smoothing, filtering, and prediction are solved once the model is identified.

11.4.3 Regularization Approach

11.4.3.1 Regularization and Constrained Least Squares Method

The inference problem (*a*) can also be solved from quite a different approach from the previous approach. The approach is called the regularization approach or constrained least squares approach. Although the mathematical forms of the models treated in both approaches, that is, innovation approach and regularization approach, are similar, the mathematical meaning of the variables and computational methods for inference are quite different. Not many textbooks pay much attention to the difference of the two approaches since the final numerical results are very similar, at least for most examples of state space modeling. However, when the state dimension becomes high and the state dynamics become complex and nonlinear, the difference between the two approaches becomes significant. In the present section, we see the similarities and dissimilarities of the "mathematical implication" of the two approaches using some simple examples.

When the r-dimensional observation data, $x_1, x_2, ..., x_N$ are considered as observations of unobserved k-dimensional vector variables, $z_1, z_2, ..., z_N$, with Gaussian observation errors, $\varepsilon_1, \varepsilon_2, ..., \varepsilon_N$, with a known observation matrix C, the unobserved variables, $z_1, z_2, ..., z_N$, can be usually estimated by the least squares method, which gives the solution, $\hat{z}_1, \hat{z}_2, ..., \hat{z}_N$, minimizing the sum of squares of the errors, that is,

$$\underset{z_1, z_2, ..., z_N}{\text{Min}} \sum_{t=1}^{N} (x_t - Cz_t)^2.$$

For the solution of the problem to exist, the dimension of the unobserved variable z_t needs to be smaller or equal to the dimension of the observed variable x_t. However, in many scientific applications, being expected to produce some solution, scientists are requested to produce a solution even though this condition is not satisfied. A typical example is the EEG inverse problem where we have 20-channels of EEG observations x_t, and people want to estimate 3000 dimensional current density vectors z_t in the brain (see Pascual-Marqui et al. 1994, Yamashita et al. 2004). The matrix C is given from the physical model of the brain and the 20-channel electrodes on the surface of the head and is called the lead field matrix. A more simple example is the seasonal adjustment method, where people want to explain a scalar time series x_t as a sum of trend component T_t, seasonal periodic component S_t, and a random noise component ε_t.

The problem is called an ill-posed problem for the obvious reasons. The key for converting the ill-posed problem into a numerically solvable problem lies in the introduction of some constraint for z_t. The most well-known constraint is the smoothness of z_t, which can be described by the condition that the sum of squares of the difference of z_t is small, that is, $\sum_{2}^{N} (z_t - z_{t-1})^2$ is small. If we write $z_t - z_{t-1} = w_t$, and $x_t - Cz_t = \varepsilon_t$, the whole setting of the constrained least squares method can be written as

$$\begin{aligned} z_t &= z_{t-1} + w_t \\ x_t &= Cz_t + \varepsilon_t. \end{aligned} \tag{11.142}$$

Here we are expected to estimate $z_1, z_2, \ldots, z_N$ from the observed data $x_1, x_2, \ldots, x_N$ with the constraint that $z_t - z_{t-1}$ is small. The actual solution is obtained by minimizing

$$\sum_{t=1}^{N} \|x_t - Cz_t\|^2 + d^2 \sum_{t=1}^{N} \|z_t - z_{t-1}\|^2, \tag{11.143}$$

where $\|.\|^2$ is the Euclidean norm. The balancing parameters d^2 and C are specified in advance. The solution, $\hat{z} = (\hat{z}_1, \hat{z}_2, \ldots, \hat{z}_N)'$, is a kind of constrained least squares estimate of z. The method was introduced by Tikhonov (1963) and is called the Tikhonov regularization method.

In statistics, the method is also known as ridge regression. It is related to the Levenberg-Marquardt algorithm for nonlinear least squares problems. The standard approach for solving an overdetermined system of linear equations given as

$$Cz = x$$

is known as linear least squares and seeks to minimize the residual $\|Cz - x\|^2$. However, the matrix C may be ill-conditioned or singular yielding a large number of solutions. In order to give preference to a particular solution with desirable properties, the regularization term $\|\Gamma z\|^2$ is included in this minimization as

$$\|Cz - x\|^2 + d^2 \|\Gamma z\|^2,$$

with some suitably chosen Tikhonov matrix Γ, and with a so-called balancing parameter d^2. In many cases, this matrix Γ is chosen as the identity matrix $\Gamma = I$. This regularization improves the conditioning of the problem, thus enabling a numerical solution. An explicit solution, denoted by $\hat{z}$ is given by

$$\hat{z} = (C^T C + d^2 \Gamma^T \Gamma)^{-1} C^T x.$$

For $\Gamma = 0$, this reduces to the nonregularized least squares solution provided that $(C'C)^{-1}$ exists.

The advantage of the use of the regularization method is that the solution of

$$\operatorname*{Min}_{x} \|Cz - x\|^2 + d^2 \|\Gamma z\|^2 \tag{11.144}$$

is explicitly given by

$$\hat{z} = (C^T C + d^2 \Gamma^T \Gamma)^{-1} C^T x$$

without using an iterative numerical optimization method if d^2 is given and the model is linear, that is, C and Γ are specified by matrices. Then the method yields an alternative solution to the inference problem-(a) for a simple state space model,

$$z_t = z_{t-1} + w_t$$
$$x_t = Cz_t + \varepsilon_t.$$

Here d^2 is an important parameter balancing the two penalties, $\|Cz - x\|^2$ and $\|\Gamma z\|^2$, in (11.144).

For the user of the regularization method, how to choose d^2 is a delicate and a critical problem affecting the final results. This is classified as a topic in the inference problem (b). The solution to this problem cannot be obtained in the framework of regularization by itself. It can be solved by the introduction of a Bayesian interpretation of the regularization method (Akaike 1980), which we see in Section 12.4.

11.4.3.2 Examples

Before going to the discussion of the choice of d^2, let us see some more examples of state space models of the constrained least squares approach.

Example 1. Trend-Seasonal Decomposition
A typical example of regularization is the Akaike (1980)'s BAYSEA method for seasonal adjustment, where we try to decompose a scalar time series x_t into a trend component T_t and cyclic seasonal component s_t. Here the time series is decomposed as the sum of the three components, a trend component T_t, a cyclic component s_t, and an irregular noise component I_t, $x_t = T_t + s_t + I_t$. Constraints introduced are follows:

1. $T_t - T_{t-1}$ should be small, that is, the trend should be changing smoothly in time.
2. $s_t - s_{t-p}$ should be small, that is, the cyclic component (of cycle p) should be changing smoothly in time.
3. $s_t + s_{t-1} + \cdots + s_{t-p+1}$ should be small, that is, the cyclic component should be oscillating around the base line in order to ensure that the cyclic component does not include the trend component.

Computationally, the BAYSEA method is formulated as a constrained least squares problem,

$$\underset{T,s}{\text{Min}} \left[\sum_{t=1}^{N} (x_t - T_t - s_t)^2 + d_1^2 \sum_{t=1}^{N} (T_t - T_{t-1})^2 + d_2^2 \sum_{t=1}^{N} \left(\sum_{j=0}^{11} s_{t-j} \right)^2 + d_3^2 \sum_{t=1}^{N} (s_t - s_{t-12})^2 \right],$$

where d_1^2, d_2^2, and d_3^2 are specified in advance by the analyst.

The state space version of the BAYSEA is introduced as the following $(p + 1)$-dimensional state space representation (Ozaki 1998a,b),

$$z_t = Az_{t-1} + Bw_t$$
$$x_t = Cz_t + \varepsilon_t, \tag{11.145}$$

where

$$A = \begin{pmatrix} 1 & O_{1 \times p} \\ O_{p \times 1} & A^{(s)} \end{pmatrix}, \quad B = \begin{pmatrix} 1 & 0 \\ O_{p \times 1} & B^{(s)} \end{pmatrix}, \quad C = \begin{pmatrix} 1 & C^{(s)} \end{pmatrix}$$

$$z_t = (T_t, s_{t-1}, s_{t-2}, \ldots, s_{t-p})', \quad w_t = \begin{pmatrix} w_t^{(T)} \\ w_t^{(s)} \end{pmatrix}$$

$$A^{(s)} = \begin{pmatrix} \phi - 1 & \phi - 1 & \cdots & \phi - 1 & \phi \\ 1 & 0 & \cdots & 0 & 0 \\ 0 & 1 & \cdots & 0 & 0 \\ \cdots & \cdots & \cdots & \cdots & \cdots \\ 0 & 0 & \cdots & 1 & 0 \end{pmatrix}, \quad B^{(s)} = \begin{pmatrix} 1 \\ 0 \\ 0 \\ \cdots \\ 0 \end{pmatrix},$$

and

$$C^{(s)} = (1, 0, 0, \ldots, 0).$$

Here the dynamic model for the trend is a simple random walk model,

$$T_t = T_{t-1} + w_t^{(T)}.$$

The dynamic model for the cyclic component is described by AR(p) type model,

$$s_t = (\phi - 1)s_{t-1} + (\phi - 1)s_{t-2} + \cdots + (\phi - 1)s_{t-p+1} + \phi s_{t-p} + w_t^{(s)},$$

which comes from the idea of making the weighted sum

$$(1 - \phi)(s_t + s_{t-1} + s_{t-2} + \cdots + s_{t-p+1}) + \phi(s_t - s_{t-p}) = w_t^{(s)}$$

small, where the two terms are balanced by the parameter ϕ.

Note that the state space model (11.142) is composed of two independent dynamic models, one for the trend and another for the cycle (see Figure 11.3). The model is an example of a parallel type structural model (see Section 3.3).

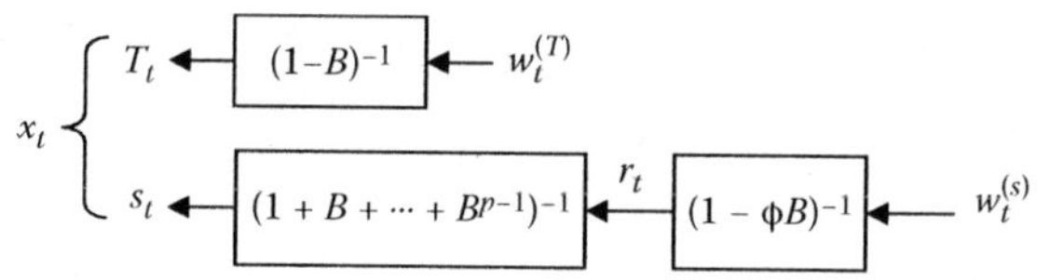

FIGURE 11.3
Structural interpretation of the regularization model for trend-seasonal model.

In fact, most state space models discussed in Chapter 3 can be interpreted from the regularization point of view, where the state dynamic equations are regarded as constraints on the observed time series.

Example 2. EEG rhythm decomposition:
A parallel type compartment state space model

$$z_t = Az_{t-1} + w_t$$

$$x_t = Cz_t + \varepsilon_t$$

for EEG rhythm decomposition is regarded as a state space model version of the regularization model. Here we have

$$z_t = (z_t^{(1)}, z_{t-1}^{(1)}, z_t^{(2)}, z_{t-1}^{(2)}, z_t^{(3)}, z_{t-1}^{(3)})',$$

$$w_t = (n_t^{(1)}, 0, n_t^{(2)}, 0, n_t^{(3)}, 0)',$$

$$A = \begin{pmatrix} \phi_1^{(\alpha)} & \phi_2^{(\alpha)} & 0 & 0 & 0 & 0 \\ 1 & 0 & 0 & 0 & 0 & 0 \\ 0 & 0 & \phi_1^{(\beta)} & \phi_2^{(\beta)} & 0 & 0 \\ 0 & 0 & 1 & 0 & 0 & 0 \\ 0 & 0 & 0 & 0 & \phi_1^{(\gamma)} & \phi_2^{(\gamma)} \\ 0 & 0 & 0 & 0 & 1 & 0 \end{pmatrix}$$

and

$$C = (1, 0, 1, 0, 1, 0).$$

With this model, we must find alpha, beta, and gamma components, $z_i^{(\alpha)}$, $z_i^{(\beta)}$, $z_i^{(\gamma)}$ in the observed scalar EEG time series x_t under the constraints:

1. $z_t^{(\alpha)} - \phi_1^{(\alpha)} z_{t-1}^{(\alpha)} - \phi_2^{(\alpha)} z_{t-2}^{(\alpha)}$ is small.
2. $z_t^{(\beta)} - \phi_1^{(\beta)} z_{t-1}^{(\beta)} - \phi_2^{(\beta)} z_{t-2}^{(\beta)}$ is small.
3. $z_t^{(\gamma)} - \phi_1^{(\gamma)} z_{t-1}^{(\gamma)} - \phi_2^{(\gamma)} z_{t-2}^{(\gamma)}$ is small.

All the coefficients, $\phi_1^{(\alpha)}$, $\phi_2^{(\alpha)}$, $\phi_1^{(\beta)}$, $\phi_2^{(\beta)}$, $\phi_1^{(\gamma)}$, and $\phi_2^{(\lambda)}$ are specified in advance by taking care of each main frequency, α, β, or γ, and the sampling interval.

The estimation of these components, that is, inferential problem-(*a*), is formulated as the following constrained least squares problem,

$$\underset{z^{(\alpha)}, z^{(\beta)}, z^{(\gamma)}}{\text{Min}} \left[\sum_{i=3}^{N} (x_i - z_i^{(\alpha)} - z_i^{(\beta)} - z_i^{(\gamma)})^2 + d_\alpha^2 \sum_{i=3}^{N} (\phi^{(\alpha)}(B) z_i^{(\alpha)})^2 \right.$$

$$\left. + d_\beta^2 \sum_{i=3}^{N} (\phi^{(\beta)}(B) z_i^{(\beta)})^2 + d_\gamma^2 \sum_{i=3}^{N} (\phi^{(\gamma)}(B) z_i^{(\gamma)})^2 \right].$$

Example-3 Lorenz Chaos

Note that the constraint does not need to be linear. The nonlinear state space model for the Lorenz chaos can be also formulated as a constrained least squares problem. Suppose we observe one of the variables, say $z^{(1)}$ of the three-dimensional Lorenz chaos model

$$\frac{dz^{(1)}}{dt} = -az^{(1)} + az^{(2)}$$

$$\frac{dz^{(2)}}{dt} = -z^{(3)}z^{(1)} + rz^{(1)}z^{(2)}$$

$$\frac{dz^{(3)}}{dt} = z^{(1)}z^{(2)} - bz^{(3)}.$$

The inference problem (*a*), that is, the estimation of the three variables, $z^{(1)}$, $z^{(2)}$, $z^{(3)}$, from the observation noise contaminated time series x_t, is formulated in a state space model, when we assume that the discrete-time dynamic models for the three variables are specified as

$$z_t^{(1)} = f^{(1)}\left(z_{t-1}^{(1)}, z_{t-1}^{(2)}\right)$$

$$z_t^{(2)} = f^{(2)}\left(z_{t-1}^{(1)}, z_{t-1}^{(2)}, z_{t-1}^{(3)}\right) \qquad (11.146)$$

$$z_t^{(3)} = f^{(3)}\left(z_{t-1}^{(1)}, z_{t-1}^{(2)}, z_{t-1}^{(3)}\right).$$

We can write this in state space model form as

$$z_t^{(1)} = f^{(1)}\left(z_{t-1}^{(1)}, z_{t-1}^{(2)}\right)$$

$$z_t^{(2)} = f^{(2)}\left(z_{t-1}^{(1)}, z_{t-1}^{(2)}, z_{t-1}^{(3)}\right)$$

$$z_t^{(3)} = f^{(3)}\left(z_{t-1}^{(1)}, z_{t-1}^{(2)}, z_{t-1}^{(3)}\right) \qquad (11.147)$$

$$x_t = (1, 0, 0) \begin{pmatrix} z_t^{(1)} \\ z_t^{(2)} \\ z_t^{(3)} \end{pmatrix} + \varepsilon_t.$$

By interpreting the three nonlinear dynamic models of (11.146), as constraints, the estimation problem of the three variables is formulated as the following constrained least squares problem:

$$
\min_{z_i^{(1)}, z_i^{(2)}, z_i^{(3)}} \left[\sum_{i=2}^{N} \left(x_i - z_i^{(3)} \right)^2 + d_{(1)}^2 \sum_{i=2}^{N} \left\{ z_i^{(1)} - f^{(1)} \big(z_{i-1}^{(1)}, z_{i-1}^{(2)} \big) \right\}^2 \right.
$$

$$
+ d_{(2)}^2 \sum_{i=2}^{N} \sum_{i=2}^{N} \left\{ z_i^{(2)} - f^{(2)} \big(z_{i-1}^{(1)}, z_{i-1}^{(2)}, z_{i-1}^{(3)} \big) \right\}^2
$$

$$
\left. + d_{(3)}^2 \sum_{i=2}^{N} \left\{ z_i^{(3)} - f^{(3)} \big(z_{i-1}^{(1)}, z_{i-1}^{(2)}, z_{i-1}^{(3)} \big) \right\}^2 \right].
$$

Kostelich and Yorke (1988) tried this approach for shadowing chaos dynamics. Farmer and Sidorowich (1991) tried a similar constrained least squares method for the smoothing of the chaos trajectory with a definite constraint for the equalities of (11.146) using the Lagrange multiplier method. Their results show that the constrained least squares method leads to numerical difficulties, which, they suspect, are caused by the complex nonlinearities of the constraints.

Incidentally, we note that stiff dynamical systems such as chaos are the examples where the difference between the two approaches, regularization approach and innovation approach, becomes clearly visible. Both inference problem-(a) and inference problem-(b) for chaos models are shown to be solved, without much numerical difficulty, by the innovation-based approach to state space modeling (Ozaki 1994b, Ozaki et al. 2004).

11.5 Discussions

Solutions for the inference problem (a) that we have seen so far are classified into two types, one is a "state space model based" solution and another is a "regularization based" solution. The two are different from each other in many ways, and looking into the differences helps in understanding the inference problems associated with state space modeling from time series.

11.5.1 Model Formulations

The state space model approach explicitly formulates the problem in a state space model with a dynamic model for the state equation and an observation equation that specifies the relation between the state and the observed time series. Typical examples are

$$\begin{cases} z_t = Az_{t-1} + w_t \\ x_t = Cz_t + \varepsilon_t, \end{cases}$$

and

$$\begin{cases} z_t = a(z_{t-1}) + w_t \\ x_t = Cz_t + \varepsilon_t. \end{cases}$$

On the other hand, the regularization-based approach formulates the problem with an observation equation and a constraint equation. The two methods are quite similar because the above two examples of state space models are formulated also in the regularization-based method as

$$\underset{z}{\text{Min}} \left[\sum_{t=1}^{N} \|x_t - Cz_t\|^2 + d^2 \sum_{t=1}^{N} \|z_t - Az_{t-1}\|^2 \right]$$

$$\underset{z}{\text{Min}} \left[\sum_{t=1}^{N} \|x_t - Cz_t\|^2 + d^2 \sum_{t=1}^{N} \|z_t - a(z_{t-1})\|^2 \right].$$

The reverse is also true as long as the constraint is set for the dynamics of the state variables. For example, the EEG inverse problem is formulated in LORETA (Pascual-Marqui et al. 1994) by using the regularization approach as

$$\underset{J}{\text{Min}} \left[\sum_{t=1}^{N} \|x_t - KJ_t\|^2 + d^2 \sum_{t=1}^{N} \|LJ_t\|^2 \right], \tag{11.148}$$

where
   x_t is an r-dimensional EEG time series measurement
   J_t is an m-dimensional primary current density of the cortices
   K is the $r \times m$ lead field matrix relating J_t and the measured data x_t

Here L is the Laplacian operator, such that

$$LJ_t^{(i,j,k)} = J_t^{(i,j,k)} - \frac{1}{6}(J_t^{(i-1,j,k)} + J_t^{(i+1,j,k)} + J_t^{(i,j-1,k)} + J_t^{(i,j+1,k)} + J_t^{(i,j,k-1)} + J_t^{(i,j,k+1)}).$$

The first term of (11.148) corresponds to the goodness of fit of the solution to the observations, and the second term imposes the spatial smoothness constraint on the solution. Since the constraint does not concern the dynamics of J_t, (11.148) cannot be transformed to a state space model.

When we consider the dynamic aspect of the current density J_t, a natural constraint to the temporal smoothness will make $\|LJ_t - LJ_{t-1}\|^2$ small also. Then a natural state space formulation of the regularization formulation (11.148) is to include the temporal smoothness of J_t by

$$\min_J \left\{ \| Y_t - KJ_t \|^2 + \lambda^2 \| (1-a)LJ_t + a(LJ_t - LJ_{t-1}) \|^2 \right\} \qquad (11.149)$$

instead of (11.148), where a balances the weights between minimizing two constraints, $\|LJ_t\|^2$ and $\|LJ_t - LJ_{t-1}\|^2$. Since $\min_J \{ \| Y_t - KJ_t \|^2 + \lambda^2 \| (1-a)LJ_t + a(LJ_t - LJ_{t-1}) \|^2 \} = \min_J \{ \| Y_t - KJ_t \|^2 + \lambda^2 \| LJ_t - aLJ_{t-1} \|^2 \}$, we have, from (11.149), a state space model formulation of the EEG inverse problem as

$$LJ_t = aLJ_{t-1} + w_t \quad w_t \sim N(0, \sigma_w^2 I_{m \times m})$$

$$x_t = KJ_t + \varepsilon_t \quad \varepsilon_t \sim N(0, \sigma_\varepsilon^2 I_{r \times r}).$$

Different constraints leads to a different state space model such as that of Schumitt et al. (2001), where their temporal smoothness constraint yields the following random walk-like state space model,

$$LJ_t = LJ_{t-1} + w_t \quad w_t \sim N(0, \sigma_\varepsilon^2 I_{m \times m})$$

$$x_t = KJ_t + \varepsilon_t. \quad \varepsilon_t \sim N(0, \sigma_\varepsilon^2 I_{r \times r}).$$

For more detailed discussions of the regularization-based approach to the "Dynamic EEG inverse problem" we refer to Yamashita et al. (2004), and for more detailed discussion of the state space model based approach to "Dynamic EEG inverse problem," we refer to Galka et al. (2004).

We note, however, that constraints commonly used in the regularization approach are limited to rather simple artificial ones such as smoothness constraints (Harrison and Stevens 1976). There is not much chance of considering such A's and C that do not satisfy the observability condition (see Sections 11.2.5 and 12.5.1). On the other hand, there are many examples of dynamic constraints treated in the state space model based approach, including continuous-time state models such as

$$\begin{cases} \dot{z} = a(z) + w(t) \\ x_t = Cz_t + \varepsilon_t \end{cases}$$

An important advantage of the explicit state space modeling approach over the regularization approach is that a more general dynamic model can be designed, in a fairly easy way, by considering more sophisticated dynamics of the state space.

11.5.2 "Independence of Errors" or "Homogeneity of Errors"

Obviously, the two approaches are also numerically quite different. The state space approach pays attention to the temporal independence of the prediction errors and is concerned about the independence of the prediction error v_t and the predictor given by the conditional mean $x_{t|t-1}$ at each time step. On the contrary, the regularization approach is concerned only with the minimization of the two types of errors, the sum of squares of errors of data fitting, and the sum of squares of errors of the constraints for the state. We note that in the regularization approach, the temporal homogeneity of the errors is implicitly assumed, while not much attention is paid to the local-in-time independence of the prediction errors. Thus, the difference between the two approaches could be characterized by whether the method cares about "independence of errors" or "homogeneity of the errors."

Since the regularization method is born out of the least squares method, it essentially inherits many of the implicit assumptions of the least squares method, where no explicit mechanism for assuring the "temporal independence" of errors is built in. In the state space model–based approach, on the contrary, the innovations are treated as "independent" at each time step, and there is no explicit mechanism forcing the errors to be "temporally homogeneous."

The difference affects the estimation results of state space models, especially for nonlinear dynamic cases, in many ways. A further detailed discussion follows in Sections 13.2, 13.4, and 13.6.

12

Inference Problem (b) for State Space Models

12.1 Introduction

As we saw in Section 11.1.3, two types of inferential problems arise from the nonlinear dynamic state space models:

(a) The estimation problem of the unobserved state variables from the observed time series (possibly contaminated with observation errors) assuming that the given state space model is true

(b) The identification problem of the model of the state equation (i.e., dynamical system model) from the partially observed time series, where some of the variables of the state equation are not observed, and the observed variables are possibly contaminated by observation errors

Here in problem (b), the true state space model is unknown, and we have 2 possible cases:

(b − 1) The model structure (i.e., the dimension of the state and parametric form specifying the state space model) is known, but precise parameter values are unknown.

(b − 2) The model structure is also unknown.

In (b − 1), we need to estimate the parameters of the state space model from time series. In (b − 2), we need to identify the model structure from time series, estimate the model parameters of the identified model structure from time series, and estimate the unobserved states of the identified model from time series.

The Kalman filtering method is useful only for solving the problem (a). In order to solve the problem (b), we need to develop a method for identifying the model structure (i.e., the dimension of the state etc.) and estimating model parameters of the identified nonlinear state space model from the given time series data. The guiding principle for the identification of linear or nonlinear dynamic models, that is, the innovation approach, discussed in Chapter 10, is still valid for the current problem of the identification of the state space models. In order to take advantage of the tools discussed in Chapter 10, we need to derive a log-likelihood function for the state space models.

12.2 Log-Likelihood of State Space Models in Continuous Time

We have seen in Chapter 10, that whitening of the time series data into prediction errors (i.e., innovations) with Markov models provides us with a powerful tool for the estimation, identification, and assessment of dynamic models generating the time series data. The innovation-based Gaussian maximum likelihood method is still useful for the identification of linear or nonlinear state space models.

Suppose we have a dynamic model,

$$\dot{z}(t) = f(z) + w^{(1)}(t),\tag{12.1}$$

for time series data, $x_1, x_2, \ldots, x_N$. In general, people discuss the estimation problem of the model (12.1) from continuous observation data, $x(t'), 0 \le t' \le t$ with an observation scheme,

$$x(t) = Cz(t) + b(t)w^{(2)}(t).\tag{12.2}$$

However, we must not forget that, in practice, the data are observed at discrete-time points, $0 \le t_1 \le t_1 \le, \ldots, \le t_n = T$, although N, the number of time points, can be increased in some cases, leading to finer partitions of the observation interval $(0,T)$. In addition, we must not forget that the observation noise $w^{(2)}(t)$, being a unit Gaussian continuous-time white noise, is a formal derivative of a Brownian motion $W^{(2)}(t)$, which does not exist in the strict mathematical sense. The same problem arises when we think of the driving noise $w^{(1)}(t) = \sigma_{w^{(1)}} \dot{W}^{(1)}(t)$ for the dynamic model (12.2), where $\dot{W}^{(1)}(t)$ is a unit Gaussian continuous-time white noise. Both observation process and system dynamics should be expressed in terms of increments; that is, the model (12.1) with (12.2) should be rewritten as the following state space model in continuous-time,

$$dz = f(z)dt + \sigma_{w^{(1)}} dW^{(1)}(t)$$

$$dx = Cz\,dt + b(t)dW^{(2)}(t).$$

If a finite sample $dx(t_1), dx(t_2), \ldots, dx(t_N)$ is taken from this model, it is difficult to write down the probability density $p(dx(t_1), dx(t_2), \ldots, dx(t_N))$, since we do not know the distribution of $z(t)$ when $f(.)$ is nonlinear, but we can at least write the log-density as a sum of the logarithms of conditional densities,

$$\log p(dx_1, \ldots, dx_n) = \sum \log p(dx_i | dx_{i-1}, \ldots, dx_1).$$

According to the innovation theorem of Frost and Kailath (see Section 7.3.4), we can replace the expression

$$dx(t) = Cz(t)dt + b(t)\,dW^{(2)}(t), \quad 0 \le t < T,$$

by

$$dx(t) = C\tilde{z}(t)dt + b(t)dU(t), \quad 0 \le t < T,$$

without affecting the problem in any way. Here, $U(t)$ is the innovation martingale and $dU(t)$ is a Gaussian white noise according to the Frost-Kailath theorem.

Now if we examine the conditional distribution of

$$dx(t_k) = C\tilde{z}(t_k)dt_k + u(t_k), \tag{12.3}$$

we realize that $C\tilde{z}(t_k)$ (the conditional expectation of $Cz(t_k)$ given the past $dx(t')$, $0 < t' < t_k$) will act as a constant, whereas $u(t_k)$ $(=dU(t_k))$ being independent of the past, will have the same distribution as $b(t_k)dW^{(2)}(t_k)$. It follows that the conditional distribution of $dx(t_k)$ will then be Gaussian with mean $C\tilde{z}(t_k)dt_k$ and variance $b(t_k)^2dt_k$. For a more formal proof, see the original paper by Frost and Kailath (1971). Since the prediction error of the observed time series is a sum of a Gaussian observation noise plus a prediction error of a continuous Markov process for an interval Δt, the theorem may look intuitively obvious to some readers.

If the process is discretized at fixed time points, model (12.3) becomes (replacing t_k by k for simplicity)

$$x_k = Cz_{k|k-1} + u_k.$$

Here, $dx(t_k)$ is approximated by $\Delta x(t_k) = x(t_k) - x(t_{k-1})$ (which we write x_k for simplicity), and $C\tilde{z}(t_k)dt_k$ is replaced by $Cz_{k|k-1}$, where

$$Cz_{k|k-1} = E[Cz_k | x_{k-1}, \ldots, x_1],$$

the conditional variance $b(t_k)^2dt_k$ is replaced by $\sigma^2_{k|k-1} = E[(x_k - Cz_{k|k-1})^2]$ and u_k is a discrete-time Gaussian white noise with variance $\sigma^2_{k|k-1}$. The conditional density of u_k is

$$p(u_k | x_{k-1}, \ldots, x_1) = \frac{1}{\sqrt{2\pi\sigma^2_{k|k-1}}} \exp\left[\frac{-\{x_k - Cz_{k|k-1}\}^2}{2\sigma^2_{k|k-1}}\right].$$

Then the (–2)log-likelihood is given by

$$(-2)\log p(x_1, \ldots, x_N) = \sum (-2)\log p(x_k|x_{k-1}, \ldots, x_1)$$

$$= \sum \left[\log \sigma_{k|k-1}^2 + \frac{\{x_k - Cz_{k|k-1}\}^2}{\sigma_{k|k-1}^2} \right] + N \log 2\pi, \qquad (12.4)$$

where we use the relation

$$\sum (-2)\log p(x_k|x_{k-1}, \ldots, x_1) = \sum (-2)\log p(u_k|x_{k-1}, \ldots, x_1).$$

An important problem in practical applications is how we estimate the conditional mean $z_{k|k-1}$ and the conditional variance $\sigma_{k|k-1}^2$. The problem is closely related to the approximation of the continuous-time state space model by a discrete-time state space model, where estimates of the conditional mean $Cz_{k|k-1}$ and the conditional variance $\sigma_{k|k-1}^2$ are obtained by applying a nonlinear Kalman filtering technique of Chapter 11 to an observed time series.

If the state vector of the multivariate continuous-time dynamic model is observed directly without observation noise, that is, $dx(t_k) = z(t_k)$, we can of course calculate the log-likelihood function (12.4) without using the Kalman filtering techniques of Chapter 11. Here the conditional mean $z_{k|k-1}$ and the conditional variance $\sigma_{k|k-1}^2$ can be approximated explicitly as functions of the observed series.

12.2.1 Linear Models and Nonlinear Models: Examples

12.2.1.1 Log-Likelihood of Tank Models

A typical example of a continuous-time linear state space model is the tank model of Section 5.3.3. A tank model, composed of k sequential tanks, is represented by a state space model (12.5) with a k-dimensional state vector, $Z(t) = (d^{k-1}z(t)/dt^{k-1}, \ldots, dz(t)/dt, z(t))'$,

$$\dot{Z}(t) = AZ(t) + Bs(t)$$

$$x_t = CZ(t) + \varepsilon_t, \qquad (12.5)$$

where

$$A = \begin{pmatrix} -\lambda & 0 & \cdots & 0 & 0 \\ 1 & -\lambda & \cdots & 0 & 0 \\ \cdots & \cdots & \cdots & \cdots & \cdots \\ 0 & 0 & \cdots & -\lambda & 0 \\ 0 & 0 & \cdots & 1 & -\lambda \end{pmatrix}, \quad B = \begin{pmatrix} b \\ 0 \\ \cdots \\ 0 \\ 0 \end{pmatrix}$$

and

$$C = \begin{pmatrix} 0 & 0 & \dots & 0 & 1 \end{pmatrix}.$$

By a standard discretization method for multivariate linear differential equation models, the continuous-time state space model (12.5) can be converted to a linear discrete-time state space model,

$$Z_t = \exp(A\Delta t)Z_{t-\Delta t} + A^{-1}(\exp(A\Delta t) - I)Bs_{t-\Delta t}$$

$$x_t = CZ(t) + \varepsilon_t.$$

$$(12.6)$$

Here we assume that x_t and s_t are observed, so that we have two sets of time series data, $x_{\Delta t}, x_{2\Delta t}, x_{3\Delta t}, \dots, x_{N\Delta t}$ and $s_{\Delta t}, s_{2\Delta t}, s_{3\Delta t}, \dots, s_{N\Delta t}$. Then, we can whiten the observed time series data, $x_{\Delta t}, x_{2\Delta t}, \dots, x_{N\Delta t}$, into $v_{\Delta t}, v_{2\Delta t}, \dots, v_{N\Delta t}$, using the linear Kalman filter scheme for a linear model with exogenous input. Thus, we can calculate the likelihood of the model and estimate the model parameters by the maximum likelihood method. The parameters of the model to estimate are b, λ, and the observation noise variance σ_ε^2.

12.2.1.2 Nonlinear Continuous-Time State Space Model

The prediction error–based Gaussian maximum likelihood method is useful for the identification of the nonlinear continuous-time state space model as well. The model could be written, with a "well-proposed" set of $a(z)$, $b(z)$ and C, as

$$\begin{cases} dz = a(z)dt + b(z)dW \\ x_t = Cz_t + \varepsilon_t \end{cases} \tag{12.7}$$

or

$$\begin{cases} \dot{z} = a(z) + b(z)w(t) \\ x_t = Cz_t + \varepsilon_t. \end{cases} \tag{12.8}$$

The term "well-posed" for a state space model means that improperly designed models, such as nonobservable state space models or explosive state models, are excluded from the model families of concern. This topic is discussed in detail in Section 12.5.

Note that a linear state space model is included as a special case of (12.7), where we have $a(z) = Az$, with a constant transition matrix A. The justification for the innovation-based Gaussian maximum likelihood approach, for the state space model identification, comes from the theorem of Frost and Kailath (1971), which states that the prediction error v_t

$$v_t = x_t - E[x_t | x_{t-\Delta t}, x_{t-2\Delta t}, \ldots, \theta]$$

converges to a Gaussian white noise for $\Delta t \to 0$. This theorem implies that if the time series data, x_1, x_2, x_3, ..., are sampled with a sufficiently small interval (i.e., the main dynamics of the process $x(t)$ $[0 \le t \le T]$ is captured as a sufficiently smooth trajectory in the discrete-time data, $x_1, x_2, x_3, \ldots, x_N$), the prediction errors, $v_1, v_2, v_3, \ldots, v_N$, are approximately Gaussian white noise.

The likelihood of the linear/nonlinear state space model (12.7) can be written as

$$(-2)\log p(x_1, \ldots, x_N | \theta) = (-2)\sum_k \log p(x_k | x_{k-1}, \ldots, x_1, \theta)$$

$$= (-2)\sum_k \log p(v_k | x_{k-1}, \ldots, x_1, \theta),$$

where θ is the parameter vector of the model. When the sampling interval of the data is sufficiently small, $p(v_k | x_{k-1}, \ldots, x_1, \theta)$ has the form of Gaussian density distribution. Then the likelihood of the state space model (12.7) can be written as

$$(-2)\log p(x_1, \ldots, x_N | \theta)$$

$$= (-2)\sum_k \log p(v_k | x_{k-1}, \ldots, x_1, \theta)$$

$$= \left(\begin{array}{l} \sum_{t=1}^{N} \left\{ \log \sigma_{v_t}^2 + \dfrac{v_t^2}{\sigma_{v_t}^2} \right\} + N \log 2\pi : \text{for scalar data } x_1, \ldots, x_N. \\[2em] \sum_{t=1}^{N} \left\{ \log |\Sigma_{v_t}| + v_t' \Sigma_{v_t}^{-1} v_t \right\} + N \log 2\pi : \text{for vector data } x_1, \ldots, x_N. \end{array} \right.$$

Actual computation of the innovations

$$v_t = x_t - E[x_t | x_{t-1}, x_{t-2}, \ldots, x_1, \theta] \quad (t = 1, 2, \ldots, N)$$

and innovation variances

$$\sigma_{v_t}^2 \ (t = 1, 2, \ldots, N) \text{ or } \Sigma_{v_t} \ (t = 1, 2, \ldots, N)$$

are realized by applying the linear/nonlinear Kalman filter, in Sections 11.2 and 11.3. The maximum likelihood estimate of the model parameter vector θ is obtained by using a numerical nonlinear optimization method.

We need to be aware in advance, while we are calculating the log-likelihood, that we always have a possibility of numerical difficulties in calculating the prediction error v_t of x_t while applying the Kalman filter to the model (12.7) or (12.8). We can eliminate these difficulties by setting some conditions for the structure of $a(z)$, $b(z)$, and C of the state space model. This topic is discussed in more detail in Chapter 13.

12.2.1.3 Hodgkin–Huxley Model with Exogenous Inputs

As seen in Section 5.4, many nonlinear dynamic models, such as the Hodgkin–Huxley model, the Zetterberg Model, etc., are useful as a dynamic model explaining the dynamics of a scalar time series within the framework of state space formulations. For example, the Hodgkin–Huxley equations ((5.43) through (5.46) in Section 5.4.3) are rewritten by a four-dimensional nonlinear dynamic state model,

$$\dot{z} = f(z) + f_0 + I_z(t)$$
$$x_t = (1,0,0,0)z_t + \varepsilon_t,$$

(12.9)

with an exogenous input $I_z(t)$, where

$$I_z(t) = \begin{pmatrix} \dfrac{1}{C_M} I(t) \\ 0 \\ 0 \\ 0 \end{pmatrix},$$

$$z(t) = \{V(t), m(t), h(t), n(t)\}',$$

$$f(z) = \begin{pmatrix} -\dfrac{1}{C_M}\{\bar{g}_{Na}m(t)^3 h(t) + \bar{g}_K n(t)^4 + \bar{g}_l\}V(t) \\ -\{\alpha_m(V(t)) + \beta_m(V(t))\}m(t) + \alpha_m(V(t)) \\ -\{\alpha_h(V(t)) + \beta_h(V(t))\}h(t) + \alpha_h(V(t)) \\ -\{\alpha_n(V(t)) + \beta_n(V(t))\}n(t) + \alpha_n(V(t)) \end{pmatrix},$$

and

$$f_0 = \begin{pmatrix} \dfrac{1}{C_M}(\bar{g}_{Na}m^3 hE_{Na} + \bar{g}_K n^4 E_K + \bar{g}_l E_l) \\ 0 \\ 0 \\ 0 \end{pmatrix}.$$

The exogenous input $I_z(t)$ could be treated as either a Gaussian white noise or an unknown current input to the voltage $V(t)$. When we regard the unknown current $I_z(t)$ as a Gaussian white noise, the time series x_t is transformed into a Gaussian white noise, $v_t = x_t - E[x_t | x_{t-1}, x_{t-2}, \ldots]$ using the LL filter, and we can estimate the unknown parameters, $\bar{g}_{NA}, \bar{g}_K, \ldots$ of the Hodgkin–Huxley model using the innovation-based maximum likelihood method.

When we consider $I_z(t)$ as a sum of unobserved current and a Gaussian white noise such as

$$I_z(t) = I_e(t) + w_V(t),$$

we could explain the data by a six-dimensional state space model,

$$\dot{z} = f(z) + f_0 + w_z(t)$$
$$x_t = (1,0,0,0)z_t + \varepsilon_t,$$

(12.10)

with a six-dimensional state vector,

$$z(t) = \{\dot{I}_e(t), I_e(t), V(t), m(t), h(t), n(t)\}'.$$

The function vector $f(z)$ and the constant input vector f_0 are also redefined, respectively, by a six-dimensional function vector,

$$f(z) = \begin{pmatrix} 0 \\ \dot{I}_e(t) \\ -\dfrac{1}{C_M}\{\bar{g}_{Na}m(t)^3 h(t) + \bar{g}_K n(t)^4 + \bar{g}_l\}V(t) \\ -\{\alpha_m(V(t)) + \beta_m(V(t))\}m(t) + \alpha_m(V(t)) \\ -\{\alpha_h(V(t)) + \beta_h(V(t))\}h(t) + \alpha_h(V(t)) \\ -\{\alpha_n(V(t)) + \beta_n(V(t))\}n(t) + \alpha_n(V(t)) \end{pmatrix},$$

and a six-dimensional constant vector,

$$f_0 = \begin{pmatrix} 0 \\ 0 \\ \dfrac{1}{C_M}(\bar{g}_{Na}m^3 h E_{Na} + \bar{g}_K n^4 E_K + \bar{g}_l E_l) \\ 0 \\ 0 \\ 0 \end{pmatrix}.$$

Here the temporal evolution of $I_e(t)$ is modeled by a dynamic model driven by a Gaussian white noise $w_I(t)$ as

$$d^2 I_e(t)/dt^2 = w_I(t).$$

Note that in discrete-time, this model corresponds to the second-order stochastic difference equation model for $I_{e,t}$ driven by a discrete-time Gaussian white noise $w_{I,i}$; that is,

$$I_{e,t} - 2I_{e,t-1} + I_{e,t-2} = w_{I,t}.$$

The model is represented as a 2-dimensional discrete-time state space model,

$$\begin{pmatrix} I_{e,t} \\ I_{e,t-1} \end{pmatrix} = \begin{pmatrix} 2 & -1 \\ 1 & 0 \end{pmatrix} \begin{pmatrix} I_{e,t-1} \\ I_{e,t-2} \end{pmatrix} + \begin{pmatrix} 1 \\ 0 \end{pmatrix} w_{I,t}.$$

By combining this simple two-dimensional discrete-time dynamic model and the discretized four-dimensional L.L. (locally linearized) Hodgkin–Huxley model, Kawai et al. (2008) ingeniously formulated the estimation problem of unknown current input $I_e(t)$ into an estimation problem of a six-dimensional stochastic Hodgkin–Huxley state space model with stochastic inputs. Here the model is driven by two different system noise, $w_V(t)$ and $w_I(t)$. With the six-dimensional state space model, the data $x_1, \ldots, x_N$ are transformed into innovations, $v_1, \ldots, v_N$, with the LL filter, and the model parameters are estimated using the innovation-based maximum likelihood method. For those who are interested in numerical problems related to Hodgkin–Huxley model estimation, we refer Kawai et al. (2008), where detailed numerical studies are shown with interesting discussions. Applications of the LL filter to more sophisticated neuro-mechanical models are found in Riera et al. (2005, 2007).

12.2.2 How It Works with Innovations: The Case of Lorenz Chaos

We have seen in the previous discussion that the inference problem (b) of the state space model can be solved using the innovation-based maximum likelihood method under the condition that the model is "well posed."

Inference problem (b) of many famous examples of complex nonlinear dynamical system models, known to be difficult to estimate, in applied mathematics, are in fact solved with the present method. Examples are nonlinear oscillation models including the Duffing model and the van der Pol model (Ozaki 1989, 1992b, 1994a, Shoji 1998), Lorenz chaos for the hydrodynamics in meteorology (Ozaki 1994b, Ozaki et al. 1999), Rikitake chaos for dynamics of a 2 disc dynamo in geophysics (Ozaki et al. 2000), the Zetterberg model for the dynamics of EEG alpha rhythm (Valdes et al. 1999), and the Hodgkin–Huxley model for neural dynamics (Kawai et al. 2008).

Leaving the detailed discussion of the numerical problems related to real neuroscience data analysis to individual studies, we see, in this section, how the innovations and the innovation-based log-likelihood method with the local linearization filter work for solving the inference problem (*b*), using two examples, Lorenz chaos (Lorenz 1963) and Rikitake chaos (Rikitake 1958). The two models are known to be typical highly nonlinear dynamic models that are difficult to estimate from time series data (Farmer and Sidorowich 1991, Ito 1980). We see that, with our innovation-based method, estimation of unobserved state variables, estimation of the model parameters, and estimation of the unobserved initial state variables of these chaos models are fairly easy and straightforward.

12.2.2.1 State Space Model for Lorenz Chaos

In this section we first see whether the likelihood method is really useful for the estimation of parameters of the continuous-time nonlinear state space models, using simulated data of the Lorenz chaos model, which is defined by three equations:

$$\dot{\xi} = -a\xi + a\eta$$

$$\dot{\eta} = -\xi\varsigma + r\xi - \eta$$

$$\dot{\varsigma} = \xi\eta - b\varsigma,$$

where *a* is called the Prandtl number and *r* is called the Rayleigh number. All the parameters, *a*, *r*, *b*, are positive, but usually *a* and *b* are set to be *a* = 10, *b* = 8/3, and *r* is varied. The system exhibits chaotic behavior for *r* = 28 but displays knotted periodic orbits for other values of *r*. For example, with *r* = 99.96 it becomes a torus knot. The attractor itself, and the equations from which it is derived, were introduced by Lorenz (1963), who derived it from a partial differential equation describing the convection roll in fluid dynamics.

Since the variable η is one of the most interesting variables in Lorenz (1963)'s analysis, we try to estimate the model from the observation of scalar time series data of the variable η. Let $z = (\xi, \eta, \varsigma)'$ be a state vector. Then if we observe only η, we have an observation matrix, $C = (0,1,0)$, leading to a state space model,

$$\dot{z} = f(z) + w(t)$$

$$x_t = Cz_t + \varepsilon_t,$$

where

$$f(z) = \begin{pmatrix} -a\xi + a\eta \\ -\xi\varsigma + r\xi - \eta \\ \xi\eta - b\varsigma \end{pmatrix},$$

and *w(t)* is a 3-dimensional Gaussian white noise with a variance matrix $\Sigma_w = \sigma_w^2 I_{3\times3}$. The variance σ_w^2 could be zero, which means the present state space model for Lorenz chaos includes deterministic Lorenz chaos as a special case.

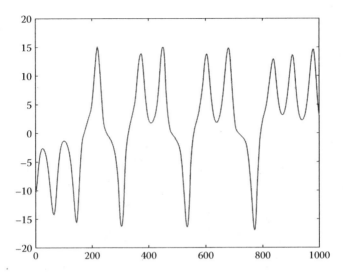

FIGURE 12.1
Lorenz data.

To obtain chaotic data we simulated the model with the same conditions as Lorenz (1963), except that we added a small observation noise to $\eta(t)$ in order to make the estimation problem more realistic and difficult. Thus, $\sigma_w^2 = 0$, $a = 10.0$, $b = 28.0$, $r = 8/3$ (≈ 2.666666667), and $(\xi, \eta, \zeta)_0' = (0, 1, 0)'$. The variance of the observation noise is set to be $\sigma_\varepsilon^2 = 0.1 \times 10^{-3}$. Then 1000 data points were taken from the 1501-th to the 2500-th data point with Gaussian independent observation noise being added to each point. The data are shown in Figure 12.1.

We are interested in seeing whether the likelihood function is really chaotic in the neighborhood of the true parameters $\theta = (a, b, r, \sigma_\varepsilon^2, \sigma_w^2)'$.

To see the behavior of the likelihood function near the true parameter values, we fix the parameters, a, b, r and σ_ε^2 to the true values and see how the likelihood function behaves when the system noise variance varies between small and large values. Here we assume that the system noise variance is a three-dimensional Gaussian white noise with a variance–covariance matrix $\sigma_w^2 I_{3\times 3}$. Table 12.1 shows the calculated (-2)log-likelihood values $lk(\sigma_w^2 | a, b, r, \sigma_\varepsilon^2)$ for 11 different values of σ_w^2 from $\sigma_w^2 = 0$ to $\sigma_w^2 = 10$.

In this example, the (-2)log-likelihood function does not take its minimum at $\sigma_w^2 = 0$, but it takes its minimum at $\sigma_w^2 = 0.01$. The function is smooth near the minimum as we expect for ordinary statistical modeling. We suspect that the estimate of the variance σ_w^2 of the system noise is closely related to, and can be partly generated by, the computational method used in simulating the data, where the data are generated by approximating and integrating the state equation for $z(t)$, whose dynamics are defined by an analytically unsolvable nonlinear differential equation. It is also related to the model we

TABLE 12.1

The Calculated (−2)log-Likelihood
Values $lk(\sigma_w^2|a,b,r,\sigma_\varepsilon^2)$ of the Lorenz
Chaos Model for 11 Different Values
of σ_w^2 from 0 to 10

σ_w^2	(−2)log-Likelihood
0.	$0.1890 \times 10^{+07}$
0.1×10^{-09}	$0.1750 \times 10^{+07}$
0.1×10^{-08}	$0.1030 \times 10^{+07}$
0.1×10^{-07}	$0.5843 \times 10^{+05}$
0.1×10^{-06}	$-0.2282 \times 10^{+04}$
0.1×10^{-05}	$-0.3684 \times 10^{+04}$
0.1×10^{-04}	$-0.4427 \times 10^{+04}$
0.1×10^{-03}	$-0.5631 \times 10^{+04}$
0.1×10^{-02}	$-0.6927 \times 10^{+04}$
0.1×10^{-01}	$-0.7121 \times 10^{+04}$
0.1×10^{-00}	$-0.6354 \times 10^{+04}$
$0.1 \times 10^{+01}$	$-0.4535 \times 10^{+04}$

used for the variance of the system noise. It can be shown that a more refined
approximate model for the variance–covariance matrix, that is,

$$\begin{pmatrix} \sigma_{w_x}^2 & 0 & 0 \\ 0 & \sigma_{w_y}^2 & 0 \\ 0 & 0 & \sigma_{w_z}^2 \end{pmatrix},$$

yields smaller estimates $\hat{\sigma}_{w_x}^2$, $\hat{\sigma}_{w_y}^2$, and $\hat{\sigma}_{w_z}^2$ with a better likelihood.

We want to see the behavior of the likelihood function as a function of other
parameters. In Table 12.2, we see $lk(\sigma_\varepsilon^2|a, b, r, \sigma_w^2)$, the likelihood function, as a
function of the observation noise variance, σ_ε^2, when we fix other parameters
to the true values, $a = 10$, $b = 28$, $r = 8/3$, and roughly the estimated value
$\sigma_w^2 = 0.01$ from the previous analysis.

Here we see that the minimum of the likelihood function is attained at $\sigma_\varepsilon^2 = 0.1 \times 10^{-03}$ out of the eight values between 0.1×10^{-06} and 1 (see Table 12.2). From
the table, we can see that the function is smooth in the neighborhood of the
true value $\sigma_\varepsilon^2 = 0.1 \times 10^{-03}$. We also calculated $lk(a|b, r, \sigma_\varepsilon^2, \sigma_w^2), lk(b|a, r, \sigma_\varepsilon^2, \sigma_w^2)$
and $lk(r|a, b, \sigma_\varepsilon^2, \sigma_w^2)$. The results clearly show that the (−2)log-likelihood func-
tions are smooth near the true values of parameters and attain their minima
in the neighborhood of the true values (see Ozaki 1994b and Ozaki et al. 2000).

12.2.2.2 Prediction Errors of Lorenz Chaos

We have seen that the likelihood function favors a nonzero system noise
variance even though the original data were generated by a deterministic
chaos model. The natural question to this may be "why is the stochastic

TABLE 12.2

(−2)Likelihood Values of the Lorenz
Chaos Model for Various σ_ε^2's

σ_w^2	(−2)log-Likelihood
0.1×10^{-06}	$-0.6481 \times 10^{+04}$
0.1×10^{-05}	$-0.6522 \times 10^{+04}$
0.1×10^{-04}	$-0.6804 \times 10^{+04}$
0.1×10^{-03}	$-0.7121 \times 10^{+04}$
0.1×10^{-02}	$-0.6034 \times 10^{+04}$
0.1×10^{-01}	$-0.4248 \times 10^{+04}$
0.1×10^{-00}	$-0.2198 \times 10^{+04}$
$0.1 \times 10^{+01}$	$0.3836 \times 10^{+02}$

version $z_t = f(z_{t-1}|\theta) + w_t$ superior for prediction when the data were simulated from the deterministic model $z_t = f(z_{t-1}|\theta)$?" We know that if a model is favored by the likelihood function, then the model explains the probabilistic behavior of the data better than other less-favored models. We can see this point by looking, more carefully, at the behavior of the innovations, which are the prediction errors $x_{k+1} - C\hat{z}_{k+1|k}$ at each time point k where $k = 0, 1, 2, \ldots, 999$. The prediction errors (i.e., innovations derived from the nonlinear Kalman filter) of the model with true parameters $a, b, r, \sigma_\varepsilon^2$, and $\sigma_w^2 = 0.1 \times 10^{-07}$ are shown in Figure 12.2 together with prediction errors of other similar

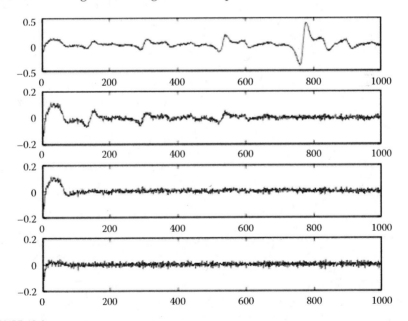

FIGURE 12.2
Lorenz: prediction errors of $\sigma_w^2 = 0.1 \times 10^{-07}$, $\sigma_w^2 = 0.1 \times 10^{-06}$, $\sigma_w^2 = 0.1 \times 10^{-04}$, and $\sigma_w^2 = 0.1 \times 10^{-01}$, from the top to the bottom.

models with $\sigma_w^2 = 0.1 \times 10^{-06}$, $\sigma_w^2 = 0.1 \times 10^{-04}$, and $\sigma_w^2 = 0.1 \times 10^{-01}$. The scale of prediction errors with $\sigma_w{}^2 = 0$ is much larger than those of $\sigma_w{}^2 = 0.1 \times 10^{-07}$.

Four figures in Figure 12.2 clearly shows that the likelihood-favored model gives better predictions than the deterministic model. One natural explanation for this is seen by considering the nonlinear Kalman filtering scheme (11.111), where the state prediction $z_{k|k-1}$ is corrected to $z_{k|k}$ by

$$z_{k|k} = z_{k|k-1} + K_k(x_k - Cz_{k|k-1}),$$

where

$$K_k = P_{k|k-1}C'(CP_{k|k-1}C' + \sigma_\varepsilon^2)^{-1}.$$

If the system is deterministic, then $P_{k|k-1}$ becomes 0, and the filtered state estimation and the one step ahead prediction are always the same, that is, $z_{k|k} = z_{k|k-1}$. This means that, for the deterministic system, the state estimation method would not take account of information drawn from recently obtained data x_k even in the case where it gives an alarmingly large prediction error $v_k = x_k - Cz_{k|k-1}$. This is like insisting on predicting tomorrow's weather with last month's weather record and ignoring today's weather.

A further closer look at the prediction errors in Figure 12.2 together with the data plot of Figure 12.1 tells us that prediction for the deterministic model is particularly bad when the data swing from the negative area to the positive area. This means that prediction errors are more likely to be large and more likely to be accumulating in those active periods than in other quiet periods. These observations strongly imply that the prediction errors have something to do with the local stability and instability of the nonlinear dynamic state equation.

In our calculation of the likelihood function, we have used a nonlinear Kalman filtering technique, that is, the local linearization filter, where the prediction of the state is approximately given locally by $z_{k+1|k} = A(z_{k|k})z_{k|k}$. Hence, if we assume that the system is deterministic and if we have a small estimation error e_k in the estimate $z_{k|k}$ of the true state z_k, that is, $z_{k|k} = z_k + e_k$, e_k is inflated or deflated (depending on the eigen-values of $A(z_{k|k})$) by the transition matrix $A(z_{k|k})$ into $A(z_{k|k})e_k$, because

$$z_{k+1|k+1} = z_{k+1|k} = A(z_{k|k})z_{k|k} = A(z_{k|k})z_k + A(z_{k|k})e_k.$$

This means that the eigen-values of the transition matrix $A(z_{k|k})$ at each time point k provides us with useful information on the local instability. Especially the maximum of the absolute value of the eigen-values is a good index of this. When the maximum is greater than 1, the system is instantaneously unstable and it contributes to inflating the previous state estimation error, while if the maximum is less than 1, it suppresses the previous error. The log of the maximum of the absolute value of eigen-values at each time point is plotted against time in Figure 12.3.

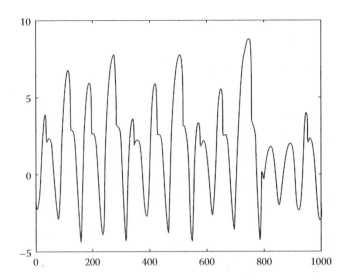

FIGURE 12.3
Local maximum eigen values.

By looking at the prediction errors in Figure 12.2 together with the data plot in Figure 12.1 and the plot of the log of the maximum absolute values of eigen-values in Figure 12.3, we can clearly see that the prediction errors are inflated during the period when the maximum eigen-values stay outside the unit circle (in continuous-time case, this means the maximum of the real part of the eigen-values stays positive). This shows that if we have a parametric chaos model like the Lorenz chaos model, qualitative information about the model, such as the local stability and the instability, may be obtained from the eigen-values of the local transition matrix, approximated from the original continuous-time model. This is far more powerful and convincing than the nonparametric estimate of the Lyapunov exponent often used in conventional chaos studies.

12.2.3 What We Win by Maximizing the Log-Likelihood: The Case of Rikitake Chaos

In the analysis of the simulation data of Lorenz chaos, as can be noticed from looking at the plots of Figure 12.2, we have inhomogeneous prediction errors. The inhomogeneity comes from the following two sources:

(i) An inappropriate initial state $z_{0|0} = (\xi_{0|0}, \eta_{0|0}, \varsigma_{0|0})'$ and an inappropriate variance–covariance matrix $P_{0|0}$ of the state estimation error $(z_0 - z_{0|0})'$

(ii) Inhomogeneous system noise because of the strong nonlinearity causing the transition matrix $A(z_{t|t})$ of the state to be temporarily enlarging with local eigen-values outside the unit circle for some $z_{t|t}$

The second source of the inhomogeneity is possible even when we use the true parameters a and r if σ_w^2 is not made sufficiently large. The inhomogeneity can, of course, be suppressed by using a larger σ_w^2 as we saw in Figure 12.2. The first source of inhomogeneity implies that $z_{0|0} = (\xi_{0|0}, \eta_{0|0}, \varsigma_{0|0})'$ and $P_{0|0}$ need to be adjusted, which can be realized by minimizing the (-2)log-likelihood.

As the initial effect usually dies out after a certain period, we can find reasonable system parameters by discarding the initial data affected by transient effects when the number of data points N is large. That means we can use, with some integer $m > 0$, the approximate likelihood function,

$$(-2)\log p(z_{m+1}, \ldots, z_N | \theta) \approx \sum_{k=m+1}^{N} \left[\log \sigma_{k|k-1}^2 + \frac{(x_k - Cz_{k|k-1})^2}{\sigma_{k|k-1}^2} \right] + (N - m)\log 2\pi,$$

(12.11)

instead of the full likelihood function,

$$(-2)\log p(z_1, \ldots, z_N | \theta) = \sum_{k=1}^{N} \left[\log \sigma_{k|k-1}^2 + \frac{(x_k - Cz_{k|k-1})^2}{\sigma_{k|k-1}^2} \right] + N \log 2\pi.$$

However, the effect of initial values sometimes remains for a long period in an unexpected way. We see this in the estimation of the unobserved state with the nonlinear Kalman filtering method used in our maximum likelihood calculations for the Rikitake data of Figure 12.4. Here, in a similar way

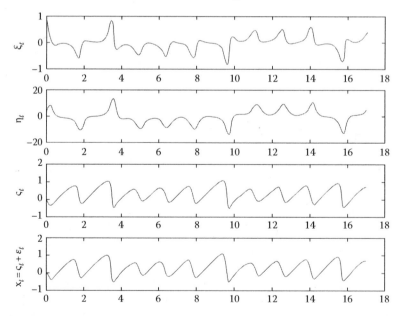

FIGURE 12.4
Rikitake chaos data.

as for Lorenz chaos, the observed scalar time series data, x_t was generated from the following stochastic Rikitake model:

$$\frac{d\xi}{dt} = -\theta_1\xi + \varsigma\eta + w(t)$$

$$\frac{d\eta}{dt} = \theta_2\xi - \theta_1\eta + \varsigma\xi \qquad (12.12)$$

$$\frac{d\varsigma}{dt} = 1 - \xi\eta,$$

where $w(t)$ is a scalar continuous-time Gaussian white noise with the variance σ_w^2.

The original deterministic model was introduced by Rikitake (1958) to explain geomagnetic polarity reversals. An extensive detailed theoretical study of the model is given by Ito (1980), where chaotic dynamics are generated from the original deterministic model with $\theta_1 = 5.00$, $\theta_2 = 124.8$. Here we use the stochastic chaos model with the parameters $\theta_1 = 5.00$, $\theta_2 = 124.8$, $\sigma_w^2 = 0.05$, to generate a time series, from $\tau = 0$ to $\tau = 17$, with observation noise variance $\sigma_\varepsilon^2 = 0.001$. The sampling interval is $\Delta t = 0.005$, and so the number of data points is $N = 3400$. The initial value of the state vector is $(\xi, \eta, \zeta)_0' = (1, 0, 0)'$. The generated state series and the observed time series data $x_i = \varsigma_i + \varepsilon_i$ ($i = 1, ..., N$) are given in Figure 12.4. First we see how the unobserved state series $\xi_1, ..., \xi_N, \eta_1, ..., \eta_N$ and $\varsigma_1, ..., \varsigma_N$ can be correctly estimated from the contaminated observations $x_1, ..., x_N$ if only we know the true model and true initial state vector z_0 and the initial state error variance–covariance matrix $P_{0|0}$ is properly adjusted. We used the exact theoretical parameter values for θ_1 and θ_2 and applied our nonlinear Kalman filter scheme to get the filtered estimates, $\xi_{1|1}, \xi_{2|2}, ..., \xi_{N|N}$ and $\eta_{1|1}, \eta_{2|2}, ..., \eta_{N|N}$, as well as $\varsigma_{1|1}, \varsigma_{2|2}, ..., \varsigma_{N|N}$. The results are plotted in Figure 12.5a together with the true state values $\xi_1, \xi_2, ..., \xi_N \eta_1, \eta_2, ..., \eta_N$, and $\varsigma_1, \varsigma_2, ..., \varsigma_N$. The difference between the true states and the estimated states is hardly noticeable. This clearly shows that if only the model is properly identified the data smoothing problem of the observed data and the estimation problem for the unobserved states (i.e., estimation of $(\xi_t, \eta_t, \varsigma_t)'$ from the observed data, $x_t = \varsigma_t + \varepsilon_t$, ($t = 1, 2, ..., N$), are solved at the same time. However, it is not obvious what will happen if we use the approximate likelihood function as (12.11) and discard the first m data points that are affected by the initial transient effects. It turns out that even in this case we can still get a reasonably good estimate of the model parameters.

For the estimation of the unobserved state, however, things are more complicated. Figure 12.5b shows the estimated states with the true model but with unadjusted poor initial values. With this model, of course the prediction errors are very large at the beginning, but the size of errors decreases to a normal level after about 700 steps. Although the prediction errors are very small after about 700 steps, the estimates of the unobserved states are

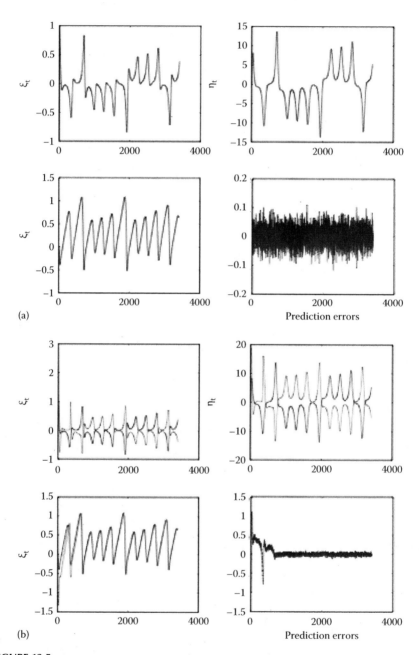

FIGURE 12.5
(a) Filtered state using theoretical initial values. (b) Filtered estimate using poor initial values.

TABLE 12.3

Parameters of the Rikitake Model: True Values, Initial Values, and the Estimated Values with the Rikitake Data

	θ_1	θ_2	σ_w^2	ξ_0	η_0	ζ_0
Theoretical	5.00	124.8	0.05	1.00	0.0	0.0
Initial	5.00	124.8	0.05	0.015	0.31	0.0
Estimated	4.95	123.1	0.078	1.002	0.08	0.003

completely incorrect and show no sign of improvement for the whole period. This means that, even with the true system parameters, we might obtain very poor estimators of the states when the initial values are inappropriate.

From these results, people may rush to the conclusion that this effect is because of the special nature of the chaos model, that is, the initial value sensitivity, which has attracted the limelight in recent times because of its contribution to the uncertainty of the deterministic chaos (Ozaki 1990, Ruelle 1987, Sugihara 1995, Takens 1981). We will see, however, that this conclusion is entirely incorrect. We can find a proper initial state z_0, which maximizes the likelihood by using a numerical optimization technique. Table 12.3 shows the results of the maximum likelihood method applied to the present case.

In this example, the other system parameters, θ_1, θ_2, and σ_w^2 are also estimated at the same time. The initial values for ξ_0, η_0, and ζ_0, in the numerical searching method are the ones used in calculating the very poor state estimates shown in Figure 12.5b. The state estimates obtained by using the estimated initial values $\xi_0 = 1.002$, $\eta_0 = 0.08$, and $\zeta_0 = 0.003$ are shown in Figure 12.5a together with the true states. In the figure, the difference between the estimated state and the true state is very small and unnoticeable (compare them with Figure 12.5b). Here we fix $P_{0|0}$ to be small at 1.0×10^{-6}. The figure of innovations in Figure 12.5a obtained from the estimated model clearly shows that this method has improved the prediction errors over the period of the initial part of the series, which can be confirmed quantitatively by the decrease in AIC.

12.3 Log-Likelihood of State Space Models in Discrete Time

12.3.1 General Linear State Space Models

The prediction error–based Gaussian maximum likelihood method is valid for the identification problem of a well-presented linear state space model:

$$\begin{cases} z_t = Az_{t-1} + Bw_t \\ x_t = Cz_t + \varepsilon_t. \end{cases} \tag{12.13}$$

The prediction error v_t, given from x_t by

$$v_t = x_t - x_{t|t-1},$$

is a Gaussian white noise. Here $x_{t|t-1} = E[x_t|x_{t-1}, x_{t-2}, ..., \theta]$ is calculated using the Kalman filter scheme (11.49) through (11.55), in Chapter 11.

We need to know that we always have a possibility of facing numerical difficulties in calculating the prediction errors of x_t while applying the Kalman filter to the model (12.13). We can eliminate these difficulties by using a "well-posed" structure of A, B, and C of the state space model in advance. The topic is discussed in more detail in Section 12.5.

Using the Gaussian innovations, $v_1, ..., v_N$, the likelihood of the linear state space model (12.13) can be written (Schweppe (1965)) as

$$
\begin{aligned}
(-2)\log p(x_1, ..., x_N|\theta) &= (-2)\sum_k \log p(x_k|x_{k-1}, ..., x_1, \theta) \\
&= (-2)\sum_k \log p(v_k|x_{k-1}, ..., x_1, \theta) \\
&= \begin{cases} \sum_{t=1}^{N}\left\{\log \sigma_{v_t}^2 + \dfrac{v_t^2}{\sigma_{v_t}^2}\right\} + N\log 2\pi: \text{ for scalar data } x_1, ..., x_N. \\[2ex] \sum_{t=1}^{N}\left\{\log |\Sigma_{v_t}| + v_t'\Sigma_{v_t}^{-1}v_t\right\} + Nr\log 2\pi: \text{ for } r\text{-dimensional vector data } x_1, ..., x_N. \end{cases}
\end{aligned}
$$

The maximum likelihood estimate of the parameter θ of the model is obtained by numerically optimizing the log-likelihood function in terms of the parameter vector θ. The consistency and the asymptotic normality of the maximum likelihood estimates of the state space models have been studied under reasonable conditions (see Caines and Rissanen 1974, Ljung and Caines 1979, Ljung 1999).

It is important to know that different types of computational difficulties may arise while we are numerically optimizing the log-likelihood function in terms of the parameter vector θ. The difficulty we encounter here is not a Kalman filter related problem, but it is a kind of redundant parameterization problem of the state space model using "well-posed" state space models also ensures that these badly designed models are excluded from our potential candidate models for the prediction of time series. Several conditions are introduced to ensure that the method works properly. The topic is discussed in Chapter 13.

12.3.2 Spectral Decomposition Model

A typical example of a linear state space model useful in neuroscience is the spectral decomposition model in discrete time, which is the parallel type structural model given by

$$\begin{cases} z_t = A z_{t-1} + \eta_t \\ x_t = C z_t + \varepsilon_t \end{cases}, \qquad (12.14)$$

where

$$z_t = (z_t^{(1)} \quad z_{t-1}^{(1)} \quad \cdots \quad z_t^{(r)} \quad z_{t-1}^{(r)} \quad z_t^{(2r+1)} \quad \cdots \quad z_t^{(p)})',$$

$$A = \begin{pmatrix} 2\,\mathrm{Re}(\lambda_1) & -|\lambda_1|^2 & \cdots & 0 & 0 & 0 & \cdots & 0 \\ 1 & 0 & \cdots & 0 & 0 & 0 & \cdots & 0 \\ \cdots & \cdots & \cdots & \cdots & \cdots & \cdots & \cdots & \cdots \\ 0 & 0 & \cdots & 2\,\mathrm{Re}(\lambda_r) & -|\lambda_r|^2 & 0 & \cdots & 0 \\ 0 & 0 & \cdots & 1 & 0 & 0 & \cdots & 0 \\ 0 & 0 & \cdots & 0 & 0 & \lambda_{2r+1} & \cdots & 0 \\ \cdots & \cdots & \cdots & \cdots & \cdots & \cdots & \cdots & \cdots \\ 0 & 0 & \cdots & 0 & 0 & 0 & \cdots & \lambda_p \end{pmatrix},$$

$$C = (1, 0, \ldots, 1, 0, 1, \ldots, 1),$$

$$\eta_t = (\varepsilon_t^{(1)} \quad 0 \quad \cdots \quad \varepsilon_t^{(r)} \quad 0 \quad \varepsilon_t^{(2r+1)} \quad \cdots \quad \varepsilon_t^{(p)})',$$

so that

$$\eta_t \sim N(0, Q),$$

where

$$Q = \begin{pmatrix} \sigma_1^2 & 0 & \cdots & 0 & 0 & 0 & \cdots & 0 \\ 0 & 0 & \cdots & 0 & 0 & 0 & \cdots & 0 \\ \cdots & \cdots & \cdots & \cdots & \cdots & \cdots & \cdots & \cdots \\ 0 & 0 & \cdots & \sigma_r^2 & 0 & 0 & \cdots & 0 \\ 0 & 0 & \cdots & 0 & 0 & 0 & \cdots & 0 \\ 0 & 0 & \cdots & 0 & 0 & \sigma_{2r+1}^2 & \cdots & 0 \\ \cdots & \cdots & \cdots & \cdots & \cdots & \cdots & \cdots & \cdots \\ 0 & 0 & \cdots & 0 & 0 & 0 & \cdots & \sigma_p^2 \end{pmatrix}.$$

Here the state equation represents a parallel type compartment model, which is composed of a few submodels. Each compartment model has either a second-order AR model structure with complex conjugate eigen-values λ_j and $\bar{\lambda}_j$ such that $|\lambda_j| < 1$ or a first-order AR model structure with a real eigen-value λ_j such that $|\lambda_j| < 1$. In the previous example, we have r submodels of AR(2) structure,

$$z_t^{(j)} = 2\,\mathrm{Re}(\lambda_j)z_{t-1}^{(j)} - |\lambda_j|^2 z_{t-2}^{(j)} + \varepsilon_t^{(j)}, \quad (j = 1, \ldots, r),$$

and $(p\text{-}2r)$ sub-models of AR(1) structure,

$$\xi_t^{(k)} = \lambda_k \xi_{t-1}^{(k)} + \varepsilon_t^{(k)}, \quad (k = 2r+1, \ldots, p).$$

The parameters to be estimated are

$$\theta = (\lambda_1, \bar{\lambda}_1, \ldots, \lambda_r, \bar{\lambda}_r, \lambda_{2r+1}, \ldots, \lambda_p, \sigma_1^2, \ldots, \sigma_r^2, \sigma_{2r+1}^2, \ldots, \sigma_p^2, \sigma_\varepsilon^2)'.$$

The prediction error, $v_t = x_t - CZ_{t|t-1}$, that we need for computing the likelihood is obtained by applying the linear Kalman filter scheme with the previous state space model (12.14).

12.3.3 Nonlinear Discrete-Time State Space Model

Suppose the nonlinear state space model

$$\begin{cases} z_t = A(z_{t-1})z_{t-1} + B(z_{t-1})w_t \\ x_t = Cz_t + \varepsilon_t \end{cases} \tag{12.15}$$

is well posed, that is, the state process z_t is of finite variance. Then the prediction error v_t, given from x_t by $v_t = x_t - x_{t|t-1}$, is a Gaussian white noise. Here, $x_{t|t-1} = E[x_t | x_{t-1}, x_{t-2}, \ldots, \theta]$ is calculated using one of the Discrete-Discrete type nonlinear Kalman filter schemes such as EKF-3 in Section 11.3.2. Then the (−2) log-likelihood of the model is given by using the Gaussian likelihood of the innovations. The maximum likelihood estimate of the model is obtained by minimizing the (−2)log-likelihood, in terms of the parameter θ, by using some numerical nonlinear optimization method such as the Newton–Raphson method or the Davidon–Fletcher–Powell method (Fletcher and Powell 1963).

The nonlinear Kalman filter scheme the for Discrete-Discrete type filtering problem is meaningful only for the finite-variance discrete-time nonlinear state space model. A nonlinear state space model such as

$$\begin{aligned} z_t &= a_1 z_{t-1} - a_2 z_{t-1}^3 + w_t \\ x_t &= z_t + \varepsilon_t \end{aligned} \tag{12.16}$$

is out of our concern since this is not a well-posed state space model. The state process z_t of (12.16) is an explosive process, for any $a_2 \neq 0$, when the system noise w_t is Gaussian white noise. The model is an ill-posed model no matter how small its variance σ_w^2 is (see Ozaki 1993a).

On the other hand a nonlinear state space model

$$z_t = \left(a_1 - a_2 e^{-z_{t-1}^2}\right) z_{t-1} + w_t$$
$$x_t = z_t + \varepsilon_t$$

is a well-posed state space model because the state process z_t is a stationary finite-variance Markov chain process when a_1 and a_2 satisfy certain conditions no matter how large σ_w^2 is. The likelihood of the model is written down using the innovations, $v_1, \ldots, v_N$, as

$$(-2)\log p(x_1, \ldots, x_N | \theta) = (-2) \sum_k \log p(v_k | x_{k-1}, \ldots, x_1, \theta)$$

$$= \left(\begin{array}{l} \displaystyle\sum_{t=1}^{N} \left\{ \log \sigma_{v_t}^2 + \frac{v_t^2}{\sigma_{v_t}^2} \right\} + N \log 2\pi : \text{ for scalar data } x_1, \ldots, x_N. \\[4mm] \displaystyle\sum_{t=1}^{N} \left\{ \log |\Sigma_{v_t}| + v_t' \Sigma_{v_t}^{-1} v_t \right\} + N \log 2\pi : \text{ for vector data } x_1, \ldots, x_N. \end{array} \right.$$

Actual computation of the innovations

$$v_t = x_t - E[x_t | x_{t-1}, x_{t-2}, \ldots, x_1, \theta] \quad (t = 1, 2, \ldots, N)$$

and innovation variances

$$\sigma_{v_t}^2 \ (t = 1, 2, \ldots, N) \quad \text{or} \quad \Sigma_{v_t} \ (t = 1, 2, \ldots, N)$$

is realized by applying the Kalman filter. The maximum likelihood estimate of the model parameter vector θ is obtained by using a numerical nonlinear optimization method.

12.3.4 RBF-AR Type State Space Model

A typical example of a useful discrete-time nonlinear state space model is the one that is derived from an RBF-AR model

$$z_t = \sum_{j=1}^{d} \phi_j(Z_{t-1}) z_{t-j} + w_t, \tag{12.17}$$

with a state vector $Z_{t-1} = (z_{t-1}, \Delta z_{t-1}, \Delta^2 z_{t-1}, \ldots, \Delta^{r-1} z_{t-1})'$ and centers $\varsigma_l = (\varsigma_l^{(1)}, \varsigma_l^{(2)}, \ldots, \varsigma_l^{(r)})'$ ($l = 1, 2, \ldots, m$). Here we have

$$\phi_j(Z_{t-1}) = c_{j,0} + \sum_l^m c_{j,l} \exp\left\{-\|Z_{t-1} - \varsigma_l\|_{H_l}^2\right\} \tag{12.18}$$

and

$$\|Z_{t-1} - \varsigma_l\|_{H_l}^2 = \sum_{k=1}^r \frac{(\Delta^{k-1} z_{t-j} - \varsigma_l^{(k)})^2}{h_l^{(k)}}, \tag{12.19}$$

where $h_l^{(k)}$ is a scaling parameter.

The stability of the process z_t defined by the RBF-AR model (12.17) is guaranteed if $c_{1,0}, c_{2,0}, \ldots, c_{p,0}$ satisfy the following conditions:

"eigen-values, $\lambda_1, \ldots, \lambda_p$ of the matrix A_0,

$$A_0 = \begin{pmatrix} c_{1,0} & c_{2,0} & \cdots & c_{p-1,0} & c_{p,0} \\ 1 & 0 & \cdots & 0 & 0 \\ \cdots & \cdots & \cdots & \cdots & \cdots \\ 0 & 0 & \cdots & 0 & 0 \\ 0 & 0 & \cdots & 1 & 0 \end{pmatrix}$$

lie inside the unit circle."

When the process z_t is observed with an observation error ε_t, that is, $x_t = z_t + \varepsilon_t$, the RBF-AR model (12.17) can be built in a state space model. A simple example is

$$\begin{pmatrix} z_t \\ z_{t-1} \\ \cdots \\ z_{t-p+2} \\ z_{t-p+1} \end{pmatrix} = \begin{pmatrix} \phi_1(Z_{t-1}) & \phi_2(Z_{t-1}) & \cdots & \phi_{p-1}(Z_{t-1}) & \phi_p(Z_{t-1}) \\ 1 & 0 & \cdots & 0 & 0 \\ \cdots & \cdots & \cdots & \cdots & \cdots \\ 0 & 0 & \cdots & 0 & 0 \\ 0 & 0 & \cdots & 1 & 0 \end{pmatrix} \begin{pmatrix} z_{t-1} \\ z_{t-2} \\ \cdots \\ z_{t-p+1} \\ z_{t-p} \end{pmatrix} + \begin{pmatrix} 1 \\ 0 \\ \cdots \\ 0 \\ 0 \end{pmatrix} w_t$$

$$x_t = z_t + \varepsilon_t$$

$$\tag{12.20}$$

If $\|Z_{t-1} - \varsigma_{t-1}\|_h^2$ becomes larger, the transition matrix approaches A_0. Therefore, if the coefficients $c_{1,0}, c_{2,0}, \ldots, c_{p,0}$ satisfy the stability condition, the state process defines a nonexplosive finite-variance state process z_t. This means that the earlier-defined state space model (12.20) is a well-posed nonlinear discrete-time state space model, and we can apply the discrete-time nonlinear filter and the innovation-based maximum likelihood method to estimate the model parameters, σ_w^2, σ_ε^2, $c_{j,k}$ ($j = 0, 1, \ldots, p, l = 1, \ldots, m$).

12.4 Regularization Approach and Type II Likelihood

In this section, we consider an alternative approach to the inference problem-(b). We studied, in Section 11.4.3, a regularization approach to the inference problem-(a). In order to make the regularization approach to be valid and useful for the inference problem-(b) in practice, we need to solve the associated problem, that is, "How to take the balance of $\|Cz - x\|^2$ and $\|\Gamma z\|^2$ by d^2 in the objective function, $J = \|Cz - x\|^2 + d^2\|\Gamma z\|^2$, where C and Γ are assumed to be known."

12.4.1 How to Choose d^2?

For the given time series data $x = (x_1, \ldots, x_N)'$ and $C = (c_{i,j})$ ($i = 1, \ldots, N, j = 1, \ldots, K$), the method of least squares for $z^{(j)}$ ($j = 1, \ldots, K$) leads to the minimization of

$$L(z) = \sum_{i=1}^{N} \left[x_i - \sum_{j=1}^{K} c_{i,j} z^{(j)} \right]^2 . \tag{12.21}$$

The solution of (12.21) behaves badly when K is large compared with N. This type of ill-conditioned problem can be solved when we introduce some preference on the values of the parameters such as that z is close to z_0 and try to reformulate the problem as a constrained least squares problem, where we try to minimize

$$L(z) + d^2 \|z - z_0\|_R^2 ,$$

with some constant d^2. The difficulty in the application of the constrained least squares method is in the choice of the value d^2. To solve this, Akaike (1980) suggests transforming the problem into the maximization of

$$l(z) = \left[\exp - \frac{1}{2\sigma^2} \left\{ L(z) + d^2 \|\Gamma z\|^2 \right\} \right],$$

where temporarily we assume σ^2 is known. Since we have

$$l(z) = \exp\left\{ -\frac{L(z)}{2\sigma^2} \right\} \exp\left\{ -\frac{d^2 \|\Gamma z\|^2}{2\sigma^2} \right\},$$

we realize that the solution of the constrained least squares problem, for a given pair of σ^2 and d^2, is now regarded as the mean of the posterior distribution of z defined by the data distribution

$$f(x|\sigma^2,z) = \left(\frac{1}{2\pi\sigma^2}\right)^{N/2} \exp\left\{-\frac{L(z)}{2\sigma^2}\right\} \tag{12.22}$$

and the prior distribution of z

$$\pi(z|d) = \left(\frac{1}{2\pi\sigma^2}\right)^{K/2} \exp\left\{-\frac{d^2\|\Gamma z\|^2}{2\sigma^2}\right\}. \tag{12.23}$$

The problem of the choice of d^2 in the constrained least square problem is then translated into a problem of the choice of σ^2 and d^2 in Bayesian modeling with the data distribution $f(x|\sigma^2,z)$ and the prior distribution $\pi(z|d)$ for z. The marginal likelihood of (d,σ^2) is defined by

$$L(d,\sigma^2) = \int f(x|\sigma^2,z)\pi(z|d)dz.$$

Since $f(x|\sigma^2,z))$ and $\pi(z|d)$ are given by (12.22) and (12.23), we get

$$L(d,\sigma^2) = \left(\frac{1}{2\pi\sigma^2}\right)^{N/2} \exp\left\{-\frac{\|s(z*|d)\|^2}{2\sigma^2}\right\} \times \frac{\|d^2\Gamma'\Gamma\|^{1/2}}{\|d^2\Gamma'\Gamma+C'C\|^{1/2}}, \tag{12.24}$$

where $\|s(z*|d)\|^2$ denotes the minimum of $\|s(z|d)\|^2$. Akaike (1980) proposed a procedure that suggests choosing σ^2 and d^2, using the minimum of $(-2)\log$ of marginal likelihood, which is given from equation (12.24) by

$$ABIC = (-2)\log L(d,\sigma_d^2)$$

$$= N\log\left\{1/N\|s(z*|d)\|^2\right\} + \log\|d^2\Gamma'\Gamma+C'C\| - \log\|d^2\Gamma'\Gamma\| + \text{const.}$$

The minimum ABIC method is equivalent to Good (1965)'s type II maximum likelihood method, which suggests choosing a model with the maximum marginal likelihood. Since we are familiar with the "log-scale" by the use of minus twice the log-likelihood for non-Bayesian models, ABIC is preferable to the type II maximum likelihood, for the Bayesian model identification problem.

12.4.2 Computational Cost and EM Algorithm

Note that the Bayesian Regularization approach (for the maximum likelihood estimation of the state space model and the state estimation) does not pay attention to the recursive nature of the optimal property (of the state esti-mate as the minimum-variance estimate) of the innovation approach. In the innovation approach, the optimality of the state estimate, that is, minimum variance of the state estimation error and the whiteness of the innovation at each time point, is guaranteed, whereas in the Bayesian Regularization approach, these local-in-time optimal properties are not guaranteed. That means in the Bayesian Regularization approach, all the state estimates at all time points need to be adjusted simultaneously in the computational stage of calculating the likelihood function for a set of parameters temporarily fixed during the stage of numerical optimization of the likelihood function. This is extraordinarily time consuming and inefficient compared with the innova-tion approach as a method of solving inference problem-*(b)*.

The EM algorithm is supposed to improve this computational inefficiency. However, according to Shumway and Stoffer (1982), it has been recognized that the convergence near the optimal point is very slow and inefficient. They recommend to switch from the EM algorithm to a Newton–Raphson type iterative method in the later stage of the likelihood maximization pro-cedure to speed up numerical convergence. The topic is discussed again in Chapter 13.

12.4.3 Jazwinski Scheme and General Nonlinear Non-Gaussian Filter

When x_t is non-Gaussian, some people may think that the simple analytical expression

$$\left(\frac{-1}{2}\right)\sum_{t=1}^{N}\left\{\log \sigma_{v_t}^2 + \frac{v_t^2}{\sigma_{v_t}^2}\right\}$$

of the Gaussian likelihood should be replaced by non-Gaussian likelihood, which idea goes back to the original general form of Jazwinski (1970) (see (11.70) through (11.72) in Section 11.2.8).

Since the assumptions of linearity, stationarity, and Gaussianity are approx-imations to reality, general nonlinear and non-Gaussian filter schemes have attracted the attention of many applied scientists. The General nonlinear non-Gaussian Kalman filter is mathematically well defined by Jazwinski's recursive scheme

$$p(x_t|x_{t-1}, \ldots, x_1, \theta) = \int p(x_t|z_t)p(z_t|x_{t-1}, \ldots, x_1, \theta)dz_t$$

$$p(z_t|x_{t-1}, \ldots, x_1, \theta) = \int p(z_t|z_{t-1})p(z_{t-1}|x_{t-1}, \ldots, x_1, \theta)dz_{t-1}$$

$$p(z_{t-1}|x_{t-1}, \ldots, x_1, \theta) = \frac{p(x_{t-1}|z_{t-1})p(z_{t-1}|x_{t-2}, \ldots, x_1, \theta)}{\int p(x_{t-1}|\varsigma_{t-1})p(\varsigma_{t-1}|x_{t-2}, \ldots, x_1, \theta)d\varsigma_{t-1}}$$

for time series, x_1, x_2, ..., x_t, x_{t+1}, ... with equal sampling intervals. Here, linearity of the dynamics and the Gaussianity of the distribution are not assumed. Only the Markov property of the process and the observation system are assumed to be given by $p(z_t|z_{t-1})$ and $p(x_t|z_t)$, respectively. Note that these equations explain the conceptual relations and do not mean that the densities are analytically obtainable. Actually, their analytical derivation is impossible for the general non-Gaussian case. Jazwinski (1970) presented the scheme for discrete nonlinear filtering, but the scheme is also true and valid for a continuous-time model with discrete-time data points. The scheme only shows the conceptual recursive relations between the conditional distributions at the discrete-time points.

The linear Kalman filter scheme can solve this Jazwinski scheme efficiently only when $p(x_t|z_t)$, $p(z_t|z_{t-1})$, and $p(x_{t-1}|z_{t-1})$ are Gaussian. When they are non-Gaussian, these conditional densities cannot be obtained in analytical form. However, even though the densities are not available in an analytical form, they are obtainable, in principle, by numerical methods with modern fast computers. Many empirical studies have been proposed and tried out with the help of computers and Monte Carlo simulation techniques. However, the computational cost for solving the Jazwinski scheme with these fully numerical methods is enormous compared with the conventional linear Kalman filter even when all the model parameters, such as the system noise variance of the state, observation noise variance, and the state transition parameters are assumed to be known. If we need to search for both model parameters and the optimal filtering as well, the required amount of computation is huge, even for a modestly low-dimensional case.

Instead of these fully numerical approaches, the possibility of more efficient parametric model-based approaches has been investigated and developed in the last few decades within the space engineering and control engineering communities. They are computationally much more efficient than the numerical integration approach based on the Jazwinski scheme. The only disadvantage of these model-based approaches has been the computational instability problem, which was solved recently by the introduction of A-stable discretization schemes such as LL and NLL. Since neuroscience data intrinsically involves high dimensional dynamics and many models are Markov by nature, we strongly recommend this parametric model–based approach to nonlinear filtering with the LL or the NLL discretization scheme in as discussed Chapter 9.

12.5 Identifiability Problems

From the very beginning of the 1960s when people started applying the Kalman filter to actual engineering problems such as space engineering, they have recognized several difficulties in various stages of the implementation of the Kalman filter algorithm. Several conditions that guarantee the eligible solution of the algorithm have been presented since.

Whether the concerned model is identifiable or not is a very important question for the users of the state space modeling method. Unfortunately, many discussions and theories given in textbooks are not very use- ful for the users. Jazwinski (1970)'s asymptotic stability condition (see Jazwinski 1970, Section 6) is indirect and may not be easy to check for the practitioners.

Typical (but not all) difficulties for nonlinear state space model identifica- tion arise from the following three:

(D-1) Conditions for the matrices, A, B, and C:
The state $z(t)$ cannot be fully determined from the knowledge of the observed data, $x_1, x_2, \ldots, x_N$, if certain conditions for the matrices, A, B, and C, are not satisfied. For example, if the pair of the observation matrix C and the tran- sition matrix A are not properly chosen, some of the state variable, say $z_t^{(i)}$, of the k-dimensional state vector $z_t = (z_t^{(1)}, z_t^{(2)}, \ldots, z_t^{(k)})'$ cannot be estimated by the Kalman filter. Also if the pair of the matrices (A, B) does not satisfy the controllability (reachability) condition, some of the state variable, say $z_t^{(i)}$, of the k-dimensional state vector $z_t = (z_t^{(1)}, z_t^{(2)}, \ldots, z_t^{(k)})'$ stays constant for the whole period, $-\infty < t < \infty$, even though the state vector is driven by a Gaussian white noise. The topic of how to design the most compact and stan- dard state space model from the observed time series $x_1, x_2, \ldots, x_N$ is called the "realization problem" (a part of the related theories we saw in Sections 3.4 and 4.4).

(D-2) Numerical instability:
Optimality (such as the minimum variance of the state estimate) of the filter- ing does not guarantee the stability of its dynamics, and the actual filtering scheme occasionally diverges even though the filtering equation is supposed to be optimal. This is called the "instability problem." The problem can cause serious difficulty and becomes especially important when the dynamics of the state equation is nonlinear and some of the instantaneous eigen-values of the transition matrix stay temporarily outside the unit circle.

(D-3) Overparameterization:
Parameters of the state space models cannot always be estimated by the maximum likelihood method, even though the matrices, A, B, and C satisfy the observability and controllability (reachability) conditions, if the param- eterization is inappropriate. This problem is called the "parameterization problem."

We have seen that solutions for the inference problem (b) are given by the two approaches: one is the "state space model based" approach and another is the "regularization based" approach. The state space model approach explic- itly formulates the problem in a state space model with a dynamic model for the state equation and an observation equation, which specifies the relation between the state and the observed time series. Typical examples are

$$
\begin{cases}
z_t = A z_{t-1} + w_t \\
x_t = C z_t + \varepsilon_t
\end{cases}
\tag{12.25}
$$

and

$$
\begin{cases}
z_t = a(z_{t-1}) + w_t \\
x_t = C z_t + \varepsilon_t
\end{cases},
\tag{12.26}
$$

where the models are expected to be identified from the time series data, $x_1, x_2, \ldots, x_N$.

These three types of computational difficulties associated with the state space model identification have been discussed mostly for the state space model based approach. There is not much systematic discussion of the numerical difficulties associated with the model structure for the "regularization based" approach. However, in principle, they share the same difficulties, because as we saw in the previous section, Section 12.4, the state space models (12.25) and (12.26) can be translated to a regularization model as

$$
\underset{z}{\mathrm{Min}} \left[\sum_{t=1}^{N} \| x_t - C z_t \|^2 + d^2 \sum_{t=1}^{N} \| z_t - A z_{t-1} \|^2 \right]
\tag{12.27}
$$

and

$$
\underset{z}{\mathrm{Min}} \left[\sum_{t=1}^{N} \| x_t - C z_t \|^2 + d^2 \sum_{t=1}^{N} \| z_t - a(z_{t-1}) \|^2 \right].
\tag{12.28}
$$

The reverse is also true, and a regularization model may be translated to a state space model as long as the constraint of the regularization model is for the dynamics of the state variables.

The fact that there is not much discussion of the difficulties of the regularization model does not necessarily mean that the method is free from difficulty. It simply means that the users of the regularization based approach are mostly interested in simple model structures that are free from the difficulties associated with observability, instability, or overparameterization. When they start using more sophisticated dynamic model structures as a constraint of the unobserved state variable, they will naturally need discussions of the same identifiability problems as the state space model based approach.

12.5.1 Gramians

We have seen in Section 3.2.1.5 that the two conditions, observability condition and controllability (reachability) condition, are essential for the estimation of the unobserved state variables and for the Kalman filter to work in the inferential problem (*a*).

To understand the role of these conditions in the inferential problems of the state space model, it is sometimes useful to use the following two Gramians, one being the observability Gramian and another the controllability (reachability) Gramian. The observability Gramian, denoted by G_O, is defined by

$$G_O = \sum_{k=0}^{\infty} (A')^k C' C A^k = O'O.$$

Since it holds that "*rank O = k* if and only if $O'O = G_O > 0$," the observability condition (see Section 11.2.5) is equivalent to the observability Gramian G_O being positive definite. Also from these relations we can see that (A,C) is observable if and only if $C\xi = 0$ implies $\xi = 0$ for any eigen-vector ξ of A.

The controllability (reachability) Gramian, denoted by G_C, is defined by

$$G_C = \sum_{k=0}^{\infty} A^k BB' (A')^k = CC'.$$

Since it holds that "*rank C = n* if and only if $CC' = G_C > 0$," the controllability (reachability) condition (see Section 11.2.5) is equivalent to the controllability (reachability) Gramian G_C being positive definite. From the relations here we can see that (A,B) is controllable (reachable) if and only if $\xi' B = 0$ implies $\xi = 0$ for any row eigen-vector ξ' of A.

These two properties guarantee that the system behaves normally, that is, it exhibits no pathological properties. For example, the state vector has singular variance when the system is not reachable. The singularity of the variance of the k-dimensional state simply means that some element of the state vector stays constant for the whole period even though the state vector is driven by the system noise w_t.

This implies that we can eliminate the offending element from the vector of the k-dimensional state variable and reformulate a new state space model with a $(k - 1)$-dimensional state vector. If this state is one of the important variables, and estimation of this state is one of your main objectives, the non-reachability and the singularity of the state variance simply means that the model is not properly designed.

When the system is observable, the least squares estimate of the state is well defined and consistent. If the model is unobservable, at least one variable in the state vector is isolated and is not affected by the present and past observation data, so that we cannot estimate the state variable from the data. Such a variable in the state vector is not necessary for the description of the generating mechanism of the time series data, and we can eliminate those variables from the state vector and reformulate the model into a new state space model. If the estimation of the state (from the current observed time series) is one of your main purposes, unobservability simply means that the current state space model is not properly designed for the purpose.

12.5.2 Instability Problem

As for the instability problem, D-2, there are many ways of setting conditions for guaranteeing the proper stable solutions. However, most of them are not easy to check in practice, and often another way of setting conditions is replacing the previous set of difficult-to-check conditions with other conditions with the same degree of difficulty (Kalman and Bertran 1960a, 1960b). Since we assume that the readers of the present book are mostly applied scientists and engineers, we don't go into the details of these mathematical discussions. Instead, we point out that there are two common types of instability problems in the state space model identification:

1. The state equation is properly presented so that the state process is stable and of finite variance, but the discretized state model is unstable and of infinite variance.

2. The state equation is improperly presented so that the state process is intrinsically unstable and explosive.

The stability of linear systems is rather easy to check. It depends on the eigenvalues of the transition matrix A. If the eigen-values of A are within the unit circle, the norm of the process z_t is always smaller than the norm of z_{t-1}, that is, $\|z_t\|^2 = \|Az_{t-1}\|^2 \leq \|z_{t-1}\|^2$. All the stationary linear time series models are transformed into a state space model with observation errors. For example, the AR(p) model is transformed into a state space model

$$
\begin{pmatrix} z_t \\ z_{t-1} \\ \cdots \\ z_{t-p} \end{pmatrix} = \begin{pmatrix} \phi_1 & \phi_2 & \cdots & \phi_p \\ 1 & 0 & \cdots & 0 \\ \cdots & \cdots & \cdots & \cdots \\ 0 & 0 & \cdots & 0 \end{pmatrix} \begin{pmatrix} z_{t-1} \\ z_{t-2} \\ \cdots \\ z_{t-p-1} \end{pmatrix} + \begin{pmatrix} w_t \\ 0 \\ \cdots \\ 0 \end{pmatrix},
$$

$$
x_t = z_t + \varepsilon_t
$$

where the observation error $\varepsilon_t = 0$.

It is not so easy to check the stability of a nonlinear system as it is to check a linear system, especially when the systems are in discrete time.

A simple example of an improperly presented nonlinear discrete-time state space model is

$$z_t = -z_{t-1}^3 + w_t$$

$$x_t = z_t + \varepsilon_t.$$

(12.29)

The state process z_t defined by the state equation is explosive when w_t is Gaussian white noise, and z_t does not satisfy the finite-variance property. It does not make sense to consider a computationally stable filter for such an infinite-variance state space model.

An example of a well-posed nonlinear state space model is

$$dz/dt = -z^3 + w(t)$$

$$x_t = z_t + \varepsilon_t.$$

(12.30)

The state process $z(t)$ of (12.30) is known to be of finite variance, and so it makes sense to consider the filtering and estimation of the state $z(t)$ from time series observations.

Even though the original continuous-time state space model is stable, its discretized version does not necessarily yield to a stable well-posed state space model. For example, z_t of the following discrete-time state space model is not stable:

$$z_t = z_{t-1} + (\Delta t)z_{t-1}^3 + \sqrt{\Delta t}w_t,$$

$$x_t = z_t + \varepsilon_t$$

(12.31)

but z_t of the following state space model,

$$z_t = z_{t-1} + (1/3)\left\{\exp(-3z_{t-1}^2\Delta t) - 1\right\}z_{t-1} + \sqrt{\Delta t}w_t$$

$$x_t = z_t + \varepsilon_t,$$

(12.32)

is stable and well posed. Note that both (12.31) and (12.32) converge to the same continuous-time state space model (12.30) for $\Delta t \to 0$.

In the calculation of the innovations by a nonlinear filtering scheme, an A-stable discretization scheme such as the LL method in Chapter 9 plays a key role. In the model identification procedure, since numerical optimization of the likelihood function itself needs quite complicated computational techniques to make it work properly without explosion, it is essential for the users to use a discretization scheme that inherits the finite-variance property of the original continuous-time model in the discretized model, so that they can concentrate on the optimization work by separating the two difficult problems, that is, search for a proper discretization and search for an appropriate model parameter vector in the searching space.

12.5.3 Instability and Variable Transformations

It is often useful in model identification, if we know that a variable transformation can sometimes solve the instability problem. One example is the trend rhythm decomposition problem, where the data are decomposed into a trend component and oscillatory rhythm component. Both components could be affected by some exogenous input variable and in an increasing (multiplicative) mode. A simple dynamic model of increasing trend is expressed in the following (Stratonovich type) stochastic differential equation model

$$\dot{T}(t) = a(u_t)T(t) + T(t)w_T(t),$$

where the exogenous variable u_t controls the increase (if $a(u_t) > 0$) and decrease (if $a(u_t) < 0$) in the trend.

A simple dynamic model for cyclic oscillations with increasing (or decreasing) amplitude with $T(t)$ is given by $T(t)s(t)$, where $s(t)$ follows the oscillation model

$$\ddot{s}(t) + bs(t) = w_s(t).$$

Here the frequency of the oscillation of $s(t)$ is determined by $b(>0)$.

Suppose a set of time series data, $x_1, x_2, \ldots, x_N$, is obtained from this dynamic model for trend and rhythm. When the trend is in an increasing mode (i.e., $a(u_t) = a_0 > 0$, the model is written as

$$\dot{T}(t) = a(u_t)T(t) + T(t)w_T(t)$$

$$\ddot{s}(t) = -bs(t) + w_s(t) \tag{12.33}$$

$$x_t = T_t + T_t s_t,$$

which can be rewritten as

$$\dot{z}(t) = Az(t) + Bw(t)$$
$$x_t = c(z_t), \tag{12.34}$$

where

$$z(t) = (T(t), \dot{s}(t), s(t))', \quad w_t = (w_T(t), w_S(t))',$$

$$c(z_t) = T_t + T_t s_t,$$

$$A = \begin{pmatrix} a_0 & 0 & 0 \\ 0 & 0 & -b \\ 0 & 1 & 0 \end{pmatrix}, \quad B = \begin{pmatrix} 1 & 0 \\ 0 & 1 \\ 0 & 0 \end{pmatrix}.$$

We cannot apply Kalman filters to whiten the time series x_t into an innovation, $v_t = x_t - E[x_t|x_{t-1}, x_{t-2}, ...]$ with the model (12.34) even though we know the true model with the exact model parameters, a_0 and the noise variances of $w_T(t)$ and $w_s(t)$. This is because one of the eigen-values of the transition matrix A is real and positive. Accordingly, the transition matrix is an enlarging map of the state and the state estimation error, $z_t - z_{t|t}$, is inflated at each transition. Therefore, the initial state estimation error $z_t - z_{t|t}$ at time point $t = 1$ will increase to ∞ for $t \to \infty$, no matter how small it is at the beginning.

However, this instability problem can be solved if we convert the state space model for $x_1, x_2, ..., x_N$ into a state space model for the log-transformed data, $\log x_1, \log x_2, ..., \log x_N$. For the transformed data, we can rewrite an observation equation as

$$\log x_t = \log T_t + \log (1 + s_t).$$

A dynamic model for $\log T_t$ is obtained from the dynamic model for T_t in (12.34) as

$$\frac{d(\log T_t)}{dt} = a + w_T(t).$$

We can use the same dynamic model for s_t in (12.34). Then we have a state space model for the log-transformed data as

$$\frac{d(\log T(t))}{dt} = a + w_T(t)$$
$$\dot{S}(t) = A_s S(t) + B_s w_s(t) \tag{12.35}$$
$$\log x_t = \log T_t + \log (1 + s_t).$$

Here,

$$S(t) = \begin{pmatrix} \dot{s}(t) \\ s(t) \end{pmatrix}, \quad A_s = \begin{pmatrix} 0 & -b \\ 1 & 0 \end{pmatrix}, \quad B_s = \begin{pmatrix} 1 \\ 0 \end{pmatrix}.$$

The transition matrix of the model (12.35) is not an inflating map of the state since the eigen-values of the transition matrix do not stay outside the unit circle. Here, the state dynamics of the state space model (12.34) is linear, but the observation equation is nonlinear. A discrete-time version of state space modeling of inflating process similar to (12.35) was used in the X-11 type trend-seasonal adjustment of economic time series (see Ozaki and Thomson 2002).

12.5.4 Overparameterization Problem

As for the overparameterization problem, D-3, there is no systematic theory for avoiding the redundant parameterization of the state space models except empirical expertise. Inappropriate parameterization often ends up with situations where the so-called regularity conditions for the maximum likelihood estimation are not satisfied. It yields ill-conditioned Hessian, which makes the nonlinear numerical optimization of the log-likelihood impossible. This problem could often happen when people start developing their own new nonlinear dynamic models for their specific problems.

If the parameterization is proper, and if some regularity conditions are satisfied, the maximum likelihood estimates of the state space model are known to be consistent and asymptotically normally distributed (Ljung and Caines 1979). However, the only realistic way for checking this is a simulation study where we use artificially generated data with known model parameters and apply the maximum likelihood estimation method to the artificially generated data. With a modern fast computer, this is quite an easy task.

In the parameterization of state space modeling, we should be careful that the state space representation is not unique. This may sometimes cause a confusion to the modelers. Suppose we have a k-dimensional state space model

$$\begin{cases} y_t = Ay_{t-1} + B\varepsilon_t \\ x_t = Cy_{t-1} \end{cases}. \tag{12.36}$$

Here we observe only x_t. Since we don't observe the state z_t, we may freely to choose any z_t satisfying $z_t = R^{-1}y_t$ with some non-singular matrix R. Then with the new state variable z_t we can have an alternative state space representation:

$$\begin{cases} z_t = A'z_{t-1} + B'\varepsilon_t \\ x_t = C'z_{t-1} \end{cases}, \tag{12.37}$$

where $z_t = R^{-1}y_t$, $A' = R^{-1}AR$, $B' = R^{-1}B$, $C' = CR$. However, nonuniqueness does not mean that the model identification method is meaningless. As long as the state space model is well posed, we can estimate the model parameters without difficulty using our innovation-based maximum log-likelihood method for the state space model.

Here the problem is not only in the choice of the state vector. Even though we fix the state vector to be y_t or z_t and go with the associated state space model, we cannot estimate A, B, and C of the model (12.36) or A', B', and C' of the model (12.37), if we treat all the elements of the matrices as free parameters. The parameterization of the state space model needs to be more compact. We can see this in the following simple example of the state space modeling of an ARMA process. The question is what is the most compact state space representation for the time series x_t that is generated from the ARMA(3,2) model

$$x_t = \phi_1 x_{t-1} + \phi_2 x_{t-2} + \phi_3 x_{t-3} + \theta_1 w_{t-1} + \theta_2 w_{t-2} + w_t. \qquad (12.38)$$

We have already seen an answer to this, which is a special case of Akaike's general solution (the canonical representation) to the multivariate ARMA model identification method in Chapter 4. The state space model is given by

$$\begin{aligned} z_t &= A z_{t-1} + b w_t \\ x_t &= c z_t, \end{aligned} \qquad (12.39)$$

where

$$A = \begin{pmatrix} 0 & 1 & 0 \\ 0 & 0 & 1 \\ \phi_3 & \phi_2 & \phi_1 \end{pmatrix}, \quad c = \begin{pmatrix} 1 & 0 & 0 \end{pmatrix}$$

and

$$b = \begin{pmatrix} 1 \\ h_1 \\ h_2 \end{pmatrix}.$$

$h_1, h_2, \ldots$ is the impulse response function of the ARMA(3,2) of (12.38) given by

$$\begin{pmatrix} 1 \\ h_1 \\ h_2 \end{pmatrix} = \begin{pmatrix} 1 & 0 & 0 \\ -\phi_1 & 1 & 0 \\ -\phi_2 & -\phi_1 & 1 \end{pmatrix}^{-1} \begin{pmatrix} 1 \\ \theta_1 \\ \theta_2 \end{pmatrix}.$$

Note that the number of independent parameters of the ARMA(3,2) model of (12.38) is 6, and the number of free parameters in the Akaike canonical representation (12.39) is also 6. Naturally, we have other similar canonical representations that define equivalent processes to the original ARMA(3,2) process of (12.38) (see Section 11.3.6.2). They are

$$
\begin{pmatrix} z_t^{(1)} \\ z_t^{(2)} \\ z_t^{(3)} \end{pmatrix} = \begin{pmatrix} \phi_1 & 1 & 0 \\ \phi_2 & 0 & 1 \\ \phi_3 & 0 & 0 \end{pmatrix} \begin{pmatrix} z_{t-1}^{(1)} \\ z_{t-1}^{(2)} \\ z_{t-1}^{(3)} \end{pmatrix} + \begin{pmatrix} 1 \\ \theta_1 \\ \theta_2 \end{pmatrix} w_t
$$

$$
x_t = (1 \quad 0 \quad 0) \begin{pmatrix} z_t^{(1)} \\ z_t^{(2)} \\ z_t^{(3)} \end{pmatrix},
$$

(12.40)

$$
\begin{pmatrix} z_t^{(1)} \\ z_t^{(2)} \\ z_t^{(3)} \end{pmatrix} = \begin{pmatrix} 0 & 0 & \phi_3 \\ 1 & 0 & \phi_2 \\ 0 & 1 & \phi_1 \end{pmatrix} \begin{pmatrix} z_{t-1}^{(1)} \\ z_{t-1}^{(2)} \\ z_{t-1}^{(3)} \end{pmatrix} + \begin{pmatrix} 1 \\ 0 \\ 0 \end{pmatrix} w_t
$$

$$
x_t = (1 \quad h_1 \quad h_2) \begin{pmatrix} z_t^{(1)} \\ z_t^{(2)} \\ z_t^{(3)} \end{pmatrix},
$$

(12.41)

and

$$
\begin{pmatrix} z_t^{(1)} \\ z_t^{(2)} \\ z_t^{(3)} \end{pmatrix} = \begin{pmatrix} \phi_1 & \phi_2 & \phi_3 \\ 1 & 0 & 0 \\ 0 & 1 & 0 \end{pmatrix} \begin{pmatrix} z_{t-1}^{(1)} \\ z_{t-1}^{(2)} \\ z_{t-1}^{(3)} \end{pmatrix} + \begin{pmatrix} 1 \\ 0 \\ 0 \end{pmatrix} w_t
$$

$$
x_t = (1 \quad \theta_1 \quad \theta_2) \begin{pmatrix} z_t^{(1)} \\ z_t^{(2)} \\ z_t^{(3)} \end{pmatrix}.
$$

(12.42)

In control engineering these models, (12.39) through (12.42) are called the observability canonical form, observer canonical form, controllability canonical form, and controller canonical form, respectively (Kailath 1980).

13

Art of Likelihood Maximization

We learned in Section 10.1 that with the innovation approach, we can write down the likelihood of any nonlinear state space model by transforming the time series, whether Gaussian or non-Gaussian, into white Gaussian innovations. Writing down the likelihood function of the Gaussian innovations may be easy, but it does not mean that its maximization with respect to the parameters is easy. In fact, numerical maximization of the likelihood of state space models is not so simple and easy. It needs some experience and skill even for the case of linear state space models. Numerical maximization of the likelihood is an important part of the work of the state space modeling of time series. What is difficult to find in the most textbooks of time series analysis are "useful instructions" for obtaining proper maximum likelihood estimates after writing down the likelihood, that is, useful instructions and guidance for obviating numerical difficulties during the stage of numerical optimization of the likelihood of the state space model.

Unfortunately, most applied mathematicians and numerical analysts are not much interested in numerical optimization problems of a specific objective function related to a specific model structure, such as the state space model identification problems. Statisticians may be interested in the asymptotic behavior of the maximum likelihood estimates in general, but they are not so much interested in numerical problems associated with maximization of the likelihood for finite sample data. For this reason, we will specially focus in this chapter on these problems of numerical optimization of the likelihood of state space models and the numerical computational problems associated with the model identifications. In fact, the numerical problems and statistical parameterization problems are closely connected (see Section 12.5). Numerical difficulties often come from incorrect parameterizations. A typical example is the redundant parameterization, which is not so easy to notice when the dynamic model is formulated in a state space representation. Obviously, redundant parameterization leads to a singular Hessian matrix, which makes numerical optimization of the likelihood function impossible.

13.1 Introduction

The maximum likelihood method for state space modeling has been introduced in control engineering since the late 1960s (see Mehra 1969, 1971, Astrom and Kallstrom 1973, Mehra and Tyler 1974) and has been used in many applications in control engineering. The models used in these applications range from very simple linear models to very complex nonlinear dynamic models such as the Hodgkin-Huxley model and Lorenz chaos. As the dynamics of the state space model became more complex, nonlinear, and high dimensional, various difficulties have been recognized in the actual numerical maximization of the likelihood.

The likelihood of the state space model is not only written using the present innovation approach. As we saw in Section 12.4 some state space models proposed in the regularization framework are estimated by maximizing the Type-II likelihood. The difference is subtle and people often ignore it. The Bayesian EM algorithm extended by Schumway and Stoffer (1982) for the state space model estimation is such an example. Durbin and Koopman (2001) also take the intermediate stance between the classic innovation-based approach and the Bayesian approach. This mixed attitude is acceptable for the estimation of models with simple state dynamics such as the random walk; however, it starts causing difficulties when we face an estimation problem of more complex nonlinear dynamic models for the state equation. Leaving this topic to a later discussion, we see first the most important numerical problems associated typically with the maximization of the likelihood of the state space model.

The typical difficulties associated with the likelihood maximization of the state space models are the following two:

1. Initial value problem
2. Slow convergence of the optimization

We first consider the computational difficulties coming from the initial value problem. Traditionally in the estimation of conventional time series models, such as AR models and ARMA models, the initial value problem is usually ignored with the assumption that the number of observations is sufficiently large. However, in recent extensions of the time series methods to various types of state space models, it has become clear that the initial values of state space models are not in fact negligible. On the contrary, it has turned out that this can cause serious damage the validity of parameter estimation in some nonlinear state space modeling cases.

Since the problem of the initial value effect was not recognized to be very important, there are many ways of treating the initial values in the conventional methods for the computation of the maximum likelihood estimation of the state space model. We will check this problem, in the following section, from the point of view of the innovation approach.

13.2 Initial Value Effects and the Innovation Likelihood

Our guideline principle, that is, the Prediction Error Approach or the Innovation Approach, is still useful for the computations of the maximum likelihood estimate of state space models, where we can transform the temporally dependent time series x_t into independent Gaussian white noise, $v_t = x_t - x_{t|t-1}$, by the Kalman filter. By transforming the time series into Gaussian white innovations, $v_t = x_t - x_{t|t-1}$ for $t = 2, ..., N$, we can write down the (−2)log-likelihood of a state space model

$$z_t = Az_{t-1} + w_t, \quad w_t: \mathcal{N}(0, R_1)$$

$$x_t = Cz_t + \varepsilon_t, \quad \varepsilon_t: \mathcal{N}(0, R_2)$$

as

$$(-2)\log p(x_1, x_2, ..., x_N | \theta) = (-2)\log p(x_1 | \theta) + (-2)\sum_{t=2}^{N} \log p(x_t | x_1, ..., x_{t-1}, \theta)$$

$$= (-2)\log p(x_1 | \theta) + (-2)\sum_{t=2}^{N} \log p(v_t | x_1, ..., x_{t-1}, \theta),$$

(13.1)

where θ is the parameter vector of the model to be estimated. In the old days, people just introduced assumptions such as the number of data points, N, is sufficiently large so that (−2)log $p(x_1 | \theta)$ is negligibly small and we can eliminate it from the objective function when optimizing. However, the effect of the initial value is not necessarily small when we deal with state space models. Typical examples are state space models for the seasonal adjustment (Ozaki and Thomson 2002) and state space models for the Rikitake and Lorenz chaos (Ozaki et al. 2000). Recently, people have started paying more attention to the initial value effect when they calculate the maximum likelihood for parameter estimations. The treatment of initial value is slightly different in each method, but the differences of the results between them are not as slight as we expect. They are rather significant and quite large.

13.2.1 Innovation Approach

For the specification of the (-2)log-likelihood of the initial value, $(-2)\log p(x_1|\theta)$, Ozaki (1992b) suggests using

$$(-2)\log p(x_1|\theta) = (-2)\log p(v_1|z_{0|0}, P_{0|0}, \theta)$$

$$= \log \sigma_{v_1}^2 + \frac{v_1^2}{\sigma_{v_1}^2}, \tag{13.2}$$

where we discuss the scalar time series case for simplicity, although the discussion also applies for the multivariate case without any essential change.

Here $z_{0|0}$, $P_{0|0}$ are the state filtered estimate and its error variance, respectively. They play an important role in initiating the recursive computation of the innovation v_t and the innovation variance $\sigma_{v_t}^2$ in the Kalman filter for $t = 1, 2, \ldots, N$. At the start of the recursive computation, we have

$$v_1 = x_1 - Cz_{0|0}$$

and

$$\sigma_{v_t}^2 = C(AP_{0|0}A' + R_1)C' + R_2.$$

Then the (-2)log-likelihood is

$$(-2)\log p(x_1, x_2, \ldots, x_N|\theta) = \log p(x_1|\theta) + \sum_{t=2}^{N} \log p(x_t|x_1, \ldots, x_{t-1}, \theta)$$

$$= \sum_{t=1}^{N} \left(\log \sigma_{v_t}^2 + \frac{v_t^2}{\sigma_{v_t}^2} \right) + N \log 2\pi. \tag{13.3}$$

In the innovation approach, $z_{0|0}$ and $P_{0|0}$ can be treated simply as extra parameters and are estimated by maximizing the log-likelihood (Ozaki 1992b).

We know intuitively that prediction error is better to be small. In fact finding θ giving smaller innovations always contributes to attaining θ of larger log-likelihood (i.e., smaller (-2)log-likelihood). This is also seen in the structure of the innovation-based (-2)log-likelihood representation (13.4), which is written as

$$(-2)\log p(x_1, x_2, \ldots, x_N|\theta) = \log \sigma_{v_1}^2 + \frac{v_1^2}{\sigma_{v_1}^2} + \sum_{t=2}^{N} \left(\log \sigma_{v_t}^2 + \frac{v_t^2}{\sigma_{v_t}^2} \right) + N \log 2\pi,$$

$$\tag{13.4}$$

where

$$v_1 = x_1 - Cz_{0|0} \tag{13.5}$$

and

$$\sigma_{v_1}^2 = C(AP_{0|0}A' + BR_1B')C' + R_2 \tag{13.6}$$

Here, v_1 is affected by a bad choice of the initial state $z_{0|0}$ from

$$v_1 = x_1 - CAz_{0|0}.$$

The large v_1 affects the next innovation v_2 by

$$v_2 = x_2 - CAz_{1|1} = x_2 - CAz_{1|0} + CAK_1v_1.$$

The choice of $P_{0|0}$ affects the value of v_2 through the Kalman gain K_1 (see (11.55) and (11.53) in Section 11.2). This is one of the typical characteristics of the Kalman filter. It takes a while for innovations to reach a stationary level. Although they are made into temporally independent sequences, they are not necessarily homogeneous in time.

In some extreme situations, even though N is very large, say 1000, more than 90% of the size of the (-2) log-likelihood could be explained by the bad choice of the initial state $z_{0|0}$. In such a situation, the effect of adjusting the parameter θ of the state space model in the numerical optimization procedure of the (-2) log-likelihood is negligibly small. The numerical optimization fails to find a proper value for the parameter θ.

The solution for this comes out easily if we go back to the original idea of making prediction errors small with the model in our hand. If the innovations near the initial value are very large because of the bad choice of the initial state $z_{0|0}$, we could choose $z_{0|0}$, which leads to small innovations and large likelihood. Numerically, this can be implemented simply by including $z_{0|0}$ as one of the variables to adjust in the optimization procedure.

Of course, the innovations are affected by the choice of $P_{0|0}$ through the Kalman gain. If $P_{0|0}$ is too small, it takes a longer time for innovations to reach a stationary level (see Figure 12.2 in Section 12.2.3), whereas if $P_{0|0}$ is chosen to be too large it does not help in reducing the (-2)log-likelihood (see 13.4 and 13.6). However, since its effect to (-2)log-likelihood is not as drastic as the effect of $z_{0|0}$, we could fix it first and concentrate on eliminating the bad effect of the inappropriate choice of $z_{0|0}$.

In principle, the minimization of (-2)log-likelihood is attained by adjusting all of θ, $z_{0|0}$, and $P_{0|0}$. However, this requires rather heavy computation especially when the dimension of the state vector is high. A practical solution could be splitting the computation into the following three stages:

First stage:

1. Choose some initial values for θ, $z_{0|0}$, and $P_{0|0}$.
2. Calculate innovations and plot the innovation with respect to the time axis.
3. Choose, by looking at the plot, the time point N_0 where the innovations reach a more or less stationary level.
4. Fix $z_{0|0}$ and $P_{0|0}$, and find an optimal θ using the (-2)log-likelihood for the time period from N_0 to N.

Second stage:

1. Fix $P_{0|0}$ and the optimized θ, and find an optimal $z_{0|0}$ using the (-2) log-likelihood for the whole time period from 1 to N.

Third stage:

1. Optimize the (-2)log-likelihood for the whole time period from 1 to N with respect to all the parameters, θ, $z_{0|0}$, and $P_{0|0}$.

Empirical checking of the sensitivity of (-2)log-likelihood to $z_{0|0}$ and $P_{0|0}$ is worth mentioning. If we optimize the (-2)log-likelihood in terms of both $z_{0|0}$ and $P_{0|0}$, while a properly optimized θ is fixed, the innovations near the initial point keep decreasing as $z_{0|0}$ converges to some point and $P_{0|0}$ approaches zero. The innovation plot of this in time axis reveals an interesting phenomena. The size of the innovations is very small, even smaller than the size at the stationary level, at the beginning, and then gradually it increases to the stationary level. In practice, we could fix $P_{0|0}$ to be some small value such as 0.01 and concentrate on finding an optimal θ and $z_{0|0}$. Since most of the variability of the initial innovations comes from $z_{0|0}$, we don't get much disadvantage from fixing $P_{0|0}$ to a constant as long as it is not too large.

13.2.2 Durbin–Koopman Approach

Durbin and Koopman (2001) suggests to fix μ_{x_1} arbitrarily and assume a Gaussian distribution, with mean μ_{x_1} and variance $\sigma_{x_1}^2$, for $p(x_1)$, so that for the univariate case,

$$(-2)\log p(x_1|\theta) = (-2)\log\left\{\frac{1}{\sqrt{2\pi\sigma_{x_1}^2}}\exp\left(\frac{-(x_1-\mu_{x_1})^2}{2\sigma_{x_1}^2}\right)\right\}$$

$$= \log 2\pi + \log \sigma_{x_1}^2 + \frac{(x_1-\mu_{x_1})^2}{\sigma_{x_1}^2}.$$

Then, the (−2)log-likelihood is

$$(-2)\log p(x_1, x_2, \ldots, x_N | \theta) = (-2)\left\{ \log p(x_1|\theta) + \sum_{t=2}^{N} \log p(x_t | x_1, \ldots, x_{t-1}, \theta) \right\}$$

$$= \log \sigma_{x_1}^2 + \frac{(x_1 - \mu_{x_1})^2}{\sigma_{x_1}^2} + \sum_{t=2}^{N}\left(\log \sigma_{v_t}^2 + \frac{v_t^2}{\sigma_{v_t}^2} \right) + N \log 2\pi.$$

$$(13.7)$$

Durbin and Koopman (2001, p. 31) says, "All terms in (13.7) remain finite as $\sigma_{x_1}^2 \to \infty$ with x fixed except the term for $t = 1$. It is thus reasonable to remove the influence of $\sigma_{v_t}^2$ as $\sigma_{x_1}^2 \to \infty$ by defining the diffuse log–likelihood L_d as

$$(-2)\log L_d = (-2)\lim_{\sigma_{x_1}^2 \to \infty}\left(\sum_{t=1}^{N} \log p(x_t | x_{t-1}, \ldots, x_1, \theta) + \frac{1}{2}\log \sigma_{x_1}^2 \right)$$

$$= N \log(2\pi) + \sum_{t=2}^{N}\left(\log \sigma_{v_t}^2 + \frac{v_t^2}{\sigma_{v_t}^2} \right).\text{"}$$

By this treatment, they may feel good at removing the influence of $\sigma_{v_t}^2$ and making the maximum likelihood estimate dependent solely on observed data, as it is free from the arbitrarily assigned initial state value.

However, this is at the cost of sacrificing innovations near the initial value. With the choice of an arbitrary initial value for the unobserved state, the innovations will remain very bad for the beginning part of the time series. In order to improve the likelihood (i.e., to make the likelihood large), we need to reduce the size of innovations. Making the variance of the initial state large does not help in improving the likelihood. Although a detailed discussion of the consistency of this approach to Bayesians theory (see Rosenberg 1973, de Jong 1991) is given by Durbin and Koopman (2001), it does not help in solving the awkward problem of large innovations.

13.2.3 Schumway–Stoffer Approach

Schumway and Stoffer (1982) proposed an alternative approach based on the EM algorithm, different from the innovation-based definition of the likelihood of the state space model. They suggest the use of the so-called complete data likelihood of a state space model, which is given by

$$p(z_1, z_2, \ldots, z_N, x_1, x_2, \ldots, x_N) = p_0(z_0)\prod_{t=1}^{n} p_z(z_t | z_{t-1})\prod_{t=1}^{n} p_x(x_t | z_t).$$

Under the Gaussian assumption and ignoring constants, the complete data likelihood can be written as

$$(-2)\log p(z_1, z_2, \ldots, z_N, x_1, x_2, \ldots, x_N|\theta)$$

$$= \log|\Sigma_0| + (z_0 - \mu)'\sum_0^{-1}(z_0 - \mu) + \log|R_1| + \sum_{t=1}^{N}(z_t - Az_{t-1})'R_1^{-1}(z_t - Az_{t-1})$$

$$+ \log|R_2| + \sum_{t=1}^{N}(x_t - Cz_t)'R_2^{-1}(x_t - Cz_t), \tag{13.8}$$

where parameters to estimate are, mean μ_0 and variance Σ_0 of the initial state value z_0, and the variances R_1 and R_2, of the system noise w_t and the observation noise ε_t, respectively. Thus, we write the parameter vector for the state space model,

$$z_t = Az_{t-1} + w_t, \quad w_t: \mathcal{N}(0, R_1)$$

$$x_t = Cz_t + \varepsilon_t, \quad \varepsilon_t: \mathcal{N}(0, R_2)$$

by $\theta = (\mu_0, \Sigma_0, A, R_1, R_2)'$.

Note that maximization of the complete data likelihood is essentially equivalent to the Constrained Least Squares Method and maximization of the type-II log-likelihood (Good 1965, Akaike 1980). Here, two kinds of errors, $(z_t - Az_{t-1})$ and $(x_t - Cz_t)$, are different from $(z_{t|t} - Az_{t-1|t-1})$ and $(x_t - Cz_{t|t})$ of the Kalman filter, respectively. These errors in (13.8) are implicitly regarded as time homogeneous in the minimization procedure, where their temporal independence is not counted as in the Kalman filter. This means that the objective function to minimize in the EM algorithm and the objective function to minimize in the classic innovation approach are different. They suggest using the EM algorithm at the beginning of optimization computation and switch to the ordinary Newton–Raphson type numerical optimization procedure at a later stage in order to avoid the slow speed of the convergence of the EM algorithm. Here they mix the tools for the innovation approach (i.e., Kalman filter) and the tools for the Bayesian approach. This could induce some extra numerical problems when the state space model becomes more complex, nonlinear, and high dimensional.

Schumway and Stoffer (2000) also recognize the difficulty of choosing initial value of the state. They recommend calculating the smoothed estimate $z_{0|N}$ of the initial state z_0 using the filtered estimate and replacing the initial state value $z_{0|0}$ by the smoothed estimate $z_{0|N}$ and then repeating the computation of the likelihood with the replaced initial state value. They recommend repeating the back-and-forward procedure until the likelihood stabilizes.

This method relies on the optimistic expectation that the iterated smoothed estimate of the initial state estimate converges to an optimal point where the likelihood is maximized. Unfortunately, this expectation does not have any theoretical support. It is easy to check in a numerical experiment (good test data will be either Rikitake and Lorenz chaos data of Ozaki et al. 2000 or seasonal data of Ozaki and Thomson 2002) that the smoothed estimate of the initial value cannot be free from the initial poor choice of $z_{0|0}$. In other words, the smoothed estimate $z_{0|N}$ of the initial state still leads to very bad innovations overall when the initial choice $z_{0|0}$ of the initial state value produces terrible innovations at the beginning period of the recursive computation of the innovations.

If we repeat the back and forward procedure of filter and smoother, we have a sequence of the smoothed estimates of the initial state values as $z_{0|N}^{(1)}, z_{0|N}^{(2)}, z_{0|N}^{(3)}, \ldots$ We may have some improvement in the innovations and the likelihood, at the first and second trials of the back and forward computations of the filter and the smoother. However, the improvement will soon die away after a few (say k) steps, even though the innovations generated from the k-times smoothed initial state $z_{0|N}^{(k)}$ still take very large values near the initial time point. An important point here is that $z_{0|N}^{(k)}$ still depends on the initial bad choice of $z_{0|0}$. In some cases, the repeatedly smoothed estimate of a badly chosen initial state value can be much worse than other unsmoothed initial state values $z_{0|0}$. Thus, the smoother method of the EM algorithm cannot be a solution to the initial state value problem, whereas in the classic innovation approach, the solution comes simply by tuning the $z_{0|0}$ and $P_{0|0}$ directly in the likelihood.

A comparison of the already mentioned three typical computational methods, that is, the EM algorithm, the Durbin–Koopman (D–K) method, and the Innovation Approach (I.A.), for a linear state space model

$$z_t = Az_{t-1} + w_t, \quad w_t: \mathcal{N}(0, R_1)$$

$$x_t = Cz_t + \varepsilon_t, \quad \varepsilon_t: \mathcal{N}(0, R_2),$$

is given in Tables 13.1 and 13.2.

13.3 Slow Convergence Problem

13.3.1 Ill Conditions near Optimal Point

The slow convergence problem is one of the most commonly experienced problems for the users of the maximum likelihood method for state space modeling. Both Schumway and Stoffer (1982) and Durbin and Koopman

TABLE 13.1

Comparison of the Three Methods, EM, Durbin–Koopman's Diffuse Likelihood, and the Innovation Approach

	Treatment of the Initial Value x_1	Actual Objective Function
EM	Introduce parameters μ_0 and Σ_0 for specifying $p_0(z_0) = \mathcal{N}(z_0, \Sigma_0)$ to give $p(x_1) = \int p_0(z_0) p(x_1 \mid z_0) dz_0$. In practice, use an arbitrary $z_{0\mid0}$ & $P_{0\mid0}$, and replace them by the smoothed estimates, $z_{0\mid N}$ & $P_{0\mid N}$, so that $p(x_1) = \mathcal{N}(CAz_{0\mid N}, C(AP_{0\mid N}A' + R_1)C')$	$\log p(x_1, \ldots, x_n \mid \theta) =$ $-\dfrac{n}{2}\log(2\pi) - \dfrac{1}{2}\sum_{t=1}^{n}\left(\log \sigma_t^2 + \dfrac{v_t^2}{\sigma_t^2}\right)$ (Even though Markov structure is neglected and not exploited in EM)
D–K	Fix an arbitrary ξ_1 and assume $p(x_1) = \mathcal{N}(\xi_1, P_1)$ Here we let $P_1 \to \infty$.	Diffuse log-likelihood $\log L_d$ $= \lim_{P_1 \to \infty}\left(\sum_{t=1}^{n}\log p(x_t \mid x_{t-1}, \ldots, x_1, \theta) + \dfrac{1}{2}\log P_1\right)$ $= -\dfrac{n}{2}\log(2\pi) - \dfrac{1}{2}\sum_{t=2}^{n}\left(\log \sigma_t^2 + \dfrac{v_t^2}{\sigma_t^2}\right)$
I.A.	Regard z_{00} and P_{00} as parameters for specifying $p(x_1)$, i.e., $p(x_1) = \mathcal{N}(CAz_{0\mid0}, C(AP_{0\mid0}A' + R_1)C')$	$\log p(x_1, \ldots, x_n \mid \theta)$ $= \sum_{t=1}^{n}\log p(x_t \mid x_{t-1}, \ldots, x_1, \theta)$ $= -\dfrac{n}{2}\log(2\pi) - \dfrac{1}{2}\sum_{t=1}^{n}\left(\log \sigma_t^2 + \dfrac{v_t^2}{\sigma_t^2}\right)$

TABLE 13.2

Comparison of the Three Methods, EM, Durbin–Koopman's Diffuse Likelihood (D–K), and the Innovation Approach (I.A.)

	Parameters to Estimate	Iterative Numerical Optimization	Whitening Innovations
EM	$\theta = (A, R_1, R_2)$	No (Although K-filter and smoother are used in EM iteration)	No
D–K	$\theta = (A, R_1, R_2)$	Yes (Newton–Raphson type)	Yes (Kalman filter)
I.A.	$\theta = (z_{0\mid0}, P_{0\mid0}, A, R_1, R_2)$	Yes (Newton–Raphson type)	Yes (Kalman filter)

(2001) recommend switching from the EM algorithm to the classic innovation-based maximum likelihood method with a Newton–Raphson type optimization scheme when the convergence becomes slow. However, there could be many reasons for the slow convergence. Some problems cannot be solved simply by switching from the EM algorithm to a numerical optimization of the innovation-based log-likelihood.

One of the common reasons for the slow convergence of the maximum likelihood computation is the parameter redundancy at the optimal point of the likelihood function. When we try to minimize the (−2)log-likelihood with the given data using some iterative numerical optimization method, the parameter vector may start converging to the true point. However as the parameter approaches to the true value, convergence always becomes slower and slower and the Hessian matrix, the second derivative of the objective function, approaches the singular matrix. This is because there exists a relation (dependency) between the parameters at the minimized point (if not at the neighbors) of the (−2)log-likelihood.

This is true even for the simple first-order AR model

$$x_t = \phi x_{t-1} + \varepsilon_t.$$

We may write down the (−2)log-likelihood ($=l(\phi, \sigma^2)$) of the model as

$$l(\phi, \sigma^2) = N \log \sigma^2 + \sum_{t=1}^{N} \frac{\{\varepsilon_t(\phi)\}^2}{\sigma^2} + N \log 2\pi$$

where

$$\varepsilon_t(\phi) = x_t - \phi x_{t-1},$$

and σ^2 is the variance of ε_t. Suppose the (−2)log-likelihood is minimal at the point $\sigma^2 = \sigma_0$; then we have

$$\frac{\partial l(\phi, \sigma^2)}{\partial \sigma^2} = \frac{N}{\sigma^2} - \frac{\sum_{t=1}^{N} \{\varepsilon_t(\phi)\}^2}{\sigma^4} = 0.$$

This means that at the optimal point of σ^2 we have a relation

$$\sigma^2 = \frac{1}{N} \sum_{t=1}^{N} \{\varepsilon_t(\phi)\}^2. \tag{13.9}$$

To avoid the numerical difficulty arising from the redundant parameterization at the converged point, people usually drop σ^2 out of the set of the independent parameters of the objective function by using the relation (13.9) satisfied at the minimized point from the start and try to minimize

$$\sum_{t=1}^{N} \{\varepsilon_t(\phi)\}^2$$

with respect to ϕ only. The maximum likelihood estimates of σ^2 are given, after the likelihood function is optimized, using the optimized ϕ by the relation (13.9). Essentially the same approach is used in most time series modeling (see Box and Jenkins 1970).

This special redundancy near the optimal point can be avoided by the same method in ordinary time series modeling; however, the problem is not so simple in the state space modeling case. Jones (1980) gave a smart way to reduce one parameter, that is, the system noise variance, in order to accelerate the speed of the convergence of the numerical optimization near the optimal point in the state space modeling for ARMA processes with some assumptions. However, it has some limitations, that is, the method is valid only for stationary ARMA models.

If we do not remove one of the parameter variables from the whole parameters to be estimated, it affects the estimation method and makes the Hessian matrix very close to a singular matrix. However, the innovation-based maximum likelihood method seems to be less vulnerable, than the least squares type method, to the parameter redundancy of including both the noise variances. This delicate difference of the least squares method and the maximum likelihood method for the estimation of the state space model with both noise variances is shown with numerical results for simulation data generated from stochastic nonlinear oscillation models in Ozaki (1994a).

Here we discuss how to eliminate the observation error variance out of the set of parameters to be optimized in the maximum likelihood computation for general state space models. Suppose we have a state space model

$$\begin{cases} z_t = Az_{t-1} + Bw_t \\ x_t = Cz_t + \varepsilon_t \end{cases}, \tag{13.10}$$

where

x_t is a scalar time series
z_t is a k-dimensional state
w_t is an r-dimensional system noise

Suppose all the parameters in the matrices, A, B, and C, are put together in the parameter vector θ. Apart from θ, the model has some more parameters to estimate, that is, the system noise variance matrix R_1, observation noise variance R_2, the initial state vector $z_{0|0}$, and the initial state error variance $P_{0|0}$. Thus, the parameter to be optimized is $\Theta = \{\theta, R_1, R_2, z_{0|0}, P_{0|0}\}$. Then, the (-2)log-likelihood is

$$(-2)\log p(x_1, x_2, \ldots, x_N | \Theta) = \log p(x_1 | \Theta) + \sum_{t=2}^{N} \log p(x_t | x_1, \ldots, x_{t-1}, \Theta)$$

$$= \sum_{t=1}^{N} \left(\log \sigma_{v_t}^2 + \frac{v_t^2}{\sigma_{v_t}^2} \right) + N \log 2\pi, \tag{13.11}$$

where we have the initial innovation and initial innovation variance from

$$v_1 = x_1 - Cz_{0|0}$$

and

$$\sigma_{v_1}^2 = C(AP_{0|0}A' + BR_1B')C' + R_2.$$

The innovations and their variances for $t = 2, \ldots, N$ are calculated by the Kalman filter scheme:

$$z_{t+1|t} = Az_{t|t}$$

$$x_{t+1|t} = Cz_{t+1|t}$$

$$v_{t+1} = x_{t=1} - x_{t+1|t}$$

$$P_{t|t} = P_{t|t-1} - K_t CP_{t|t-1}$$

$$z_{t+1|t+1} = z_{t+1|t} + K_{t+1}(x_{t+1} - x_{t+1|t})$$

$$K_{t+1} = P_{t+1|t}C'(CP_{t+1|t}C' + R_2)^{-1}$$

$$P_{t+1|t} = AP_{t|t}A' + BR_1B'$$

$$\sigma_{v_{t+1}}^2 = CP_{t+1|t}C + R_2.$$

(13.12)

Next we see whether it is possible to reduce the number of parameters to optimize in the maximum likelihood computation for the state space modeling. One way to do this is to rewrite the Kalman filter representation of the recursive computation of innovations and innovation variances by normalizing the scale of the "observation errors." For example, if we rewrite the innovation variance by

$$\sigma_{v_{t+1}}^2 = R_2\Sigma_{t+1},$$

and then the updating equation of the innovation variance in the Kalman filter scheme (13.12) will be written as

$$R_2\Sigma_{t+1} = R_2(Cp_{t+1|t}C + 1)$$

where we have $P_{t+1|t} = R_2 p_{t+1|t}$. With this new variable $p_{t+1|t}$ for the state prediction error variance and with $r_2 = \left(\dfrac{1}{R_2}\right) R_1$, we can rewrite the whole Kalman filter scheme as follows:

$$z_{t+1|t} = A z_{t|t}$$

$$x_{t+1|t} = C z_{t+1|t}$$

$$v_{t+1} = x_{t=1} - x_{t+1|t}$$

$$p_{t|t} = p_{t|t-1} - K_t C p_{t|t-1}$$

$$z_{t+1|t+1} = z_{t+1|t} + K_{t+1}(x_{t+1} - x_{t+1|t}) \tag{13.13}$$

$$K_{t+1} = p_{t+1|t} C'(C p_{t+1|t} C' + 1)^{-1}$$

$$p_{t+1|t} = A p_{t|t} A' + B r_1 B'$$

$$\Sigma_{t+1} = (C p_{t+1|t} C + 1)$$

$$\sigma^2_{v_{t+1}} = R_2 \Sigma_{t+1}.$$

Note that the equation for the innovation variance is in fact

$$R_2 \Sigma_{t+1} = R_2 (C p_{t+1|t} C + 1).$$

Therefore, Σ_{t+1} in the recursive scheme of (13.13) needs to be renormalized back to the original scale by the last equation of (13.13), that is,

$$\sigma^2_{v_{t+1}} = R_2 \Sigma_{t+1}.$$

Then, the (−2)log-likelihood is

$$(-2)\log p(x_1, \ldots, x_N | \Theta)$$

$$= (-2) \sum_t \{\log p(v_t | x_{t-1}, \ldots, x_1, \Theta)$$

$$= \sum_{t=1}^{N} \log \sigma^2_{v,t} + \sum_{t=1}^{N} \frac{v_t^2}{\sigma^2_{v,t}} + N \log 2\pi$$

$$= \sum_{t=1}^{N} \log R_1 \Sigma_t + \sum_{t=1}^{N} \frac{v_t^2}{R_1 \Sigma_t} + N \log 2\pi$$

$$= N \log R_1 + \sum_{t=1}^{N} \log \Sigma_t + \frac{1}{R_1} \sum_{t=1}^{N} \frac{v_t^2}{\Sigma_t} + N \log 2\pi. \tag{13.14}$$

Since the gradient of the (−2)log-likelihood (=*l*) is zero at the optimal point, we have

$$\frac{\partial l}{\partial R_1} = 0$$

and

$$\frac{N}{R_1} - \frac{1}{R_1^2} \sum_{t=1}^{N} \frac{v_t^2}{\Sigma_t} = 0.$$

That means we can eliminate *R* from the likelihood representation (13.14) by

$$R_1 = \frac{1}{N} \sum_{t=1}^{N} \frac{v_t^2}{\Sigma_t}. \tag{13.15}$$

Then, we have

$$(-2)\log p(x_1, \ldots, x_N \mid \Theta) = N \log \sum_{t=1}^{N} \frac{v_t^2}{\Sigma_t} + \sum_{t=1}^{N} \log \Sigma_t + N + N \log 2\pi.$$

Thus, we can eliminate R_1 from the parameters for the likelihood maximization. The estimate of R_1 is calculated from (13.15) after the optimization is completed. Note that in the present optimization, replacing R_1 by (13.15) is implicitly assuming that the innovation is stationary.

An alternative way to do this is to rewrite the Kalman filter representation of the recursive computation of innovations and innovation variances by normalizing the scale of the "system noise." This is possible when the dimension *r* of the system noise w_t is one. For example, if we rewrite the innovation variance by

$$\sigma_{v_{t+1}}^2 = R_2 \Sigma_{t+1},$$

and then the updating equation of the innovation variance in the Kalman filter scheme (13.12) will be written as

$$R_2 \Sigma_{t+1} = R_2 (C p_{t+1|t} C + 1),$$

where we have $P_{t+1|t} = R_2 p_{t+1|t}$. With this new variable $p_{t+1|t}$ for the state prediction error variance, we can rewrite the whole Kalman filter scheme (13.12) as the following:

$$z_{t+1|t} = A z_{t|t}$$

$$x_{t+1|t} = C z_{t+1|t}$$

$$v_{t+1} = x_{t=1} - x_{t+1|t}$$

$$p_{t|t} = p_{t|t-1} - K_t C p_{t|t-1}$$

$$z_{t+1|t+1} = z_{t+1|t} + K_{t+1}(x_{t+1} - x_{t+1|t}) \tag{13.16}$$

$$K_{t+1} = p_{t+1|t} C'(C p_{t+1|t} C' + r_2)^{-1}$$

$$p_{t+1|t} = A p_{t|t} A' + B1B'$$

$$\Sigma_{t+1} = C p_{t+1|t} C + r_2$$

$$\sigma_{v_{t+1}}^2 = Q \Sigma_{t+1},$$

where $r_2 = R_2/R_1$. Note that the innovation variance needs to be renormalized back to the original scale by the last equation of (13.16), that is,

$$\sigma_{v_{t+1}}^2 = R_1 \Sigma_{t+1}.$$

Then, the (−2)log-likelihood is

$$(-2)\log p(x_1, \ldots, x_N | \Theta)$$

$$= \sum_{t=1}^{N} \log R_1 \Sigma_t + \sum_{t=1}^{N} \frac{v_t^2}{R_1 \Sigma_t} + N \log 2\pi$$

$$= N \log R_1 + \sum_{t=1}^{N} \log \Sigma_t + \frac{1}{R_1} \sum_{t=1}^{N} \frac{v_t^2}{\Sigma_t} + N \log 2\pi.$$

Using the same logic as for (13.14), we have

$$R_1 = \frac{1}{N} \sum_{t=1}^{N} \frac{v_t^2}{\Sigma_t}, \tag{13.17}$$

and we can eliminate R_1 from the parameters for optimizing the likelihood

$$(-2)\log p(x_1, \ldots, x_N | \Theta) = N \log \sum_{t=1}^{N} \frac{v_t^2}{\Sigma_t} + \sum_{t=1}^{N} \log \Sigma_t + N + N \log 2\pi.$$

Note that, in both (13.15) and (13.17), the temporal homogeneity of innovations is not assumed. The equations are compatible with the violent behavior of innovations near the starting time point influenced by a bad choice of the initial state value $z_{0|0}$ and the initial state error variance $P_{0|0}$.

13.3.2 Ill-Conditioned Optimizations

Ill-conditioned optimization is one of the main causes for the slow convergence of the optimization computation. In general, we always need to be careful about redundant parameterization or overparameterization if we want to avoid the ill conditions in the maximization of the likelihood for the state space modeling.

Rigorously speaking, redundant parameterization and overparameterization are not the same. In state space modeling of nonartificial real data, most parameterization is not really redundant but is very likely to be overparameterized if the number of parameters is large. A problem with the overparameterized case is that the Hessian matrix becomes very close to being singular, so that some of the eigen-values of the Hessian become very close to zero. This leads to unreliable estimates of the updating step for the optimization, yielding a slow speed of convergence. Similar phenomena are seen in the ARMA modeling of real data with very high lag orders.

Both Schumway and Stoffer (2000) and Durbin and Koopman (2001) noticed the slow convergence of the EM algorithm and recommended switching to the classic approach of the numerical optimization method together with the innovation-based likelihood. However, as we see in the previous discussions, this cannot be a solution. It does not solve the problem of parameter redundancy near the optimal point and the problem of ill-conditioned parameterization inherent to the state space models.

In general, the ill-conditioned situation of the state space model and the slow speed of convergence can be improved if we divide the parameters into three groups and optimize the likelihood with respect to parameters in one group while the parameters in other two groups are fixed. The three groups of parameters are the parameters related to the initial value, $\theta_1 = (z_{0|0}, P_{0|0})$, the parameters related to the dynamics of the model, $\theta_2 = A$, and the parameters related to the noise variances, $\theta_3 = (R_1, R_2)$. In the third group, the parameters are further divided into two in order to avoid the redundancy at the optimal point. If we do this partial optimization with respect to the parameters of each group iteratively, the speed of convergence improves drastically. After certain iterative optimizations, we can finally maximize the likelihood with respect to all the parameters.

13.4 Innovation-Based Approach versus Innovation-Free Approach

Historically, two different approaches to the state space modeling have been developed independently. One is the classic innovation-based approach and another is the innovation-free approach. A typical example of the innovation-free approach is the regularization method. In the regularization method, for example, they may try to estimate the data x_t using explanatory variables $z_t = (z_t^{(1)}, z_t^{(2)}, \dots z_t^{(k)})'$, with some fixed observation matrix C such as $C = (1, 1, \dots, 1)$ by minimizing the sum of squares of the error:

$$\sum_{t=1}^{N} (x_t - Cz_t)^2.$$

However, this equation cannot be solved when $\dim(x_t) < \dim(z_t)$. Such a problem is called an ill-posed problem.

The regularization approach suggests solving the ill-posed problem by introducing a constraint on z_t. One of the commonly used constraints is the smoothness constraint, that is, $z_t - z_{t-1}$ is small. This leads to the following constrained least squares method with fixed C and d, where the solution is given as

$$(z_1, \dots, z_N)' = \arg \left\{ \underset{z_1, \dots, z_N}{\text{Min}} \sum_{t=1}^{N} \|x_t - Cz_t\|^2 + d^2 \sum_{t=1}^{N} \|z_t - z_{t-1}\|^2 \right\}. \qquad (13.18)$$

Note that the constraint in the regularization can be considered as a dynamic model, $z_t = z_{t-1} + w_t$, driven by the error w_t.

A constraint can be written, in general, with a k×k matrix A as $z_t = Az_{t-1} + w_t$. Then, the regularization method with the Bayesian interpretation of the constrained least squares method in Section 12.4 leads to a solution of the estimation of the states and the parameters of the following state space model:

$$z_t = Az_{t-1} + w_t$$

$$x_t = Cz_t + \varepsilon_t.$$

Many users of state space modeling do not pay much attention to the difference of the classic innovation approach and the regularization approach. They often compute the estimates of the state variables and the model parameters (or hyperparameters), by mixing the logic and computational tools of the two approaches in applications.

The difference in the logical base of the two approaches is obvious, but as long as they produce almost the same results in actual numerical computations, this should not be a great problem at least to the users in applied sciences. Unfortunately, however, the two approaches lead to very different computational results. We have already seen two typical examples of the differences between the two approaches, that is, about the initial value problem and about the parameter redundancy problem near the optimal point.

13.4.1 Temporal Homogeneity of Errors

The most significant difference between the classic innovation-based likelihood approach and the innovation-free approach, say the EM algorithm or regularization approach, is in the assumption (whether explicit or implicit) of the independence and temporal homogeneity of the three kinds of errors, prediction errors, system errors (i.e., system noise) and observation errors. Remember that in the EM algorithm approach it suggests minimizing

$$(-2)\log p(z_1, z_2, \ldots, z_N, x_1, x_2, \ldots, x_N | \theta)$$

$$= \log |\Sigma| + (z_0 - \mu)' \Sigma^{-1}(z_0 - \mu) + \log |R_1| + \sum_{t=1}^{N} (z_t - Az_{t-1})' R_1^{-1} Q^{-1}(z_t - Az_{t-1})$$

$$+ \log |R_2| + \sum_{t=1}^{N} (x_t - Cz_t)' R_2^{-1}(x_t - Cz_t) \tag{13.19}$$

without any assumption for the whiteness of both the errors $(z_t - Az_{t-1})$ $(t = 1, 2, \ldots, N)$ and $(x_t - Cz_t)$ $(t = 1, 2, \ldots, N)$. On the other hand, as we saw in the initial value problem for the optimization of the likelihood, this approach implicitly expects the temporal homogeneity of the system errors $(z_t - Az_{t-1})$ and observation errors $(x_t - Cz_t)$. They are simply trying to minimize the sum of squares of these errors with constant variances R_1 and R_2 in (13.19). Note that the estimate of z_t, which is obtained by maximizing

$$\log p(z_t | x_1, x_2, \ldots, x_t),$$

and the estimate of z_t, which is obtained by maximizing the joint density,

$$\log p(z_0, z_1, \ldots, z_t | x_1, x_2, \ldots, x_t)$$

is not the same, unless the conditional density distributions are "unimodal and symmetric" (see Jazwinski 1970, p. 156).

In the regularization approach or in the EM algorithm, innovations don't play any explicit role. There is no way for the temporally inhomogeneous innovations to affect the likelihood through the two errors $(z_t - Az_{t-1})$ and $(x_t - Cz_t)$. This shows that the regularization approach is a kind of "innovation-free approach" to the estimation of state space models. However, it does not necessarily mean it yields a superior approach to likelihood maximization. In fact, by being detached from the innovations, the innovation-free approaches are losing the chance of using one of the most useful information for improving the estimation results.

In the classic innovation-based approach, on the contrary to the regularization approach, the innovations are treated to be possibly "temporally inhomogeneous," and they play a central role in the computation of the likelihood as well as in the computation of the estimates of the system variables. The dynamic error of the state, that is, the estimate of $(z_t - Az_{t-1})$, is specified by using the innovations. It is given explicitly as a difference in the filtered estimate and the predicted estimate of the state as $(z_{t|t} - Az_{t-1|t-1})$. An estimate of another error, that is, $(x_t - Cz_t)$, is also specified explicitly using the innovations as $(x_t - Cz_{t|t})$.

13.4.2 Temporal Independence of Errors

The least squares method is a product of the estimation method for non-dynamic models for independent data, where the validity of the method is guaranteed for large number of data, N. In the time series case, the use of the idea of "large N" needs special caution. This is because the data are temporally dependent, so that any large number of data points may not be sufficiently large when the temporal dependency is very strong. For example, in some situations of ARMA modeling, the characteristic roots defined by coefficients of AR and MA parameters can go indefinitely close to the unit circle (see some examples of ARMA(2,2) model for Series B of Box–Jenkins data in Ozaki 1977). In such a case, any large number of N is not sufficiently large.

The use of exact likelihood can solve this problem. Jones (1980) gave an example how to solve this problem by using the stationary state space model, where he uses the exact likelihood function for the ARMA model. However, this is possible only for linear modeling. When the dynamics are nonlinear, we have no way to write down the non-Gaussian distribution of the initial state value and the computation of the exact likelihood is impossible.

Since the regularization method is born out of the least squares method, it essentially inherits many of the implicit assumptions of the least squares method, where no mechanism of yielding the "temporal independence" of error is built in. On the other hand, the method is implicitly in pursuit of the "temporal homogeneity of errors" by the inclusion of the two terms,

$$\sum_{t=1}^{N} (x_t - Cz_t)' R_2^{-1}(x_t - Cz_t)$$

and

$$\sum_{t=1}^{N}(z_t - Az_{t-1})'R_1^{-1}(z_t - Az_{t-1}),$$

in the objective function to be minimized. Naturally, these differences yield significantly different results, when the innovations are temporally inhomogeneous, to the results obtained by the innovation approach.

13.4.3 Innovation-Free Approach and Local Non-Gauss State Space Models

So far we have considered the maximization of the likelihood for the model with the assumption of the local Gaussianity of the discrete-time state process z_t or continuous-time state process $z(t)$. Both the innovation-based likelihood approach and innovation-free likelihood approach, such as type-II likelihood and EM algorithm, were first developed for the linear state space model

$$\begin{cases} z_t = Az_{t-1} + Bw_t \\ x_t = Cz_t + \varepsilon_t \end{cases}$$

or

$$\begin{cases} dz = A^{(c)}zdt + B^{(c)}dw(t) \\ x_t = Cz_t + \varepsilon_t \end{cases}$$

with the local Gaussianity assumption of the state process z_t.

Note that the innovation-based approach to the calculation of the likelihood is closely related to the idea of the local Gaussianity of the state process. It does not make sense to use Gaussian likelihood for the innovation if the prediction error of the state process z_t is non-Gaussian distributed, for example, with a multimodal density distribution or asymmetric density distribution. After all, the innovation is essentially a sum of the two errors, i.e., the state prediction error and the observation error.

On the other hand, the specialty of the innovation-free approach is the local non-Gaussianity assumption of the state process. It must be remembered, however, that it does not make sense and is unjustifiable, for the innovation-free approach, to use the "quadratic" cost in the total data likelihood (or the type-II log-likelihood) in the regularization method, unless the "local non-Gaussianity" assumption of the state process z_t is withdrawn.

Objective evidence of the validity of the state space modeling comes from the innovations and the likelihood function. It is easy to notice anything wrong in the modeling method when people look at the innovations, while it is not so easy to notice anything wrong by just looking at the value of the optimized likelihood. Naturally, the innovations are often terribly messy and far from Gaussian for highly nonlinear time series data.

Whether it is right or not, people started to look for a solution, for the case when innovations are badly non-Gaussian, by introducing the local non-Gaussianity assumption for the state process z_t and the observation error. In the local non-Gaussian approach, the conventional nonlinear local-Gaussian state equation

$$z_t = a(z_{t-1}) + b(z_{t-1})w_t$$

is replaced by a probability density function, $p(z_t|z_{t-1})$, which characterizes the transition probability density of the general Markov chain process z_t defined on the k-dimensional continuous Euclidian space. Here, we naturally assume that the general Markov chain state process z_t is a nonexplosive finite-variance ergodic process.

The Gaussian observation errors could also be generalized into non-Gaussian errors. Then the conventional nonlinear observation equation, $x_t = c(z_t) + \varepsilon_t$, can be replaced by a conditional density distribution, $p(x_t|z_t)$. Then for the general Markov chain state space model, defined by the two conditional densities, $p(z_t|z_{t-1}, \theta)$ and $p(x_t|z_t, \theta)$, with the parameter vector θ, we can write down the log-likelihood of the model for the observed time series data, $x_1, x_2, \ldots, x_N$, as

$$\log p(x_1, \ldots, x_N|\theta) = \sum \log p(x_t|x_{t-1}, \ldots, x_1, \theta),$$

where the conditional densities $p(x_t|x_{t-1}, \ldots, x_1, \theta)$ ($t = 2, \ldots, N$) are specified recursively by using the Jazwinski scheme (see Section 12.6.2), for the general Markov chain state space model, as

$$p(x_t|x_{t-1}, \ldots, x_1, \theta) = \int p(x_t|z_t, \theta) p(z_t|x_{t-1}, \ldots, x_1, \theta) dz_t$$

$$p(z_t|x_{t-1}, \ldots, x_1, \theta) = \int p(z_t|z_{t-1}, \theta) p(z_{t-1}|x_{t-1}, \ldots, x_1, \theta) dz_{t-1}$$

$$p(z_{t-1}|x_{t-1}, \ldots, x_1, \theta) = \frac{p(x_{t-1}|z_{t-1}, \theta) p(z_{t-1}|x_{t-2}, \ldots, x_1, \theta)}{\int p(x_{t-1}|\varsigma_{t-1}, \theta) p(\varsigma_{t-1}|x_{t-2}, \ldots, x_1, \theta) d\varsigma_{t-1}},$$

and $p(x_1 | \theta)$ is specified by the initial conditions,

$$p(x_1 | \theta) = \int p(x_1 | z_1, \theta) p(z_1 | \theta) dz_1$$

$$p(z_1 | \theta) = \int p(z_1 | z_0, \theta) p(z_0) dz_0,$$

where $p(z_0)$ could be either $\delta(z - z_0)$ or Durbin and Koopman (2001)'s diffuse prior $p(z_0)$.

These recursive integral relations (given by Jazwinski 1970) provide the basis of innovation-free approaches to the filtering and smoothing of local non-Gaussian state space models. The lately introduced filtering method, called the Monte Carlo filter (or particle filter), is one of the innovation-free methods for the filtering of the local Gauss state space models. It is a brute-force computational approach (Cappe et al. 2005) that calculates all the integrals in the Jazwinski scheme using the Monte Carlo Markov Chain method, where all the conditional densities are non-Gaussian. The maximization of the likelihood, which requires such huge computation for the evaluation of the objective function, at each stage of the iterative procedure, is even more time consuming and formidable for practical users of the maximum likelihood method for state space modeling. Here the validity of this innovation-free approach to the maximization of the likelihood solely depends on the accuracy and the speed of the numerical integration method and the huge dimensional numerical optimization method.

It must be noted that the introduction of local non-Gaussianity does not necessarily help in solving the problem of slow convergence caused by the overparameterization near the optimal point. Rather it could be harmful in some cases. The problem with the local non-Gaussian approach may be that even though we may have temporally inhomogeneous looking innovations, they are accepted to be consistent with the assumptions of the local non-Gaussianity of the general Markov chain process, so that the analyst loses motivation and incentive for improving the computational problems such as the initial value effect and the overparameterization effect. Note that a blind numerical optimization of the likelihood, without checking innovations, is very difficult even for the experienced data analysts and tends to mislead analysts to poor conclusions, especially when the dynamics of the model is highly nonlinear and the number of parameters to optimize is large. Even for the case of linear state space model estimation, we will have great difficulty in estimating parameters if the parameters are not sufficiently reduced into one of the canonical state space representations by some standard method such as Akaike's method in Chapters 3 and 4. We need to specify the minimum necessary and sufficient parameters out of all the elements of the matrices, A, B, and C. Without proper parameterization of the state equation and the observation equation, those general state space model identification methods certainly lead to the numerical problems such as the slow convergence problem and the singularity problem of the Hessian in the numerical optimization, which makes the results unreliable and doubtful in actual applications.

13.5 Innovation-Based Approach and the Local Levy State Space Models

The innovation-based approach to the likelihood maximization has been proved to be very useful for the modeling highly periodic data, where the power spectrum has sharp peaks or modeling highly nonlinear dynamics, yielding strong initial value effects. The method has been proved to be useful for many dynamic models including various structural models for trend-seasonal adjustment in time series analysis and nonlinear dynamic models such as Hodgkin–Huxley models in neurosciences.

The innovation-based approach is valid only for nonlinear state space models with local Gauss assumption, such as

$$z_t = a(z_{t-1}) + b(z_{t-1})w_t$$

$$x_k = c(z_k) + \varepsilon_t$$

and

$$\dot{z} = a^{(c)}(z) + b^{(c)}(z)w(t)$$

$$x_t = c(z_t) + \varepsilon_t,$$

where w_t, $w(t)$ and ε_t are Gaussian white noise.

If the innovations don't look as if they are Gaussian, we need to reconsider the modeling. Often people tend to look for an easy solution, that is, use the same dynamic model with a more general non-Gaussian noise model. Nonzero centered prediction errors or bi-modally distributed prediction errors sound more general and suitable than Gaussian errors, but actually this kind of idea is nonsensical. The non-Gaussian characters of the prediction errors are simply showing the inappropriateness of the dynamic model we assumed for the data.

On the other hand, we must pay attention to the mathematical fact that the prediction errors cannot be too general. We already know that the prediction errors of Markov processes are Gaussian when the process is continuous (Doob 1953, Feller 1966). When the process is not continuous with occasional discontinuous jumps in the sample path, the Levy–Ito Theorem says that the prediction error is decomposed into two mutually independent noises, that is, Gaussian white noise and the (compensated) compound Poisson process (see Sato 1999 for details).

The most sensible solution in this awkward situation for the innovation-based maximum likelihood approach may be to redesign the experiment and eliminate all the possibility of pulse-like shot noise and collect a new data set. If the shot noise is unavoidable in the experiment, an alternative natural extension of the innovation–based maximum likelihood approach may be to use a Markov diffusion type local Gaussian model with jumps, where the driving white noise of the stochastic dynamical system is replaced by the sum of Gaussian white noise and a compensated compound Poisson noise process. We could call the state space model whose state dynamic is specified by such a dynamic model driven by a mixed noise of Gaussian white noise and the compensated compound Poisson noise a "local-Levy" state space model, instead of "local non-Gauss" state space model. The problem is then whether we can have any computationally efficient practical method for calculating and maximizing the likelihood of the local-Levy state space model.

13.5.1 Likelihood of Local-Levy State Space Models

Next we introduce a practical and computationally efficient innovation-based method for calculating and maximizing the likelihood of the local-Levy state space model. Suppose we have a local-levy state space model,

$$z_t = A_{t-1} z_{t-1} + B_{t-1} n_t$$
$$x_t = C z_t + \varepsilon_t,$$

(13.20)

where n_t is a Levy noise, which is a Gaussian white noise except when a Poisson jump noise occurs. Suppose m_t is defined as

$$m_t = \begin{cases} m_t^{(0)} & \text{when there is no jump for } n_t \\ m_t^{(1)} & \text{when there is a jump for } n_t. \end{cases}$$

Let $I_t^{(i)}$ be an indicating function of m_t such that

$$I_t^{(i)} = \begin{cases} 1 & \text{if } m_t = m_t^{(i)} \\ 0 & \text{otherwise.} \end{cases}$$

By introducing the prior density $p(m_1, \ldots, m_N)$ of a sequence of events $(m_1, m_2, \ldots, m_N)$, we can write down the (–2)log-likelihood function by

$$(-2) \log p(x_1, x_2, \ldots, x_N, m_1, \ldots, m_N)$$
$$= \{(-2) \log p(x_1, x_2, \ldots, x_N \mid m_1, \ldots, m_N) + (-2) \log p(m_1, \ldots, m_N)\}.$$

With the uniform assumption of the prior density distribution $p(m_1, \ldots, m_N)$, the maximization of the likelihood is attained by finding the parameters and a sequence $(m_1, m_2, \ldots, m_N)$ that minimize the (-2)log-likelihood.

We will first consider the (-2)log-likelihood

$$(-2)\log p(x_1, x_2, \ldots, x_N | m_1, \ldots, m_N) \tag{13.21}$$

for a given sequence of $(m_1, m_2, \ldots, m_N)$. Suppose we have an estimate $z_{t-1|t-1}$ at time point t. Whether there is a jump in the noise n_t or not, the predictor of z_t is given by

$$z_{t|t-1} = A_{t-1}z_{t-1|t-1}.$$

Thus, the innovation v_t at time point t is given by

$$v_t = x_t - Cz_{t|t-1}.$$

However the filtered estimate $z_{t|t}$ is affected (see Figure 13.1) by our judgment whether $m_t = m_t^{(1)}$ or $m_t = m_t^{(0)}$, that is, a jump occurred at t or not. This is because we have

$$z_{t|t}^{(i)} = z_{t|t-1} + K_t^{(i)}v_t$$

and

$$K_t^{(i)} = P_{t|t-1}^{(i)}C'(CP_{t|t-1}^{(i)}C' + R_2)^{-1}$$
$$P_{t|t-1}^{(i)} = A_{t-1}P_{t-1|t-1}^{(i)}A'_{t-1} + B_{t-1}Q_t^{(i)}B'_{t-1}. \tag{13.22}$$

Here $K_t^{(i)}$ is dependent on our judgment for $Q_t^{(i)}$ for the variance of the Levy noise n_t at time point t. If a jump occurs, that is, if $m_t = m_t^{(1)}$ then $Q_t^{(i)} = Q$ and if $m_t = m_t^{(0)}$, then $Q_t^{(i)} = R_1$. Then m_t affects the innovation at time point $t + 1$ by

$$v_{t+1}^{(i)} = x_{t+1} - CA_{t-1}z_{t|t}^{(i)}.$$

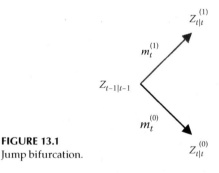

FIGURE 13.1
Jump bifurcation.

The innovation variance at time point $t + 1$ is also affected by m_t as

$$V_{t+1}^{(i)} = CP_{t|t-1}^{(i)}C' + R_2.$$

Then the minimization of (13.21) reduces to the minimization of

$$\sum_{t=1}^{N}\left[\sum_{i=0}^{1}\left\{I_t^{(i)}\log V_t^{(i)}\right\}+\sum_{i=0}^{1}\left\{I_t^{(i)}\frac{(v_t^{(i)})^2}{V_t^{(i)}}\right\}\right]+\text{Constant}. \qquad (13.23)$$

Here $(I_1^{(i)}, I_2^{(i)}, \ldots, I_N^{(i)})$ is a sequence of 0s and 1s. $I_t^{(i)} = 1$ when $m_t = m_t^{(i)}$ and $I_t^{(i)} = 0$ otherwise at the time point t. $v_t^{(i)}$ is the innovation corresponding to $m_t = m_t^{(i)}$ and $V_t^{(i)}$ is the innovation variance corresponding to $m_t = m_t^{(i)}$. Since we have 2^N possible sequences $(m_1, m_2, \ldots, m_N)$, we need to search for the optimal sequence $(m_1, m_2, \ldots, m_N)$ that optimizes (13.23) out of the 2^N possible sequences. However, this procedure is not realistic, since the computational cost of numerical minimization of (13.23) for 2^N possible sequences $(m_1, m_2, \ldots, m_N)$ is too huge.

We see next how we can transform the present approach into a more realistic approximate computationally efficient procedure for the minimization of (13.21). To get the maximum of the likelihood of the model with shot noise, we need to maximize the likelihood function for 2^N possible cases. This is because the observations $x_s(s > t)$ depend on the past state m_t. This formidable computational task can be drastically simplified if we use an approximate likelihood function of a Markov state space model. Here we use the property, that is, the dependency of the Markov state m_t dies out exponentially. In other words, we assume that the influence of m_t persists for only a short time period. Then, we have

$$\Pr(m_t = m_t^{(i)}|X^N) \approx \Pr(m_t = m_t^{(i)}|X^{t+1}),$$

where $X^t = (x_1, x_2, \ldots, x_t)$. Then the probability of getting a jump, that is, $m_t = m_t^{(1)}$ or $m_t = m_t^{(0)}$, for t is calculated by

$$\Pr(M_t = m_t^{(i)}|X^{t+1}) = \frac{\sum_{k=0}^{1}p(x_{t+1}|X^t, m_t^{(i)}, m_{t+1}^{(k)})p(x_t|X^{t-1}, m_t^{(i)})\Pr(m_t^{(i)})\Pr(m_{t+1}^{(k)})}{\sum_{k=0}^{1}\sum_{i=0}^{1}p(x_{t+1}|X^t, m_t^{(i)}, m_{t+1}^{(k)})p(x_t|X^{t-1}, m_t^{(i)})\Pr(m_t^{(i)})\Pr(m_{t+1}^{(k)})}.$$

$$(13.24)$$

Here probability densities $p(x_t | X^{t-1}, m_t^{(i)})$ $(t = 2, 3, \ldots, N)$ and $p(x_{t+1} | X^t, m_t^{(i)}, m_{t+1}^{(k)})$ $(t = 2, 3, \ldots, N)$ are obtained by using the innovations $v_t^{(i)}$ and the innovation variances $V_t^{(i)}$ obtained from the earlier mentioned modified Kalman filter scheme (13.22). For example, $p(x_{t+1} | X^t, m_t^{(i)})$ is given by

$$p(x_{t+1} | X^t, m_t^{(i)}) = p(v_{t+1}^{(i)} | X^t, m_t^{(i)})$$

$$= \frac{1}{\sqrt{2\pi V_{t+1}^{(i)}}} \exp-\left\{ \frac{1}{2} \frac{(v_{t+1}^{(i)})^2}{V_{t+1}^{(i)}} \right\}.$$

The estimated state $\hat{m}_t$ is given by

$$\hat{m}_t = \begin{cases} m_t^{(0)} & \text{if } \Pr(m_t = m_t^{(0)} | X^{t+1}) > \Pr(m_t = m_t^{(1)} | X^{t+1}) \\ m_t^{(1)} & \text{if } \Pr(m_t = m_t^{(0)} | X^{t+1}) < \Pr(m_t = m_t^{(1)} | X^{t+1}) \end{cases}, \quad (13.25)$$

where the estimate is either $m_t = m_t^{(0)}$ or $m_t = m_t^{(1)}$ according to the aforementioned inequalities of the posterior probabilities. The prior probability of having a jump at time t is λ and we have

$$\Pr(m_t^{(i)}) = \begin{cases} 1 - \lambda & \text{for } m_t = m_t^{(0)} \\ \lambda & \text{for } m_t = m_t^{(1)} \end{cases}.$$

Thus, by taking advantage of the property

$$\Pr(m_t = m_t^{(i)} | X^N) \approx \Pr(m_t = m_t^{(i)} | X^{t+1})$$

of the Markov process, we can have an efficient innovation-based method of computing the likelihood of the local-Levy state space model. With this method, the likelihood function of the model with shot noise can be calculated and maximized easily, as easily as ARMA models. While calculating the likelihood function step by step, jumps are detected on-line and recursively.

13.5.2 Recursive Computations

Next we are going to see an actual explicit representation of the recursive likelihood at each time step. In the present method, the judgment whether there is a jump in the system noise n_t or not is made one step after new data are obtained on-line.

Suppose judgment on the occurrence of jumps from time point $\tau = 1$ to $\tau = t - 1$ has been already made as

$$M^{t-1} = \{m_1^{(*)}, m_2^{(*)}, \ldots, m_{t-1}^{(*)}\},$$

where (*) denotes either 0 or 1. Here we need to judge, with newly obtained observation data x_{t+1}, whether the previous noise n_t was driven by a noise without a jump or with a jump. The likelihood of the model for the data, $x_1, x_2, \ldots, x_t, x_{t+1}$ is written as

$$(-2)\log p(x_1, x_2, \ldots, x_{t+1}, M^{t-1}) = (-2)\log p(x_1) + (-2)\sum_{\tau=2}^{t+1} \log p(x_\tau | X^{\tau-1}, M^{t-1}).$$

Note that the events of jump occurrences have been judged up to time point $\tau = t - 1$, and so the indicating function $I_\tau^{(i)}$ is specified only up to $\tau = t - 1$. Then the conditional density $p(x^\tau | X^{\tau-1}, M^{t-1})$ can be replaced by the conditional density of the innovation only up to $\tau = t - 1$. Thus, the likelihood is written as

$$(-2)\log p(x_1, x_2, \ldots, x_{t+1}, M^{t-1})$$

$$= (-2)\log p(x_1) + (-2)\left[\left\{\sum_{\tau=2}^{t-1} \log \sum_{i=0}^{1} I_\tau^{(i)} p(v_\tau^{(i)} | X^{\tau-1}, M^{t-1})\right\}\right.$$

$$\left. + \log p(x_t | X^{t-1}, M^{t-1}) + \log p(x_{t+1} | X^t, M^{t-1})\right]. \tag{13.26}$$

$v_t^{(1)}$ denotes the innovation derived from the path 1 (i.e., jump) and $v_t^{(0)}$ denotes the innovation derived from the path 0 (i.e., no jump). Similarly, $v_t^{(i,j)}$ means the innovation derived by the path i at time point $t - 1$ and path j at time point t. Thus $v_t^{(0,0)}$ denotes the innovation derived from the path 0 (i.e., no jump) at time point $t - 1$ and path 0 at time point t, while $v_t^{(1,0)}$ denotes the innovation derived from the path 1 (i.e., jump) at time point $t - 1$ and path 0 at time point t.

Here we have two paths of whitening the data x_t given X^{t-1}. One is the innovation v_t calculated by assuming a jump in n_t and another is the innovation v_t calculated by assuming no jump at n_t. Similarly we have four paths of whitening the data x_{t+1} given X^{t-1}. Since the probability of jump is specified by

$$\pi^{(0)} = \Pr(M_t = m_t^{(0)}) = 1 - \lambda$$

$$\pi^{(1)} = \Pr(M_t = m_t^{(1)}) = \lambda,$$

the probability of each path can be specified and the conditional densities are rewritten explicitly as

$$\log p(x_t | X^{t-1}, M^{t-1}) = \log\{\pi^{(0)} p(v_t^{(0)} | X^{t-1}, M^{t-1}, m_t^{(0)}) + \pi^{(1)} p(v_t^{(1)} | X^{t-1}, M^{t-1}, m_t^{(1)})\}$$

and

$$\log p(x_{t+1}|X^t, M^{t-1}) = \log \left\{ \pi^{(0)}\pi^{(0)}p(v_{t+1}^{(0,0)}|X^t, M^{t-1}, m_t^{(0)}, m_{t+1}^{(0)}) \right.$$

$$+ \pi^{(0)}\pi^{(1)}p(v_{t+1}^{(0,1)}|X^t, M^{t-1}, m_t^{(0)}, m_{t+1}^{(1)})$$

$$+ \pi^{(1)}\pi^{(0)}p(v_{t+1}^{(1,0)}|X^t, M^{t-1}, m_t^{(1)}, m_{t+1}^{(0)})$$

$$\left. + \pi^{(1)}\pi^{(1)}p(v_{t+1}^{(1,1)}|X^t, M^{t-1}, m_t^{(1)}, m_{t+1}^{(1)}) \right\}.$$

Then we have

$$\log p(x_t|X^{t-1}, M^{t-1}) = \log \left\{ \pi^{(0)}p(v_t^{(0)}|X^{t-1}, M^{t-1}, m_t^{(0)}) + \pi^{(1)}p(v_t^{(1)}|X^{t-1}, M^{t-1}, m_t^{(1)}) \right\}$$

and

$$\log p(x_{t+1}|X^t, M^{t-1}) = \log \left\{ \pi^{(0)}\pi^{(0)}p(v_{t+1}|X^t, M^{t-1}, m_t^{(0)}, m_{t+1}^{(0)}) \right.$$

$$+ \pi^{(0)}\pi^{(1)}p(v_{t+1}|X^t, M^{t-1}, m_t^{(0)}, m_{t+1}^{(1)})$$

$$+ \pi^{(1)}\pi^{(0)}p(v_{t+1}|X^t, M^{t-1}, m_t^{(1)}, m_{t+1}^{(0)})$$

$$\left. + \pi^{(1)}\pi^{(1)}p(v_{t+1}|X^t, M^{t-1}, m_t^{(1)}, m_{t+1}^{(1)}) \right\}$$

With these forms, the likelihood of (13.26) is rewritten as

$$(-2)\log p(x_1, x_2, \ldots, x_{t+1}, M^{t-1})$$

$$= (-2)\log p(x_1) + (-2)\left[\sum_{\tau=2}^{t-1} \log \left\{ \sum_{i=0}^{1} I_t^{(i)}p(v_\tau^{(i)}|X^{\tau-1}, M^{t-1}) \right\} \right.$$

$$+ \log \left\{ \pi^{(0)}p(v_t^{(0)}|X^{t-1}, M^{t-1}, m_t^{(0)}) + \pi^{(1)}p(v_t^{(1)}|X^{t-1}, M^{t-1}, m_t^{(1)}) \right\}$$

$$+ \log \left\{ \pi^{(0)}\pi^{(0)}p(v_{t+1}|X^t, M^{t-1}, m_t^{(0)}, m_{t+1}^{(0)}) + \pi^{(0)}\pi^{(1)}p(v_{t+1}|X^t, M^{t-1}, m_t^{(0)}, m_{t+1}^{(1)}) \right.$$

$$\left.\left. + \pi^{(1)}\pi^{(0)}p(v_{t+1}|X^t, M^{t-1}, m_t^{(1)}, m_{t+1}^{(0)}) + \pi^{(1)}\pi^{(1)}p(v_{t+1}|X^t, M^{t-1}, m_t^{(1)}, m_{t+1}^{(1)}) \right\} \right]$$

$$(13.27)$$

Next we make a decision for the jump at time point *t* with the jump detection rule (13.25) using the posterior probability of a jump calculated with (13.24). Suppose m_t is judged to be $m_t^{(0)}$. Then we have $M^t = \{M^{t-1}, m_t^{(0)}\}$. Also, log $p(x_t|X^{t-1}, M^t)$ and log $p(x_{t+1}|X^t, M^{t-1})$ are updated as

$$\log p(x_t|X^{t-1}, M^t) = \log p(v_t^{(0)}|X^{t-1}, M^{t-1}, m_t^{(0)}) = \log p(v_t^{(0)}|X^{t-1}, M^t) \quad (13.28)$$

and

$$\log p(x_{t+1}|X^t,M^{t-1}) = \log\left\{\pi^{(0)}p(v_{t+1}|X^t,M^t,m_{t+1}^{(0)}) + \pi^{(1)}p(v_{t+1}|X^t,M^t,m_{t+1}^{(1)})\right\}.$$

(13.29)

Finally, the likelihood function for the data, x_1, x_2, ..., x_t, x_{t+1}, is updated from (13.27) to

$$(-2)\log p(x_1,x_2,\ldots,x_{t+1},M^t)$$

$$= (-2)\log p(x_1) + \sum_{t=2}^{t}\left[\log\left\{\sum_{i=0}^{1}I_t^{(i)}V_\tau^{(i)}\right\} + \left\{\sum_{i=0}^{1}I_t^{(i)}\frac{(v_\tau^{(i)})^2}{V_\tau^{(i)}}\right\}\right]$$

$$+ (-2)\log\left\{\pi^{(0)}p(v_{t+1}^{(0)}|X_t,M^t,m_{t+1}^{(0)}) + \pi^{(1)}p(v_{t+1}^{(1)}|X_t,M^t,m_{t+1}^{(1)})\right\}$$

(13.30)

Next we go one step forward in time and consider the likelihood for the data $X^{t+2} = (X^{t+1}, x_{t+2})$, which we can obtain by adding $(-2)\log p(x_{t+2}|X^{t+1}, M^t)$ to (13.30):

$$(-2)\log p(x_1,x_2,\ldots,x_{t+2},M^t)$$

$$= (-2)\log p(x_1,x_2,\ldots,x_{t+1},M^t) + (-2)\log p(x_{t+2}|X^{t+1},M^t).$$

(13.31)

Rewriting the second term of the left-hand side of (13.27) with t replaced by $t+1$, we can rewrite (13.31) as

$$(-2)\log p(x_1,x_2,\ldots,x_{t+2},M^t)$$

$$(-2)\log p(x_1) + \sum_{t=2}^{t}\left[\log\left\{\sum_{i=0}^{1}I_t^{(i)}V_\tau^{(i)}\right\} + \left\{\sum_{i=0}^{1}I_t^{(i)}\frac{(v_\tau^{(i)})^2}{V_\tau^{(i)}}\right\}\right]$$

$$+ (-2)\log\left\{\pi^{(0)}p(v_{t+1}^{(0)}|X^t,M^t,m_{t+1}^{(0)})\right.$$

$$+ \pi^{(1)}p(v_{t+1}^{(1)}|X^t,M^t,m_{t+1}^{(1)})\right\}$$

$$+ \log\left\{\pi^{(0)}\pi^{(0)}p(v_{t+2}|X^{t+1},M^t,m_{t+1}^{(0)},m_{t+2}^{(0)})\right.$$

$$+ \pi^{(0)}\pi^{(1)}p(v_{t+2}|X^{t+1},M^t,m_{t+1}^{(0)},m_{t+2}^{(1)})$$

$$+ \pi^{(1)}\pi^{(0)}p(v_{t+2}|X^{t+1},M^t,m_{t+1}^{(1)},m_{t+2}^{(0)})$$

$$+ \pi^{(1)}\pi^{(1)}p(v_{t+2}|X^{t+1},M^t,m_{t+1}^{(1)},m_{t+2}^{(1)})\right\}.$$

(13.32)

13.5.3 Algorithm

Note that (13.32) has an equivalent form to (13.27) except that the time point t is increased by 1. This means the likelihood can be computed recursively in time as shown in Figure 13.2.

The algorithm is summarized as follows.

Step 1. Calculate $(-2)\log p(x_1)$ with the usual Kalman filter scheme for local Gaussian system noise n_1, assuming no jump for n_1. Start with $t = 1$. Let $M^{t-1} = \{m_1^{(0)}, \ldots, m_{t-1}^{(*)}\}$. Here (*) means either 0 or 1, and M^0 is empty.

Step 2. Calculate $(-2)\log p(x_1, x_2, \ldots, x_{t+1} | M^{t-1})$ using the relations

$$(-2)\log p(x_1, x_2, \ldots, x_{t+1} | M^{t-1})$$

$$= (-2)\log p(x_1) + \sum_{\tau=2}^{t-1}\left[\log \sum_{i=0}^{1} I_\tau^{(i)}V_\tau^{(i)} + \sum_{i=0}^{1} I_\tau^{(i)}\frac{(v_\tau^{(i)})^2}{V_\tau^{(i)}}\right]$$

$$+ (-2)\log p(x_t | X^{t-1}, M^{t-1})$$

$$+ (-2)\log p(x_{t+1} | X^t, M^{t-1}) + (t-2)\log 2\pi. \tag{13.33}$$

Step 3. Decide whether m_t is $m_t^{(0)}$ or $m_t^{(1)}$ using the posterior distributions as

$$\hat{m}_t = \begin{cases} m_t^{(0)} & \text{if } \Pr(m_t = m_t^{(0)}|X^{t+1}) > \Pr(m_t = m_t^{(1)}|X^{t+1}) \\ m_t^{(1)} & \text{if } \Pr(m_t = m_t^{(0)}|X^{t+1}) < \Pr(m_t = m_t^{(1)}|X^{t+1}) \end{cases}.$$

Step 4. Update the likelihood of (13.33) with the decision of Step 3. For example, if m_t is judged to be $m_t^{(1)}$, that is, $\hat{m}_t = m_t^{(1)}$, then we have $I_t^{(1)} = 1$ and $I_t^{(0)} = 0$, and terms in (13.33) are updated as

$$M^{t-1} \to M^t = \{M^{t-1}, \hat{m}_t\}$$

$$(-2)\log p(x_t|X^{t-1}, M^{t-1}) \to (-2)\log p(x_t|X^{t-1}, M^t)$$

$$= \log 2\pi + \log \sum_{i=0}^{1} I_t^{(i)}(v_t^{(i)})^2 + \sum_{i=0}^{1} I_t^{(i)}\frac{(v_t^{(i)})^2}{V_t^{(i)}}$$

$$(-2)\log p(x_{t+1}|X^t, M^{t-1}) \to (-2)\log p(x_{t+1}|X^t, M^t).$$

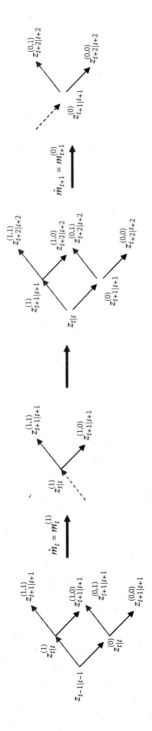

FIGURE 13.2
Sequence of jump bifurcations.

Then the likelihood of (13.33) is updated as

$$(-2)\log p(x_1, x_2, \ldots, x_{t+1} \mid M^{t-1}) \rightarrow (-2)\log p(x_1, x_2, \ldots, x_{t+1}, M^t)$$

$$= (-2)\log p(x_1) + \sum_{\tau=2}^{t} \left[\log \sum_{i=0}^{1} I_\tau^{(i)} V_\tau^{(i)} + \sum_{i=0}^{1} I_\tau^{(i)} \frac{(v_\tau^{(i)})^2}{V_\tau^{(i)}} \right]$$

$$+ (-2)\log\{p(x_{t+1} \mid X^t, M^t)\}$$

$$+ (t-1)\log 2\pi. \tag{13.34}$$

Step 5. Add $(-2)\log p(x_{t+2} \mid X^{t+1}, M^t)$ to (13.34). Increase the time point t by 1 and if $t > N - 1$ stop the computation. Otherwise go to Step 2 and continue the computation.

13.5.4 Discussions

This efficient method of likelihood computation is derived from the approximation

$$\Pr(M_t = M_t^{(i)} \mid X^N) \approx \Pr(M_t = M_t^{(i)} \mid X^{t+1}),$$

which leads to the recursive detection of jumps in the Levy noise and the recursive computation of the likelihood function for the estimated jump noise. The likelihood may improve with a more elaborate jump detection with a more elaborate approximation

$$\Pr(M_t = M_t^{(i)} \mid X^N) \approx \Pr(M_t = M_t^{(i)} \mid X^{t+2}),$$

where the decision of the occurrence of jump is made for the two lags behind the present time point $t + 2$, by comparing the posterior distributions as

$$\hat{M}_t = \begin{cases} M_t^{(0)} & \text{if } \Pr(M_t = M_t^{(0)} \mid X^{t+2}) > \Pr(M_t = M_t^{(1)} \mid X^{t+2}) \\ M_t^{(1)} & \text{if } \Pr(M_t = M_t^{(0)} \mid X^{t+2}) < \Pr(M_t = M_t^{(1)} \mid X^{t+2}) \end{cases}.$$

Here the posterior distribution $\Pr(M_t = M_t^{(i)} | X^{t+2})$ is calculated using the Bayes rule as

$$\Pr(M_t = M_t^{(i)} | X^{t+2})$$

$$= \frac{\sum\limits_{k=0}^{1}\sum\limits_{i=0}^{1}\Pr(x_{t+2}|X^{t+1}, M_t^{(i)}, M_{t+1}^{(k)}, M_{t+2}^{(l)})\Pr(x_{t+1}|X^t, M_t^{(i)}, M_{t+1}^{(k)}) \times \Pr(x_t|X^{t-1}, M_t^{(i)})\Pr(M_t^{(i)})\Pr(M_{t+1}^{(k)})\Pr(M_{t+2}^{(l)})}{\sum\limits_{l=0}^{1}\sum\limits_{k=0}^{1}\sum\limits_{i=0}^{1}\Pr(x_{t+2}|X^t, M_t^{(i)}, M_{t+1}^{(k)}, M_{t+2}^{(l)})\Pr(x_{t+1}|X^t, M_t^{(i)}, M_{t+1}^{(k)}) \times \Pr(x_t|X^{t-1}, M_t^{(i)})\Pr(M_t^{(i)})\Pr(M_{t+1}^{(k)})\Pr(M_{t+2}^{(l)})}.$$

The computation of the initial distribution, $\log p(x_1)$, may be done in the same way as the innovation-based method for the local Gauss state space model, by treating $z_{0|0}$ and $P_{0|0}$ as parameters to estimate (see Ozaki and Iino, 2001). When the dynamics of the local-Levy state space model is highly non-linear like chaos, the Hodgkin–Huxley model, and many other stiff systems, we should not forget to pay special attention to this term together with the present treatment of the jump detection at the same time.

The local non-Gauss state space model has an intrinsic problem from the start. There is no standard way to define temporally independent but mutu-ally correlated multivariate random noise except for the Gaussian case. In reality the multivariate noises could be mutually correlated, for example, between $v_t^{(i)}$ and $v_t^{(j)}$, even though they are temporally independent. However, correlation is a Gaussian concept. There is no reasonable way to define the correlation between generally distributed random variables, such as a mul-timodally distributed random variable or positive valued random variables. The Levy–Ito theorem (see Sato 1999) supports this empirical understanding from the theoretical side. When we are dealing with multivariate correlated time series, there is no way to check whether the correlation between the two time series is removed from the residuals if we give up the Gaussianity of the prediction errors.

On the other hand, as we saw in Chapter 7, the Levy–Doob Theorem shows that if the prediction errors are zero mean, sample continuous (i.e., no out-lying large prediction errors), and of finite variance (i.e., nonexplosive), the prediction errors are Gaussian white noise. It means that the only meaning-ful local non-Gaussian noise is not a generally distributed noise but a local Levy noise, where the Poisson shot noise is mixed with the Gaussian white noise. With the local Levy, we can still consider the correlation between its Gaussian noise components without any conceptual difficulty.

Our approach for time series modeling is to find a dynamic model whitening the time series into Gaussian white noise. If the prediction errors of a dynamic model do not appear to be Gaussian, we need to reconsider the dynamic model.

The following mathematical theorems are useful for diagnosing and improving our dynamic models for time series analysis:

1. Prediction errors need to have infinitely divisible probability distributions (Feller 1966).
2. For Markov and continuous (diffusion type) processes, prediction errors are Gaussian (Doob 1953).
3. For Markov diffusion processes with jumps, prediction errors decompose into the sum of Gaussian white noise and compensated Poisson processes (Levy 1956, Protter 1990).

Thus, guideline principles of the diagnostic checking could be the following:

1. If the prediction errors are non-Gaussian looking, try to detect outliers in the residuals.
2. Decompose the residuals into Gaussian white noise and a few outliers by improving the dynamic model together with the outlier detection method.
3. Find phenomenological explanations for the outlying noise and reconsider both the experiment (which the data come from) and the dynamic model (for the phenomena behind the experiment).

13.6 Heteroscedastic State Space Modeling

So far the state space models we have considered are local-Gauss state space models whose state models are either stationary linear dynamic models or stationary nonlinear dynamic models. Time series data, such as EEG record, in neuroscience applications are often nonstationary. For example, in the clinical study of subjects' EEG records, the increase and decrease in "delta rhythm" in EEG is of specific interest for monitoring an anesthetized patient in surgery in order to detect whether the patient is in awake mode or in unconscious mode. The real-time phenomenological characterization of the EEG of subjects in surgery is important for clinical doctors and surgeons, where monitoring and detection methods for the increase and decrease in various rhythms play an important role.

We have seen some examples of parametric modeling of this kind of non-stationarity time series in Section 6.7 where the time-varying noise variance is characterized adaptively by using the past prediction errors. However, the state space models we have considered so far are all homoscedastic models, where the noise variances are fixed to be constant. In some cases, the prediction errors of state space models look temporally inhomogeneous. In such cases, the likelihood and AIC could be improved by generalizing the model into a heteroscedastic model from the homoscedastic model. State space modeling is expected to be specially advantageous for real-time phenomenological modeling because of its recursive nature coming from the Kalman filter. It will be useful if the ideas of ordinary heteroscedastic modeling are generalized to the state space situation where some of the variables are not necessarily observed. In this section, we see how we can generalize an ordinary homoscedastic state space model into a heteroscedastic state space model using an example of the EEG power spectrum estimation problem. Applications of the heteroscedastic state space modeling are found in Galka et al. (2010) for the signal decomposition of EEG signals, and in Galka et al. (2004a) and Barton et al. (2009) for the EEG dynamic inverse problem.

13.6.1 A Heteroscedastic State Space Modeling for EEG Power Spectrum Estimation

Stationarity is what people usually like to assume in time series modeling. It makes the modeling procedure easy. We saw in Chapter 3 that AR models are one of the most convenient and computationally efficient model classes for estimating the power spectrum of a stationary time series. If the EEG time series (e.g., with sampling frequency, 256 Hz) is composed of five rhythms, delta, theta, alpha, beta, and gamma, the series can be approximated with a 10-th order AR model, where the five pairs of the complex conjugate eigenvalues of the model stay inside the unit circle and are distributed as is schematically shown in Figure 13.3.

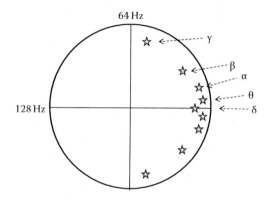

FIGURE 13.3
Eigen-values of EEG rhythms.

Here we expect that the five complex roots of the 10-th order characteristic polynomial

$$\Lambda^{10} - \phi_1 \Lambda^9 - \cdots - \phi_9 \Lambda - \cdots - \phi_{10} = 0$$

yield 5 peaks in the power spectrum. In other words, the 10-th order polynomial is factored into 5 second-order polynomials as

$$\Lambda^{10} - \phi_1 \Lambda^9 - \cdots - \phi_9 \Lambda - \cdots - \phi_{10}$$

$$= (\Lambda^2 - \phi_1^{(\delta)} \Lambda - \phi_2^{(\delta)})(\Lambda^2 - \phi_1^{(\theta)} \Lambda - \phi_2^{(\theta)})(\Lambda^2 - \phi_1^{(\alpha)} \Lambda - \phi_2^{(\alpha)})$$

$$\times (\Lambda^2 - \phi_1^{(\beta)} \Lambda - \phi_2^{(\beta)})(\Lambda^2 - \phi_1^{(\gamma)} \Lambda - \phi_2^{(\gamma)})$$

and two coefficients of each second-order polynomial produces oscillatory dynamics with a specific frequency.

A problem for the actual use of the AR power spectrum in EEG analysis is that EEG data are very nonstationary. This is natural because the human brain never stays in a simple equilibrium state. It is always working and responding to the stimulus from the surrounding environment. Accordingly, the main frequency of the EEG power spectrum changes in time.

An interesting idea of characterizing the time-varying EEG power spectrum comes from considering the time varying variances of the driving noise of each compartment model. Then the change in the driving noise variances contributes to the change in the shape of the spectrum such as the shift from delta frequency to alpha frequency. In Section 3.2.3, we saw that with a parallel type state space model the shape of the power spectrum is modified by increasing or decreasing the variance of the system noise of each compartment model, while the main frequency of each component model is fixed to a constant frequency.

Often the time-varying variance of each compartment variable provides useful information about the activity of the latent state variables related to the specific rhythm in EEG. It would be more desirable if we could let the variances of the compartment model change in time and estimate it adaptively, like in the cases for the ARCH and GARCH modeling. This approach is in contrast to another approach to the time-varying EEG power spectrum estimation method, where the coefficients of an AR model are changing smoothly in time while the noise variance of the AR model is fixed to constant.

For example, when we consider a parallel compartment type state space model with two independent compartment models,

$$x_t^{(1)} = \phi_1^{(1)} x_{t-1}^{(1)} + \phi_2^{(1)} x_{t-2}^{(1)} + \sigma^{(1)} w_t^{(1)}$$

$$x_t^{(2)} = \phi_1^{(2)} x_{t-1}^{(2)} + \phi_2^{(2)} x_{t-2}^{(2)} + \sigma^{(2)} w_t^{(2)},$$

we could formulate it into the following parallel type state space model,

$$
\begin{pmatrix} x_t^{(1)} \\ x_{t-1}^{(1)} \\ x_t^{(2)} \\ x_{t-1}^{(2)} \end{pmatrix} = \begin{pmatrix} \phi_1^{(1)} & \phi_2^{(1)} & 0 & 0 \\ 1 & 0 & 0 & 0 \\ 0 & 0 & \phi_1^{(2)} & \phi_2^{(2)} \\ 0 & 0 & 1 & 0 \end{pmatrix} \begin{pmatrix} x_{t-1}^{(1)} \\ x_{t-2}^{(1)} \\ x_{t-1}^{(2)} \\ x_{t-2}^{(2)} \end{pmatrix} + \begin{pmatrix} \sigma^{(1)} w_t^{(1)} \\ 0 \\ \sigma^{(2)} w_t^{(2)} \\ 0 \end{pmatrix}
$$

$$
x_t = \begin{pmatrix} 1 & 0 & 1 & 0 \end{pmatrix} \begin{pmatrix} x_t^{(1)} \\ x_{t-1}^{(1)} \\ x_t^{(2)} \\ x_{t-1}^{(2)} \end{pmatrix} + \varepsilon_t.
$$

Next parallel state space model is generalized into a heteroscedastic model where the system noise variances $(\sigma^{(1)})^2$ and $(\sigma^{(2)})^2$ change in time. Then it may represent a dynamically changing power spectrum as in Figure 13.4. Here we could estimate the time-varying noise variance of each compartment model adaptively, like in ARCH and GARCH models, using the "past information of the prediction errors."

The problem with the state space model is, unlike the ARCH and GARCH model, we don't observe the state variables, $x_t^{(1)}$ and $x_t^{(2)}$, directly. We observe x_t only. When x_t is observed with an observation error, we cannot update the variances of the system noise of the state variables, $x_t^{(1)}$ and $x_t^{(2)}$, using the squares of the past prediction errors of $x_t^{(1)}$ and $x_t^{(2)}$, that is, $(x_{t-1}^{(1)} - x_{t-1|t-2}^{(1)})^2$ and $(x_{t-1}^{(2)} - x_{t-1|t-2}^{(2)})^2$. Then the question is "what is the best substitute for $(x_{t-1}^{(1)} - x_{t-1|t-2}^{(1)})$ and $(x_{t-1}^{(2)} - x_{t-1|t-2}^{(2)})$?" A heuristic solution for this is to replace them by $(x_{t-1|t-1}^{(1)} - x_{t-1|t-2}^{(1)})$ and $(x_{t-1|t-1}^{(2)} - x_{t-1|t-2}^{(2)})$, respectively. Remember we have

$$
z_{t-1|t-1} = z_{t-1|t-2} + K_{t-1} v_{t-1},
$$

$p(f)$

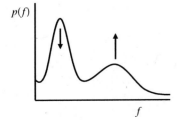

f

FIGURE 13.4
Change in spectrum of heterocompartment model.

where the Kalman gain K_{t-1} is mapping a one dimensional prediction error v_{t-1} into a four-dimensional vector $K_{t-1}v_{t-1}$. Here the Kalman filter is for the state space model,

$$z_t = Az_{t-1} + w_t$$
$$x_t = Cz_t + \varepsilon_t$$

$$z_t = \begin{pmatrix} x_t^{(1)} \\ x_{t-1}^{(1)} \\ x_t^{(2)} \\ x_{t-1}^{(2)} \end{pmatrix}, \quad A = \begin{pmatrix} \phi_1^{(1)} & \phi_2^{(1)} & 0 & 0 \\ 1 & 0 & 0 & 0 \\ 0 & 0 & \phi_1^{(2)} & \phi_2^{(2)} \\ 0 & 0 & 1 & 0 \end{pmatrix}, \quad w_t = \begin{pmatrix} \sigma^{(1)}n_t^{(1)} \\ 0 \\ \sigma^{(2)}n_t^{(2)} \\ 0 \end{pmatrix}.$$

$$C = (1 \quad 0 \quad 1 \quad 0).$$

The next problem is how to update the system noise variance, $(\sigma_t^{(i)})^2$, with the information obtained through the squares of the past information, $K_{t-1}v_{t-1}$, $K_{t-2}v_{t-2}$, $\ldots$

13.6.2 Compartment-GARCH Model

The compartment-GARCH model, composed with two compartment models for $x_t^{(1)}$ and $x_t^{(2)}$, is given as

$$z_t = Az_{t-1} + B_{t-1}w_t$$
$$x_t = Cz_t + \varepsilon_t,$$

where

$$A = \begin{pmatrix} \phi_1^{(1)} & \phi_2^{(1)} & 0 & 0 \\ 1 & 0 & 0 & 0 \\ 0 & 0 & \phi_1^{(2)} & \phi_2^{(2)} \\ 0 & 0 & 1 & 0 \end{pmatrix}, \quad B = \begin{pmatrix} \sigma_{t-1}^{(1)} & 0 \\ 0 & 0 \\ 0 & \sigma_{t-1}^{(2)} \\ 0 & 0 \end{pmatrix},$$

$$C = (1 \quad 0 \quad 1 \quad 0), \quad w_t = \left(w_t^{(1)} \quad w_t^{(2)} \right)',$$

$$z_t = \left(x_t^{(1)} \quad x_{t-1}^{(1)} \quad x_t^{(2)} \quad x_{t-1}^{(2)} \right)',$$

with some updating scheme of the system noise variances, $(\sigma_{t-1}^{(1)})^2$ and $(\sigma_{t-1}^{(2)})^2$ using, $K_{t-1}v_{t-1}, K_{t-2}v_{t-2}, \ldots$, which comes from the past innovations. The updating scheme, which we call GARCH machine, can be given, using similar ideas as for GARCH models. A few simple examples are given as follows.

Example 1 Moving average type: A moving average type GARCH machine is given, for the first compartment, as

$$\left(\sigma_t^{(1)}\right)^2 = \left(\sigma_0^{(1)}\right)^2 + \sum_{m=1}^{d} \gamma_m^{(1)}\{(K_{t-m}\upsilon_{t-m})^{(1)}\}^2. \qquad (13.35)$$

Here $(K_{t-m}\upsilon_{t-m})^{(1)}$ is the first element of the four-dimensional vector $K_{t-m}\upsilon_{t-m}$. The parameters to estimate are, for the first compartment, $\gamma_m^{(1)}$ $(m = 1, \ldots, d)$ and $(\sigma_0^{(1)})^2$. In the actual modeling procedure, however, it is recommended not to make the model too complex with high lag orders. A very useful model for a start is the following simplified MA type model,

$$\left(\sigma_t^{(1)}\right)^2 = \left(\sigma_0^{(1)}\right)^2 + \gamma^{(1)} \frac{1}{d} \sum_{m=1}^{d} \left\{(K_{t-m}\upsilon_{t-m})^{(1)}\right\}^2. \qquad (13.36)$$

Here for each compartment we have two parameters, $(\sigma_0^{(i)})^2$ and $\gamma^{(i)}$, to estimate by the maximum likelihood method.

The state space–GARCH model is written, for the simplified MA type (13.36), as

$$\begin{cases} z_t = Az_{t-1} + Bw_t \\ x_t = Cz_t + \varepsilon_t \end{cases}.$$

where

$$z_t = \begin{pmatrix} z_t^{(1)} & z_{t-1}^{(1)} & z_t^{(2)} & z_t^{(2)} \end{pmatrix}, \quad w_t = \begin{pmatrix} w_t^{(1)} & w_t^{(2)} \end{pmatrix}$$

$$A = \begin{pmatrix} \phi_1^{(1)} & \phi_2^{(1)} & 0 & 0 \\ 1 & 0 & 0 & 0 \\ 0 & 0 & \phi_1^{(2)} & \phi_2^{(2)} \\ 0 & 0 & 1 & 0 \end{pmatrix}, \quad B = \begin{pmatrix} \sigma_t^{(1)} & 0 \\ 0 & 0 \\ 0 & \sigma_t^{(2)} \\ 0 & 0 \end{pmatrix}$$

$$C = (1 \quad 0 \quad 1 \quad 0),$$

together with the GARCH machines,

$$\left(\sigma_t^{(1)}\right)^2 = \left(\sigma_0^{(1)}\right)^2 + \gamma^{(1)} \frac{1}{d} \sum_{m=1}^{d} \left\{(K_{t-m}\upsilon_{t-m})^{(1)}\right\}^2$$

and

$$\left(\sigma_t^{(2)}\right)^2 = \left(\sigma_0^{(2)}\right)^2 + \gamma^{(2)} \frac{1}{d} \sum_{m=1}^{d} \left\{ (K_{t-m} v_{t-m})^{(3)} \right\}^2 .$$

Since the state space representation of a compartment model is not unique, there could be many different variations of state space GARCH. As long as we use them in a consistent way, they are useful in characterizing heteroscedastic dynamics with latent variables.

Example 2 Autoregressive type: An autoregressive type GARCH machine is also possible. An example of the first-order AR type is given, for the *i*-th compartment, by

$$\left(\sigma_t^{(i)}\right)^2 = \left(\sigma_0^{(i)}\right)^2 + \alpha^{(i)} \left(\sigma_{t-1}^{(i)}\right)^2 + \gamma^{(i)} \left\{ (K_{t-1} v_{t-1})^{(i)} \right\}^2 , \tag{13.37}$$

where parameters to estimate are $(\sigma_0^{(i)})^2, \alpha^{(i)}$, and $\gamma^{(i)}$. An ARMA type generalization is also possible as the following Example 3.

Example 3 ARMA type: An ARMA type GARCH machine for the *i*-th compartment is given by

$$\left(\sigma_t^{(i)}\right)^2 = \left(\sigma_0^{(i)}\right)^2 + \sum_{m=1}^{p} a_m^{(i)} \left(\sigma_{t-m}^{(i)}\right)^2 + \sum_{m=1}^{q} b_m^{(i)} \left\{ (K_{t-1} v_{t-m})^{(i)} \right\}^2 . \tag{13.38}$$

Here with the GARCH machine, the number of parameters to estimate, for each compartment, is $(p + q + 1)$.

In state space GARCH modeling, we need to pay attention to the fact that the computational cost increases as the number of parameters increase. Note also that the process $(\sigma_t^{(i)})^2$ is supposed to be positive valued. This makes estimation of the parameters difficult, except for model (13.36) of Example 1 of the simplified moving average type. In order to obviate the difficulty, parameterization in the log-transformed space such as

$$\log\left(\sigma_t^{(i)}\right)^2 = \sigma_0^{(i)} + \sum_{m=1}^{p} a_m^{(i)} \log\left(\sigma_{t-m}^{(i)}\right)^2 + \sum_{m=1}^{q} b_m^{(i)} \left\{ \log(K_{t-1} v_{t-m})^{(i)} \right\}^2 \tag{13.39}$$

is often useful.

In the present method, we used $(K_{t-1} v_{t-1})^{(k)}$, that is, a *k*-th element of the vector $K_{t-1} v_{t-1}$, as an estimate for the *k*-th element of the state prediction error vector,

$$v_{t-1}^{(k)} = x_{t-1}^{(k)} - x_{t-1|t-2}^{(k)} ,$$

in order to give an approximation for the variance of $v_{t-1}^{(k)}$. However, $\left\{ (K_{t-1}v_{t-1})^{(k)} \right\}^2$ has a bias as an estimate of the variance of $v_{t-1}^{(k)}$. It has been confirmed that a small improvement in likelihood can be obtained by a bias correction in the estimation of $x_{t-1}^{(k)} - x_{t-1|t-2}^{(k)}$, leading to a better AIC (see Wong et al. 2006 for details).

13.6.3 Application to Nonstationary EEG Power Spectrum Estimation

One of the commonly used methods for estimating a nonstationary EEG spectrum is fitting the locally stationary AR models (Ozaki and Tong 1975), where the EEG record is divided into subintervals. Each of the intervals is regarded as stationary, and an ordinary p-th order AR(p) model is used for the characterization of the spectrum of the interval. A more flexible approach to the nonstationary power spectrum estimation is using the time-varying coefficient AR model introduced by Harrison and Stevens (1976). The model is one of those state space models where the state dynamics is a simple random walk model. For example, if we use a p-th order AR model structure for the power spectrum representation with a sufficiently large p, we have the following p-dimensional state space model,

$$z_t = z_{t-1} + w_t$$
$$x_t = C_{t-1}z_t + \varepsilon_t,$$

where

$$z_t = \left(\phi_t^{(1)}, \phi_t^{(2)}, ..., \phi_t^{(p)} \right)',$$

$$C_{t-1} = (x_{t-1}, x_{t-2}, ..., x_{t-p}),$$

and w_t is a p-dimensional Gaussian white noise with the variance matrix $\sigma_w^2 I_{p \times p}$. The model is rewritten as a time-varying coefficient AR model,

$$x_t = \phi_{t-1}^{(1)} x_{t-1} + \phi_{t-1}^{(2)} x_{t-2} + \cdots + \phi_{t-1}^{(p)} x_{t-p} + \varepsilon_t, \tag{13.40}$$

where the p coefficients change smoothly in time when w_t is small. The estimated time-varying coefficients AR model yields a nonstationary time-varying coefficient AR power spectrum representation,

$$p_t(f) = \frac{1}{2\pi} \left\{ \frac{\sigma_\varepsilon^2}{\left| 1 - \phi_t^{(t)} e^{-i2\pi f} - \phi_2^{(t)} e^{-i2\pi f \times 2} - \cdots - \phi_p^{(t)} e^{-i2\pi f \times p} \right|^2} \right\},$$

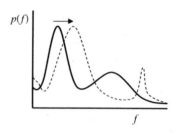

FIGURE 13.5
Change in spectrum of time-varying AR Model.

where the parameters, $\phi_t^{(1)}, \phi_t^{(2)}, \dots, \phi_t^{(p)}$ are t-dependent, while the noise variance, σ_ε^2, stays constant. An idea behind the time-varying AR approach is that the eigen-values of the time-varying AR model (13.40) defined by the transition matrix

$$\begin{pmatrix} \phi_1^{(t)} & \phi_2^{(t)} & \cdots & \phi_{p-1}^{(t)} & \phi_p^{(t)} \\ 1 & 0 & \cdots & 0 & 0 \\ 0 & 1 & \cdots & 0 & 0 \\ \cdots & \cdots & \cdots & \cdots & \cdots \\ 0 & 0 & \cdots & 1 & 0 \end{pmatrix}$$

move as time evolves (see Figure 13.5). Then they contribute to the change in the shape of the spectrum such as a shift from delta frequency to alpha frequency. The noise variance can contribute to adjust the level of the whole power, it cannot contribute to the change in the peak frequency of the spectrum shape. It has been shown that, for the estimation of time-varying EEG power spectrum, this type of time-varying AR model is not as good as the heteroscedastic compartment-GARCH state space model (Wong et al. 2006).

14

Causality Analysis

14.1 Introduction

The causality between the variables in the observed multivariate time series was first discussed by Wiener (1956) in the study of feedback systems. According to Wiener, a time series is called causal to a second if knowledge of the first series reduces the mean square prediction error of the second series (see Sinha and Kuszta 1983). Then feedback is said to be present when each of the two series is causal to the other.

Granger (1963, 1969) studied Wiener's idea further in the field of econometric time series. He introduced a few concepts related to causality, mostly in the framework of bivariate AR modeling, where the statistical significance of the autoregressive coefficients plays an important role in the interpretation of the causal and noncausal relations between the two time series. Recently, a similarity between the causality study in econometrics and neurosciences has been recognized. Although Granger discusses causality together with the multivariate feedback systems, the focus is on the causality between two time series variables or two time series vectors, which we may call pair-wise causality.

On the other hand, Wiener's original idea of causality and feedback system for multivariate time series was developed with a different approach by Akaike (1968) with special emphasis on application to the statistical identification and control of complex multivariate feedback systems in industrial engineering. Here not only the analysis of pair-wise causality between the two time series but also the total causality relations producing the complex controlled feedback system between many variables is the important problem to be solved for improving the control performance.

Akaike's method is meant for solving the feedback problem of Wiener rather than for solving the simplified pair-wise causality problem. Accordingly, it has been well recognized and used mostly in the field of large complex industrial control engineering systems such as cement rotary kiln, boiler of nuclear thermal power plant, auto-pilot navigation system of a ship in the ocean etc. (Akaike 1968, Otomo et al. 1972, Fukunishi 1977, Ohtsu et al. 1979, Nakamura and Akaike 1981). If Granger's pair-wise causality method in econometrics is relevant to the connectivity study in neurosciences, more relevant could be the method for the study of causality and feedbacks in complex industrial engineering systems, where not only pair-wise causality but also the total causality needs to be studied carefully with statistical system identification techniques.

14.2 Granger Causality and Limitations

Following Wiener (1956)'s idea of the prediction error approach, Granger introduced a definition of the causality for the pair of time series variables, x_t and y_t. Let $Var(x_t|U_s)$ be the prediction error variance of x_t given U_s. Here U_s means the information set containing all the information in the universe up to time point s, and is called the universal set. Granger causality was defined originally by Granger (1969) for the pair of time series variables, x_t and y_t as

Definition 1: *Causality.* If $Var(x_t|U_{t-1}) < Var(x_t|U_{t-1} - y_{t-1})$, y_t is causing x_t.

Definition 2: *Feedback.* If $Var(x_t|U_{t-1}) < Var(x_t|U_{t-1} - y_{t-1})$ and $Var(y_t|U_{t-1}) < Var(y_t|U_{t-1} - x_{t-1})$ feedback occurs, which is denoted $y_t \Leftrightarrow x_t$, that is, feedback occurs when y_t causes x_t and also x_t causes y_t.

Definition 3: *Instantaneous causality.* If $Var(x_t|U_{t-1}, y_t) < Var(x_t|U_{t-1})$, instantaneous causality occurs from y_t to x_t. In other words, the current value of x_t is better predicted if the value of y_t is included in the predictor.

Definition 4: *Causality lag.* If y_t causes x_t, the causality lag is defined to be the least value of lag k such that $Var(x_t|U_t - y_k) < Var(x_t|U_t - y_{k+1})$. Here, knowing y_{k+1} will be of no help in improving the prediction of x_t.

The weak points of the Granger Causality come from the following:

1. The use of the nonmathematical "universal" set U_s.
2. Causality is considered on a pair-wise base.

Although the universal set U_s could be a vector variable, the comparison of the value of the prediction error variance is on a pair-wise base between $Var(x_t|U_{t-1})$ and $Var(x_t|U_{t-1} - y_k)$. Here, the method for finding total causal relations between the many variables in a feedback system is not well established.

From the beginning, Granger himself noticed the obscurity of his definition of causality and discussed (Granger 1969, pp. 429–430) the danger of drawing out incorrect conclusions from dubious handling of three or more variables in the universal set. However, in the applied sciences, because of the strong need and expectations for a statistical method of causality analysis in real data analysis, it seems that not enough attention has been paid to the risk of relying on the method of Granger causality. We will see two typical examples of the difficulty with Granger causality after we introduce mathematical criteria for the check of Granger type causality in the next section.

14.2.1 Mathematical Criteria for Pair-Wise Causality

Granger causality is defined with a nonmathematical "Universal" set U_s. Here U_s means all the information set containing all information in the universe up to time point s and is called the universal set. We avoid this kind of nonmathematical terms in our discussion, since our interest is on causality from the viewpoint of statistical data analysis rather than philosophy, and we replace U_t by the set of time series variables available for the statistical modeling.

First, we restrict the problem to a very simplified case where the universal set is composed of only two time series x_t and y_t, that is $U_t = (x_t, y_t)$. Then the universal set is mathematically well defined, and the Granger's definition of the causality is restated as

1. Causality by Granger (1969):
   "A time series y_t Granger causes another time series x_t,

$$\text{If } Var(x_t | x_{t-}) - Var(x_t | x_{t-}, y_{t-}) > 0, \text{"}\qquad(14.1)$$

   where $x_{t-} = (x_{t-1}, x_{t-2}, \ldots)'$, $y_{t-} = (y_{t-1}, y_{t-2}, \ldots)'$.

   In Geweke's revised definition, this is generalized to

2. Causality by Geweke (1982):
   "A vector time series y_t Geweke causes another time series x_t

$$\text{If } \log|Var(x_t | x_{t-})| - \log|Var(x_t | x_{t-}, y_{t-})| > 0, \text{"}\qquad(14.2)$$

   where $x_{t-} = (x_{t-1}, x_{t-2}, \ldots)'$, $y_{t-} = (y_{t-1}, y_{t-2}, \ldots)'$.

   Here $Var(x_t | x_{t-})$ is the prediction error variance of x_t, specified by $Model^{(0)}$, which is a prediction model for x_t with x_{t-}. $Var(x_t | x_{t-}, y_{t-})$ is the prediction error variance of x_t, specified by $Model^{(1)}$, which is a prediction model for x_t using x_{t-} and y_{t-}.

   Incidentally we note that the aforementioned mathematical criteria of Granger and Geweke are essentially equivalent to the following generalized criteria, (14.3) and (14.4), based on the two models, $Model^{(0)}$ and $Model^{(1)}$

3. Causality by the log-likelihood:
   "An observed time series $x_t^{(j)}$ causes another series $x_t^{(i)}$, in the sense of "log-likelihood,"

$$\text{If } (-2)\log p^{Model^{(0)}}(x_t^{(i)}) - (-2)\log p^{Model^{(1)}}(x_t^{(i)})\} > 0. \text{"}\qquad(14.3)$$

Here $p^{Model^{(0)}}(x_t^{(i)})$ is the likelihood of the $x_t^{(i)}$ specified by the prediction model, $Model^{(0)}$, and $p^{Model^{(1)}}(x_t^{(i)})$ is the likelihood of the $x_t^{(i)}$ specified by the prediction model, $Model^{(1)}$.

If we consider that the sample data are from a true unknown density distribution $q(x)$, instead of $p^{\text{Model}^{(0)}}(x_t^{(i)})$ or $p^{\text{Model}(1)}(x_t^i)$ we can have another definition of the pair-wise causality, that is,

4. Causality by expected log-likelihood:
    "An observed time series $x_t^{(j)}$ causes another series $x_t^{(i)}$, in the sense of 'expected log-likelihood',

$$\text{if } \{(-2) \underset{q(x)}{E} \log p^{\text{Model}^{(0)}}(x_t^{(i)}) - (-2) \underset{q(x)}{E} \log p^{\text{Model}^{(1)}}(x_t^{(i)})\} > 0." \qquad (14.4)$$

14.2.2 Limitations of Granger Causality

Granger (1969) discusses an example of causality using the following model

$$\begin{aligned} x_t &= by_{t-1} + u_t^{(x)} \\ y_t &= cx_{t-2} + u_t^{(y)} \end{aligned} \qquad (14.5)$$

with the assumptions of whiteness of the noise $u_t^{(x)}$ and $u_t^{(y)}$, saying "If y is not causing x, then $b = 0$." Using this kind of examples sometimes causes confusion because such a simple model is unlikely to be chosen for real time series data analysis. A realistic situation is either that the prediction model includes many lag order terms or a simple model as (14.5) with more complex "nonwhite" noise $u_t^{(x)}$ and $u_t^{(y)}$. Such an example as (14.5) misleads some scientists to the idea of naive use of Granger causality, that is, fit an AR model,

$$\begin{aligned} x_t &= \sum_{i=}^{p_1} a_i x_{t-i} + \sum_{j=1}^{p_2} b_j y_{t-j} + u_t^{(x)} \\ y_t &= \sum_{i=}^{q_1} c_i x_{t-i} + \sum_{j=1}^{q_2} d_j y_{t-j} + u_t^{(y)}, \end{aligned}$$

to the bivariate time series, x_t and y_t, and check whether the coefficients $b_j = 0$ for $j = 1, \ldots, p_2$ without paying attention to the correlations between $u_t^{(x)}$ and $u_t^{(y)}$. Presenting a new definition of "Causality lag" does not improve the situation. It still leads to a confusing use of the original idea of Wiener (1956)'s prediction error approach to the causality analysis.

A similar confusing example is the following infinite order bivariate AR model,

$$\begin{aligned} x_t &= ax_{t-1} + bx_{t-2} + bdx_{t-3} + bd^2 x_{t-4} + \cdots + w_t^{(x)} \\ y_t &= dy_{t-1} + cy_{t-2} + cay_{t-3} + ca^2 y_{t-4} + \cdots + w_t^{(y)}, \end{aligned} \qquad (14.6)$$

where we assume that $w_t^{(x)}$ and $w_t^{(y)}$ are "colored" noise specified by

$$w_t^{(x)} = bu_{t-1}^{(y)} + bdu_{t-2}^{(y)} + bd^2u_{t-3}^{(y)} + \cdots + u_t^{(x)}$$

$$w_t^{(y)} = cu_{t-1}^{(x)} + cau_{t-2}^{(x)} + ca^2u_{t-3}^{(x)} + \cdots + u_t^{(y)}.$$

Here, x_t and y_t of (14.6) may not look causal each other. The model (14.6) is actually derived from a simple feedback system model driven by the noise $u_t^{(x)}$ and $u_t^{(y)}$ and is equivalent to

$$x_t = ax_{t-1} + by_{t-1} + u_t^{(x)}$$
$$y_t = cx_{t-1} + dy_{t-1} + u_t^{(y)}. \tag{14.7}$$

From (14.7), it appears like y_{t-1} that causes x_t when $b \neq 0$, and x_{t-1} causes y_t, when $c \neq 0$.

However, we must be careful about drawing any conclusion of causality out of the model if the model is a phenomenological time series model and is not an exact physical model. With any phenomenological time series model, we should not jump into any conclusion without confirming the whiteness and mutual independence of the driving noise $u_t^{(x)}$ and $u_t^{(y)}$. This is one of the reasons why the innovation approach becomes so important in applied time series analysis.

In general, we cannot expect that the driving noise of a feedback model is always white. If the driving noises $u_t^{(x)}$ and $u_t^{(y)}$ of (14.7) are colored noises, some more equivalent model representations are possible such as the following two scalar AR(2) type models,

$$x_t = (a+d)x_{t-1} + (bc-ad)x_{t-2} + v_t^{(x)}$$
$$y_t = (a+d)y_{t-1} + (bc-ad)y_{t-2} + v_t^{(y)}, \tag{14.8}$$

with the noise $v_t^{(x)}$ and $v_t^{(y)}$ given by

$$v_t^{(x)} = bu_{t-1}^{(y)} - du_{t-1}^{(x)} + u_t^{(x)}$$
$$v_t^{(x)} = cu_{t-1}^{(x)} - au_{t-1}^{(y)} + u_t^{(y)},$$

and an impulse response type model,

$$x_t = by_{t-1} + bay_{t-2} + ba^2y_{t-3} + \cdots + \xi_t^{(x)}$$
$$y_t = cx_{t-1} + cdx_{t-2} + cd^2x_{t-3} + \cdots + \xi_t^{(y)}, \tag{14.9}$$

with the noise $\xi_t^{(x)}$ and $\xi_t^{(y)}$, given by

$$\xi_t^{(x)} = au_{t-1}^{(x)} + a^2u_{t-2}^{(x)} + \cdots + u_t^{(x)}$$
$$\xi_t^{(y)} = du_{t-1}^{(y)} + d^2u_{t-2}^{(y)} + \cdots + u_t^{(y)}.$$

Note that bivariate driving noises,

$$\begin{pmatrix} w_t^{(x)} \\ w_t^{(y)} \end{pmatrix}, \quad \begin{pmatrix} u_t^{(x)} \\ u_t^{(y)} \end{pmatrix}, \quad \begin{pmatrix} v_t^{(x)} \\ v_t^{(y)} \end{pmatrix}, \quad \text{and} \quad \begin{pmatrix} \xi_t^{(x)} \\ \xi_t^{(y)} \end{pmatrix},$$

in (14.6) through (14.9) are all nonwhite colored noise. It shows that under the presence of colored driving noise, checking the significance (zero or non-zero) of the autoregressive coefficients of first lag order or second lag order is meaningless for the causality analysis.

14.2.3 Log-Likelihood instead of Coefficients

According to the definition of Granger causality,
    "An observed time series y_t Granger causes another series x_t

$$\text{if } a_{1,2} \neq 0,"$$

where

$$\begin{pmatrix} x_t \\ y_t \end{pmatrix} = \begin{pmatrix} a_{1,1} & a_{1,2} \\ a_{2,1} & a_{2,2} \end{pmatrix} \begin{pmatrix} x_{t-1} \\ y_{t-1} \end{pmatrix} + \begin{pmatrix} u_t^{(x)} \\ u_t^{(y)} \end{pmatrix}.$$

This may be useful for some very simple system of 2 variables, but we can easily see in the following example that it does not work well for systems with more than 2 variables.

14.2.3.1 Example A

Suppose we have a 3-dimensional system defined by

$$\begin{pmatrix} x_t \\ y_t \\ z_t \end{pmatrix} = \begin{pmatrix} a_{1,1} & a_{1,2} & 0 \\ 0 & a_{2,2} & a_{2,3} \\ a_{1,3} & 0 & a_{3,3} \end{pmatrix} \begin{pmatrix} x_{t-1} \\ y_{t-1} \\ z_{t-1} \end{pmatrix} + \begin{pmatrix} u_t^{(x)} \\ u_t^{(y)} \\ u_t^{(z)} \end{pmatrix}. \tag{14.10}$$

Here z causes y and y causes x (see Figure 14.1). Then z causes x even though $a_{1,3}$ is zero.

However, as long as the causality is checked on a pair-wise basis, there is always a case where this Geweke-Granger causality does not work well. This is clearly seen in the following example B (see Figure 14.2).

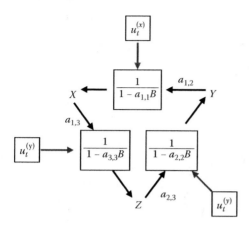

FIGURE 14.1
Example A for Granger causality.

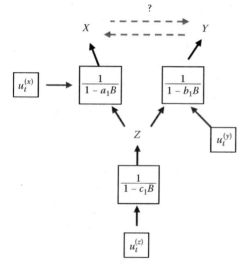

FIGURE 14.2
Example B for Granger causality.

14.2.3.2 Example B

In the 3-variables situation, defined by

$$(1 - a_1 B)x_t = a_2 z_{t-2} + u_t^{(x)}$$
$$(1 - b_1 B)y_t = b_2 z_{t-1} + u_t^{(y)} \qquad (14.11)$$
$$(1 - c_1 B)z_t = u_t^{(z)},$$

Granger–Geweke causality says "y causes x," because y_{t-1} has useful information (i.e., z_{t-2}) for predicting x_t, and we have

$$\log Var(x_t | x_{t-}) - \log Var(x_t | x_{t-}, y_{t-}) > 0.$$

However, the truth is that "z causes x and z causes y, but y does not cause x."

Geweke (1982) clarified and improved Granger's work into more general cases, where causality is defined (see (14.2)) for a pair of vector variables. However since the check of the causality of the vector variables is on a pair-wise basis, this cannot solve the difficulty, and essentially the same problem of the original Granger causality still remains.

14.3 Akaike Causality

It is certainly not safe and is inappropriate to rely simply on the pair-wise causal relationships to understand the whole picture of the causality between the variables, since the number of variables is usually very large in real-world systems such as industrial engineering systems or brain functioning systems. In order to characterize the total causality of k-variate time series, we need to characterize the dynamics of the whole variables simultaneously with a k-variate time series model, such as the AR model. The k-variate AR model for the k-variate time series may look rather too complicated to understand the tangled feedback structure between the variables with delays. When k is as large as 5 or 10, it is not realistic to check the causal path between the variables including delays by testing the significance of the coefficient $a_m^{(i,j)}$ ($m = 1, \ldots, M$) of the autoregressive coefficient A_m of the k-variate AR model,

$$x_t = A_1 x_{t-1} + \cdots + A_M x_{t-M} + w_t.$$

It may be surprising to many readers to know that the classic multivariate AR modeling, together with the frequency domain theory, can provide us with a solution to these formidable complicated problems of causality analysis of large dimensional feedback systems.

We will see how a general multivariate AR model provides us with a useful tool for checking the dynamic causality between the variables. A key for the solution here is again the idea of whitening the multivariate AR processes into innovations.

14.3.1 Innovation Contribution Ratio

Let us remember that the k-variate AR(M) model,

$$x_t = A_1 x_{t-1} + \cdots + A_M x_{t-M} + w_t,$$

has a parametric power spectrum representation (see Section 4.1.1),

$$p_{xx}(f) = \{A(f)^{-1}\}\Sigma_w\{\overline{A(f)'}\}^{-1}$$

$$= \{(a_{ij}(f))\}^{-1}(\sigma_{i,j})[\{\overline{(a_{ij}(f))}\}^{-1}]'$$

for $-1/2 \leq f \leq 1/2$, where $\Sigma_w = (\sigma_{i,j})$, $A_0 = -I$, and $A_m = 0$ for $m = -1, -2, \ldots$ and for $m > M$, so that

$$p_{i,i}(f) = \sum_{m=1}^{K} |\alpha_{i,m}(f)|^2 \, \sigma_m^2 + \sum_{m \neq n}^{K} \beta_{i.m.n}(f) \sigma_{m,n},$$

where $\beta_{i.m.n}(f)$ is determined by the elements of the matrix $A(f) = (a_{i,j}(f))$. If Σ_W is a diagonal matrix,

$$\Sigma_W = \begin{pmatrix} \sigma_1^2 & \cdots & 0 \\ \cdots & \cdots & \cdots \\ 0 & \cdots & \sigma_k^2 \end{pmatrix},$$

then we have

$$p_{i,i}(f) = \sum_{j=1}^{k} |\alpha_{i,j}(f)|^2 \, \sigma_j^2. \tag{14.12}$$

Equation (14.12) shows that for a multivariate system driven by a Gaussian white noise with a diagonal variance matrix Σ_w, the power spectrum of each variable is expressed as a weighted sum of the driving noise of all the variables, where the weight $|\alpha_{i,j}(f)|^2$ represents how much contribution the power $p_{i,i}(f)$ receives from the driving noise of the (j)-th variable. For some j this may not be zero but may be very small, or for some j this may be significantly large. The important point is that the size of the contribution is given for each frequency $-1/2 \leq f \leq 1/2$. Since the scale of $|\alpha_{i,j}(f)|^2$ depends on the scale of $p_{i,i}(f)$, Akaike (1968) introduced the normalized variable

$$r_i^{(j)}(f) = \frac{|\alpha_{i,j}(f)|^2 \, \sigma_j^2}{p_{i,i}(f)}$$

and called it the innovation contribution ratio (ICR) (or noise contribution ratio (NCR)). Then we have

$$\sum_{j=1}^{k} r_i^{(j)}(f) = \sum_{j=1}^{k} \frac{|\alpha_{i,j}(f)|^2 \, \sigma_j^2}{p_{i,i}(f)} = \frac{p_{i,i}(f)}{p_{i,i}(f)} = 1.$$

At each frequency f, the spectral power intensity is normalized to 100%, so that the ratio of the contribution from each innovation (noise) variance at each frequency can be seen clearly by plotting $r_i^{(j)}(f)$ $(j = 1, \ldots, k)$ for $0 < f < 1/2$.

Since $r_i^{(j)}(f)$ directly represents the strength of the causal effect from the j-th variable $x_t^{(j)}$ to i-th variable $x_t^{(i)}$, it may be natural to call it "innovation contribution ratio" or "noise contribution ratio" from $x_t^{(k)}$ to $x_t^{(i)}$. We note that similar methods based on the multivariate AR model have been repeatedly introduced after Akaike (1968) (see, e.g., Saito and Harashima 1981, Bernasconi and Konig 1999, Baccala and Samehsima 2001 and Kaminski et al. 2001).

14.3.2 Causality and Feedback Systems

We must admit, however, that Examples A and B, in Sections 14.2.3.1 and 14.2.3.2, are rather toy-like models. The real systems are not so clear-cut and simple. Causality relations in the real world sometimes become complicated and confusing when we deal with feedback systems. In order to take advantage of the Akaike causality equation (14.12) for multivariate feedback systems, it becomes essential that we identify a multivariate AR representation of a general feedback system from the observed time series efficiently.

We have seen in Section 4.2 that a bivariate feedback system driven by colored noise u_t and v_t, that is,

$$y_t = \sum_{m=1}^{M} a_m\, x_{t-m} + u_t \tag{14.13}$$

$$x_t = \sum_{m=1}^{M} b_m\, y_{t-m} + v_t, \tag{14.14}$$

can be estimated (consistently) by rewriting the model in AR form as

$$\begin{pmatrix} x_t \\ y_t \end{pmatrix} = \begin{pmatrix} d_1 & B_1 \\ A_1 & c_1 \end{pmatrix}\begin{pmatrix} x_{t-1} \\ y_{t-1} \end{pmatrix} + \cdots + \begin{pmatrix} d_L & B_L \\ A_L & c_L \end{pmatrix}\begin{pmatrix} x_{t-L} \\ y_{t-L} \end{pmatrix} + \begin{pmatrix} 0 & B_{L+1} \\ A_{L+1} & 0 \end{pmatrix}\begin{pmatrix} x_{t-L-1} \\ y_{t-L-1} \end{pmatrix}$$

$$+ \cdots + \begin{pmatrix} 0 & B_{L+M} \\ A_{L+M} & 0 \end{pmatrix}\begin{pmatrix} x_{t-L-M} \\ y_{t-L-M} \end{pmatrix} + \begin{pmatrix} \xi_t \\ \eta_t \end{pmatrix}. \tag{14.15}$$

Here η_t and ξ_t are white noise derived from u_t and v_t, respectively, by whitening using an AR(L) model as follows

$$u_t - \sum_{i=1}^{L} c_l\, u_{t-l} = \eta_t,$$

and

$$v_t - \sum_{i=1}^{L} d_l \, v_{t-l} = \xi_t.$$

The autoregressive coefficients, $A_1, \ldots, A_{L+M}, B_1, \ldots, B_{L+M}$ in (14.15) are specified by

$$A_1 = a_1,$$

$$A_m = a_m - \sum_{l=1}^{m-1} \tilde{c}_l a_{m-l}, \quad (m = 2, 3, \ldots, M+L)$$

$$B_1 = b_1,$$

$$B_m = b_m - \sum_{l=1}^{m-1} \tilde{d}_l b_{m-l}, \quad (m = 2, 3, \ldots, M+L),$$

where

$$a_m = 0, \quad b_m = 0, \quad (\text{for } m > M)$$

$$\tilde{c}_l = \begin{cases} c_l & \text{for } l = 1, 2, \ldots, L \\ 0 & \text{for } l > L \end{cases}$$

and

$$\tilde{d}_l = \begin{cases} d_l & \text{for } l = 1, 2, \ldots, L \\ 0 & \text{for } l > L \end{cases}.$$

This idea of identifying a feedback system model from a multivariate AR model can be extended from a bivariate feedback system to a general k-variate feedback system,

$$x_t^{(i)} = \sum_{\substack{j=1 \\ j \neq i}}^{k} \sum_{m=1}^{M} a_m^{(ij)} x_{t-m}^{(j)} + u_t^{(i)} \quad (i = 1, 2, \ldots, k),$$

where the system is driven by a k-dimensional colored noise $u_t^{(i)}$ $(i = 1, 2, \ldots, k)$ (see Figure 14.3).

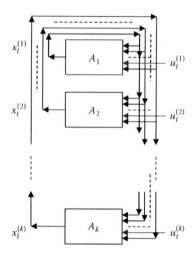

FIGURE 14.3
Causality and feedback system in Akaike's multivariate AR approach.

In the same way as for the bivariate case, we obtain a model of AR form,

$$x_t^{(i)} = \sum_{l=1}^{L} c_l^{(i)} x_{t-l}^{(i)} + \sum_{\substack{j=1 \\ j \neq i}}^{k} \sum_{m=1}^{M+L} A_{i,j,m} x_{t-m}^{(j)} + \varepsilon_t^{(i)}, \quad (i = 1, 2, \ldots, k), \quad (14.16)$$

using a whitening filter

$$u_t^{(i)} - \sum_{l=1}^{L} c_l^{(i)} u_{t-l}^{(i)} = \varepsilon_t^{(i)} \quad (i = 1, 2, \ldots, k)$$

for each noise $u_t^{(i)}$. The model (14.16) is a special form of a general k-variate AR model,

$$x_t^{(i)} = \sum_{m=1}^{M} \sum_{j=1}^{k} A_m^{(i,j)} x_{t-m}^{(j)} + \varepsilon_t^{(i)}, \quad (i = 1, 2, \ldots, k)$$

driven by a white noise $\varepsilon_t^{(i)}$ $(i = 1, 2, \ldots, k)$. Here we can estimate coefficients $A_m^{(i,j)}$ $(i = 1, \ldots, k, j = 1, \ldots, k, m = 1, \ldots, L + M)$ consistently, by either the least squares method or by the maximum likelihood method. From the estimated $A_m^{(i,j)}$'s, we can obtain consistent estimates of the coefficients $a_m^{(i,j)}$'s of the original feedback model in the same way as for the bivariate case.

14.3.3 Application to Gaze Data

In this section we see how the method of Akaike causality works in real data analysis. We have multivariate time series obtained by measuring the movement of heads of two persons in an experiment.

In the experiment, angles (yawing, pitching, and rolling) of the two subjects are measured. The two subjects are looking at each other at the beginning. A commander standing outside is giving the command. The first subject turns his head to look at an object in front of the two, when the commander gives him a sign. The second subject is supposed to watch the first subject and turn his head to look at the same object as soon as he noticed the first subject turns his head to the object. After a short interval, the commander gives another command and then the two subjects turn their heads to the original position to look at each other. This procedure is repeated several times with unequal time intervals.

Since the typical movement of the two subjects is in the yawing angle, we used the measurement of the yawing angles (out of the three angles) of the two subjects for the data analysis (see Figure 14.4). The number of data points is 2000, covering about 33 s (sampling rate is 60 Hz). The data were provided by kind permission of Professors S. Itakura (Kyoto University) and N. Sadato (National Institute of Physiological Sciences, Japan).

The Akaike causality results are, of course, affected by the model order we choose. However, changing model order does not substantially change the conclusion of the causal direction. For example, the results from the bivariate AR(5) model are shown in Figure 14.5, where the top left of the figure shows the power spectrum $p_{xx}(f)$ of the variable x and the bottom-left shows the power spectrum $p_{yy}(f)$ of the variable y. The dark area of the top right shows the ICR, $r_x^{(x)}(f)$, from x's own innovations. It shows that there is hardly any contribution of the innovations from y's to the power of the variable x. On the other hand, the dark area of the bottom-right of Figure 14.5 shows the ratio, $r_y^{(x)}(f)$, of the innovation contribution of x's innovation to the

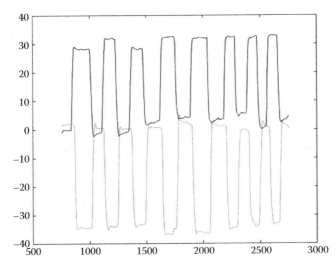

FIGURE 14.4
Gaze data.

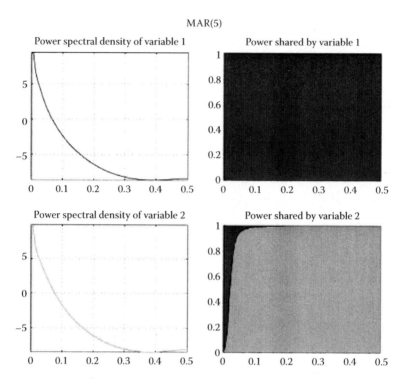

FIGURE 14.5
Innovation contribution ratio with multivariate AR(5).

power of y's, and the grey area of the bottom-right of Figure 14.5 shows the ratio, $r_y^{(y)}(f)$, of the contribution of the y's own innovations in the power of the variable y. This shows that x is causing y (especially at the low frequency band), while y is not causing x.

All the variables, $p_{xx}(f), p_{yy}(f), r_x^{(x)}(f), r_x^{(y)}(f), r_y^{(x)}(f), r_y^{(y)}(f), \sigma_1^2, \sigma_2^2$, are calculated from the 2-dimensional AR(5) model,

$$
\begin{pmatrix} x_t \\ y_t \end{pmatrix} = \begin{pmatrix} 2.5367 & -0.0125 \\ -0.0135 & 2.5328 \end{pmatrix} \begin{pmatrix} x_{t-1} \\ y_{t-1} \end{pmatrix} + \begin{pmatrix} -1.8213 & 0.0174 \\ -0.0104 & -1.8524 \end{pmatrix} \begin{pmatrix} x_{t-2} \\ y_{t-2} \end{pmatrix}
$$

$$
+ \begin{pmatrix} -0.1383 & 0.0221 \\ 0.0808 & -0.1637 \end{pmatrix} \begin{pmatrix} x_{t-3} \\ y_{t-3} \end{pmatrix} + \begin{pmatrix} 0.5645 & -0.0498 \\ -0.0595 & 0.6768 \end{pmatrix} \begin{pmatrix} x_{t-4} \\ y_{t-4} \end{pmatrix}
$$

$$
+ \begin{pmatrix} -0.1421 & 0.0230 \\ -0.0026 & -0.1978 \end{pmatrix} \begin{pmatrix} x_{t-5} \\ y_{t-5} \end{pmatrix} + \begin{pmatrix} \varepsilon_t^{(x)} \\ \varepsilon_t^{(y)} \end{pmatrix},
$$

which is estimated from the data of Figure 14.4, by assuming that the white noise $\varepsilon_t^{(x)}$ and $\varepsilon_t^{(y)}$ are uncorrelated.

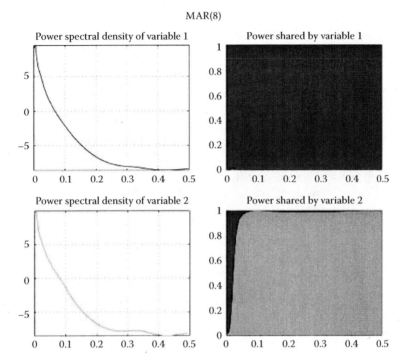

FIGURE 14.6
Innovation contribution ratio with multivariate AR(8).

The causality results are robust against the change on the order of the chosen AR model. Figure 14.6 shows the results of the Akaike innovation contribution analysis based on the AR(8) model, which is estimated from the same data of Figure 14.4, as

$$\begin{pmatrix} x_t \\ y_t \end{pmatrix} = \begin{pmatrix} 2.5230 & -0.0071 \\ -0.0221 & 2.4887 \end{pmatrix}\begin{pmatrix} x_{t-1} \\ y_{t-1} \end{pmatrix} + \begin{pmatrix} -1.7616 & 0.0139 \\ 0.0083 & -1.6884 \end{pmatrix}\begin{pmatrix} x_{t-2} \\ y_{t-2} \end{pmatrix}$$

$$+ \begin{pmatrix} -0.1598 & 0.0132 \\ 0.1095 & -0.2209 \end{pmatrix}\begin{pmatrix} x_{t-3} \\ y_{t-3} \end{pmatrix} + \begin{pmatrix} 0.3741 & -0.0676 \\ -0.1207 & 0.2654 \end{pmatrix}\begin{pmatrix} x_{t-4} \\ y_{t-4} \end{pmatrix}$$

$$+ \begin{pmatrix} 0.1009 & 0.0533 \\ -0.0829 & 0.2604 \end{pmatrix}\begin{pmatrix} x_{t-5} \\ y_{t-5} \end{pmatrix} + \begin{pmatrix} 0.0205 & 0.0564 \\ 0.1860 & 0.1245 \end{pmatrix}\begin{pmatrix} x_{t-6} \\ y_{t-6} \end{pmatrix}$$

$$+ \begin{pmatrix} 0.1608 & -0.1069 \\ -0.0744 & -0.3677 \end{pmatrix}\begin{pmatrix} x_{t-7} \\ y_{t-7} \end{pmatrix} + \begin{pmatrix} 0.0632 & 0.0449 \\ -0.0093 & 0.1334 \end{pmatrix}\begin{pmatrix} x_{t-8} \\ y_{t-8} \end{pmatrix} + \begin{pmatrix} \varepsilon_t^{(x)} \\ \varepsilon_t^{(y)} \end{pmatrix}.$$

Although the coefficients of the two models, AR(5) and AR(8), are quite different, the results of the ICR of the two models are quite similar, and this shows that Akaike's causality analysis is quite robust against the change in coefficients and model orders. Note that the data are from real phenomena, and the true model for the data is unknown. Strictly speaking, even the stationarity assumption may not be valid for this kind of situations. Naturally we cannot expect that the true model (linear or nonlinear, discrete time or continuous time) is always included in the candidate model family of bivariate AR models. All the bivariate AR models we use for the analysis are approximations. In this situation, checking the coefficients of the AR models does not help in obtaining any clear answer to the causal direction without introducing artificial assumptions for performing a statistical test, but some people may feel uncomfortable to use any hypothesis testing procedure in the present situation since there is not any true mathematical structure behind the data.

In the Akaike causality method, a causal direction is clearly shown by AR models of any lag order greater than 1, although these models are not true but approximations. Remember (see Section 10.5) that in Akaike's approach, all the statistical models are approximations to reality. We are just trying to choose the best "approximate" model among the candidate models in our hand.

14.4 How to Define Pair-Wise Causality with Akaike Method

It is easy to see that, in both Examples A and B, Akaike's innovative contribution approach takes us to the right answers. With Akaike's approach, z causes x, x causes y, and y causes z, in Example A, even though the coefficients (i.e., (1, 3)-element, (2, 1)-element, and (3, 2)-element) of the transition matrix of (14.10) are zero. This is because the driving noise for each variable is flowing into all the other variables through the feedback loop.

In Example B, y does not cause x in Akaike's sense, because the driving noise $u_t^{(y)}$ flowing into y does not flow into x. Certainly, it seems that Akaike's approach has advantage over Granger's approach when there are more than 2 variables.

On the other hand, Akaike's approach lacks a clear statement of pair-wise causality, such as Granger–Geweke causality provides, between chosen pairs of variables out of the several variables observed from the complicated feedback system. The causality in pair-wise is in fact what many scientists want to know in the early stage of the causality analysis of complex feedback systems.

14.4.1 Partial Pair-Wise Causality

It will be useful if Granger–Geweke type pair-wise causality information can be drawn out from the Akaike total causality. Remember that Akaike's total causality is derived from the power spectral representation of each variable in the multivariate AR model in terms of the innovation (i.e., prediction error) variance of each variable. For example, $p_{i,i}(f)$, that is, the power spectrum of the i-th variable $x_t^{(i)}$, is given by

$$p_{i,i}(f) = \sum_{m=1}^{k} |\alpha_{i,m}(f)|^2 \, \sigma_m^2. \tag{14.17}$$

This formulas implies that if the effect of the j-th innovation were "cut off" from the i-th variable, the power spectrum, $p_{i,i}^{(j)}(f)$ of the i-th variable would be

$$p_{i,i}^{(j)}(f) = \sum_{m=1}^{j-1} |\alpha_{i,m}(f)|^2 \, \sigma_m^2 + \sum_{m=j+1}^{k} |\alpha_{i,m}(f)|^2 \, \sigma_m^2. \tag{14.18}$$

Here we note that, according to the classic theory of Kolmogorov (1941), the power spectrum $p(f)$ and the prediction error variance σ^2 of a stationary process have the relation

$$\log \sigma^2 = \int_{-1/2}^{1/2} \log p(f)df.$$

Akaike's equation (14.17) means that $\log \sigma_i^2$, that is, the log of the prediction error variance of the i-th variable $x_t^{(i)}$ and its power spectrum $p_{i,i}(f)$ is related by

$$\log \sigma_i^2 = \int_{-1/2}^{1/2} \log p_{i,i}(f)df$$

$$= \int_{-1/2}^{1/2} \log \sum_{m=1}^{k} |\alpha_{i,m}(f)|^2 \, \sigma_m^2 df.$$

Then we can think of a virtual hypothetical process $x_t^{(i\sim j)}$, which is derived from the original k-variate AR model by cutting the flow of j-th innovations into i-th variable $x_t^{(i)}$. Then $\log \sigma_{i-j}^2$, that is, the log of the prediction error variance of the hypothetical process $x_t^{(i\sim j)}$, would be specified by

$$\log \sigma_{i-j}^2 = \int_{-1/2}^{1/2} \log p_{i,i}^{(j)}(f)df$$

$$= \int_{-1/2}^{1/2} \log \left\{ \sum_{m=1}^{j-1} |\alpha_{i,m}(f)|^2 \, \sigma_m^2 + \sum_{m=j+1}^{k} |\alpha_{i,m}(f)|^2 \, \sigma_m^2 \right\} df.$$

Let us consider the difference of $\log \sigma_{i-j}^2$ and $\log \sigma_i^2$ as Geweke (1982). The difference can be written as

$$\log \sigma_i^2 - \log \sigma_i^2 = \int \log p_{i,i}^{(j)}(f)df - \int \log p_{i,i}(f)df$$

$$= \int \log \left\{ 1 - \frac{|C_{i,j}(f)|^2 \, \sigma_j^2}{p_{i,i}(f)} \right\} df$$

$$\approx \int \frac{|C_{i,j}(f)|^2 \, \sigma_j^2}{p_{i,i}(f)} \}df$$

$$= \int_{-0.5}^{0.5} r_{i,j}(f)df.$$

It shows that the improvement of the log of the prediction error variance of the i-th variable gained by allowing the flow of the j-th innovations is approximately equal to the integration of the innovation contribution $r_{i,j}(f)$ of the j-th variable to the i-th variable with respect to the frequency, $-\frac{1}{2} < f < \frac{1}{2}$. This measure could be interpreted as Granger–Geweke type pair-wise causality measure obtained from the Akaike's total causality relations, $r_{i,j}(f)$ ($i = 1, \ldots, k, j = 1, \ldots, k$).

Then we can summarize the pair-wise causality in the sense of Granger–Geweke for each pair of the variables in the k-variate feedback system out of Akaike's method for the total causality of the system (see Table 14.1).

When we are not interested in the detail of the frequency-wise causal contribution of each variable to other variables, this kind of table will be useful to get an idea of the total picture of causal relations between the many variables of the system.

TABLE 14.1

Causality Map of k-Dimensional Time Series x_t

	$\downarrow 1$	$\downarrow 2$	...	$\downarrow k$
$1 \leftarrow$	$\int r_{1,1}(f)df$	$\int r_{1,2}(f)df$		$\int r_{1,k}(f)df$
	$(=\log \sigma_{1-1}^2 - \log \sigma_1^2)$	$(=\log \sigma_{1-2}^2 - \log \sigma_1^2)$		$(=\log \sigma_{1-k}^2 - \log \sigma_1^2)$
$2 \leftarrow$	$\int r_{2,1}(f)df$	$\int r_{2,2}(f)df$		$\int r_{2,k}(f)df$
	$(=\log \sigma_{2-1}^2 - \log \sigma_2^2)$	$(=\log \sigma_{2-2}^2 - \log \sigma_2^2)$		$(=\log \sigma_{2-k}^2 - \log \sigma_2^2)$
$k \leftarrow$	$\int r_{k,1}(f)df$	$\int r_{k,2}(f)df$		$\int r_{k,k}(f)df$
	$(=\log \sigma_{k-1}^2 - \log \sigma_k^2)$	$(=\log \sigma_{k-2}^2 - \log \sigma_k^2)$		$(=\log \sigma_{k-k}^2 - \log \sigma_k^2)$

14.4.2 Partial Causality and Total Causality

When we are interested only in the causality between the two subsets of the original set of variables, we can think of other types of causal relations between the subset variables, out of the results of the total causality of Akaike's innovation contribution equation, by considering other types of virtual processes and their power spectra.

1. For example, if we are interested in the causal relations between the two variables, $x_t^{(i)}$ and $x_t^{(j)}$ out of the set of all variables, $x_t^{(1)}, x_t^{(2)}, \ldots, x_t^{(k)}$, we can consider two virtual processes, $x_t^{(i^\wedge)}$ and $x_t^{(j^\wedge)}$, whose partial power spectra may be specified, using the results of the total causality of the whole variables, as

$$p_{(i^\wedge, i^\wedge)} = |C_{i,i}(f)|^2 \, \sigma_i^2 + |C_{i,j}(f)|^2 \, \sigma_j^2$$

$$p_{(j^\wedge, j^\wedge)} = |C_{j,j}(f)|^2 \, \sigma_j^2 + |C_{j,i}(f)|^2 \, \sigma_i^2.$$

Then a partial innovation contribution ratio, $r_{(i^\wedge, j^\wedge)}(f)$, between $x_t^{(i^\wedge)}$ and $x_t^{(j^\wedge)}$ is defined by

$$r_{(i^\wedge, j^\wedge)}(f) = \frac{|C_{i,j}(f)|^2 \, \sigma_j^2}{|C_{i,i}(f)|^2 \, \sigma_i^2 + |C_{i,j}(f)|^2 \, \sigma_j^2},$$

which represents the strength of the contribution of $x_t^{(j^\wedge)}$ to the power of $x_t^{(i^\wedge)}$. A partial causality in the reverse direction is defined by the partial innovation contribution ratio $r_{(j^\wedge, i^\wedge)}(f)$ given by

$$r_{(j^\wedge, i^\wedge)}(f) = \frac{|C_{j,i}(f)|^2 \, \sigma_i^2}{|C_{j,j}(f)|^2 \, \sigma_j^2 + |C_{j,i}(f)|^2 \, \sigma_i^2},$$

which represents the strength of the contribution of $x_t^{(i^\wedge)}$ to the power of $x_t^{(j^\wedge)}$. Naturally, the partial pair-wise contribution of $x_t^{(j^\wedge)}$ to the reduction of the log of prediction error variance of $x_t^{(i^\wedge)}$ is given by $\int_{-1/2}^{1/2} r_{(i^\wedge, j^\wedge)}(f) df$, and the partial pair-wise contribution of $x_t^{(i^\wedge)}$ to the reduction of the log of prediction error variance of $x_t^{(j^\wedge)}$ is given by $\int_{-1/2}^{1/2} r_{(j^\wedge, i^\wedge)}(f) df$.

2. Another way of using the partial innovation contribution is when we want to see whether the causality relations, suggested by using the original set of variables, $x_t^{(1)}, x_t^{(2)}, \ldots, x_t^{(k)}$, are distorted and affected by the causality results obtained by an introduction of extra variables,

$x_t^{(k+1)}, x_t^{(k+2)}, \ldots, x_t^{(k+l)}$. If the results are distorted and changed, we may want to see how. We can check these simply by comparing the ICR, $r_{(i,j)}(f)$, $(1 \leq i, j \leq k)$, of the original (k)-variables, $x_t^{(1)}, x_t^{(2)}, \ldots, x_t^{(k)}$, derived from the first model (A) for the (k)-variables, $x_t^{(1)}, x_t^{(2)}, \ldots, x_t^{(k)}$, and the partial innovation contribution ratio, $r_{(i,j)}(f)$, $(1 \leq i, j \leq k)$, of the updated model (B) fitted to $(k+l)$-variables, $x_t^{(1)}, x_t^{(2)}, \ldots, x_t^{(k)}, x_t^{(k+1)}, \ldots, x_t^{(k+l)}$. The partial innovation contribution ratio $r_{i,j}^{(B)}(f)$ from j-th variable $x_t^{(j)}$ to i-th variable $x_t^{(i)}$ is simply given, using the power spectrum of the model (B) by

$$r_{i,j}^{(B)}(f) = \frac{|C_{i,j,B}(f)|^2 \, \sigma_{j,B}^2}{\displaystyle\sum_{j=1}^{k} |C_{i,j,B}(f)|^2 \, \sigma_{j,B}^2},$$

while the innovation contribution $r_{i,j}^{(A)}(f)$ suggested by the model (A) of the original (k)-variables, $x_t^{(1)}, x_t^{(2)}, \ldots, x_t^{(k)}$, is

$$r_{i,j}^{(A)}(f) = \frac{|C_{i,j,A}(f)|^2 \, \sigma_{j,A}^2}{\displaystyle\sum_{j=1}^{k} |C_{i,j,A}(f)|^2 \, \sigma_{j,A}^2}$$

Here the index A in $C_{i,j,A}(f)$ and $\sigma_{j,A}^2$ mean that they are correspond to the model A. Similarly, $C_{i,j,B}(f)$ and $\sigma_{j,B}^2$ correspond to the model B.

We have seen that the information of Granger–Geweke type pair-wise causality between 2 vector variables can be obtained from the assessment of the total causality between many scalar variables. However, the reverse is not true, and we cannot obtain the total causality information from the collection of the pair-wise causalities between the component variables. This means, in the study of complex multivariate feedback systems, the identification of "total causality" is the most important and should have more priority than the identification of the "pair-wise causality."

14.5 Identifying Power Spectrum for Causality Analysis

From the previous discussions, it is now clear that the answer to the question of the pair-wise causality comes from the total causality of the k-variate time series, and the total causality information comes from the power spectrum of the k-variate time series. The power spectrum of the k-variate time series comes from the prediction model of the time series, and the prediction model is determined using one of the criteria, that is, (14.1) through (14.4) in Section 14.2.1.

Before moving to the discussion of the inferential part of the model estimation for the causality, we remember an important mathematical structure behind the prediction models (see Chapter 3), that is, the relations between (1) the prediction model, (2) the power spectrum of the model, and (3) the auto covariance function of the prediction model. If one of them is determined, the other two are also determined.

We know, from the discussion in Chapter 3, that the power spectrum or the auto covariance function of the time series is completely specified by linear models such as AR, MA, ARMA, and state space models. Each model family has its own unique way of characterizing the power spectrum or auto covariances from a simple structure to a more complex structure by increasing the order of the model. For example, with the parametric linear models, AR, MA, ARMA and state space models, we can have four different ways of characterizing the general linear process representation,

$$x_t = w_t + \sum_{k=1}^{\infty} H_k w_{t-k}.$$

Then with H_k ($k = 1, 2, \ldots$), we have a power spectrum representation,

$$p(f) = [I + H(f)]\Sigma_w \overline{[I + H(f)]}',$$

for each linear model, where

$$H(f) = \sum_{m=-\infty}^{\infty} e^{-i2\pi fm} H_m.$$

This means the following 5 missions for the k-variate time series are equivalent:

M1. Identification of the best explaining total causality of the time series

M2. Identification of the best prediction model of the time series

M3. Identification of the best explaining power spectrum of the time series

M4. Identification of the best explaining auto covariance function of the time series

M5. Identification of the best general linear process representation

Note that in Akaike 1968's original method, the idea of innovative contribution was introduced for the k-variate AR model of a given k-variate time series. Hence the results of total causality are derived from the power spectrum $p(f)$ of the k-dimensional AR model. However, the previous discussion shows that the essence of the causality is not in the use of AR model for the predictor. The essential information of the total causality is in the power spectrum, $\{p(f), -1/2 < f < 1/2\}$, or in the auto covariance function, $\{\ldots, \gamma_{-2}, \gamma_{-1}, \gamma_0, \gamma_1, \gamma_2, \ldots\}$, or in $\{H_1, H_2, H_3, \ldots\}$ of the general linear process representation,

$$x_t = w_t + H_1 w_{t-1} + H_2 w_{t-2} + \cdots.$$

This means that any k-variate prediction model that gives rise to a general linear process representation can be a model for explaining the total causality, and any model that predicts the data best could be chosen as the most appropriate model. The question is what is the best model for the prediction of a given time series.

14.5.1 Mathematical Criteria for Total Causality

We have seen, in Section 14.2.1, two criteria for the Granger causality, one is (14.1) by Granger (1969) and another is (14.2) by Geweke (1982), which are mathematically equivalent. If the identification of causality is considered in the framework of model identification of prediction models for the given k-variate time series, naturally we have some more criteria, including Granger's and Geweke's. The following five mathematical criteria are essentially equivalent, although differences between them become critical, when they are used in an inferential stage.

(Criterion 1): *In the sense of Granger's prediction error variance*:
   The model satisfying

$$\underset{Model \in U}{\text{Min}} \; Var \big| (X_t \mid Model) \big| \tag{14.19}$$

is chosen as the best explaining model, in the model set U, in the Mean Squares sense.

(Criterion 2): *In the sense of Geweke's log of prediction error variance*:
   The model satisfying

$$\underset{Model \in U}{\text{Min}} \; \log \big| Var(X_t \mid Model) \big| \tag{14.20}$$

is chosen as the best explaining model, in the model set U, in the log of prediction error variance sense.

(Criterion 3): *In the sense of Log-likelihood*:
   The model satisfying

$$\underset{Model \in U}{\text{Min}} \; \{ -\log p(X_t \mid Model) \} \tag{14.21}$$

is chosen as the best explaining model, in the model set U, in the log-likelihood sense.

(Criterion 4): *In the sense of expected log-likelihood*:
The model satisfying

$$\underset{Model \in U}{\text{Min}} \left\{ - \underset{q(x)}{E} \log p(X_t \mid Model) \right\} \tag{14.22}$$

is chosen as the best explaining model, in the model set U, in the $E[L$(expected-Log-likelihood)] sense.

(Criterion 5): *In the sense of Boltzmann–Kullback–Leibler entropy*:
The model satisfying

$$\underset{Model \in U}{\text{Min}} [\, KL\{q(X_t), p(X_t \mid Model)\}] \tag{14.23}$$

is chosen as the best explaining model, in the model set U, in the Boltzmann–Kullback–Leibler sense.

Here the model set U could be defined as "all" the possible prediction models like the Universal set in Granger's definition. However, to make our procedure realistic, U should be fixed, in advance before the analysis, to be a finite set of candidate models such as $U = [\{AR(i)\ i = 1, ..., p\}, (MA(j)\ j = 1, ..., q), (ARMA(i, j)\ i = 1, ..., p, j = 1, ..., q), (State\ Space\ models)]$.

We note that, in (14.22), the expectation is taken in terms of the true distribution $q(x)$ of the data. In addition, we note that the distribution $q(x)$ is unknown. In reality, it may be that $q(x) \neq p^{Model^{(0)}}(x)$ and $q(x) \neq p^{Model^{(1)}}(x)$, since all the models for the real data are approximations to the unknown true model.

Of course, we expect our models are close and not far from the truth, but we must accept the reality that no one is sure whether the set of our models includes the true model.

14.5.2 Statistical Criteria for Total Causality

In order to use the criteria in actual data analysis, we need to replace the mathematical quantities by statistical estimates through the estimated models with the finite sample time series data, and the estimated models need to be assessed and justified by a statistical method.

In other words, when we apply the aforementioned five mathematical criteria for the identification of the total causality in real data analysis, we need to replace the mathematical quantities such as the variance of prediction errors of a model, log-likelihood of a model, expected log-likelihood, and Boltzmann–Kullback–Leibler information of a model by statistical estimations.

For example, in Criterion 1 and Criterion 2, the $Var(X_t|Model^{(i)})$ should be replaced by a least squares estimate of the prediction error variance, $\widehat{Var}(X_t \mid Model^{(i)})$, in Criterion 3, the log $p(X_t|Model^{(i)})$ should be replaced by the maximum likelihood log $\hat{p}(X_t|Model^{(i)})$ of the $Model^{(i)}$, and in Criterion 4 and Criterion 5, the expected log-likelihood $\underset{q(x)}{E} \log p\,(X_t|Model^{(i)})$ and the Kullback–Leibler distance $KL\{q(X_t), p(X_t|Model^{(i)})\}$ should be replaced by AIC of the $Model^{(i)}$.

When statistical criteria are used in the modeling of actual time series, over fitting the model appears to be always better in the sense of minimizing prediction error variance or maximizing the log-likelihood. Here it is a common practice to combine a statistical method of significance test in order to prevent this overfitting. For example, Geweke (1982) suggests using the least squares criterion together with a chi-squared significance test, even though none of the models in our hand is the true model $q(x)$. For Criterion 4 and Criterion 5, however, the stopping mechanism of overfitting exists from the beginning through the introduction of the idea of the expectation, $\underset{q(x)}{E}[.]$, in terms of the true distribution $q(x)$. Here, we do not need to add extra statistical procedure such as the significance test, for the least squares estimator of the prediction error variance, to avoid the overfitting problem.

What we should do in the actual data analysis is to use Criterion 4 or equivalently Criterion 5, that is, to calculate AIC of all the possible candidate models in the candidate model set U, and choose the model giving the minimum AIC out of U. Which models to consider for U depends on the analyst's preference. If the analyst wants to avoid heavy computational load, he or she could choose all the k-variate AR models of order less than say 20. If he or she is concerned about the detailed relations of the total causality structures between the k-variables, he or she could include, apart from the AR models, ARMA models and state space models.

14.6 Instantaneous Causality

It has been widely recognized that the Akaike method of innovation contribution analysis, although useful, has a serious limitation that it can be used only for the case where the variance matrix of the innovation of the AR model is "diagonal." When the innovation variance matrix is nondiagonal, the power spectrum of each variable cannot have such a clear decomposition (14.12) in terms of the innovation variances of all the variables. In such cases, Akaike simply suggests not to use his method of the ICR analysis and suggests, instead, to improve the sampling of time series data, that is, to reconsider the design of measurement of the concerned dynamical system since this kind of instantaneously correlated innovations are mostly caused by some hidden driving force in the system (Akaike 1972).

On the other hand, in Geweke 1982's method, the concept of instantaneous causality is introduced, and a measure of assessing the strength of the instantaneous causality is presented. According to Geweke (1982), Y_t "instantaneously" causes X_t,

$$\text{if } F_{X \cdot Y} = \log \left\{ \frac{\det[Var(\tilde{X} \mid X^-, Y^-)]}{\det[Var(\tilde{X} \mid X^-, Y^+)]} \right\} > 0.$$

Although this brings us some useful information, we want to know more than just the existence of the instantaneous causality, which is more or less obvious from looking at the nonzero off-diagonal element of the innovation variance matrix. It would be useful if Akaike's method could be generalized to cases where the innovation variance matrix is nondiagonal, so that we can see in which frequency the instantaneous causality is strong. With recent advanced techniques of state space modeling, we can show that this can be actually done using a hidden latent variable, which is flowing into more than one time series variable, and the contribution ratio of the innovation of the latent variable can be measured like the case where the innovation variance matrix is diagonal. We will see this in the following.

14.6.1 Introduction of a Latent Noise Variable

Let us consider, as an example, a case of bivariate time series data for simplicity. Suppose bivariate AR models are applied to the data, and the p-th order AR model

$$\begin{pmatrix} x_t^{(1)} \\ x_t^{(2)} \end{pmatrix} = \begin{pmatrix} \phi_{11}^{(1)} & \phi_{12}^{(1)} \\ \phi_{21}^{(1)} & \phi_{22}^{(1)} \end{pmatrix} \begin{pmatrix} x_{t-1}^{(1)} \\ x_{t-1}^{(2)} \end{pmatrix} + \cdots + \begin{pmatrix} \phi_{11}^{(p)} & \phi_{12}^{(p)} \\ \phi_{21}^{(p)} & \phi_{22}^{(p)} \end{pmatrix} \begin{pmatrix} x_{t-p}^{(1)} \\ x_{t-p}^{(2)} \end{pmatrix} + \begin{pmatrix} w_t^{(1)} \\ w_t^{(2)} \end{pmatrix}$$

is found as the best explaining model for the data by AIC, but the variance matrix of innovations shows a large instantaneous correlation between $w_t^{(1)}$ and $w_t^{(2)}$. Then we can think of a latent state variable ξ_t together with uncorrelated white noise $u_t^{(1)}$ and $u_t^{(2)}$ driving this bivariate AR(p) model. The latent variable ξ_t may be driven by a third noise $u_t^{(3)}$ as in Figure 14.7.

Here we could assume that $u_t^{(3)}$ is uncorrelated with the other two noise, $u_t^{(1)}$ and $u_t^{(2)}$, which are also uncorrelated to each other, so that the variance matrix of the 3-dimensional noise $(u_t^{(1)}, u_t^{(2)}, u_t^{(3)})'$ is diagonal. This type of dynamical system can be realized by a state space model where the observed time series is 2 dimensional but the driving noise of the state, that is, system

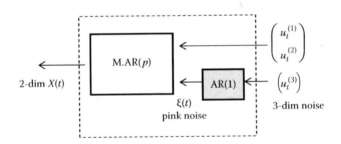

FIGURE 14.7
State space modeling.

noise is 3 dimensional. One example of such a state space model is given
here by (14.24):

$$
\begin{pmatrix} x_{t|t}^{(1)} \\ x_{t|t}^{(2)} \\ \cdots \\ x_{t+p-1|t}^{(1)} \\ x_{t+p-1|t}^{(2)} \\ \xi_t^{(1)} \end{pmatrix} =
\begin{pmatrix}
a_{11}^{(1)} & a_{12}^{(1)} & \cdots & 0 & 0 & 1 \\
a_{21}^{(1)} & a_{22}^{(1)} & \cdots & 0 & 0 & c \\
\cdots & \cdots & \cdots & \cdots & \cdots & \cdots \\
a_{11}^{(p)} & a_{12}^{(p)} & \cdots & 0 & 0 & 0 \\
a_{21}^{(p)} & a_{22}^{(p)} & \cdots & 0 & 0 & 0 \\
0 & 0 & \cdots & 0 & 0 & \phi^{(1)}
\end{pmatrix}
\begin{pmatrix} x_{t-1|t-1}^{(1)} \\ x_{t-1|t-1}^{(2)} \\ \cdots \\ x_{t+p-2|t-1}^{(1)} \\ x_{t+p-2|t-1}^{(2)} \\ \xi_{t-1}^{(1)} \end{pmatrix} +
\begin{pmatrix}
1 & 0 & 0 \\
0 & 1 & 0 \\
\cdots & \cdots & \cdots \\
0 & 0 & 0 \\
0 & 0 & 0 \\
0 & 0 & 1
\end{pmatrix}
\begin{pmatrix} u_t^{(1)} \\ u_t^{(2)} \\ u_t^{(3)} \end{pmatrix}
$$

$$
\begin{pmatrix} x_t^{(1)} \\ x_t^{(2)} \end{pmatrix} =
\begin{pmatrix} 1 & 0 & \cdots & 0 & 0 & 0 \\ 0 & 1 & \cdots & 0 & 0 & 0 \end{pmatrix}
\begin{pmatrix} x_{t|t}^{(1)} \\ x_{t|t}^{(2)} \\ \cdots \\ x_{t+p-1|t}^{(1)} \\ x_{t+p-1|t}^{(2)} \\ \xi_t^{(1)} \end{pmatrix}
$$

$$(14.24)$$

We can rewrite this with $Z_t = (x_{t|t}^{(1)}, x_{t|t}^{(2)}, \sqrt{a^2 + b^2}, \ldots, x_{t+p-1|t}^{(1)}, x_{t+p-1|t}^{(2)}, \xi_t^{(1)})'$ and
$U_t = (u_t^{(1)}, u_t^{(2)}, u_t^{(3)})'$ as

$$
\begin{cases} Z_t = AZ_{t-1} + BU_t \\ X_t = CZ_t \end{cases},
$$

$$(14.25)$$

where

$$
\mathrm{cov} \begin{pmatrix} u_t^{(1)} \\ u_t^{(2)} \\ u_t^{(3)} \end{pmatrix} = \Sigma_u =
\begin{pmatrix} \sigma_1^2 & 0 & 0 \\ 0 & \sigma_2^2 & 0 \\ 0 & 0 & \sigma_3^2 \end{pmatrix},
$$

$$A = \begin{pmatrix} a_{11}^{(1)} & a_{12}^{(1)} & \cdots & 0 & 0 & 1 \\ a_{21}^{(1)} & a_{22}^{(1)} & \cdots & 0 & 0 & c \\ \cdots & \cdots & \cdots & \cdots & \cdots & \cdots \\ a_{11}^{(p)} & a_{12}^{(p)} & \cdots & 0 & 0 & 0 \\ a_{21}^{(p)} & a_{22}^{(p)} & \cdots & 0 & 0 & 0 \\ 0 & 0 & \cdots & 0 & 0 & \phi^{(1)} \end{pmatrix}, \quad B = \begin{pmatrix} 1 & 0 & 0 \\ 0 & 1 & 0 \\ \cdots & \cdots & \cdots \\ 0 & 0 & 0 \\ 0 & 0 & 0 \\ 0 & 0 & 1 \end{pmatrix}, \quad z_t = \begin{pmatrix} x_{t|t}^{(1)} \\ x_{t|t}^{(2)} \\ \cdots \\ x_{t+p-1|t}^{(1)} \\ x_{t+p-1|t}^{(2)} \\ \xi_t^{(1)} \end{pmatrix},$$

$$C = \begin{pmatrix} 1 & 0 & \cdots & 0 & 0 & 0 \\ 0 & 1 & \cdots & 0 & 0 & 0 \end{pmatrix},$$

and

$$X_t = \begin{pmatrix} x_t^{(1)} \\ x_t^{(2)} \end{pmatrix}.$$

14.6.2 Innovation Contribution through Latent Variable

As we saw in Section 14.5, we have from the state space model (14.24) or (14.25) the power spectrum representation

$$p_{XX}^{(SS)}(f) = \{I + H^{(SS)}(f)\}\Sigma_u^{(SS)}\overline{[I + H^{(SS)}(f)]}' = C(f)\Sigma_u^{(SS)}\overline{C(f)}' \tag{14.26}$$

through the general linear process representation

$$X_t = w_t^{(SS)} + \sum_{k=1}^{\infty} H_k^{(SS)} w_{t-k}^{(SS)}$$

of the state space model (14.24). Here the power spectrum $p_{XX}^{(SS)}(f)$ is a 2×2 matrix,

$$p_{XX}^{(SS)}(f) = \begin{pmatrix} p_{1,1}(f) & p_{1,2}(f) \\ p_{2,1}(f) & p_{2,2}(f) \end{pmatrix}$$

but the variance matrix of the system noise is a 3×3 diagonal matrix,

$$\Sigma_u^{(SS)} = \begin{pmatrix} \sigma_1^2 & 0 & 0 \\ 0 & \sigma_2^2 & 0 \\ 0 & 0 & \sigma_3^2 \end{pmatrix},$$

so that $C(f)$ of (14.26) is a 2×3 matrix,

$$C(f) = \begin{pmatrix} C_{1,1}(f) & C_{1,2}(f) & C_{1,3}(f) \\ C_{2,1}(f) & C_{2,2}(f) & C_{2,3}(f) \end{pmatrix},$$

and the power spectra $p_{1,1}(f)$ and $p_{2,2}(f)$ are expressed as weighted sums of the variances, σ_1^2, σ_2^2, and σ_3^2, of the innovations, $u_t^{(1)}$, $u_t^{(2)}$, and $u_t^{(3)}$, respectively, as the following:

$$p_{1,1}(f) = |C_{1,1}(f)|^2\,\sigma_1^2 + |C_{1,2}(f)|^2\,\sigma_2^2 + |C_{1,3}(f)|^2\,\sigma_3^2$$

$$p_{2,2}(f) = |C_{2,1}(f)|^2\,\sigma_1^2 + |C_{2,2}(f)|^2\,\sigma_2^2 + |C_{2,3}(f)|^2\,\sigma_3^2.$$

ICR from the latent variable to the first variable is given by

$$r_{1,3}(f) = \frac{|C_{1,3}(f)|^2\,\sigma_3^2}{P_{1,1}(f)},$$

and the ICR from the latent variable to the second variable is given by

$$r_{2,3}(f) = \frac{|C_{2,3}(f)|^2\,\sigma_3^2}{P_{2,2}(f)}.$$

The contribution to the reduction of log-prediction error variance of the first variable by the latent variable is defined by

$$\log \sigma_{1-3}^2 - \log \sigma_1^2 = \int \log p_{1-3}(f)df - \int \log p_{1,1}(f)df = \int r_{1,3}(f)df$$

and the contribution for the reduction of log-prediction error variance of the second variable is defined by

$$\log \sigma_{2-3}^2 - \log \sigma_2^2 = \int \log p_{2-3}(f)df - \int \log p_{2,2}(f)df = \int r_{2,3}(f)df.$$

The total causality for the state space model with a latent variable is summarized in a 2×3 causality map (see Table 14.2).

TABLE 14.2

Causality Map of the State Space Model with a Latent Variable

	↓ 1	↓ 2	↓ 3
$1 \leftarrow$	$\int r_{1,1}(f)df$	$\int r_{1,2}(f)df$	$\int r_{1,3}(f)df$
	$(=\log \sigma_{1-1}^2 - \log \sigma_1^2)$	$(=\log \sigma_{1-2}^2 - \log \sigma_1^2)$	$(=\log \sigma_{1-3}^2 - \log \sigma_1^2)$
$2 \leftarrow$	$\int r_{2,1}(f)df$	$\int r_{2,2}(f)df$	$\int r_{2,3}(f)df$
	$(=\log \sigma_{2-1}^2 - \log \sigma_2^2)$	$(=\log \sigma_{2-2}^2 - \log \sigma_2^2)$	$(=\log \sigma_{2-3}^2 - \log \sigma_2^2)$

14.7 Application to fMRI Data

We applied the generalized Akaike method for the causality study of fMRI data. The data selected as an example to illustrate the method are the same data treated in Yamashita et al.(2005) (see Figure 14.8), that is, the time series measurement of 3 well-known regions of the brain, the primary visual cortex (V1), the visual motion-detection area (V5), and the posterior parietal cortex (PP), under visual stimulus under certain conditions. These areas have been reported to be activated during attention to visual motion (Buchel and Friston 1997). The data were provided by kind permission of Dr. O. Yamashita and Prof. Sadato. A detailed description of the experiment and the preprocessing procedure is in Yamashita et al. (2005).

Yamashita et al. (2005) analyzed the data with multivariate AR (MAR) models and found strong causality, by the Akaike method, from V1 to V5

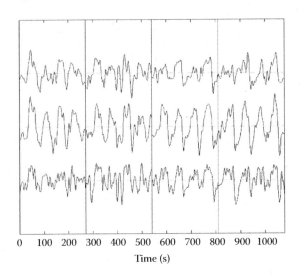

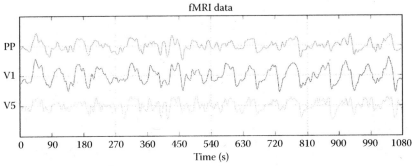

FIGURE 14.8
3-D fMRI data.

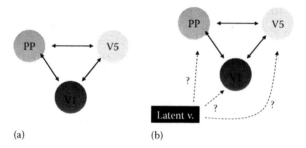

FIGURE 14.9
(a) Relations: V1-V5-pp. (b) Relations: V1-V5-PP and latent variable.

and from V5 to PP in a period of 60 s (see Figure 14.9a), which is assuring since this is equivalent to the time between the onsets of two consecutive stimuli. The only problem could be that these causality results are obtained under the assumption of the diagonal innovation variance of the MAR models, while the estimated innovation variance matrices are not really diagonal. The correlation coefficient between the innovations of the time series of V1 and the innovations of the time series of V5 is rather large (close to 0.3 at the highest). It will be interesting to see whether this problem can be solved and similar consistent results are obtainable by using the generalized Akaike method for instantaneous causality.

Our interest here is to see whether the causal effect of a latent variable to the three observed variables can be identified by the generalized Akaike method (See Figure 14.9b). First, we fit the state space model (14.25) to the 3-dimensional fMRI data of Figure 14.8. The parameters of the model (14.25) can be estimated by using the maximum likelihood method given in Chapters 10, 12, and 13. The AIC of the state space model (14.25) is 965.3.

Next we fit an MAR(4) model with a constraint that the innovation variance matrix is diagonal. AIC of the MAR(4) model was 1452.7. This shows that the total causality implied by the model (14.25) with a latent variable explains the data better than the MAR(4) model with the diagonal innovation variance.

AIC of the unconstrained MAR(4) model improves, when the constraint of the diagonal innovation variance is removed, where the largest value of the off-diagonal element of the estimated correlation matrix of the innovations is 0.3. However, the AIC is not as good as the state space model with the latent variable (Wong and Ozaki 2007).

In Figure 14.10a, we show the power spectra of PP, V1, and V5, based on the estimated state space model, where the spectrum of each variable, PP, V1, and V5, is decomposed into four components, light grey (V5), grey (PP), dark grey (V1), and black (Common), each corresponding to the four system noises of the four state variables of the state space model, V5, PP, V1, and the variable of the common source, respectively. Through (a), (b), and (c), the four noises make a distinctive contribution the time series, as shown by the model spectra.

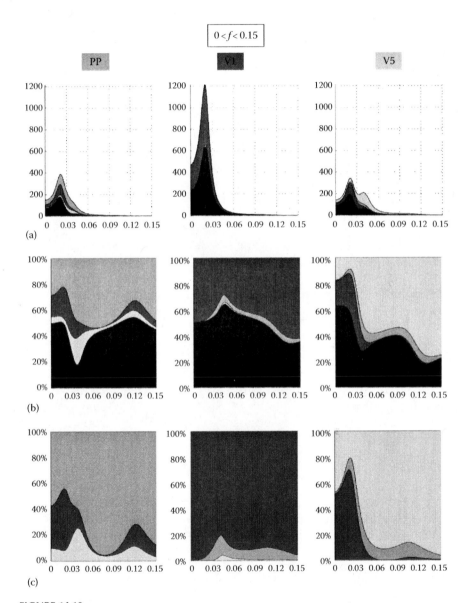

FIGURE 14.10

(a) Innovation contribution ratio of S.S.-1. (b) Innovation contribution ratio of S.S.-2. (c) Innovation contribution ratio of S.S.-3.

Among the three, in Figure 14.10a, the spectrum of V1 has the highest power intensity at around 0.02, corresponding to an oscillation with a period of 50 s, which can also be seen clearly in the data.

In Figure 14.10b we show the ICR, which are obtained by normalizing the spectra in Figure 14.10a. The area of the darkest color represents the

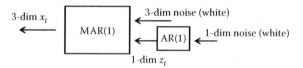

FIGURE 14.11
Structure of the state space causality model.

contribution from the innovation, which, by assumption, is driving the three time series simultaneously. We can see that in all the three spectra this common source explains more than 50% of the spectral power within the region of lower frequency.

Note that this common source has been introduced in the state space model through an idea that the 3-variate AR(1) process is driven by an almost white pink noise, which is represented by a scalar AR(1) model with a coefficient of 0.05 (see Figure 14.11).

In Figure 14.10c we show the partial ICR causality maps, where the innovation contribution between V1, V5, and PP are elucidated. The contributions of the common source (darkest color) to V1 and V5 (see Figure 14.10b) are disregarded. These remaining colors in Figure 14.10c show the causality between the three observed time series after eliminating the influence of the latent variable.

In all maps, V1 is showing up around the low frequency region, implying that causality from V1 to PP and to V5 is significant. PP is causing V1 and V5 only to a little extent, mostly around the frequency of 0.05 (corresponding to oscillations of period 20–25 s), and at the same time, V5 is causing PP to a little extent and V1 only negligibly.

Now we compare these results with the causality results, obtained by modeling the data by a pure MAR model, with the innovation variance matrix constrained to be diagonal; the model was estimated by the least square method. The AIC of the optimal MAR(4) model assumes a value of 1452.7, a value much greater than the AIC of the state space model, meaning that this MAR(4) model is a less suitable model for the data.

In Figure 14.12a and b, we show the power spectra and the ICR causality maps, respectively, of PP, V1, and V5, based on the estimated MAR(4) model. To our surprise, these ICR causality maps are very similar to those of Figure 14.10c. On the one hand, this shows that the common source component contributes for reducing the sum of squares of the residuals, while not distorting any characteristics of the causal dynamics in the data, by the imposition of our model assumptions, and on the other hand we can be assured that our results obtained from state space modeling are consistent with MAR modeling.

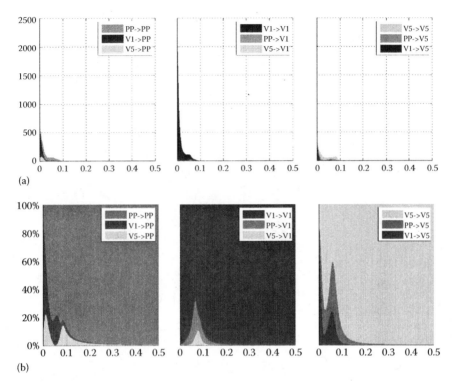

(a)

(b)

FIGURE 14.12
(a) Innovation contribution ratio-1 of MAR(4). (b) Innovation contribution ratio-2 of MAR(4).

14.8 Discussions

It may be possible to generalize the original Akaike method, for the case where the innovation variance of the AR model is nondiagonal, without generalizing the model to a state space model. For example, if the error variance matrix,

$$\Sigma = \begin{pmatrix} \sigma_1^2 & \sigma_{12} \\ \sigma_{21} & \sigma_2^2 \end{pmatrix},$$

of the AR(1) model,

$$\begin{pmatrix} x_t \\ y_t \end{pmatrix} = \begin{pmatrix} a_{1,1} & a_{1,2} \\ a_{2,1} & a_{2,2} \end{pmatrix} \begin{pmatrix} x_{t-1} \\ y_{t-1} \end{pmatrix} + \begin{pmatrix} w_t^{(1)} \\ w_t^{(2)} \end{pmatrix}, \tag{14.27}$$

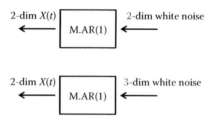

FIGURE 14.13
Comparing two types of model structure, I-O system with two-dimensional noise and I-O system with three-dimensional noise.

is nondiagonal, that is, $\sigma_{12} \neq 0$, we could think of an alternative AR(1) model (see Figure 14.13),

$$
\begin{pmatrix} x_t \\ y_t \end{pmatrix} = \begin{pmatrix} a_{1,1} & a_{1,2} \\ a_{2,1} & a_{2,2} \end{pmatrix} \begin{pmatrix} x_{t-1} \\ y_{t-1} \end{pmatrix} + \begin{pmatrix} 1 & 0 & 1 \\ 0 & 1 & c \end{pmatrix} \begin{pmatrix} u_t^{(1)} \\ u_t^{(2)} \\ u_t^{(3)} \end{pmatrix}, \tag{14.28}
$$

driven by 3-dimensional noise $(u_t^{(1)}, u_t^{(2)}, u_t^{(3)})'$, whose variance matrix is

$$
Var \begin{pmatrix} u_t^{(1)} \\ u_t^{(2)} \\ u_t^{(3)} \end{pmatrix} = \begin{pmatrix} s_1^2 & 0 & 0 \\ 0 & s_2^2 & 0 \\ 0 & 0 & s_3^2 \end{pmatrix}.
$$

Then the variance matrix of the system noise of the state space model (14.25) is given by

$$
\Sigma = Var \begin{pmatrix} u_t^{(1)} + u_t^{(3)} \\ u_t^{(2)} + c u_t^{(3)} \end{pmatrix} = \begin{pmatrix} s_1^2 + s_3^2 & c s_3^2 \\ c s_3^2 & s_2^2 + c^2 s_3^2 \end{pmatrix}. \tag{14.29}
$$

Unfortunately, this parameterization does not work. For example, for a given Σ,

$$
\Sigma = \begin{pmatrix} \sigma_1^2 & \sigma_{12} \\ \sigma_{21} & \sigma_2^2 \end{pmatrix} = \begin{pmatrix} 2 & 0.42 \\ 0.42 & 1 \end{pmatrix}, \tag{14.30}
$$

we can have many different c's that produce one and the same Σ. Some examples of a combination of four variables, c, s_1^2, s_2^2, s_3^2, yielding to the same Σ of (14.30) are shown in Table 14.3.

TABLE 14.3

Indeterminacy of the Four Variables, c, s_1^2, s_2^2, s_3^2, of the Two-Dimensional AR with Three-Dimensional Noise

c	s_1	s_2	s_3
1.6	1.73	0.32	0.27
1.4	1.70	0.41	0.30
1.2	1.65	0.49	0.35
1.9	1.58	0.58	0.42
0.8	1.47	0.66	0.53
0.6	1.29	0.74	0.70
0.4	0.94	0.83	1.08
0.3	0.59	0.87	1.41

Note that the strength of the innovation contribution of the common noise $u_t^{(3)}$ is determined by s_3^2. This means that we cannot determine the ICR of the power spectrum of x_t and y_t in terms of the three noise sources.

Note that the innovation variance matrix (14.29) of the AR model (14.28) has 4 parameters, s_1^2, s_2^2, s_3^2, and c. This is a typical example of redundant parameterization, where different combinations of 4 variables lead to the same innovation variance matrix. We may be able to write down the likelihood of the model, but if we try to maximize the likelihood in terms of the four variables, c, s_1^2, s_2^2 and s_3^2, certainly we will have numerical difficulty coming from the ill-posed parameterization. We cannot obtain the estimation of the model (14.28) without an arbitrary assumption for c. Some people may suggest choosing $c = \sqrt{\sigma_2^2/\sigma_1^2}$ with $s_3^2 = \sigma_{1,2}$ with $s_1^2 = \sigma_1^2$, $s_2^2 = \sigma_2^2$. However, there is no statistical reason why c should take that specific value. The value c should be chosen on the statistical ground such as the maximum likelihood principle or Akaike's entropy maximization principle. As long as we stick to the framework of the AR modeling, this cannot be realized.

Acknowledgment

Figures 14.4 through 14.6, 14.8, 14.10 and 14.12 were calculated by Dr. K.F.K. Wong of MGH, Harvard University.

15

Conclusion: The New and Old Problems

In the early days of time series analysis, researchers (Wold, Kolmogorov, Wiener, etc.) thought that smoothing the observed data would yield useful information for identifying the dynamics behind the data. In fact, it was Wiener (1949) who first considered the observation noise explicitly in the inferential problem of time series. Wiener tried to filter out the observation noise from the observed time series of dynamic phenomena (the trajectory of a flying object) in order to better predict the future orbit. In this respect, he was concerned about two inferential problems, (a) (the estimation of noise-contaminated state variables) and (b) (the identification of a dynamics model, i.e., a whitening operator), at the same time. The mathematical techniques he used in his method (the Wiener filter) showed that the most essential problem of temporally correlated time series can be solved finding a "whitening operator" of the time series. This leads the statisticians and scientists of this age to the study of the estimation of the color of time series (i.e., spectral analysis) during the process of solving the inferential problems (a) and (b).

Whitening a time series means transforming the temporally dependent series into independent prediction errors (i.e., innovations). To whiten a given time series into white noise, we need to identify the whitening operator from the sample time series data. In this respect, finding a whitening operator is classified as inference problem (b).

Unfortunately, Wiener did not pay much attention to the modeling of dynamic phenomena. He tried to formulate and solve inference problem (b) using a very general approach where all the dynamics are represented by impulse response functions. Wiener tried to estimate the impulse response function, h_t ($t = 0, 1, 2, 3, \ldots$), and the whitening operator from finite sample time series using estimated auto covariance functions without reference to any specific dynamic model. Although the method works, it is rather inefficient, since it is necessary to estimate so many variables, $h_1, h_2, h_3, \ldots$, from finite sample data. What is missing here is the idea of the parsimonious parameterization of the impulse response function through modeling.

Kalman took a very different approach to Wiener. He employed a state space model for characterizing the impulse response function and pointed out the importance of the role played by the state space models for the prediction, filtering, and smoothing of time series, especially for the inference problem (a), and gave an elegant alternative solution to Wiener's filtering problem. Here he explicitly separates the two inference problems (a) (the estimation of unobserved state variables from the data using a known model) and (*b*) (the identification of the model from the data) and takes full advantage of the Markov structure of the state space models in his reformulation of the Wiener problem.

What Kalman showed is that if a true state space model is known for a given Gaussian time series, the optimal smoother, optimal filter, and optimal predictor are uniquely obtainable from the model. This means that the problems of smoothing, filtering, and prediction are solved if the model is identified from the given time series data. In Kalman's paradigm, the optimal filtered estimate of the state is derived from the optimal identified model, and the optimal smoothed estimate of the state is derived from the optimal filtered estimate of the state. Note that his solution is given not for (*b*) but only for (*a*). He assumes that the state space model for the time series is "known." Similarly, the impulse response of the Wiener filter is assumed to be known.

What we need then is an identification of the optimal model for the given time series, that is, a solution for the inference problem (*b*). After solving problem (*a*), Kalman tried to develop a method for solving the inference problem (*b*) through using realization methods in his mathematical system theory (Kalman et al. 1969) without much success because of the lack of statistical criterion for the assessment of the realized model from the time series data.

One reliable criterion for the adequateness of a statistical model is the likelihood function, which is essentially equivalent to the prediction error variance in the case of dynamic models. The likelihood of a state space model is obtained through the calculated filtered state estimates and the innovations (prediction errors) of the state space model calculated from the time series.

On the other hand, a well-known shortcoming of the likelihood principle for the system identification has been recognized among the applied time series analysts who have to decide which time series model to choose among many candidate models with different lag orders. For time series data of finite length, a model with a larger number of parameters always wins over a model with fewer parameters, in the sense of maximizing the likelihood. Here the likelihood maximization principle cannot solve the problem properly. Akaike (1973a) solved this problem by introducing AIC, which is a kind of generalized likelihood, with the idea of measuring and comparing the deviation of a model from the true model using the Boltzmann-Kullback entropy. AIC is an unbiased estimate of the expected Boltzmann-Kullback entropy. Including too many parameters may result in an increase in the deviation of the model from the true model, while at the same time the reduction of the (-2)log-likelihood is very small. On the other hand, including an insufficient number of parameters may result in a significant increase in the (-2)log-likelihood, which would lead to an the increase in the deviation of the model from the true model. Akaike showed that AIC is asymptotically equivalent to the FPE (Final Prediction Error) criterion, which was introduced by Akaike (1969), for the order determination of AR models for prediction. The essential point in Akaike's finding is that "prediction" and "system identification" are intrinsically equivalent. This has proved to be

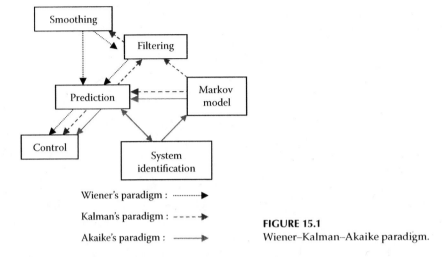

FIGURE 15.1
Wiener–Kalman–Akaike paradigm.

the most essential idea leading to a solution of Wiener's and Kalman's inference problem (*b*) (see Figure 15.1).

The work of Wiener–Kalman–Akaike had a great impact on science and engineering. The Wiener–Kalman–Akaike paradigm (the innovation approach) provides a perfect framework and useful solutions for the problems related to predictions associated with the dynamic phenomena. Modeling of dynamic phenomena using observed time series data has attracted more and more attention from scientists and engineers, but while significant benefits are reported from the application of the paradigm to the solution of inference problems (*a*) and (*b*), some drawbacks and difficulties have been recognized since the 1980s.

One of the common criticisms of the Wiener–Kalman–Akaike method is that the models treated are all linear Gaussian models. Although most dynamic phenomena can be approximately modeled by a linear Gaussian model, there are some dynamic phenomena showing very nonlinear and non-Gaussian nature. These nonlinear and non-Gaussian phenomena may not be as common as linear Gaussian phenomena, but they tend to attract more attention from scientists and engineers. This tendency has become more apparent as the measurement techniques of dynamic phenomena improved significantly after the 1980s, and we could see more details of the nature of dynamic processes behind the observed time series data.

One of the important questions is whether the Wiener–Kalman–Akaike paradigm is extendable to a more general framework such as

1. Linear → Nonlinear
2. Gaussian → non-Gaussian
3. Homoscedastic → Heteroscedastic
4. Univariate → Multivariate → Spatial

Unfortunately, many scientists and engineers took a negative attitude to this question. Many books and papers on nonlinear or non-Gaussian time series methods have been published since the late 1970s, in which the discussions start from denying the traditional approach and traditional linear time serie models such as AR and ARMA models. The present author thinks that this is not a good idea. Most nonlinear or non-Gaussian models are, as is shown in Part II, considered to be locally linear or locally Gaussian so that prediction and control problems can be discussed and solved on the same grounds of a local (in-time) linear or local (in-time) Gaussian framework (which is the essence of the Markov process theory), where the Wiener–Kalman–Akaike paradigm can be easily extended. What is important for science and engineering is not a discussion of a general model nor a general theory of model assessment criteria. What is important is to elucidate something unexplainable by the conventional approach (linear or Gaussian model) and present an actual improved model, where the improvement is assessed on the same ground of predictability, that is, the "innovation approach."

References

Ait-Sahalia, Y. (2002) Maximum likelihood estimation of discretely sampled diffusions, *Econometrica*, 70, 223–262.

Akaike, H. (1968) On the use of a linear model for the identification of feedback systems, *Ann. Inst. Stat. Math.*, 20, 425–439.

Akaike, H. (1969) Fitting autoregressive models for prediction, *Ann. Inst. Stat. Math.*, 21, 243–247.

Akaike, H. (1971) Autoregressive model fitting for control, *Ann. Inst. Stat. Math.*, 23, 163–180.

Akaike, H. (1972) *A Statistical Analysis and Control of Dynamical Systems*, Saiensusha, Tokyo (in Japanese) Its translated version (1988, in English) is available from Kluwer Academic Publishers.

Akaike, H. (1973a) Information theory and an extension of maximum likelihood principle. In *Proceeding of 2nd International Symposium on Information Theory*, eds. B.N. Petrov and F. Csaki, Akademiai-Kiado, Budapest, Hungary, pp. 267–281.

Akaike, H. (1973b) Maximum likelihood identification of Gaussian autoregressive moving average models, *Biometrika*, 60, 255–265.

Akaike, H. (1974a) Markovian representation of stochastic processes and its application to the analysis of autoregressive moving average processes, *Ann. Inst. Stat. Math.*, 26, 363–387.

Akaike, H. (1974b) Stochastic theory of minimal realization, *IEEE Trans. Automat. Control*, AC-19, 667–674.

Akaike, H. (1974c) A new look at the statistical model identification, *IEEE Trans. Automat. Control*, AC-19, 6, 716–723.

Akaike, H. (1975) Markovian representation of stochastic processes by canonical variables, *SIAM J. Control*, 13, 162–173.

Akaike, H. (1977) On entropy maximization principle. In *Application of Statistics*, ed. P.R. Krishnaiah, North-Holland, Amsterdam, the Netherlands, pp. 27–41.

Akaike, H. (1978) Time series analysis and control through parametric models. In *Applied Time Series Analysis*, ed. D.F. Findley, Academic Press, New York, pp. 1–23.

Akaike, H. (1980) Seasonal adjustment by a Bayesian modelling, *J. Time Ser. Anal.*, 1, 1–13.

Akaike, H. (1985) Prediction and entropy. In *A Celebration of Statistica, The ISI Centenary Volume*, eds. A.C. Atkinson and S.E. Fienberg, Springer-Verlag, New York, pp. 1–24.

Akaike, H. and Nakagawa, T. (1972) *Statistical Analysis and Control of Dynamic Systems*, Saiensusha, Tokyo (in Japanese) English translation has been published in 1989 by Kluwer Academic Publish with the same title.

Aoki, M. (1990) *State Space Modeling of Time Series*, Springer-Verlag, New York.

Astrom, K.J. and Kallstrom, C.G. (1973) Application of system identification techniques to the determination of ship dynamics. In *Identification and System Parameter Estimation*, ed. P. Ehkoff, North-Holland publishing Co., Amsterdam, the Netherlands, pp. 415–424.

Baccala, L.A. and Sameshima, K. (2001) Partial directed coherence: A new concept in neural structure determination, *Biol. Cybernet.*, 84, 463–474.

Balakrishnan, A.V. (1980) Nonlinear white noise theory. In *Multivariate Analysis*, ed. P.R. Krishnaiah, North-Holland, Amsterdam, the Netherlands, pp. 97–109.

Barndorff-Nielsen, O. (1978) *Information and Exponential Families in Statistical Theory*, Wiley, New York.

Bartlett, M. (1990) Chance and chaos, a discussion paper, *J. Roy. Stat. Soc. Ser. A.*, 158, 321–347.

Barton, M.J., Robinson, P.A., Kumar Soresh, Galka, A., Durrant-Whyte, H.F., Guivant, J. and Ozaki, T. (2009) Evaluating the performance of Kalman-Filter-Based EEG Source Localization, *IEEE Trans. of Biomedical Engineering*, 56, 122–136.

Bellman, R. (1957) *Dynamic Programming*, Princeton University Press, Princeton, NJ, Dover paper version edition (2003).

Bergstrom, A.R. (1966) Non-recursive models as discrete approximations to systems of stochastic differential equations, *Econometrica*, 34, 173–182.

Bernasconi, C. and Konig, P. (1999) On the directionality of cortical interactions studied by structural analysis of electrophysiological recordings, *Biol. Cybernet.*, 81, 199–210.

Biscay, R., Jimenez, J.C., Riera, J.J., and Valdes, P.A. (1995) Local linearization method for the numerical solution of stochastic differential equations, *Ann. Inst. Stat. Math.*, 48, 631–644.

Boltzmann, L. (1877) Uber die Beziehung zwischen dem zweiten Huptsatze der mechanischen Warmetheorie und der Wahrscheinkeitsrechnung respective den Satzen uber das Warmegleichgewicht, *Wiener Berichte*, 76, 373–435.

Box, G.E.P. and Jenkins, G.M. (1970) *Time Series Analysis, Forecasting and Control*, Holden-Day, San Francisco, CA.

Brockwell, P.J. and Davis, R.A. (2006) *Time Series: Theory and Methods*, Springer, New York.

Brown, P.E., Karesen, K.F., Roberts G.O., and Tonellato, S. (2000) Blur-generated non-separable space-time models, *J. R. Stat. Soc. B*, 62(4), 847–860.

Bryson, A.E. and Frazier, M. (1963) Smoothing for linear and nonlinear dynamic systems. In *Proceedings of Optimum System Synthesis Conference, Wright-Patterson AFB, OH, 1962*, pp. 354–363. (Tech. Doc. Rep. No. ASD-TDR-63-119, Aeronautical Systems Division, Wright-Patterson AFB, pp. 395).

Buchel, C. and Friston, K.J. (1997) Modulation of connectivity in visual pathways by attention: Cortical interactions evaluated with structural equation modeling and fMRI, *Cereb. Cortex*, 7, 768–778.

Burg, J.P. (1967) Maximum entropy spectral analysis. In *37th Annual International S.E.G. Meeting*, Oklahoma City, OK.

Buxton, R.B., Wong, E.C., and Frank, L.R. (1998) Dynamics of blood flow and oxygenation changes during brain activation: The balloon model, *Magn. Reson. Med.*, 39, 855–864.

Caines, P.E. and Rissanen, J. (1974) Maximum likelihood estimation of parameters in multivariate Gaussian stochastic processes, *IEEE Trans. Info. Theory*, IT-20, 102–104.

Cappe, O., Moulines, E., and Ryden, T. (2005) *Inference in Hidden Markov Models*, Springer-Verlag, New York.

Carbonell, F., Jimenez, J.C., and Biscay, R. (2002) A numerical method for the computation of the Lyapunov exponents of nonlinear ordinary differential equations, *Appl. Math. Comput.*, 131, 21–37.

Carroll, L. (1872) The Walrus and the carpenter. In *Through the Looking-Glass and What Alice Found There*, MacMillan & Co, London, U.K.

Casdagli, M., Eubank, S., Farmer, J., and Gibson, J. (1991) State space reconstruction in the presence of noise, *Physica. D*, 51, 52–98.

Caughey, T.K. (1963) Derivation and application of the Fokker-Planck equation to discrete nonlinear dynamic systems subjected to white random excitation, *J. Acoust. Soc. Am.*, 35, 1683–1692.

Cichoski, A. and Unbehauen, R. (1993) *Neural Networks for Optimization and Signal Processing*, John Wiley & Sons, Chichester, U.K.

Cramer, H. (1960) On some classes of nonstationary stochastic processes. In *Proceeding of the Fourth Berkeley Symposium on Stochastics and Probability*, Vol. 11, pp. 57–78.

Cronin, J. (1987) *Mathematical Aspects of Hodgkin-Huxley Neural Theory*, Cambridge University Press, Cambridge, U.K.

De Jong, P. (1991) The diffuse Kalman filter, *Ann. Stat.*, 19, 1073–1083.

De la Cruz, H., Biscay, R.J., Carbonell, F.M., Jimenez, J.C., and Ozaki, T. (2006) Local Linearization-Runge Kutta (LLRK) methods for solving ordinary differential equations, *Lecture Notes in Computer Sciences 3991*, Springer-Verlag, Berlin, Germany, pp. 132–139.

De la Cruz, H., Biscay, R.J., Carbonell, F., Ozaki, T., and Jimenez, J.C. (2007) A higher order local linearization method for solving ordinary differential equations, *Appl. Math. Comp.*, 185, 197–212.

Doeblin, W. (1940) Elements d'une theorie general des chaines simple constantes de Markoff, *Ann. Sci. Ecole Norm. Sup., III. Ser.*, 57, 61–111.

Doob, J.L. (1953) *Stochastic Processes*, John Wiley & Sons, New York.

Duffing, G. (1918) *Erzwungene Schingungen bei veranderlicher Eigenfrequenz*, F. Viewag & Sohn, Braunschweig, Germany.

Durbin, J. (1960) The fitting of time series models, *Rev. Int. Inst. Stat.*, 28, 233–244.

Durbin, J. and Koopman, S.J. (2001) *Time Series Analysis by State Space Methods*, Oxford University Press, Oxford, U.K.

Dyn, N. (1989a) Interpolation and approximation by radial and related functions. In *Approximation Theory*, Vol. VI, London Academic Press, London, U.K., pp. 211–234.

Dyn, N. (1989b) Interpolation by piece-wise radial basis functions, *J. Approx. Theory*, 59, 202–223.

Ehrenfest, P. and Ehrenfest, T. (1912) The conceptual foundation of the statistical approach in mechanics. In *German Encyclopedia of Mathematical Sciences*, English translation (by M.J. Moravcsik), Dover Publishing Co., New York, 1959.

Einstein, A. (1905) Uber die von der molekularkinetischen Theorie der Warme geforderte Bewegung von in ruhenden Flussigkeiten suspendierten Teilchen, *Annalen der Physik*, 17, 549–560.

Einstein, A. (1956) *Investigation of the Theory of Brownian Movement*, Dover, New York.

Farmer, J.D. and Sidorowich, J.J. (1991) Optimal shadowing and noise reduction, *Physica D*, 47, 373–392.

Feller, W. (1966) *An Introduction to Probability Theory and Its Applications*, John Wiley & Sons, New York.

Fisher, R.A. (1956) *Statistical Methods and Scientific Inference*, Oliver and Boyd, Edinburgh.

Fletcher, R. and Powell, M.J.D. (1963) A rapidly convergent descent method for minimization, *Computer J.*, 6, 163–168.

Freeman, W.J. (1975) *Mass Action in the Nervous System*, Academic Press, New York.

Friston, K.J. (1994) Functional and effective connectivity in neuroimaging: A synthesis, *Hum. Brain Mapp.*, 2, 56–78.

Friston, K.J. (1995) Statistical parametric mapping: Ontology and current issues, *J. Cereb. Blood Flow Metab.*, 15, 361–370.

Friston, K.J. (2003) Functional connectivity, in R.S.J. Frackowiak, K.J. Friston, C. Frith, R. Dolan, C.J. Price, S. Zeki, J. Ashburner, and W.D. Penny, editors, Human Brain Function, Academic Press, 2nd edition.

Friston, K.J., Harrison, L., and Penny, W. (2003) Dynamic causal modelling, *Neuroimage*, 19, 1273–1302.

Friston, K.J., Holmes, A.P., Worsley, K.J., Poline, J.B., Frith, C., and Frackowiak R.S.J. (1995) Statistical parametric maps in functional imaging: A general linear approach, *Hum. Brain Mapp.*, 2, 189–210.

Friston, K.J., Jezzard, P., and Turner, R. (1994) Analysis of functional MRI time series, *Hum. Brain Map.*, 1, 153–171.

Friston, K.J., Mechelli, A., Turner, R., and Price, C.J. (2000) Nonlinear responses in fMRI: The balloon model, voltera kernels, and other hemodynamics, *Neuroimage*, 12, 466–477.

Frost, P.A. and Kailath, T. (1971) An innovation approach to least squares estimation-part III: Nonlinear estimation in white Gaussian noise, *IEEE Trans. Automat. Contr.*, AC-16, 217–226.

Fukunishi, K. (1977) Diagnostic analysis of a nuclear power plant using multivariate autoregressive processes, *Nucl. Sci. Eng.*, 62, 215–225.

Galka A., Ozaki, T., Bosch Bayard, J., and Yamashita, O. (2006) Whitening as a tool for estimating mutual information in spatiotemporal data sets, *J. Stat. Phys.*, 124(5), 1275–1315.

Galka, A., Wong, K.F.K., and Ozaki, T. (2010) Generalized state space models for modeling non-stationary EEG time series. In *Modeling Phase Transition in the Brain*, Springer Series in Computational Neuroscience, Vol. 4, eds. D.A. Syein-Ross and M. Stein-Ross, Springer, New York, pp. 27–52.

Galka, A., Yamashita, O., and Ozaki, T. (2004a) GARCH modelling of covariance in dynamical estimation of inverse solutions, *Phys. Lett. A*, 333, 261–268.

Galka, A., Yamashita, O., Ozaki, T., Biscay, R., and Valdes-Sosa, P. (2004b) A solution to the dynamical inverse problem of EEG generation using spatiotemporal Kalman filtering, *Neuroimage*, 23, 435–453.

Gallant, A.R. and Long, J.R. (1997) Estimating stochastic differential equations efficiently by minimum chi-squared, *Biometrika*, 84, 125–141.

Gallant, A.R. and Tanchen, G. (1997) Estimation of Continuous-time Models for Stock Returns and Interest Rates, *Macroeconomic Dynamics*, 1, 135–168.

Geweke, J. (1982) Measurement of linear dependence and feedback between multiple time series, *J. Am. Stat. Assoc.*, 77, 304–313.

Gibbs, J.W. (1902) *Elementary Principles in Statistical Mechanics*, C. Scribner's Sons, New York.

Goel, N.S. and Richter-Dyn, N. (1974) *Stochastic Models in Biology*, Academic Press, New York.

Good, I.J. (1965) *The Estimation of Probabilities: An Essay on Modern Bayesian Methods*, MIT Press, Cambridge, MA.

Granger, C.W.J. (1963) Economic processes involving feedback, *Inf. Control*, 6, 28–48.

Granger, C.W.J. (1969) Investigating causal relations by econometric models and cross-spectral methods, *Econometrika*, 37, 424–438.

Gribbin, J. (1995) *Schrodinger's Kitten*, Weidenfeld & Nicholson, London, U.K.

Haggan, V. and Ozaki, T. (1981) Modelling nonlinear random vibrations using an amplitude-dependent autoregressive time series model, *Biometrika*, 68, 189–196.

Haggan-Ozaki, V., Ozaki, T., and Toyoda, Y. (2009) An Akaike state space controller for RBF-ARX models, *IEEE Trans. Contr. Syst. Tech.*, 17, 191–198.

Hannan, E.J. (1960) *Time Series Analysis*, Methuen & Co., Ltd, London, U.K.

Hannan, E.J. (1969a) Identification of vector mixed autoregressive-moving average systems, *Biometrika*, 56, 223–225.

Hannan, E.J. (1969b) The estimation of mixed moving average autoregressive systems, *Biometrika*, 56, 579–593.

Hannan, E.J. and Deistler, M. (1988) *The Statistical Theory of Linear Systems*, John Wiley, New York.

Hanner, O. (1949) Deterministic and nondeterministic stationary random processes, *Ark. Mat.*, 1, 161–177.

Hansen, L.P. (1982) Large sample properties of generalized method of moments estimators, *Econometrica*, 50, 547–557.

Harrison, P.J. and Stevens, C.F. (1976) Bayesian forecasting (with discussion), *J. R. Stat. Soc. Ser. B*, 38, 205–247.

Harvey, A.C. (1989) *Forecasting, Structural Time Series Models and the Kalman Filter*, Cambridge University Press, Cambridge, U.K.

Henrici, P. (1962) *Discrete Variables Methods in Ordinary Differential Equations*, Wiley, New York.

Ho, Y.C. and Lee, R.C.K. (1964) A Baysian approach to problems in stochastic estimation and control, *IEEE Trans. Automat. Control*, 9, 333–339.

Hodgkin, A.L. and Huxley, A.F. (1952) A quantitative description of membrane current and its application to conduction and excitation in nerve, *J. Physiol.*, 117, 500–544.

Ishiguro, M., Sakamoto, Y., and Kitagawa, G. (1994) Bootstrapping log-likelihood and EIC, an extension of AIC, Research Memo. No. 532, The Institute of Statistical Mathematics, Tokyo, Japan.

Ito, K. (1942) On stochastic processes, I (Infinitely divisible laws of probability), *Japan. J. Math.*, 18, 261–301. (Reprinted in *Kiyoshi Ito Selected Papers*, Springer, New York, 1987).

Ito, K. (1944) Stochastic integral, *Proc. Imp. Acad. Tokyo*, 20, 519–524.

Ito, K. (1951) On stochastic differential equations, *Mem. Am. Math. Soc.*, 4, 1–51. College Press, London, U.K., pp. 64–76.

Ito, K. (1980) Chaos in the Rikitake two-disc dynamo system, *Earth Planet. Sci. Lett.*, 51, 451–456.

Ito, K. (1984) Infinite dimensional Ornstein-Uhlenbeck processes. In *Taniguchi Symp. SA, Katata, 1982*, ed. K. Ito, North-Holland, Amsterdam, the Netherlands, pp. 197–224.

Jaynes, E.T. (1988) How does the Brain do plausible reasoning? In *Maximum Entropy and Bayesian Methods in Science and Engineering*, Vol. 1: Foundations, ed. G.J. Erickson and C.R. Smith, Kluwer Academic Publishers, Dordrecht, the Netherland.

Jazwinski, A.H. (1970) *Stochastic Processes and Filtering Theory*, Academic Press, New York.

Jimenez, J.C. (2012) *Local Linearization Methods for the Integration and Estimation of Differential Equations*, to appear.

Jimenez, J.C., Biscay, R., and Ozaki, T. (2006) Inference methods for discretely observed continuous-time stochastic volatility models: A commented overview, *Asia Pacific Financial Markets*, 12, 109–141.

Jimenez, J.C. and Ozaki, T. (2002) Linear estimation of continuous-discrete state space models with multiplicative noise, *Sys. Control Lett.*, 47, 91–101.

Jimenez, J.C. and Ozaki, T. (2003) Local linearization filters for non-linear continuous-discrete state space models with multiplicative noise, *Int. J. Contr.*, 76, 1159–1170.

Jimenez, J.C. and Ozaki, T. (2005) An approximate innovation method for the estimation of diffusion processes from discrete data, *J. Time Ser. Anal.*, 27, 77–97.

Jimenez, J.C., Shoji, J., and Ozaki, T. (1999) Simulation of stochastic differential equations through the local linearization method. A comparative study, *J. Stat. Phys.*, 94, 587–602.

Jones, D.A. (1978) Nonlinear autoregressive processes, *Proc. Roy. Soc. London Ser. A*, 360, 71–95.

Jones, R.H. (1980) Maximum likelihood fitting of ARMA models to time series with missing observations, *Technometrics*, 22, 389–395.

Kailath, T. (1968) An innovation approach to least squares estimation Part I: linear filtering in additive white noise, *IEEE Trans. Auomat. Control*, AC-13, 646–655.

Kailath, T. (1969) A generalized likelihood-ratio formula for random signals in Gaussian noise, *IEEE Trans. Inform. Theory*, IT-15, 350–361.

Kailath, T. (1980) *Linear Systems*, Prentice-Hall.

Kailath, T. (1981) *Lectures on Wiener and Kalman Filtering*, Springer Verlag, Wien – New York.

Kalman, R.E. (1960) A new approach to linear filtering and prediction problems, trans. ASME, series D, *J. Basic Eng.*, 82, 35–45.

Kalman, R.E. and Bertran, J.E. (1960a) Control system analysis and design via the Second Method of Lyapunov, I. continuous-time systems, *Trans. ASME, Ser. D: J. Basic Eng.*, 82, 371–393.

Kalman, R.E. and Bertran, J.E. (1960b) Control system analysis and design via the Second Method of Lyapunov, II. discrete-time systems, *Trans. ASME, Ser. D: J. Basic Eng.*, 82, 394–400.

Kalman, R.E. and Bucy, R.S. (1961) New results in linear filtering and prediction theory, *Trans. ASME*, 83D, 15–108.

Kalman, R.E., Falb, P.L., and Arbib, M.A. (1969) *Topics in Mathematical System Theory*, McGraw-Hill, New York.

Kaminski, M., Ding, M., Truccolo, W.A., and Bressler, S.L. (2001) Evaluating causal relations in neural systems: Granger causality, directed transfer function and statistical assessment of significance, *Biol. Cybern.*, 85, 145–157.

Karhunen, K. (1949) Uber die struktur stationarer zufallinger funktionen, *Ark. Mat.*, 1, 141–160.

Kato, H. and Ozaki, T. (2002) Adding data process feedback to the nonlinear autoregressive model, *Signal Processing*, 82, 1189–1204.

Kawai, S., Ishiguro, M., and Oku, Y. (2008) Simultaneous estimations of parameters and external current in Hodgkin-Huxley model, *J. Soc. Instr. Control Eng.*, 44, 838–845. (in Japanese).

Kingman, J. (1963) Ergodic properties of continuous-time Markov processes and their discrete skeletons, *Proc. London Math. Soc.*, 13, 593–604.

Kloeden, P.E. and Platen, E. (1995) *Numerical Solution of Stochastic Differential Equations*, Springer, New York.

Kolmogorov, A.N. (1931) On analytical methods in probability theory, *Math. Ann.*, 104, 415–458. (Selected works of A.N. Kolmogorov, 1992). In *Probability Theory and Mathematical Statistics*, Vol. II, ed. A.N. Shiryayev, Kluwer Academic Publishers, Dordrecht, the Netherlands, pp. 62–108.

Kolmogorov, A.N. (1939) Sur l'interpolation et l'extrapolation des suites stationnaires, *C. R. Acad. Sci. Paris*, 208, 2043–2045. (A translation has been published by RAND Corp., Santa Monica, CA as Memo -3090-PR.as RM.)

Kolmogorov, A.N. (1941) Stationary sequences in Hilbert space, *Bull. Math. Univ. Moscou.*, 5, 3–14.

Konishi, S. and Kitagawa, G. (1996) Generalized information criteria in model selection, *Biometrika*, 83, 875–890.

Kostelich, E.J. and Yorke, J.A. (1988) Noise reduction in dynamical systems, *Phys. Rev. A*, 38, 1649–1652.

Kullback, S. (1959) *Information and Statistics*, John Wiley & Sons, New York.

Kushner, H.J. (1962) On the differential equations satisfied by conditional probability densities of Markov processes, with applications, *J. SIAM Control, Ser. A*, 2, 106–119.

Kushner, H.J. (1967) Approximations to optimal nonlinear filters, *IEEE Trans. Automat. Control*, AC-12, 546–556.

Kutoyantz, Y.A. (1984) *Parameter Estimation for Stochastic Processes*, (Translated and edited by B.L.S. Prakasa Rao), Heldermann, Berlin, Germany.

Lange, N. and Zeger, S.L. (1997) Non-linear Fourier time series analysis for human brain mapping by functional magnetic resonance imaging (with discussion), *Appl. Stat.*, 46, 1–29.

Lapedes, J. and Farber, R. (1987) Neural network works. In *Neural Information Systems*, Academic Institute of Physics, New York, pp. 442–456.

Laplace, P.S. (1814) *A Philosophical Essay on Probabilities*, translated from the 6th French edition by F.W. Truscott and F.L. Emory (1951), Dover Publications, New York.

Larson, H.J. and Shubert, B.O. (1979) *Probabilistic Models in Engineering Sciences*, Vol. II, John Wiley & Sons, New York.

Leonard, J. and Kramer, M. (1991) Radial basis function networks for classifying process faults, *IEEE Control Syst. Mag.*, 11, 31–38.

Levinson, N. (1947) The Wiener RMS (Root Mean Square) error criterion in filter design and prediction, *J. Math. Phys.*, XXV, 261–278.

Levy, P. (1956) Sur une classe de courbes de l'espace de Hilbert et sur une equation integrale non lineaire, *Ann. Sci. Ecole Norm. Sup.*, 73, 121–156.

Liang, D.F. (1983) Comparisons of nonlinear recursive filters for systems with non-negligible nonlinearities. In *Control and Dynamical Systems, Advances in Theory and Applications*, Vol. 20: Nonlinear and Kalman Filtering Techniques, eds. C.T. Leondis, Academic Press, New York, pp. 341–401.

Ljung, L. (1987) *System Identification: Theory for the User*, Prentice-Hall Inc., Upper Saddle River, NJ.

Ljung, L. and Caines, P.E. (1979) Asymptotic normality of prediction error estimators for approximate system models, *Stochastics*, 3, 29–46.

Lopes da Silva, F.H., van Rotterdam, A., Barts, P., van Heusden, E., and W. Burr (1976b) Models of neuronal populations: The basic mechanisms. In *Perspective in Brain Research*, eds. M.A. Corner and D.F. Swaab, *Prog. Brain Res.*, 45, 281–308.

Lorenz, E. (1963) Deterministic non-periodic flow, *J. Atmos. Sci.*, 20, 130–141.

Lutkepohl, H. (2006) *New Introduction to Multiple Time Series Analysis*, Springer, Berlin, Germany.

Masani, P. and Wiener, N. (1959) Nonlinear prediction. In *Probability and Statistics*, ed. V. Grenander, John Wiley & Sons, New York, pp. 190–212.

May, R.M. (1976) Simple mathematical models with very complicated dynamics, *Nature*, 261, 459–467.

Meditch, J.S. (1967) Orthogonal projection and discrete optimal linear smoothing, *J. SIAM Contr.*, 5, 74–89.

Mehra, R.K. (1969) Identification of stochastic linear dynamic systems, *Proc. IEEE Symp. Adapt. Process. Decis. Contr.*, Atlanta, GA, November 17–19, pp. 6-f-1–6-f-4.

Mehra, R.K. (1971) Identification of stochastic linear dynamical systems, *AIAA J.*, 9, 28–31.

Mehra, R.K. and Tyler, J.S. (1974) Case studies in aircraft parameter identification. In *Proceeding 3rd IFAC Symposium on Identification and System Parameter Estimation*, The Hague, the Netherlands, pp. 117–145.

Milstein, G.N. (1974) Approximate integration of stochastic differential equations, *Theory Prob. Appl.*, 23, 396–401.

Milstein, G.N. (1994) *Numerical Integration of Stochastic Differential Equations*, Kluwer Academic Publishers, Dordrecht, the Netherlands.

Moody, J. and Darken, J.C. (1989) Fast learning in networks of locally tuned processing units, *Neural Comput.*, 1, 281–294.

Mortensen, R.E. (1969) Mathematical problems of modeling stochastic nonlinear dynamic systems, *J. Stat. Phys.*, 1, 271–296.

Nakamura, H. and Akaike, H. (1981) Statistical identification for optimal control of supercritical thermal power plants, *Automatica*, 17, 143–155.

Narendra, K.S. and Gallman, P.G. (1966) An iterative method for the identification of nonlinear systems using a Hammerstein model, *IEEE Trans. Automat. Control*, AC–11, 546–550.

Newbold, P. (1974) The exact likelihood function for a mixed autoregressive-moving average process, *Biometrika*, 61, 423–426.

Nunetz, P.L. (1981) *Electrical Fields of the Brain*, Oxford University Press, New York.

Ohtsu, K., Horigome, M., and Kitagawa, G. (1979) A new ship's auto pilot design through a stochastic model, *Automatica*, 15, 255–268.

Otomo, T., Nakagawa, T., and Akaike, H. (1972) Statistical approach to computer control of cement rotary kilns, *Automatica*, 8, 35–48.

Ozaki, T. (1977) On the order determination of ARIMA models, *Appl. Stat.*, 26, 290–301.

Ozaki, T. (1980) Nonlinear time series models for nonlinear random vibrations, *J. Appl. Prob.*, 17, 84–93.

Ozaki, T. (1981a) Nonlinear threshold autoregressive models for nonlinear random vibrations, *J. Appl. Prob.*, 18, 443–451.

Ozaki, T. (1981b) Nonlinear phenomena and time series models, Invited paper, 43rd Session of the International Statistical Institute, Buenos Aires, Argentina, *Bull. Int. Stat. Inst.*, 49, Book 3, 1193–1210.

Ozaki, T. (1982) Statistical analysis of perturbed limit cycle processes using nonlinear time series models, *J. Time Ser. Anal.*, 3, 29–41.

Ozaki, T. (1985a) Nonlinear time series models and dynamical systems. In *Handbook of Statistics*, Vol. 5, eds. E.J. Hannan et al., North-Holland, Amsterdam, the Netherlands, pp. 25–83.

Ozaki, T. (1985b) Statistical identification of storage models with application to stochastic hydrology, *Water Resour. Bull.*, 21, 663–675.

Ozaki, T. (1986) Local Gaussian modelling of stochastic dynamical systems in the analysis of nonlinear random vibrations. In *Essays on Time Series and Allied Processes, Festschrift in honour of Prof. E.J. Hannan*, Applied Probability Trust, U.K., pp. 241–255.

Ozaki, T. (1989) Statistical identification of nonlinear random vibration systems, *J. Appl. Mech.*, 56, 186–191.

Ozaki, T. (1990) Contribution to the discussion of M.S. Bartlett's paper, "Chance and Chaos" *J. Roy. Stat. Soc. Ser. A*, 153, 330–346.

Ozaki, T. (1992a) A bridge between nonlinear time series models and nonlinear stochastic dynamical systems: A local linearization approach, *Stat. Sinica*, 2 (1), 113–135.

Ozaki, T. (1992b) Identification of nonlinearities and non-Gaussianities in time series. In *New Directions in Time Series Analysis, Part I*, eds. D.R. Brillinger et al., IMA Volumes in Mathematics and Its Application, Vol. 45, Springer Verlag, New York/ Berlin, pp. 227–264.

Ozaki, T. (1993a) A local linearization approach to nonlinear filtering, *Int. J. Contr.*, 57 (1), 75–96.

Ozaki, T. (1993b) Non-Gaussian characteristics of exponential autoregressive processes. In *Developments in Time Series Analysis, Festshrift in honour of Prof. Maurice Priestley*, ed. T. Sabba Rao, Chapman & Hall, London, U.K., pp. 257–273.

Ozaki, T. (1994a) The local linearization filter with application to nonlinear system identifications. In *Proceeding of the First US/Japan Conference on the Frontiers of Statistical Modeling: An Informational Approach*, ed. H. Bozdogan, Kluwer Academic Publishers, Dordrecht, the Netherlands, pp. 3217–3240.

Ozaki, T. (1994b) A likelihood analysis of chaos. In *Bulletin of the International Statistical Institute, 50th Session of the ISI*, Beijing, China.

Ozaki, T. (1998a) Dynamic X11 and nonlinear seasonal adjustment 2, *Proc. Inst. Stat. Math.*, 45, 287–300 (in Japanese).

Ozaki, T. (1998b) Dynamic X11 and nonlinear seasonal adjustment 1, *Proc. Inst. Stat. Math.*, 45, 265–285 (in Japanese).

Ozaki, T. (2008) Innovation approach to the modeling of spatial time series with application to fMRI connectivity studies, *Research Memo No. 1052*, Institute of Statistical Mathematics, Tokyo.

Ozaki, T. (in press) Spatial time series modelling for fMRI data analysis in neurosciences. To appear in *Handbook of Statistics*, 30, eds. T.S. Rao and C.R. Rao, Elsevier, Amsterdam, the Netherlands.

Ozaki, T. and Iino, M. (2001) An innovation approach to non-Gaussian time series analysis, *J. Appl. Prob. Trust*, 38A, 78–92.

Ozaki, T., Jimenez, J.C., and Ozaki, V.H. (2000) Role of the likelihood function in the estimation of chaos models, *J. Time Ser. Anal.*, 21, 363–387.

Ozaki, T., Jimenez, J.C., Peng, H., and Ozaki, V.H. (2004) The innovation approach to the identification of nonlinear causal models in time series analysis. IMA volume on *Time Series Analysis and Applications to Geophysical Systems*, Springer, New York, pp. 195–226.

Ozaki, T. and Oda, H. (1978) Nonlinear time series model identification by Akaike's information criterion. In *Information and Systems*, ed. Dubuisson, Pergamon Press, U.K.

Ozaki, T. and Thomson, P.J. (1992) A dynamical system approach to X–11 type seasonal adjustment, *Research Memorandum*, Institute of Statistical Mathematics, Tokyo, Japan, No. 498.

Ozaki, T. and Thomson, P.J. (2002) A nonlinear dynamic model for multiplicative seasonal-trend decomposition, *J. Forecast.*, 21, 107–124.

Ozaki, T. and Tong, H. (1975) On the fitting of non-stationary autoregressive models in time series analysis. In *Proceeding of 8-th Hawaii International Conference on System Sciences*, Hawaii, Western Periodical Co., pp. 224–226.

Ozaki-Haggan, V., Toyoda, Y., and Ozaki, T. (2012) *RBF-AR Time Series Modeling for Predictive Control: Case Studies in Power Plant Control*, to appear.

Ozaki, T., Valdes, P., and Ozaki, V.H. (1997) Reconstructing nonlinear dynamics from time series: With application to epilepsy data analysis, Technical Report 1997/02, University of Manchester, Manchester Centre for Statistical Science, U.K.

Ozaki, T., Valdes-Sosa, P., and Ozaki, V. (1999) Reconstructing the nonlinear dynamics of epilepsy data using nonlinear time series analysis, *J. Sig. Process.*, 3, 153–162.

Park, J. and Sandberg, I. (1991) Universal approximation using radial basis function networks, *Neural Comput.*, 3, 247–257.

Park, J. and Sandberg, I. (1993) Approximation and radial basis networks, *Neural Comput.*, 5, 305–306.

Pascual-Marqui, R.D., Michel, C.M., and Lehmann, D. (1994) Low resolution electromagnetic tomography: A new method for localizing electrical activity in the brain, *Int. J. Psychophysiol.*, 18, 49–65.

Pearson, K. (1900) On the criterion that a given system of deviations from the probable in the case of a correlated system of variables is such that it can be reasonably supposed to have arisen from random sampling. *Philosopical Magazine*, 158–175.

Pearson, R.K. (1999) *Discrete-Time Dynamic Models*, Oxford University Press, New York.

Peng, H., Ozaki, T., Haggan-Ozaki, V., and Toyoda, Y. (2002) Structured parameter optimization method for the radial basis function-based state-dependent autoregressive model, *Int. J. Syst. Sci.*, 33, 1087–1098.

Peng, H., Ozaki, T., Toyoda, Y., Shioya, H., Nakano, K., Haggan-Ozaki, V., and Mori, M. (2003) RBF-ARX model based nonlinear system modeling and predictive control with application to a NOx decomposition process, *(IFAC) Contr. Eng. Pract.*, 12, 191–203.

Poincare, H. (1881) Memoir sur les courbes definies par une equation differentielle, *J. Math.*, 7 (3), 375–422. (Oeuvres 1, 3–84.)

Prakasa Rao, B.L.S. (1999) *Statistical Inference 4 for Diffusion Type Processes*, Arnold, London (co-published in USA by Oxford University Press, New York).

Planck, M. (1904) Überdiemechanische Bedeutung der Température und Eutropie. In Festschrift f. L. Boltzmann, Barth, Leipzig, pp. 113–122.

Priestley, M.B. (1981) *Spectral Analysis and Time Series*, Vol. 1 and 2, Academic Press, London, U.K.

Priestley, M.B. (1988) *Non-Linear and Non-Stationary Time Series Analysis*, Academic Press, London, U.K.

Protter, P. (1990) *Stochastic Integration and Differential Equations*, Springer-Verlag, Berlin, Germany.

Rauch, H.E., Tung, F., and Striebel, C.T. (1965) Maximum likelihood estimates of linear dynamic systems, *AIAA J.*, 3, 1445–1450.

Rayleigh, L., (Strutt, J.W.) (1894) *The Theory of Sounds*, MacMillan, London, Reprinted (1945) Dover, New York.

Riera, J., Aubert, E., Iwata, K., Kawashima, R., Wan, X., and Ozaki, T. (2005) Fusing EEG and fMRI based on a bottom-up model: Inferring activation and effective connectivity in neural masses, *Phil. Trans. R. Soc. Biol. Sci.*, 360 (1457), 1025–1041.

Riera, J.J., Jimenez, J.C., Wan, X., Kawashima, R., and Ozaki, T. (2007) Nonlinear Local Coupling. II; From data to neuronal masses, *Hum. Brain Mapp.*, 28, 335–354.

Riera, J., Watanabe, J., Kazuki, I., Naoki, M., Aubert, E., Ozaki, T., and Kawashima, R. (2004a). A state-space model of the hemodynamic approach: Nonlinear filtering of BOLD signals, *Neuroimage*, 21, 547–567.

Riera, J., Yamashita, O., Kawashima, R., Sadato, N., Okada, T., and Ozaki, T. (2004b) fMRI activation maps based on the NN-ARX model, *Neuroimage*, 23, 680–697.

Rikitake, T. (1958) Oscillations of a system of disc dynamos, *Proc. Camb. Philos. Soc.*, 54, 89–105.

Roberts, G.O. and Tweedie, R.L. (1996) Exponential convergence of Langevin diffusions and their discrete approximations, *Bernuilli*, 2, 341–363.

Roberts, O.D. and Stramer, O. (2001) On inference for partially observed nonlinear diffusion models using the Metropolis-Hastings algorithm, *Biometrika*, 88, 603–621.

Robinson, P.M. (1976) The estimation of linear differential equations with constant coefficients, *Econometrica*, 44, 751–764.

Rosenberg, B. (1973) Random coefficients models: The analysis of cross-section of time series by stochastically convergent parameter regression, *Ann. Econo. Social Measurement*, 2, 399–428.

Ruelle, D. (1987) Diagnosis of dynamical systems with fluctuating parameters, *Proc. Roy. Soc. Lond. A*, 413, 5–8.

Ruelle, D. and Takens, F. (1971) On the nature of turbulence, *Commun. Math. Phys.*, 20, 167–192.

Rumelhart, D. and McClelland, J. (1986) *Parallel Distributed Processing*, MIT Press, Cambridge, MA.

Sage, A.P. and Melsa, J.L. (1971) *Estimation Theory with Application to Communication and Control*, McGraw-Hill, New York. (or Sage, A.P. and Melsa, J.L. (1971) System Identification, Academic Press, New York.)

Saito, Y. and Harashima, H. (1981) Tracking of information within multi-channel EEG record–causal analysis in EEG. In *Recent Advances in EEG and MEG Data Processing*, eds. N. Yamaguchi and K. Fujisawa, Elsevier, Amsterdam, the Netherlands, pp. 133–146.

Sato, K. (1999) *Levy Processes and Infinitely Divisible Distributions*, University Press, Cambridge, U.K.

Schumitt, U., Louis, A.K., Darvas, F., Buchner, H., and Fuchs, M. (2001) Numerical aspects of spatio-temporal current density reconstruction from EEG-/MEG-dta, *IEEE Trans. Med. Imaging*, 20, 314–324.

Schurz, H. (2002) Numerical analysis of SDE without tears. In *Handbook of Stochastic Analysis and Applications*, eds. D. Kannan and V. Lakshmikantham, Marcel Dekker, Basel, Switzerland, pp. 237–359.

Schwartz, L. and Stear, E.B. (1968) A computational comparison of several nonlinear filters, *IEEE Trans. Automatic Control*, AC-13, 83–86.

Schumway, R.H. and Stoffer, D.S. (1982) An approach to time series smoothing and forecasting using the EM algorithm, *J. Time Ser. Anal.*, 3, 253–264.

Schumway, R.H. and Stoffer, D.S. (2000) *Time Series Analysis and Its Application*, Springer, New York.

Schweppe, H. (1965) Evaluation of likelihood functions for Gaussian signals, *IEEE Trans. Autom. Control*, AC-11, 61–70.

Shannon, C.E. (1948) A mathematical theory of communication, *Bell Syst. Tech. J.*, 27, 379–423 and 623–656.

Shi, Z., Tamura, Y., and Ozaki, T. (1999) Nonlinear time series modeling by radial basis function-based state-dependent autoregressive model, *Int. J. Sys. Sci.*, 30, 717–727.

Shi, Z., Tamura, Y., and Ozaki, T. (2001) Monitoring the stability of BWR oscillation by nonlinear time series modeling, *Ann. Nucl. Energy*, 28, 953–966.

Shoji, I. (1998) Approximation of continuous time stochastic processes by a local linearization method, *Mathematics of Computation*, 67, 287–298.

Shoji, I. (1998) A comparative study of maximum likelihood estimators for nonlinear dynamical system models, *Int. J. Control*, 71, 391–404.

Shoji, I. and Ozaki, T. (1997) Comparative study of estimation methods for continuous time stochastic processes, *J. Time Ser. Anal.*, 18, 485–506.

Shoji, I. and Ozaki, T. (1998) A statistical method of estimation and simulation for systems of stochastic differential equations, *Biometrika*, 85, 240–243.

Shumway, R.H. and Stoffer, D.S. (1982) An approach to time series smoothing and forecasting using the EM algorithm, *J. Time Ser. Anal.*, 3, 253–264.

Shyskin, J., Young, H., and Musgrave, J.C. (1967) The X-11 variant of the census method II seasonal adjustment program, Technical Report 15, U.S. department of Commerce, Bureau of the Census, Washington, DC.

Singer, H. (2002) Parameter estimations of nonlinear stochastic differential equations, *J. Comp. Graph. Stat.*, 11, 972–995

Sinha, N.K. and Kuszta, B. (1983) *Modeling and Identification of Dynamical Systems*, Van Nostrand Reinhold Company, New York.

Sorenson, H.W. (1966) Kalman filtering technique. In *Advances in Control Systems, Theory and Applications*, 3, Academic Press, London, U.K., pp. 219–292.

Sorensen, H. (2004) Parametric inference for diffusion processes observed at discrete points in time: A survey, *Int. Stat. Rev.*, 72, 337–354.

Stewart, I. (1989) *Does God Play Dice, The Mathematics of Chaos*, Basil Blackwell, Oxford, U.K.

Stoker, J.J. (1950) *Nonlinear Vibrations*, Interscience, New York.

Stramer, O. and Tweedie, R.L. (1999) Langevin-type models II: Self-targeting candidates for MCMC algorithms, *Methodolo. Comput. Appl. Probab.*, 1, 307–328.

Stratonovich, R.L. (1966) A new representation for stochastic integrals and equations, *SIAM J. Control*, 4, 362–371.

Sugawara, M. (1962) On the analysis of runoff structure about several Japanese rivers, *Japanese J. Geophysics*, 2, 1–76.

Sugihara, G. (1995) Prediction as a criterion for classifying natural time series. In *Chaos and Forecasting, Proceedings of the Royal Statist. Soc. Discussion Meeting*, ed. H. Tong, World Scientific, River Edge, NJ, pp. 269–294.

Takens, F. (1981) Detecting strange attractors in turbulence, *Lect. Notes Math.*, 898, 366–381.

Takens, F. (1994) Analysis of nonlinear time series with noise, Technical Report, Department of Mathematics, Groningen University, Groningen, the Netherlands.

Talay, D. (1995) Simulation and numerical analysis of stochastic differential systems: A review. In *Probabilistic Methods in Applied Physics*, eds. P. Kree and W. Wedig, Vol. 451 of Lecture Notes in Physics, Chapter 3, Springer Verlag, Berlin, Germany, pp. 63–106.

Terasvirta, T. (1994) Specification, estimation and evaluation of smooth transition autoregressive models, *J. Amer. Stat. Assoc.*, 89, 208–218.

Tikhonov, A.N. (1963) Solution of incorrectly formulated problems and the regularization method, *Soviet Math. Dokl.*, 4, 1035–1038 (English translation of Dokl Akad Nauk SSSR 151, 1963, 501–504).

Tong, H. (1994) A personal view of nonlinear time series analysis from a chaos perspective, *Scandinan J. Stat.*, 22, 399–455.

Tong, H. and Lim, K.S. (1980) Threshold autoregression, limit cycles and cyclical data (with discussion), *J. R. Statist. Soc., Ser. B*, 42, 245–292.

Toyoda, Y., Oda, K., and Ozaki, T. (1997) The nonlinear system identification method for advanced control of the fossil power plants. In *Proceeding of 11th IFAC Symposium on System Identification*, July 1997, Kitakyusyu, Fukuoka, Japan, pp. 8–11.

Tribus, M. (1988) An engineer looks at Bayes. In *Maximum Entropy and Bayesian Methods in Science and Engineering*, Volume 1: Foundations, eds. G.J. Erickson and C.R. Smith, Kluwer Academic Publishers, Dordrecht, the Netherlands.

Tsay, R. (2010) *Analysis of Financial Time Series*, John Wiley & Sons, New York.

Tweedie, R.L. (1975) Sufficient conditions for ergodicity and stationarity of Markov chains on a general state space, *Stoch. Proc. Their Appl.*, 3, 385–403.

Tweedie, R.L. (1983) The Existence of Moments and Stationary Markov Chains, *J. Appl. Prob.*, 20, 191–196.

Valdes, P., Jimenez, J.C., Riera, J., Biscay, R., and Ozaki, T. (1999) Nonlinear EEG analysis based on a neural mass model, *Biol. Cybern.*, 81, 415–424.

van der Pol, B. (1927) Forced oscillations in a circuit with nonlinear resistance, *Phil. Mag.*, 3, 65–80.

Vesin, J. (1993) An amplitude-dependent autoregressive signal model based on a radial basis function expansion, *Proc. Internat. Conf. Acoust. Speech Signal Process*, Minnesota, Vol. 3, 129–132.

West, M. and Harrison, J. (1997) *Bayesian Forecasting and Dynamic Models*, Springer-Verlag, New York.

Whittle, P. (1953) The analysis of multiple stationary time series, *J. Roy. Statist. Soc.*, B15, 125–139.

Whittle, P. (1962) Gaussian estimation in stationary time series, *Bulletine l'Institute International de Statistique*, 39 (2), 105–129.

Whittle, P. (1963a) On the fitting of multivariate autoregressions and the approximate canonical factorization of a spectral density matrix, *Biometrika*, 50, 129–134.

Whittle, P. (1963b) *Prediction and Regulation by Linear Least Squares Methods*, English University Press, London, U.K.

Whittle, P. (1996) *Optimal Control*, John Wiley & Sons, New York.

Wiener, N. (1949) *Extrapolation, Interpolation and Smoothing of Stationary Time Series with Engineering Applications*, Wiley, New York. (Originally issued as a classified report by MIT Radiation Lab., Cambridge, MA, February 1942.)

Wiener, N. (1956) The theory of prediction. In *Modern Mathematics for Engineers*, ed. E.F. Beckenback, McGraw-Hill, New York, Chapter 2, pp. 165–190.

Wiener, N. (1958) *Nonlinear Problems in Random Theory*, MIT Press, Cambridge, MA.

Wiener, N. and Masani, P. (1957) The predictiontheory of multivariate stochastic processes, I, *Acta Math.*, 98, 111–150.

Wold, H. (1938) *A Study on the Analysis of Stationary Time Series*, Almqvist and Wicksell, Uppsala, Sweden.

Wong, E. (1963) The construction of a class of stationary Markoff process, *Proc. Amer. Math. Soc. Symp. Appl. Math.*, 16, 264–276.

Wong, K.F.K., Galka, A., Yamashita, O., and Ozaki, T. (2006) Modelling non-stationary variance in EEG time series by state space GARCH model, *Comput. Biol. Med.*, 36, 1327–1335.

Wong, K.F.K. and Ozaki, T. (2007) Akaike causality in state space—Instantaneous causality between visual cortex in fMRI time series, *Biol. Cybern.*, 97, 151–157.

Worsley, K.J., Liao, C.H., Aston, J., Petre, V., Duncan, G.H., Morales, F., and Evans, A.C. (2002) A general statistical analysis for fMRI data, *Neuroimage*, 15, 1–15.

Yamashita, O., Galka, A., Ozaki, T., Biscay, R., and Valdes-Sosa, P. (2004) Recursive penalized least squares solution for dynamical inverse problems of EEG generation, *Hum. Brain Mapp.*, 21, 221–235.

Yamashita, O., Sadato, N., Okada, T., and Ozaki, T. (2005) Evaluating frequency-wise directed connectivity of BOLD signals applying relative power contribution with the linear multivariate time series models, *Neuroimage*, 25, 478–490.

Yoshida, N. (1992) Estimation for diffusion processes from discrete observation, *J. Multivar. Anal.*, 41, 220–242.

Zeeman, E.C. (1976) Duffing equation in brain modelling, *Bull. Inst. Math. Appl.*, 12, 107–214.

Zetterberg, L.H., Kristiansson, L., and Mossberg, K. (1978) Performance of a model for a local neural population, *Biol. Cybern.*, 31, 15–26.

Index